Svetlin G. Georgiev

Time Scales Analysis

Also of Interest

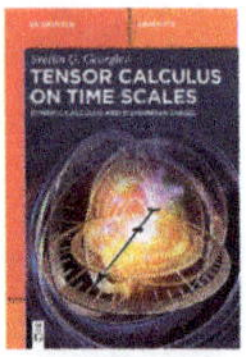

Tensor Calculus on Time Scales
Dynamic Calculus and Riemannian Spaces
Svetlin G. Georgiev, 2026
ISBN 978-3-11-222848-7, e-ISBN (PDF) 978-3-11-222849-4

Differential and Integral Calculus
Implicit Functions, Stieltjes Integrals and Curvilinear Integrals
Svetlin G. Georgiev, Khaled Zennir, 2025
ISBN 978-3-11-914462-9, e-ISBN (PDF) 978-3-11-221808-2

Multiple Integrals in Calculus
Improper Integrals, Line Integrals, Surface Integrals
Svetlin G. Georgiev, Khaled Zennir, 2025
ISBN 978-3-11-914345-5, e-ISBN (PDF) 978-3-11-221960-7

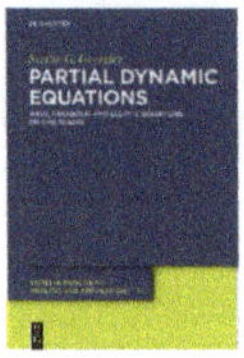

Partial Dynamic Equations
Wave, Parabolic and Elliptic Equations on Time Scales
Svetlin G. Georgiev, 2025
ISBN 978-3-11-163551-4, e-ISBN (PDF) 978-3-11-163614-6

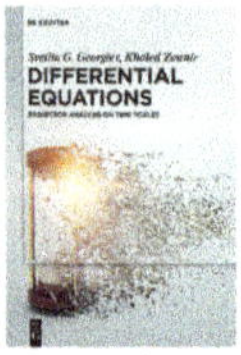

Differential Equations
Projector Analysis on Time Scales
Svetlin G. Georgiev, Khaled Zennir, 2024
ISBN 978-3-11-137509-0, e-ISBN (PDF) 978-3-11-137715-5

Differential Geometry
Frenet Equations and Differentiable Maps
Muhittin E. Aydin, Svetlin G. Georgiev, 2024
ISBN 978-3-11-150089-8, e-ISBN (PDF) 978-3-11-150185-7

Svetlin G. Georgiev

Time Scales Analysis

Differentiation, Integration, Elementary Functions

DE GRUYTER

Mathematics Subject Classification 2020
Primary: 39Axx, 39A10, 39A12, 39A13, 65Lxx, 65L05; Secondary: 65L10, 65L15

Author
Dr. Svetlin G. Georgiev
Sorbonne University
1 Rue Victor Hugo
70005 Paris
France
svetlingeorgiev1@gmail.com

ISBN 978-3-11-223207-1
e-ISBN (PDF) 978-3-11-223208-8
e-ISBN (EPUB) 978-3-11-223209-5

Library of Congress Control Number: 2025951308

Bibliographic information published by the Deutsche Nationalbibliothek
The Deutsche Nationalbibliothek lists this publication in the Deutsche Nationalbibliografie; detailed bibliographic data are available on the Internet at http://dnb.dnb.de.

Cover image: photosaint / iStock / Getty Images Plus
Typesetting: VTeX UAB, Lithuania

www.degruyterbrill.com
Questions about General Product Safety Regulation:
productsafety@degruyterbrill.com

Contents

1 Time scales

In this chapter, we define time scales and give some examples. Forward jump operators, backward jump operators, forward graininess functions, and backward graininess functions are defined, and some of their properties are given. The induction and dual induction principles are introduced.

1.1 Definition. Examples

Definition 1.1. A time scale is an arbitrary nonempty closed subset of the real numbers.

We will denote a time scale by $\mathbb{T}$. We suppose that a time scale $\mathbb{T}$ has the topology inherited from the real numbers with standard topology.

Example 1.1. The sets $[-1,4]$, $\mathbb{R}$, $\mathbb{Z}$, $\mathbb{N}$,

$$\left\{-2,\ -1,\ -\frac{1}{2},\ 0,\ \frac{1}{4},\ \frac{1}{3},\ 2,\ 3,\ 6\right\},$$

and

$$\{1\} \cup \left\{\frac{1}{n}+1\right\}_{n\in\mathbb{N}} \cup \{3\} \cup \left\{\frac{4}{n^2}+3\right\}_{n\in\mathbb{N}} \cup \{9\} \cup \left\{\frac{7}{n^4}+9\right\}_{n\in\mathbb{N}}$$

are time scales.

Example 1.2. The sets $(-3,7)$, $[0,5)$, $(1,7]$, and $\{\frac{5}{n}+3\}_{n\in\mathbb{N}}$ are not time scales.

Example 1.3. Let $a, b > 0$. The sets

$$P_{a,b} = \bigcup_{k=0}^{\infty} [k(a+b), k(a+b)+a]$$

are time scales.

Example 1.4. The set of harmonic numbers

$$H_0 = 0,$$
$$H_n = \sum_{k=1}^{n} \frac{1}{k}, \quad n \in \mathbb{N},$$

is a time scale.

Example 1.5. Let $\{a_n\}_{n\in\mathbb{N}_0}$ be a sequence of positive real numbers. Define

$$t_n = \sum_{k=0}^{n-1} a_k, \quad n \in \mathbb{N}.$$

https://doi.org/10.1515/9783112232088-001

Then the set

$$\mathbb{T} = \{t_n : n \in \mathbb{N}\}$$

is a time scale.

Example 1.6 (Cantor set). Consider the interval $K_0 = [0,1]$. We obtain a subset K_1 of K_0 by removing the open "middle third" of K_0, i. e., the open interval $(\frac{1}{3}, \frac{2}{3})$ from K_0. The set K_2 is obtained by removing the two open middle thirds of K_1, i. e., the two open intervals $(\frac{1}{9}, \frac{2}{9})$ and $(\frac{7}{9}, \frac{8}{9})$, from K_1. Proceeding in this manner, we obtain a sequence $\{K_n\}_{n\in\mathbb{N}_0}$ of subsets of the interval $[0,1]$. Figure 1.1 shows the sets K_0, K_1, K_2, K_3, and so forth. The Cantor set is now defined as

$$C = \bigcap_{n=0}^{\infty} K_n.$$

The Cantor set is a time scale. Any $x \in [0,1]$ can be represented in its ternary expansion as

$$x = \sum_{j=1}^{\infty} \frac{a_j}{3^j}, \quad \text{where } a_j \in \{0,1,2\},\ j \in \mathbb{N}.$$

This expansion is unique unless x is of the form $p3^{-k}$ for some integers p and k. In this case, x has two expansions:

1. $a_j = 0$ for $j > k$.
2. $a_j = 2$ for $j > k$.

Assume that p is not divisible by 3. One of these expansions will have $a_k = 1$, and the other will have $a_k = 0$ or $a_k = 2$. We have that

$$a_1 = 1 \quad \text{if and only if} \quad \frac{1}{3} < x < \frac{2}{3}$$

and

$$a_1 \neq 1 \quad \text{and} \quad a_2 = 1 \quad \text{if and only if} \quad \frac{1}{9} < x < \frac{2}{9} \quad \text{or} \quad \frac{7}{9} < x < \frac{8}{9},$$

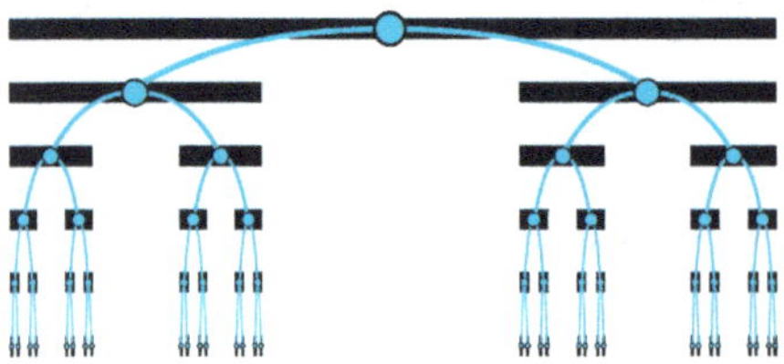

Figure 1.1: Expansion of the Cantor Set.

and so forth. If

$$x = \sum_{j=1}^{\infty} \frac{a_j}{3^j} \quad \text{and} \quad y = \sum_{j=1}^{\infty} \frac{b_j}{3^j},$$

then $x < y$ if and only if there exists $n \in \mathbb{N}$ such that $a_n < b_n$ and $a_j = b_j$ for $j < n$. Thus the Cantor set C is the set of all $x \in [0, 1]$ that have a base-3 expansion

$$x = \sum_{j=1}^{\infty} \frac{a_j}{3^j} \quad \text{with } a_j \neq 1 \text{ for all } j.$$

Example 1.7. The set

$$\mathbb{T} = \left\{ t_n = -\frac{1}{n} : n \in \mathbb{N} \right\} \cup \mathbb{N}_0$$

is a time scale.

Example 1.8. The set

$$[0,1] \cup \left\{ 1 + \frac{1}{n} \right\}_{n \in \mathbb{N}} \cup (2,3] \cup \left\{ 3 + \frac{1}{n} \right\}_{n \in \mathbb{N}}$$

is a time scale.

Example 1.9. Let

$$U = \left\{ \frac{1}{2^n} : n \in \mathbb{N}_0 \right\}.$$

Then the set

$$\{0\} \cup U \cup (1 - u) \cup (1 + U) \cup (2 - U) \cup (2 + U) \cup (3 - U) \cup (3 + U) \cup \{1, 2, 3, 4\}$$

is a time scale.

Exercise 1.1. Check if the following sets are time scales:

1. $2^{\mathbb{N}_0}$.
2. $(-1, 1] \cup [2, 3] \cup [4, 8]$.
3. $\{-\frac{1}{2n} : n \in \mathbb{N}\} \cup 2\mathbb{N}_0$.
4. $U \cup (2 - U) \cup (2 + U)$, $U = \{\frac{1}{4^n} : n \in \mathbb{N}_0\}$.
5. $[0, 2] \cup \{2 + \frac{1}{n}\}_{n \in \mathbb{N}} \cup (3, 5] \cup 7^{\mathbb{N}_0}$.

1.2 Forward jump operators, backward jump operators and graininess functions

We start by defining the forward jump operator.

Definition 1.2. Let $\mathbb{T}$ be a time scale. For $t \in \mathbb{T}$, we define the forward jump operator $\sigma : \mathbb{T} \to \mathbb{T}$ as follows:

$$\sigma(t) = \inf\{s \in \mathbb{T} : s > t\}.$$

In this definition, we put $\inf \emptyset = \sup \mathbb{T}$. Then $t = \sigma(t)$ if t is a maximum of $\mathbb{T}$.

Note that $\sigma(t) \geq t$ for all $t \in \mathbb{T}$.

Example 1.10. Let $\mathbb{T} = h\mathbb{Z}$, $h > 0$. Take arbitrary $t \in \mathbb{T}$. Then there is $n \in \mathbb{Z}$ such that $t = hn$. Applying the definition of forward jump operators, we find

$$\begin{aligned}\sigma(t) &= \inf\{s = hp, p \in \mathbb{Z} : hp > hn\} \\ &= h(n+1) \\ &= hn + h \\ &= t + h.\end{aligned}$$

Example 1.11. Let $\mathbb{T} = 3^{\mathbb{N}_0}$. Take arbitrary $t \in \mathbb{T}$. Then, there is $n \in \mathbb{N}_0$ such that $t = 3^n$. Applying the definition of forward jump operators, we find

$$\begin{aligned}\sigma(t) &= \inf\{3^s, s \in \mathbb{N}_0 : 3^s > 3^n\} \\ &= 3^{n+1} \\ &= 3 \cdot 3^n \\ &= 3t.\end{aligned}$$

Example 1.12. Let $\mathbb{T} = \mathbb{N}_0^k$, where $k \in \mathbb{N}$ is fixed. Take arbitrary $t \in \mathbb{T}$. Then there is $n \in \mathbb{N}_0$ such that $t = n^k$. Hence $n = \sqrt[k]{t}$. Now applying the definition of forward jump operators, we arrive at

$$\begin{aligned}\sigma(t) &= \inf\{s^k, s \in \mathbb{N}_0 : s^k > n^k\} \\ &= (n+1)^k \\ &= (\sqrt[k]{t} + 1)^k.\end{aligned}$$

Example 1.13. Let $\mathbb{T} = \{H_n : n \in \mathbb{N}_0\}$, where H_n, $n \in \mathbb{N}_0$, are the harmonic numbers. Take arbitrary $n \in \mathbb{N}_0$. Then applying the definition of forward jump operators, we find

$$\sigma(H_n) = \inf\{H_s, s \in \mathbb{N}_0 : H_s > H_n\}$$

$$
\begin{aligned}
&= \inf\left\{H_s, s \in \mathbb{N}_0 : \sum_{k=1}^{s} \frac{1}{k} > \sum_{k=1}^{n} \frac{1}{k}\right\} \\
&= \sum_{k=1}^{n+1} \frac{1}{k} \\
&= H_{n+1}.
\end{aligned}
$$

Example 1.14. Let $\mathbb{T} = P_{1,3}$. Then

$$
\begin{aligned}
\mathbb{T} &= \bigcup_{k=0}^{\infty} [4k, 4k+1] \\
&= [0,1] \cup [4,5] \cup [8,9] \cup [12,13] \cup \dots .
\end{aligned}
$$

If $t \in [0,1)$, then applying the definition of forward jump operators, we find

$$
\begin{aligned}
\sigma(t) &= \inf\{s \in \mathbb{T} : s > t\} \\
&= t.
\end{aligned}
$$

If $t = 1$, then

$$
\begin{aligned}
\sigma(1) &= \inf\{s \in \mathbb{T} : s > 1\} \\
&= 4.
\end{aligned}
$$

Now take arbitrary $k \in \mathbb{N}$. If $t \in [4k, 4k+1)$, then we have

$$
\begin{aligned}
\sigma(t) &= \inf\{s \in \mathbb{T} : s > t\} \\
&= t.
\end{aligned}
$$

If $t = 4k + 1$, then

$$
\begin{aligned}
\sigma(t) &= \inf\{s \in \mathbb{T} : s > 4k+1\} \\
&= 4(k+1) \\
&= 4k + 4 \\
&= 4k + 1 + 3 \\
&= t + 3.
\end{aligned}
$$

Therefore

$$
\sigma(t) = \begin{cases} t & \text{if } t \in \bigcup_{k=0}^{\infty} [4k, 4k+1), \\ t + 3 & \text{if } t \in \bigcup_{k=0}^{\infty} \{4k+1\}. \end{cases}
$$

Exercise 1.2. Let $\mathbb{T} = P_{a,b}$, where $a, b > 0$. Find $\sigma(t)$, $t \in \mathbb{T}$.

Example 1.15. Let $\mathbb{T} = C$, where C is the Cantor set. We will find $\sigma(t)$ for $t \in \mathbb{T}$. For this aim, let C_1 be the set of all left-hand end points of the open intervals that are removed. Then

$$C_1 = \left\{ \sum_{k=1}^{m} \frac{a_k}{3^k} + \frac{1}{3^{m+1}} : m \in \mathbb{N},\ a_k \in \{0, 2\} \text{ for all } 1 \le k \le m \right\}.$$

Let C_2 be the set of all right-hand end points of the open intervals that are removed. We have

$$C_2 = \left\{ \sum_{k=1}^{m} \frac{a_k}{3^k} + \frac{2}{3^{m+1}} : m \in \mathbb{N},\ a_k \in \{0, 2\} \text{ for all } 1 \le k \le m \right\}.$$

Take arbitrary $t \in C$. We have the following cases.

1. $t \in C_1$. Then

$$t = \sum_{k=1}^{m} \frac{a_k}{3^k} + \frac{1}{3^{m+1}}.$$

   Hence we obtain

$$\begin{aligned} \sigma(t) &= \inf\{s \in \mathbb{T} : s > t\} \\ &= \sum_{k=1}^{m} \frac{a_k}{3^k} + \frac{2}{3^{m+1}} \\ &= \sum_{k=1}^{m} \frac{a_k}{3^k} + \frac{1}{3^{m+1}} + \frac{1}{3^{m+1}} \\ &= t + \frac{1}{3^{m+1}}. \end{aligned}$$

2. $t \in C_2$. Then

$$t = \sum_{k=1}^{m} \frac{a_k}{3^k} + \frac{2}{3^{m+1}}.$$

   Hence

$$\begin{aligned} \sigma(t) &= \inf\{s \in \mathbb{T} : s > t\} \\ &= t. \end{aligned}$$

3. $t \in \mathbb{T} \backslash (C_1 \cup C_2)$. Then

$$\begin{aligned} \sigma(t) &= \inf\{s \in \mathbb{T} : s > t\} \\ &= t. \end{aligned}$$

Consequently,

$$\sigma(t) = \begin{cases} t + \frac{1}{3^{m+1}} & \text{if } t \in C_1, t = \sum_{k=1}^{m} \frac{a_k}{3^k} + \frac{1}{3^{m+1}}, \\ t & \text{if } t \in \mathbb{T}\backslash C_1. \end{cases}$$

Example 1.16. Let $\{a_n\}_{n \in \mathbb{N}_0}$ be a sequence of positive real numbers, and let

$$t_n = \sum_{k=0}^{n-1} a_k, \quad n \in \mathbb{N},$$

and

$$\mathbb{T} = \{t_n : n \in \mathbb{N}\}.$$

We will find $\sigma(t)$, $t \in \mathbb{T}$. Take arbitrary $n \in \mathbb{N}$. Then

$$\begin{aligned} \sigma(t_n) &= \inf\left\{x \in \mathbb{T} : s = \sum_{k=0}^{n-1} a_k, \ s > t_n\right\} \\ &= \sum_{k=0}^{n} a_k \\ &= \sum_{k=0}^{n-1} a_n \\ &= t_n + a_n. \end{aligned}$$

Example 1.17. Let

$$\mathbb{T} = \left\{t_n = -\frac{1}{n} : n \in \mathbb{N}\right\} \cup \mathbb{N}_0.$$

We will find $\sigma(t)$, $t \in \mathbb{T}$. Take arbitrary $n \in \mathbb{N}$. Then

$$n = -\frac{1}{t_n},$$

and

$$\begin{aligned} \sigma(t_n) &= \inf\left\{s \in \mathbb{T} : s = -\frac{1}{m}, m \in \mathbb{N}, s > t_n\right\} \\ &= -\frac{1}{n+1} \\ &= -\frac{1}{-\frac{1}{t_n} + 1} \\ &= -\frac{t_n}{t_n - 1}. \end{aligned}$$

Next, if $t \in \mathbb{N}_0$, then

$$\begin{aligned}\sigma(t) &= \inf\{s \in \mathbb{T} : s > t\}\\ &= t+1.\end{aligned}$$

Consequently,

$$\sigma(t) = \begin{cases} -\frac{t}{t-1} & \text{if } t \in \{t_n = -\frac{1}{n} : n \in \mathbb{N}\}, \quad t = t_n,\\ t+1 & \text{if } t \in \mathbb{N}_0.\end{cases}$$

Example 1.18. Let

$$\mathbb{T} = \left\{t_n = \left(\frac{1}{2}\right)^{2^n} : n \in \mathbb{N}_0\right\} \cup \{0,1\}.$$

We will find $\sigma(t)$, $t \in \mathbb{T}$. Take arbitrary $n \in \mathbb{N}$. Then

$$\begin{aligned}\sigma(t_n) &= \inf\{s \in \mathbb{T} : s > t_n\}\\ &= \left(\frac{1}{2}\right)^{2^{n-1}}\\ &= \left(\frac{1}{2}\right)^{2^n \cdot \frac{1}{2}}\\ &= \left(\left(\frac{1}{2}\right)^{2^n}\right)^{\frac{1}{2}}\\ &= \sqrt{t_n}.\end{aligned}$$

Next,

$$t_0 = \frac{1}{2} \quad \text{and} \quad \sigma(t_0) = 1,$$

and

$$\begin{aligned}\sigma(0) &= 0,\\ \sigma(1) &= 1.\end{aligned}$$

Consequently,

$$\sigma(t) = \begin{cases} \sqrt{t} & \text{if } t \in \{t_n = (\frac{1}{2})^{2^n} : n \in \mathbb{N}\},\\ 1 & \text{if } t = \frac{1}{2},\\ 0 & \text{if } t = 0,\\ 1 & \text{if } t = 1.\end{cases}$$

Example 1.19. Let $U = \{\frac{1}{2^n} : n \in \mathbb{N}\}$, and let

$$\mathbb{T} = U \cup (1 - U) \cup (1 + U) \cup (2 - U) \cup (2 + U) \cup \{0, 1, 2\}.$$

We will find $\sigma(t)$, $t \in \mathbb{T}$. We have the following cases.

1. $t = 0$. Then
$$\sigma(0) = 0.$$
2. $t = \frac{1}{2}$. Then
$$\sigma\left(\frac{1}{2}\right) = \frac{3}{4}.$$
3. $t = 1$. Then
$$\sigma(1) = 1.$$
4. $t = \frac{3}{2}$. Then
$$\sigma\left(\frac{3}{2}\right) = \frac{7}{4}.$$
5. $t = 2$. Then
$$\sigma(2) = 2.$$
6. $t = \frac{5}{2}$. Then
$$\sigma\left(\frac{5}{2}\right) = \frac{5}{2}.$$
7. $t \in U\backslash\{\frac{1}{2}\}$. Then
$$t = \frac{1}{2^n}$$
for some $n \in \mathbb{N}$, and
$$\begin{aligned} \sigma(t) &= \frac{1}{2^{n-1}} \\ &= \frac{2}{2^n} \\ &= 2t. \end{aligned}$$

8. $t \in (1-U)\backslash\{\frac{1}{2}\}$. Then

$$t = 1 - \frac{1}{2^n}$$

for some $n \in \mathbb{N}$, and

$$\frac{1}{2^n} = 1 - t.$$

Hence

$$\begin{aligned}\sigma(t) &= 1 - \frac{1}{2^{n+1}}\\ &= 1 - \frac{1}{2}\cdot\frac{1}{2^n}\\ &= 1 - \frac{1-t}{2}\\ &= \frac{2-1+t}{2}\\ &= \frac{1+t}{2}.\end{aligned}$$

9. $t \in (1+U)\backslash\{\frac{3}{2}\}$. Then

$$t = 1 + \frac{1}{2^n}$$

for some $n \in \mathbb{N}$. Hence

$$\frac{1}{2^n} = t - 1,$$

and

$$\begin{aligned}\sigma(t) &= 1 + \frac{1}{2^{n-1}}\\ &= 1 + \frac{2}{2^n}\\ &= 1 + 2(t-1)\\ &= 2t - 1.\end{aligned}$$

10. $t \in (2-U)\backslash\{\frac{3}{2}\}$. Then

$$t = 2 - \frac{1}{2^n}$$

for some $n \in \mathbb{N}$, and

$$\frac{1}{2^n} = 2 - t.$$

Hence

$$\begin{aligned}\sigma(t) &= 2 - \frac{1}{2^{n+1}}\\ &= 2 - \frac{1}{2}\cdot\frac{1}{2^n}\\ &= 2 - \frac{2-t}{2}\\ &= \frac{t+2}{2}.\end{aligned}$$

11. $t \in (2+U)\backslash\{\frac{5}{2}\}$. Then

$$t = 2 + \frac{1}{2^n}$$

for some $n \in \mathbb{N}$, and

$$\frac{1}{2^n} = t - 2.$$

Hence

$$\begin{aligned}\sigma(t) &= 2 + \frac{1}{2^{n-1}}\\ &= 2 + \frac{2}{2^n}\\ &= 2 + 2(t-2)\\ &= 2(t-1).\end{aligned}$$

Consequently,

$$\sigma(t) = \begin{cases} 0 & \text{if } t = 0,\\ \frac{3}{4} & \text{if } t = \frac{1}{2},\\ 1 & \text{if } t = 1,\\ \frac{7}{4} & \text{if } t = \frac{3}{2},\\ 2 & \text{if } t = 2,\\ \frac{5}{2} & \text{if } t = \frac{5}{2},\\ 2t & \text{if } t \in U\backslash\{\frac{1}{2}\},\\ \frac{1+t}{2} & \text{if } t \in (1-U)\backslash\{\frac{1}{2}\},\\ 2t-1 & \text{if } t \in (1+U)\backslash\{\frac{3}{2}\},\\ \frac{t+2}{2} & \text{if } t \in (2-U)\backslash\{\frac{3}{2}\},\\ 2(t-1) & \text{if } t \in (2+U)\backslash\{\frac{5}{2}\}.\end{cases}$$

Exercise 1.3. Find $\sigma(t)$, $t \in \mathbb{T}$, where
1. $\mathbb{T} = h\mathbb{Z} + k$, $h > 0$, $k \in \mathbb{R}$.
2. $\mathbb{T} = (-2\mathbb{N}_0) \cup 3^{\mathbb{N}_0}$.
3. $\mathbb{T} = P_{3,7} \cup [4, 6]$.
4. $\mathbb{T} = 11^{\mathbb{N}_0} \cup \{0\}$.
5. $\mathbb{T} = [1, 2] \cup [3, 4] \cup [7, 8] \cup 9^{\mathbb{N}}$.

Definition 1.3. Define the forward graininess function $\mu : \mathbb{T} \to \mathbb{R}$ as

$$\mu(t) = \sigma(t) - t, \quad t \in \mathbb{T}.$$

Example 1.20. Let $\mathbb{T} = h\mathbb{Z}$, where $h > 0$. Then using the computations in Example 1.10, we find

$$\begin{aligned} \mu(t) &= \sigma(t) - t \\ &= t + h - t \\ &= h, \quad t \in \mathbb{T}. \end{aligned}$$

Example 1.21. Let $\mathbb{T} = 3^{\mathbb{N}_0}$. Then using the computations in Example 1.11, we get

$$\begin{aligned} \mu(t) &= \sigma(t) - t \\ &= 3t - t \\ &= 2t, \quad t \in \mathbb{T}. \end{aligned}$$

Example 1.22. Let $\mathbb{T} = \mathbb{N}_0^k$, where $k \in \mathbb{N}$. Then using the computations in Example 1.12, we find

$$\begin{aligned} \mu(t) &= \sigma(t) - t \\ &= (\sqrt[k]{t} + 1)^k - t \\ &= t + \binom{k}{1}\sqrt[k]{t^{k-1}} + \binom{k}{2}\sqrt[k]{t^{k-2}} + \cdots + \binom{k}{k-1}\sqrt[k]{t} + 1 - t \\ &= \binom{k}{1}\sqrt[k]{t^{k-1}} + \binom{k}{2}\sqrt[k]{t^{k-2}} + \cdots + \binom{k}{k-1}\sqrt[k]{t} + 1, \quad t \in \mathbb{T}. \end{aligned}$$

Example 1.23. Let $\mathbb{T} = \{H_n : n \in \mathbb{N}_0\}$, where H_n, $n \in \mathbb{N}_0$, are the harmonic numbers. Then using the computations in Example 1.13, we find

$$\begin{aligned} \mu(H_n) &= \sigma(H_n) - H_n \\ &= H_{n+1} - H_n \\ &= \frac{1}{n+1}, \quad n \in \mathbb{N}. \end{aligned}$$

Example 1.24. Let $\mathbb{T} = P_{1,3}$. Then using the computations in Example 1.14, we find

$$\begin{aligned}\mu(t) &= \sigma(t) - t\\ &= \begin{cases}0 & \text{if } t \in \bigcup_{k=0}^{\infty}[4k, 4k+1),\\ 3 & \text{if } t \in \bigcup_{k=0}^{\infty}\{4k+1\}.\end{cases}\end{aligned}$$

Example 1.25. Let $\mathbb{T} = C$, where C is the Cantor set. Then using the computations in Example 1.15, we find

$$\begin{aligned}\mu(t) &= \sigma(t) - t\\ &= \begin{cases}\frac{1}{3^{m+1}} & \text{if } t \in C_1, \quad t = \sum_{k=1}^{m}\frac{a_k}{3^k} + \frac{1}{3^{m+1}},\\ 0 & \text{if } t \in \mathbb{T}\backslash C_1.\end{cases}\end{aligned}$$

Example 1.26. Let $\{a_n\}_{n\in\mathbb{N}}$ be a sequence of positive real numbers, let

$$t_n = \sum_{k=0}^{n-1} a_k, \quad n \in \mathbb{N}, \quad \text{and let} \quad \mathbb{T} = \{t_n : n \in \mathbb{N}\}.$$

Then using the computations in Example 1.16, we find

$$\begin{aligned}\mu(t_n) &= \sigma(t_n) - t_n\\ &= t_{n+1} - t_n\\ &= a_n, \quad n \in \mathbb{N}.\end{aligned}$$

Example 1.27. Let $\mathbb{T} = \{t_n = -\frac{1}{n} : n \in \mathbb{N}\} \cup \mathbb{N}_0$. Then using the computations in Example 1.17, we find

$$\begin{aligned}\mu(t) &= \sigma(t) - t\\ &= \begin{cases}-\frac{t^2}{t-1} & \text{if } t \in \{t_n = -\frac{1}{n} : n \in \mathbb{N}\}, \quad t = t_n,\\ 1 & \text{if } t \in \mathbb{N}_0.\end{cases}\end{aligned}$$

Example 1.28. Let $\mathbb{T} = \left\{t_n = (\frac{1}{2})^{2^n} : n \in \mathbb{N}_0\right\} \cup \{0, 1\}$. Then using the computations in Example 1.18, we find

$$\begin{aligned}\mu(t) &= \sigma(t) - t\\ &= \begin{cases}\sqrt{t} - t & \text{if } t \in \left\{t_n = (\frac{1}{2})^{2^n} : n \in \mathbb{N}\right\},\\ \frac{1}{2} & \text{if } t = \frac{1}{2},\\ 0 & \text{if } t \in \{0, 1\}.\end{cases}\end{aligned}$$

Example 1.29. Let $U = \{\frac{1}{2^n} : n \in \mathbb{N}\}$, and let

$$\mathbb{T} = \{0\} \cup U \cup (1 - U) \cup (1 + U) \cup (2 - U) \cup (2 + U) \cup \{1, 2\}.$$

Then using the computations in Example 1.19, we find

$$\begin{aligned}\mu(t) &= \sigma(t) - t \\ &= \begin{cases} 0 & \text{if } t = 0, \\ \frac{1}{4} & \text{if } t = \frac{1}{2}, \\ 0 & \text{if } t = 1, \\ \frac{1}{4} & \text{if } t = \frac{3}{2}, \\ 0 & \text{if } t = 2, \\ 0 & \text{if } t = \frac{5}{2}, \\ t & \text{if } t \in U \backslash \{\frac{1}{2}\}, \\ \frac{1-t}{2} & \text{if } t \in (1 - U) \backslash \{\frac{1}{2}\}, \\ t - 1 & \text{if } t \in (1 + U) \backslash \{\frac{3}{2}\}, \\ \frac{2-t}{2} & \text{if } t \in (2 - U) \backslash \{\frac{3}{2}\}, \\ t - 2 & \text{if } t \in (2 + U) \backslash \{\frac{5}{2}\}. \end{cases}\end{aligned}$$

Exercise 1.4. Find $\mu(t)$, $t \in \mathbb{T}$, where $\mathbb{T}$ are the time scales in Exercise 1.3.

Definition 1.4. The backward jump operator $\rho : \mathbb{T} \to \mathbb{T}$ is defined as

$$\rho(t) = \sup\{s \in \mathbb{T} : s < t\}.$$

In this definition, we put $\sup \emptyset = \inf \mathbb{T}$. Then $\rho(t) = t$ if t is a minimum of $\mathbb{T}$.

Note that $\rho(t) \leq t$ for all $t \in \mathbb{T}$.

Example 1.30. Let $\mathbb{T} = h\mathbb{Z}$, $h > 0$. Take arbitrary $t \in \mathbb{T}$. Then there is $n \in \mathbb{Z}$ such that $t = hn$. Hence, applying the definition of backward jump operators, we find

$$\begin{aligned}\rho(t) &= \sup\{s \in \mathbb{T} : s = hm,\ m \in \mathbb{Z},\ s < hn\} \\ &= h(n - 1) \\ &= hn - h \\ &= t - h.\end{aligned}$$

Example 1.31. Let $\mathbb{T} = 3^{\mathbb{N}_0}$. Take arbitrary $t \in \mathbb{T}$. We have the following cases.

1. $t = 1$. Then

$$\rho(1) = 1.$$

2. $t > 1$. Then $t = 3^l$ for some $l \in \mathbb{N}$. Hence, applying the definition of backward jump operators, we find

$$\begin{aligned}\rho(t) &= \sup\{s \in \mathbb{T} : s = 3^k,\ k \in \mathbb{N}_0,\ s < 3^l\}\\ &= 3^{l-1}\\ &= \frac{3^l}{3}\\ &= \frac{t}{3}.\end{aligned}$$

Consequently,

$$\rho(t) = \begin{cases} 1 & \text{if } t = 1,\\ \frac{t}{3} & \text{if } t > 1.\end{cases}$$

Example 1.32. Let $\mathbb{T} = \mathbb{N}_0^k$, $k \in \mathbb{N}$. Take arbitrary $t \in \mathbb{T}$. We have the following cases.
1. $t = 0$. Then

$$\rho(0) = 0.$$

2. $t > 0$. Then there is $n \in \mathbb{N}$ such that $t = n^k$. Hence $n = \sqrt[k]{t}$. Now applying the definition of backward jump operators, we find

$$\begin{aligned}\rho(t) &= \sup\{s \in \mathbb{T} : s = l^k,\ l \in \mathbb{N}_0,\ s < n^k\}\\ &= (n-1)^k\\ &= (\sqrt[k]{t} - 1)^k.\end{aligned}$$

Consequently,

$$\rho(t) = \begin{cases} 0 & \text{if } t = 0,\\ (\sqrt[k]{t} - 1)^k & \text{if } t > 0.\end{cases}$$

Example 1.33. Let $\mathbb{T} = \{H_n : n \in \mathbb{N}_0\}$, where H_n, $n \in \mathbb{N}_0$, are the harmonic numbers. Take arbitrary $n \in \mathbb{N}_0$. We have the following cases.
1. $n = 0$. Then

$$\rho(H_0) = H_0.$$

2. $n \geq 1$. Then applying the definition of backward jump operators, we find

$$\begin{aligned}\rho(H_n) &= \sup\{s \in \mathbb{T} : s = H_l,\ l \in \mathbb{N}_0,\ s < H_n\}\\ &= H_{n-1}.\end{aligned}$$

Consequently,

$$\rho(t) = \begin{cases} H_0 & \text{if } n = 0, \\ H_{n-1} & \text{if } n \geq 1. \end{cases}$$

Example 1.34. Let $\mathbb{T} = P_{1,3}$. Take arbitrary $t \in \mathbb{T}$. We have the following cases.

1. $t = 0$. Then

$$\rho(0) = 0.$$

2. $t > 0$. Then we have the following subcases.
   a. $t \in (0,1] \cup \bigcup_{k=1}^{\infty}(4k, 4k+1]$. Then

$$\rho(t) = t.$$

   b. $t = 4k$, $k > 0$. Then $k = \frac{t}{4}$, and

$$\begin{aligned} \rho(t) &= 4(k-1) + 1 \\ &= 4k - 4 + 1 \\ &= 4k - 3 \\ &= t - 3. \end{aligned}$$

Consequently,

$$\rho(t) = \begin{cases} t & \text{if } t \in [0,1] \cup \bigcup_{k=1}^{\infty}(4k, 4k+1], \\ t - 3 & \text{if } t \in \bigcup_{k=1}^{\infty}\{4k\}. \end{cases}$$

Example 1.35. Let $\mathbb{T} = C$, where C is the Cantor set. Let C_2 be the set of all right-hand end points of the intervals that are removed from the interval $[0,1]$, i. e.,

$$C_2 = \left\{ \sum_{k=1}^{m} \frac{a_k}{3^k} + \frac{2}{3^{m+1}} : m \in \mathbb{N},\ a_k \in \{0,2\},\ 1 \leq k \leq m \right\}.$$

Take arbitrary $t \in \mathbb{T}$. Then we have the following cases.

1. $t \in C_2$. Then

$$t = \sum_{k=1}^{m} \frac{a_k}{3^k} + \frac{2}{3^{m+1}},$$

   and

$$\rho(t) = \sum_{k=1}^{m} \frac{a_k}{3^k} + \frac{1}{3^{m+1}}$$

$$= \sum_{k=1}^{m} \frac{a_k}{3^k} + \frac{2}{3^{m+1}} - \frac{1}{3^{m+1}}$$
$$= t - \frac{1}{3^{m+1}}, \quad \text{where } a_k \in \{0, 2\},\ 1 \le k \le m.$$

2. $t \in \mathbb{T} \setminus C_2$. Then $\rho(t) = t$.

Consequently,

$$\rho(t) = \begin{cases} t - \frac{1}{3^{m+1}} & \text{if } t \in C_2,\ t = \sum_{k=1}^{m} \frac{a_k}{3^k} + \frac{2}{3^{m+1}}, \\ t & \text{if } t \in \mathbb{T} \setminus C_2. \end{cases}$$

Example 1.36. Let $\mathbb{T} = \{t_n = \sum_{k=0}^{n-1} a_k : n \in \mathbb{N}\}$, where $a_n > 0$, $n \in \mathbb{N}_0$. Take arbitrary $t_n \in \mathbb{T}$ for some $n \in \mathbb{N}$. We have the following cases.

1. $n = 1$. Then

$$\rho(t_1) = t_1.$$

2. $n > 1$. Then

$$\rho(t_n) = \sup\left\{ t_m = \sum_{k=0}^{m-1} a_k,\ m \in \mathbb{N}, : t_m < t_n \right\}$$
$$= \sum_{k=0}^{n-2} a_k$$
$$= \sum_{k=0}^{n-1} a_k - a_{n-1}$$
$$= t_n - a_{n-1}.$$

Consequently,

$$\rho(t_n) = \begin{cases} t_1 & \text{if } n = 1, \\ t_n - a_{n-1} & \text{if } n \ge 2. \end{cases}$$

Example 1.37. Let $\mathbb{T} = \{t_n = -\frac{1}{n} : n \in \mathbb{N}\} \cup \mathbb{N}_0$. Take arbitrary $t \in \mathbb{T}$. We have the following cases.

1. $t = t_n$ for some $n \in \mathbb{N}$.
   a. If $n = 1$, then

$$\rho(t_1) = t_1.$$

b. If $n \geq 2$, then $n = -\frac{1}{t_n}$, and

$$\begin{aligned}\rho(t_n) &= -\frac{1}{n-1}\\ &= -\frac{1}{-\frac{1}{t_n}-1}\\ &= \frac{t_n}{t_n+1}.\end{aligned}$$

2. Let $t \in \mathbb{N}_0$.
   a. If $t = 0$, then

$$\rho(0) = 0.$$

   b. If $t \geq 1$, then

$$\rho(t) = t - 1.$$

Consequently,

$$\rho(t) = \begin{cases} t & \text{if } t \in \{-1, 0\},\\ \frac{t}{1+t} & \text{if } t \in \{t_n = -\frac{1}{n} : n \in \mathbb{N},\ n \geq 2\},\\ t-1 & \text{if } t \in \mathbb{N}.\end{cases}$$

Example 1.38. Let $\mathbb{T} = \{t_n = (\frac{1}{2})^{2^n} : n \in \mathbb{N}_0\} \cup \{0, 1\}$. Take arbitrary $t \in \mathbb{T}$. We have the following cases.

1. $t = (\frac{1}{2})^{2^n}$ for some $n \in \mathbb{N}_0$. Then

$$\begin{aligned}\rho(t) &= \left(\frac{1}{2}\right)^{2^{n+1}}\\ &= \left(\frac{1}{2}\right)^{2^n \cdot 2}\\ &= \left(\left(\frac{1}{2}\right)^{2^n}\right)^2\\ &= t^2.\end{aligned}$$

2. $t = 0$. Then

$$\rho(0) = 0.$$

3. $t = 1$. Then

$$\rho(1) = \frac{1}{2}.$$

Consequently,

$$\rho(t) = \begin{cases} t^2 & \text{if } t \in \{t_n = (\frac{1}{2})^{2^n} : n \in \mathbb{N}_0\}, \\ 0 & \text{if } t = 0, \\ \frac{1}{2} & \text{if } t = 1. \end{cases}$$

Example 1.39. Let $U = \{\frac{1}{2^n} : n \in \mathbb{N}\}$, and let

$$\mathbb{T} = \{0\} \cup U \cup (1 - U) \cup (1 + U) \cup (2 - U) \cup (2 + U) \cup \{1, 2\}.$$

We will find $\rho(t)$, $t \in \mathbb{T}$. Take arbitrary $t \in \mathbb{T}$. We have the following cases.

1. $t = \frac{1}{2^n}$ for some $n \in \mathbb{N}$. Then

$$\begin{aligned} \rho(t) &= \frac{1}{2^{n+1}} \\ &= \frac{1}{2} \cdot \frac{1}{2^n} \\ &= \frac{t}{2}. \end{aligned}$$

2. $t \in (1 - U)$.
   a. If $t = \frac{1}{2}$, then

$$\rho\left(\frac{1}{2}\right) = \frac{1}{4}.$$

   b. If

$$t = 1 - \frac{1}{2^n}, \quad n \geq 2,$$

   then

$$\frac{1}{2^n} = 1 - t, \quad n \geq 2,$$

   and

$$\begin{aligned} \rho(t) &= 1 - \frac{1}{2^{n-1}} \\ &= 1 - \frac{2}{2^n} \\ &= 1 - 2(1 - t) \\ &= 2t - 1. \end{aligned}$$

3. $t \in (1+U)$ and $t = 1 + \frac{1}{2^n}$ for some $n \in \mathbb{N}$. Then

$$\frac{1}{2^n} = t - 1,$$

and

$$\begin{aligned} \rho(t) &= 1 + \frac{1}{2^{n+1}} \\ &= 1 + \frac{1}{2} \cdot \frac{1}{2^n} \\ &= 1 + \frac{t-1}{2} \\ &= \frac{t+1}{2}. \end{aligned}$$

4. $t \in (2-U)$.
   a. If $t = \frac{3}{2}$, then

$$\rho\left(\frac{3}{2}\right) = \frac{5}{4}.$$

   b. If $t = 2 - \frac{1}{2^n}$ for some $n \in \mathbb{N}$, $n \geq 2$, then

$$\frac{1}{2^n} = 2 - t,$$

   and

$$\begin{aligned} \rho(t) &= 2 - \frac{1}{2^{n-1}} \\ &= 2 - \frac{2}{2^n} \\ &= 2 - 2(2-t) \\ &= 2t - 2. \end{aligned}$$

5. $t \in (2+U)$ and $t = 2 + \frac{1}{2^n}$, $n \in \mathbb{N}$. Then

$$\frac{1}{2^n} = t - 2,$$

and

$$\begin{aligned} \rho(t) &= 2 + \frac{1}{2^{n+1}} \\ &= 2 + \frac{1}{2} \cdot \frac{1}{2^n} \\ &= 2 + \frac{t-2}{2} \\ &= \frac{t+2}{2}. \end{aligned}$$

6. $t = 0$. Then
$$\rho(0) = 0.$$

7. $t = 1$. Then
$$\rho(1) = 1.$$

8. $t = 2$. Then
$$\rho(2) = 2.$$

Consequently,

$$\rho(t) = \begin{cases} \frac{1}{2}t & \text{if } t \in U, \\ \frac{1}{4} & \text{if } t = \frac{1}{2}, \\ 2t - 1 & \text{if } t \in (1 - U)\backslash\{\frac{1}{2}\}, \\ \frac{t+1}{2} & \text{if } t \in (1 + U), \\ \frac{5}{4} & \text{if } t = \frac{3}{2}, \\ 2(t-1) & \text{if } t \in (2 - U)\backslash\{\frac{3}{2}\}, \\ \frac{t+2}{2} & \text{if } t \in (2 + U), \\ 0 & \text{if } t = 0, \\ 1 & \text{if } t = 1, \\ 2 & \text{if } t = 2. \end{cases}$$

Exercise 1.5. Find $\rho(t)$, $t \in \mathbb{T}$, where $\mathbb{T}$ are the time scales in Exercise 1.3.

Definition 1.5. The backward graininess function $\nu : \mathbb{T} \to \mathbb{R}$ is defined as
$$\nu(t) = t - \rho(t), \quad t \in \mathbb{T}.$$

Note that $\nu(t) \geq 0$, $t \in \mathbb{T}$.

Example 1.40. Let $\mathbb{T} = h\mathbb{Z}$, where $h > 0$. Then using Example 1.30, we find
$$\begin{aligned} \nu(t) &= t - \rho(t) \\ &= t - (t - h) \\ &= h, \quad t \in \mathbb{T}. \end{aligned}$$

Example 1.41. Let $\mathbb{T} = 3^{\mathbb{N}_0}$. Then using Example 1.31, we find
$$\begin{aligned} \nu(t) &= t - \rho(t) \\ &= \begin{cases} 0 & \text{if } t = 1, \\ \frac{2t}{3} & \text{if } t > 1. \end{cases} \end{aligned}$$

Example 1.42. Let $\mathbb{T} = \mathbb{N}_0^k$, $k \in \mathbb{N}$. Then using Example 1.32, we find

$$\begin{aligned} \nu(t) &= t - \rho(t) \\ &= \begin{cases} 0 & \text{if } t = 0, \\ t - (\sqrt[k]{t} - 1)^k & \text{if } t > 0. \end{cases} \end{aligned}$$

Observe that

$$\begin{aligned} t - (\sqrt[k]{t} - 1)^k &= t - \left(t - \binom{k}{1}\sqrt[k]{t^{k-1}} + \binom{k}{2}\sqrt[k]{t^{k-2}} - \cdots + (-1)^k\right) \\ &= \binom{k}{1}\sqrt[k]{t^{k-1}} - \binom{k}{2}\sqrt[k]{t^{k-2}} + \cdots + (-1)^k. \end{aligned}$$

Consequently,

$$\nu(t) = \begin{cases} 0 & \text{if } t = 0, \\ \binom{k}{1}\sqrt[k]{t^{k-1}} - \binom{k}{2}\sqrt[k]{t^{k-2}} + \cdots + (-1)^k & \text{if } t > 0. \end{cases}$$

Example 1.43. Let $\mathbb{T} = \{H_n : n \in \mathbb{N}_0\}$, where H_n, $n \in \mathbb{N}_0$, are the harmonic numbers. Using Example 1.33, we find

$$\begin{aligned} \nu(t) &= t - \rho(t) \\ &= \begin{cases} 0 & \text{if } n = 0, \\ H_n - H_{n-1} & \text{if } n \geq 1. \end{cases} \end{aligned}$$

Example 1.44. Let $\mathbb{T} = P_{1,3}$. Using Example 1.34, we find

$$\begin{aligned} \nu(t) &= t - \rho(t) \\ &= \begin{cases} 0 & \text{if } t \in [0,1] \cup \bigcup_{k=1}^{\infty}(4k, 4k+1], \\ 3 & \text{if } t \in \bigcup_{k=1}^{\infty}\{4k\}. \end{cases} \end{aligned}$$

Example 1.45. Let $\mathbb{T} = C$, where C is the Cantor set. Using Example 1.35, we find

$$\begin{aligned} \nu(t) &= t - \rho(t) \\ &= \begin{cases} \frac{1}{3^{m+1}} & \text{if } t \in C_2,\ t = \sum_{k=1}^{k} \frac{a_k}{3^k} + \frac{2}{3^{m+1}}, \\ 0 & \text{if } t \in \mathbb{T}\backslash C_2. \end{cases} \end{aligned}$$

Example 1.46. Let

$$\mathbb{T} = \left\{ \sum_{k=0}^{n-1} a_k : n \in \mathbb{N},\ a_k > 0,\ k \in \mathbb{N} \right\}.$$

Using Example 1.36, we find

$$\nu(t) = t - \rho(t) = \begin{cases} 0 & \text{if } n = 1, \\ \alpha_{n-1} & \text{if } n \geq 2. \end{cases}$$

Example 1.47. Let $\mathbb{T} = \{t_n = -\frac{1}{n} : n \in \mathbb{N}\} \cup \mathbb{N}_0$. Using Example 1.37, we find

$$\nu(t) = t - \rho(t) = \begin{cases} 0 & \text{if } t \in \{-1, 0\}, \\ \frac{t^2}{t+1} & \text{if } t \in \{t_n = -\frac{1}{n} : n \in \mathbb{N},\ n \geq 2\}, \\ 1 & \text{if } t \in \mathbb{N}. \end{cases}$$

Example 1.48. Let $\mathbb{T} = \{(\frac{1}{2})^{2^n} : n \in \mathbb{N}_0\} \cup \{0, 1\}$. Using Example 1.38, we find

$$\nu(t) = t - \rho(t) = \begin{cases} t - t^2 & \text{if } t \in \{(\frac{1}{2})^{2^n} : n \in \mathbb{N}_0\}, \\ 0 & \text{if } t = 0 \\ \frac{1}{2} & \text{if } t = 1. \end{cases}$$

Example 1.49. Let $U = \{\frac{1}{2^n} : n \in \mathbb{N}\}$, and let

$$\mathbb{T} = \{0\} \cup U \cup (1 - U) \cup (1 + U) \cup (2 - U) \cup (2 + U) \cup \{1, 2\}.$$

Then using Example 1.39, we find

$$\nu(t) = t - \rho(t) = \begin{cases} \frac{t}{2} & \text{if } t \in U, \\ 0 & \text{if } t = 0, \\ 1 - t & \text{if } t \in (1 - U)\backslash\{\frac{1}{2}\}, \\ \frac{t-1}{2} & \text{if } t \in (1 + U), \\ \frac{1}{4} & \text{if } t = \frac{3}{2}, \\ 2 - t & \text{if } t \in (2 - U)\backslash\{\frac{3}{2}\}, \\ \frac{t-2}{2} & \text{if } t \in (2 + U), \\ 0 & \text{if } t = 0, \\ 0 & \text{if } t = 1, \\ 0 & \text{if } t = 2. \end{cases}$$

Exercise 1.6. Find $\nu(t)$, $t \in \mathbb{T}$, where $\mathbb{T}$ are the time scales in Exercise 1.3.

Example 1.50. We will show that in general σ is not continuous. Consider

$$\mathbb{T} = \left\{-\frac{1}{n} : n \in \mathbb{N}\right\} \cup \mathbb{N}_0.$$

We have

$$\begin{aligned} \sigma(0) &= 1, \\ \sigma\left(-\frac{1}{n}\right) &= -\frac{1}{n+1}, \quad n \in \mathbb{N}. \end{aligned}$$

Hence

$$\begin{aligned} \lim_{n\to\infty} \sigma\left(-\frac{1}{n}\right) &= -\lim_{n\to\infty} \frac{1}{n+1} \\ &= 0 \\ &\neq 1 \\ &= \sigma(0) \\ &= \sigma\left(\lim_{n\to\infty}\left(-\frac{1}{n}\right)\right). \end{aligned}$$

Example 1.51. We will show that in general ρ is not continuous. Consider

$$\mathbb{T} = [-2, -1] \cup \left\{\frac{1}{n}\right\}_{n\in\mathbb{N}} \cup \mathbb{N}_0.$$

Then

$$\begin{aligned} \rho(0) &= -1, \\ \rho\left(\frac{1}{n}\right) &= \frac{1}{n+1}, \quad n \in \mathbb{N}. \end{aligned}$$

Hence

$$\begin{aligned} \lim_{n\to\infty} \rho\left(\frac{1}{n}\right) &= \lim_{n\to\infty} \frac{1}{n+1} \\ &= 0 \\ &\neq -1 \\ &= \rho(0) \\ &= \rho\left(\lim_{n\to\infty} \frac{1}{n}\right). \end{aligned}$$

1.3 Classification of points

For elements of any time scale, we have the following classification.

Definition 1.6. For $t \in \mathbb{T}$, we have the following cases.

1. If $\sigma(t) > t$, then we say that t is right-scattered.
2. If $t < \sup \mathbb{T}$ and $\sigma(t) = t$, then we say that t is right-dense.

3. If $\rho(t) < t$, then we say that t is left-scattered.
4. If $t > \inf \mathbb{T}$ and $\rho(t) = t$, then we say that t is left-dense.
5. If t is both left- and right-scattered, then we say that t is isolated.
6. If t is both left- and right-dense, then we say that t is dense.

Example 1.52. Let $\mathbb{T} = h\mathbb{Z}$, $h > 0$. By Example 1.10 we have that

$$\begin{aligned}\sigma(t) &= t + h \\ &> t, \quad t \in \mathbb{T}.\end{aligned}$$

Thus all points of $\mathbb{T}$ are right-scattered. Now using Example 1.30, we get

$$\begin{aligned}\rho(t) &= t - h \\ &< t, \quad t \in \mathbb{T}.\end{aligned}$$

Therefore all points of $\mathbb{T}$ are left-scattered. Hence we conclude that all points of $\mathbb{T}$ are isolated.

Example 1.53. Let $\mathbb{T} = 3^{\mathbb{N}_0}$. Take arbitrary $t \in \mathbb{T}$. We have the following cases.

1. $t = 1$. Then by Example 1.11 we have

$$\begin{aligned}\sigma(1) &= 3 \\ &> 1,\end{aligned}$$

   i. e., $t = 1$ is right-scattered. By Example 1.31 we have

$$\rho(1) = 1.$$

   Since $1 = \inf \mathbb{T}$, we conclude that $t = 1$ is not left-dense.
2. $t > 1$. By Example 1.11 we have

$$\begin{aligned}\sigma(t) &= 3t \\ &> t.\end{aligned}$$

   Thus t is right-scattered. By Example 1.31 we get

$$\begin{aligned}\rho(t) &= \frac{t}{3} \\ &< t,\end{aligned}$$

   i. e., t is left-scattered. Hence we conclude that t is isolated.

Example 1.54. Let $\mathbb{T} = \mathbb{N}_0^k$, $k \in \mathbb{N}$. Take arbitrary $t \in \mathbb{T}$. We have the following cases.

1. $t = 0$. Then by Example 1.12 we have

$$\begin{aligned}\sigma(0) &= 1\\ &> 0,\end{aligned}$$

i. e., $t = 0$ is right-scattered. By Example 1.32 we obtain

$$\rho(0) = 0.$$

Since $0 = \inf \mathbb{T}$, we conclude that $t = 0$ is not left-dense.

2. $t > 0$. Then by Example 1.12 we get

$$\begin{aligned}\sigma(t) &= \left(\sqrt[k]{t} + 1\right)^k\\ &> t,\end{aligned}$$

i. e., t is right-scattered. By Example 1.32 we find

$$\begin{aligned}\rho(t) &= \left(\sqrt[k]{t} - 1\right)^k\\ &< t,\end{aligned}$$

i. e., t is left-scattered. Therefore t is isolated.

Example 1.55. Let $\mathbb{T} = \{H_n : n \in \mathbb{N}_0\}$, where H_n, $n \in \mathbb{N}_0$, are the harmonic numbers. Take $n \in \mathbb{N}_0$. We have the following cases.

1. $n = 0$. Then by Example 1.13 we get

$$\sigma(H_0) = H_1,$$

i. e., H_0 is right-scattered. By Example 1.33 we have

$$\rho(H_0) = H_0.$$

Since $H_0 = \inf \mathbb{T}$, we conclude that H_0 is not left-dense.

2. $n > 0$. Then by Example 1.13 we get

$$\begin{aligned}\sigma(H_n) &= H_{n+1}\\ &> H_n.\end{aligned}$$

Then H_n is right-scattered. By Example 1.33 we get

$$\begin{aligned}\rho(H_n) &= H_{n-1}\\ &< H_n,\end{aligned}$$

i. e., H_n is left-scattered. Then H_n is isolated.

Example 1.56. Let $\mathbb{T} = P_{1,3}$. Take arbitrary $t \in \mathbb{T}$. We have the following cases.

1. Let $t \in \bigcup_{k=0}^{\infty}[4k, 4k+1)$. Then by Example 1.14 we get

$$\sigma(t) = t,$$

i. e., t is right-dense. By Example 1.34 we find

$$\rho(t) = t \quad \text{if } t \neq 0,$$

i. e., t is left-dens if $t \neq 0$. Thus t is dense if $t \neq 0$.

2. $t = 0$. Then by Example 1.14 we obtain

$$\sigma(0) = 0,$$

i.e, $t = 0$ is right-dense. By Example 1.34 we find

$$\rho(0) = 0.$$

Since $0 = \inf \mathbb{T}$, we conclude that 0 is not left-dense.

3. $t \in \bigcup_{k=1}^{\infty}\{4k\}$. Then by Example 1.14 we get

$$\sigma(t) = t,$$

i. e., t is right-dense. By Example 1.34 we find

$$\begin{aligned} \rho(t) &= t - 3 \\ &< t, \end{aligned}$$

i. e., t is left-scattered.

4. $t \in \bigcup_{k=0}^{\infty}\{4k+1\}$. Then by Example 1.14 we find

$$\begin{aligned} \sigma(t) &= t + 3 \\ &> t, \end{aligned}$$

i. e., t is right-scattered. By Example 1.34 we find

$$\rho(t) = t,$$

i. e., t is left-dense.

Example 1.57. Let $\mathbb{T} = C$, where C is the Cantor set. Take arbitrary $t \in \mathbb{T}$. We have the following cases.

1. $t \in C_1$. Then by Example 1.15 we have

$$\begin{aligned}\sigma(t) &= t + \frac{1}{3^{m+1}}\\ &> t,\end{aligned}$$

   i. e., t is right-scattered. By Example 1.35 we find

$$\rho(t) = t.$$

   If $t \neq 0$, then it is left-dense. If $t = 0$, then it is not left-dense because $0 = \inf \mathbb{T}$.
2. $t \in C_2$. Then by Example 1.15 we get

$$\sigma(t) = t,$$

   i. e., t is right-dense. By Example 1.35 we find

$$\begin{aligned}\rho(t) &= t - \frac{1}{3^{m+1}}\\ &< t,\end{aligned}$$

   i. e., t is left-scattered.
3. $t \in T\backslash C_1$. We have the following subcases.
   a. $t \in C_2$. Then by Example 1.15 we find

$$\sigma(t) = t,$$

   i. e., t is right-dense. By Example 1.35 we obtain

$$\begin{aligned}\rho(t) &= t - \frac{1}{3^{m+1}}\\ &< t,\end{aligned}$$

   i. e., t is left-scattered.
   b. $t \in T\backslash C_2$. Then by Example 1.15 we arrive at

$$\sigma(t) = t,$$

   i. e., t is right-dense. By Example 1.35 we have

$$\rho(t) = t.$$

   If $t \neq 0$, then it is left-dense. If $t = 0$, then it is not left-dense.
4. $t \in \mathbb{T}\backslash C_2$. We have the following subcases.

a. $t \in C_1$. Then by Example 1.15 we find

$$\sigma(t) = t + \frac{1}{3^{m+1}} \\ > t,$$

i. e., t is right-scattered. By Example 1.35 we find

$$\rho(t) = t.$$

If $t \neq 0$, then it is left-dense. If $t = 0$, then it is not left-dense.

b. Let $t \in T\backslash C_1$. By Example 1.15, we have

$$\sigma(t) = t,$$

i. e., t is right-dense. By Example 1.35 we have

$$\rho(t) = t.$$

If $t \neq 0$, then it is left-dense and hence dense. If $t = 0$, then it is not left-dense.

Example 1.58. Let

$$\mathbb{T} = \left\{ \sum_{k=0}^{n-1} a_k : a_k > 0,\ k \in \mathbb{N}_0,\ n \in \mathbb{N} \right\}.$$

Take arbitrary $t \in \mathbb{T}$. Then there is $n \in \mathbb{N}$ such that

$$t = \sum_{k=0}^{n-1} a_k.$$

We have the following cases.

1. $n = 1$. Then by Example 1.16 we have

$$\sigma(t) = t + a_n \\ > t,$$

i. e., t is right-scattered. By Example 1.36 we arrive at

$$\rho(t) = t.$$

Since $t = \inf \mathbb{T}$, we conclude that t is not left-dense.

2. $n > 1$. Then by Example 1.16 we find

$$\begin{aligned}\sigma(t) &= \sum_{k=0}^{n} a_k \\ &> t,\end{aligned}$$

i. e., t is right-scattered. By Example 1.36 we find

$$\begin{aligned}\rho(t) &= t - a_{n-1} \\ &< t,\end{aligned}$$

i. e., t is left-scattered. Therefore t is isolated.

Example 1.59. Let $\mathbb{T} = \{t_n = -\frac{1}{n} : n \in \mathbb{N}\} \cup \mathbb{N}_0$. Take arbitrary $t \in \mathbb{T}$. We have the following cases.

1. $t = -1$. Then by Example 1.17 we have

$$\begin{aligned}\sigma(-1) &= -\frac{1}{2} \\ &> -1,\end{aligned}$$

i. e., $t = -1$ is right-scattered. By Example 1.37 we find

$$\rho(-1) = -1.$$

Since $-1 = \inf \mathbb{T}$, we conclude that $t = -1$ is not left-dense.

2. $t = 0$. Then by Example 1.17 we have

$$\begin{aligned}\sigma(0) &= 1 \\ &> 0,\end{aligned}$$

i. e., $t = 0$ is right-scattered. By Example 1.37 we obtain

$$\rho(0) = 0.$$

Since $0 > \inf \mathbb{T}$, we conclude that $t = 0$ is left-dense.

3. $t \in \{t_n = -\frac{1}{n} : n \in \mathbb{N}\} \backslash \{-1\}$. Then by Example 1.17 we have

$$\begin{aligned}\sigma(t) &= -\frac{t}{t-1} \\ &> t,\end{aligned}$$

i. e., t is right-scattered. By Example 1.37 we get

$$\begin{aligned}\rho(t) &= \frac{t}{t+1} \\ &< t,\end{aligned}$$

i. e., t is left-scattered. Thus t is isolated.

4. $t \in \mathbb{N}$. Then by Example 1.17 we have

$$\begin{aligned}\sigma(t) &= t+1\\ &> t,\end{aligned}$$

i. e., t is right-scattered. By Example 1.37 we obtain

$$\begin{aligned}\rho(t) &= t-1\\ &< t,\end{aligned}$$

i. e., t is left-scattered. Thus t is isolated.

Example 1.60. Let $\mathbb{T} = \{t_n = (\frac{1}{2})^{2^n} : n \in \mathbb{N}_0\} \cup \{0,1\}$. Take arbitrary $t \in \mathbb{T}$. We have the following cases.

1. $t = 0$. Then by Example 1.18 we have

$$\sigma(0) = 0,$$

i. e., $t = 0$ is right-dense. By Example 1.38 we get

$$\rho(0) = 0.$$

Since $0 = \inf \mathbb{T}$, we conclude that $t = 0$ is not left-dense.

2. $t = \frac{1}{2}$. Then by Example 1.18 we have

$$\begin{aligned}\sigma\left(\frac{1}{2}\right) &= 1\\ &> \frac{1}{2},\end{aligned}$$

i. e., $t = \frac{1}{2}$ is right-scattered. By Example 1.38 we get

$$\begin{aligned}\rho\left(\frac{1}{2}\right) &= \frac{1}{4}\\ &< \frac{1}{2},\end{aligned}$$

i. e., $t = \frac{1}{2}$ is left-scattered. Thus $t = \frac{1}{2}$ is isolated.

3. $t = 1$. Then by Example 1.18 we have

$$\sigma(1) = 1.$$

Since $1 = \sup \mathbb{T}$, we conclude that $t = 1$ is not right-dense. By Example 1.38 we obtain

$$\rho(1) = \frac{1}{2} < 1,$$

i. e., $t = 1$ is left-scattered.

4. $t \in \{t_n = (\frac{1}{2})^{2^n} : n \in \mathbb{N}\}$. Then there is $n \in \mathbb{N}$ such that

$$t = \left(\frac{1}{2}\right)^{2^n}.$$

By Example 1.18 we have

$$\sigma(t) = \sqrt{t} > t,$$

i. e., t is right-scattered. By Example 1.38 we obtain

$$\rho(t) = t^2 < t,$$

i. e., t is left-scattered. Thus t is isolated.

Example 1.61. Let $U = \{\frac{1}{2^n} : n \in \mathbb{N}\}$, and let

$$\mathbb{T} = \{0\} \cup U \cup (1 - U) \cup (1 + U) \cup (2 - U) \cup (2 + U) \cup \{1, 2\}.$$

Take arbitrary $t \in \mathbb{T}$. Then we have the following cases.

1. $t = 0$. Then by Example 1.19 we have

$$\sigma(0) = 0,$$

i. e., $t = 0$ is right-dense. By Example 1.39 we get

$$\rho(0) = 0.$$

Since $0 = \inf \mathbb{T}$, we conclude that $t = 0$ is not left-dense.

2. $t = \frac{1}{2}$. Then by Example 1.19 we have

$$\sigma\left(\frac{1}{2}\right) = \frac{3}{4},$$

i. e., $t = \frac{1}{2}$ is right-scattered. By Example 1.39 we get

$$\rho\left(\frac{1}{2}\right) = \frac{1}{4},$$

i. e., $t = \frac{1}{2}$ is left-scattered. Thus $t = \frac{1}{2}$ is isolated.

3. $t = 1$. Then by Example 1.19 we have

$$\sigma(1) = 1,$$

i. e., $t = 1$ is right-dense. By Example 1.39 we find

$$\rho(1) = 1,$$

i. e., $t = 1$ is left-dense. Thus $t = 1$ is dense.

4. $t = \frac{3}{2}$. Then by Example 1.19 we have

$$\sigma\left(\frac{3}{2}\right) = \frac{7}{4},$$

i. e., $t = \frac{3}{2}$ is right-scattered. By Example 1.39 we find

$$\rho\left(\frac{3}{2}\right) = \frac{5}{4},$$

i. e., $t = \frac{3}{2}$ is left-scattered. Thus $t = \frac{3}{2}$ is isolated.

5. $t = 2$. Then by Example 1.19 we have

$$\sigma(2) = 2,$$

i. e., $t = 2$ is right-dense. By Example 1.39 we get

$$\rho(2) = 2,$$

i. e., $t = 2$ is left-dense. Thus $t = 2$ is dense.

6. $t = \frac{5}{2}$. Then by Example 1.19 we get

$$\sigma\left(\frac{5}{2}\right) = \frac{5}{2}.$$

Since $\frac{5}{2} = \sup \mathbb{T}$, we conclude that $t = \frac{5}{2}$ is not right-dense. By Example 1.39 we find

$$\rho\left(\frac{5}{2}\right) = \frac{9}{4}.$$

Thus $t = \frac{5}{2}$ is left-scattered.

7. $t \in U \backslash \{\frac{1}{2}\}$. Then by Example 1.19 we have

$$\begin{aligned} \sigma(t) &= 2t \\ &> t, \end{aligned}$$

i. e., t is right-scattered. By Example 1.39 we get

$$\rho(t) = \frac{t}{2}$$

$$< t,$$

i. e., t is left-scattered. Thus t is isolated.

8. $t \in (1 - U)\backslash\{\frac{1}{2}\}$. Then by Example 1.19 we have

$$\begin{aligned}\sigma(t) &= \frac{1+t}{2}\\ &> t,\end{aligned}$$

i. e., t is right-scattered. By Example 1.39 we find

$$\begin{aligned}\rho(t) &= 2t - 1\\ &< t,\end{aligned}$$

i. e., t is left-scattered. Thus t is isolated.

9. $t \in (1 + U)\backslash\{\frac{3}{2}\}$. Then by Example 1.19 we have

$$\begin{aligned}\sigma(t) &= 2t - 1\\ &> t,\end{aligned}$$

i. e., t is right-scattered. By Example 1.39 we find

$$\begin{aligned}\rho(t) &= \frac{t+1}{2}\\ &< t,\end{aligned}$$

i. e., t is left-scattered. Thus t is isolated.

10. $t \in (2 - U)\backslash\{\frac{3}{2}\}$. Then by Example 1.19 we have

$$\begin{aligned}\sigma(t) &= \frac{2+t}{2}\\ &> t,\end{aligned}$$

i. e., t is right-scattered. By Example 1.39 we find

$$\begin{aligned}\rho(t) &= 2t - 2\\ &< t,\end{aligned}$$

i. e., t is left-scattered. Thus t is isolated.

11. $t \in (2 + U)\backslash\{\frac{5}{2}\}$. Then by Example 1.19 we have

$$\begin{aligned}\sigma(t) &= 2(t - 1)\\ &> t,\end{aligned}$$

i. e., t is right-scattered. By Example 1.39 we find

$$\begin{aligned}\rho(t) &= \frac{t+2}{2}\\ &< t,\end{aligned}$$

i. e., t is left-scattered. Thus t is isolated.

Exercise 1.7. Classify the points of $\mathbb{T}$, where $\mathbb{T}$ are the time scales in Exercise 1.3.

1.4 The topology of time scales

Definition 1.7. Let $a, b \in \mathbb{T}$, $a \leq b$. Define closed, half-open, and open time scales intervals as

$$\begin{aligned}&[a,b]_{\mathbb{T}} = \{x \in \mathbb{T} : a \leq x \leq b\},\\ &[a,b)_{\mathbb{T}} = \{x \in \mathbb{T} : a \leq x < b\}, \quad (a,b]_{\mathbb{T}} = \{x \in \mathbb{T} : a < x \leq b\},\\ &(a,b)_{\mathbb{T}} = \{x \in \mathbb{T} : a < x < b\},\end{aligned}$$

respectively.

Example 1.62. Let $\mathbb{T} = 3\mathbb{Z}$. Then

$$\begin{aligned}[-3,12]_{\mathbb{T}} &= \{-3,0,3,6,9,12\},\\ [-3,12)_{\mathbb{T}} &= \{-3,0,3,6,9\},\\ (-3,12]_{\mathbb{T}} &= \{0,3,6,9,12\},\\ (-3,12)_{\mathbb{T}} &= \{0,3,6,9\}.\end{aligned}$$

Example 1.63. Let $\mathbb{T} = 3^{\mathbb{N}_0}$. Then

$$\begin{aligned}[3,243]_{\mathbb{T}} &= \{3,9,27,81,243\},\\ [3,243)_{\mathbb{T}} &= \{3,9,27,81\},\\ (3,243]_{\mathbb{T}} &= \{9,27,81,243\},\\ (3,243)_{\mathbb{T}} &= \{9,27,81\}.\end{aligned}$$

Example 1.64. Let $\mathbb{T} = \mathbb{N}_0^3$. Then

$$\begin{aligned}[1,27]_{\mathbb{T}} &= \{1,8,27\},\\ [1,27)_{\mathbb{T}} &= \{1,8\},\\ (1,27]_{\mathbb{T}} &= \{8,27\},\\ (1,27)_{\mathbb{T}} &= \{8\}.\end{aligned}$$

Example 1.65. Let $\mathbb{T} = \{H_n : n \in \mathbb{N}_0\}$, where H_n, $n \in \mathbb{N}_0$, are the harmonic numbers. Then

$$\begin{aligned}
\left[1, \frac{147}{60}\right]_{\mathbb{T}} &= \left\{1, \frac{3}{2}, \frac{11}{6}, \frac{25}{12}, \frac{137}{60}, \frac{147}{60}\right\},\\
\left[1, \frac{147}{60}\right)_{\mathbb{T}} &= \left\{1, \frac{3}{2}, \frac{11}{6}, \frac{25}{12}, \frac{137}{60}\right\},\\
\left(1, \frac{147}{60}\right]_{\mathbb{T}} &= \left\{\frac{3}{2}, \frac{11}{6}, \frac{25}{12}, \frac{137}{60}, \frac{147}{60}\right\},\\
\left(1, \frac{147}{60}\right)_{\mathbb{T}} &= \left\{\frac{3}{2}, \frac{11}{6}, \frac{25}{12}, \frac{137}{60}\right\}.
\end{aligned}$$

Example 1.66. Let $\mathbb{T} = P_{1,3}$. Then

$$\begin{aligned}
[0, 12]_{\mathbb{T}} &= [0, 1] \cup [4, 5] \cup [8, 9] \cup \{12\},\\
[0, 12)_{\mathbb{T}} &= [0, 1] \cup [4, 5] \cup [8, 9],\\
(0, 12]_{\mathbb{T}} &= (0, 1] \cup [4, 5] \cup [8, 9] \cup \{12\},\\
(0, 12)_{\mathbb{T}} &= (0, 1] \cup [4, 5] \cup [8, 9].
\end{aligned}$$

Example 1.67. Let $\mathbb{T} = \{\sum_{k=0}^{n} k : n \in \mathbb{N}\}$. Then

$$\begin{aligned}
[0, 28]_{\mathbb{T}} &= \{0, 1, 3, 6, 10, 15, 21, 28\},\\
[0, 28)_{\mathbb{T}} &= \{0, 1, 3, 6, 10, 15, 21\},\\
(0, 28]_{\mathbb{T}} &= \{1, 3, 6, 10, 15, 21, 28\},\\
(0, 28)_{\mathbb{T}} &= \{1, 3, 6, 10, 15, 21\}.
\end{aligned}$$

Example 1.68. Let $\mathbb{T} = \{-\frac{1}{n} : n \in \mathbb{N}\} \cup \mathbb{N}_0$. Then

$$\begin{aligned}
\left[-\frac{1}{3}, 3\right]_{\mathbb{T}} &= \left\{-\frac{1}{n} : n \in \mathbb{N},\ n \ge 3\right\} \cup \{0, 1, 2, 3\},\\
\left[-\frac{1}{3}, 3\right)_{\mathbb{T}} &= \left\{-\frac{1}{n} : n \in \mathbb{N},\ n \ge 3\right\} \cup \{0, 1, 2\},\\
\left(-\frac{1}{3}, 3\right]_{\mathbb{T}} &= \left\{-\frac{1}{n} : n \in \mathbb{N},\ n \ge 4\right\} \cup \{0, 1, 2, 3\},\\
\left(-\frac{1}{3}, 3\right)_{\mathbb{T}} &= \left\{-\frac{1}{n} : n \in \mathbb{N},\ n \ge 4\right\} \cup \{0, 1, 2\}.
\end{aligned}$$

Example 1.69. Let $\mathbb{T} = \{(\frac{1}{2})^{2^n} : n \in \mathbb{N}_0\} \cup \{0, 1\}$. Then

$$\left[0, \frac{1}{2}\right]_{\mathbb{T}} = \{0\} \cup \left\{\left(\frac{1}{2}\right)^{2^n} : n \in \mathbb{N}_0\right\},$$

$$\left[0, \frac{1}{2}\right)_{\mathbb{T}} = \{0\} \cup \left\{\left(\frac{1}{2}\right)^{2^n} : n \in \mathbb{N}\right\},$$
$$\left(0, \frac{1}{2}\right]_{\mathbb{T}} = \cup \left\{\left(\frac{1}{2}\right)^{2^n} : n \in \mathbb{N}_0\right\},$$
$$\left(0, \frac{1}{2}\right)_{\mathbb{T}} = \left\{\left(\frac{1}{2}\right)^{2^n} : n \in \mathbb{N}\right\}.$$

Example 1.70. Let $U = \{\frac{1}{2^n} : n \in \mathbb{N}\}$, and let

$$\mathbb{T} = \{0\} \cup U \cup (1 - U) \cup (1 + U) \cup (2 - U) \cup (2 + U) \cup \{1, 2\}.$$

Then

$$\left[0, \frac{7}{8}\right]_{\mathbb{T}} = \{0\} \cup U \cup \left\{\frac{3}{4}, \frac{7}{8}\right\},$$
$$\left[0, \frac{7}{8}\right)_{\mathbb{T}} = \{0\} \cup U \cup \left\{\frac{3}{4}\right\},$$
$$\left(0, \frac{7}{8}\right]_{\mathbb{T}} = U \cup \left\{\frac{3}{4}, \frac{7}{8}\right\},$$
$$\left(0, \frac{7}{8}\right)_{\mathbb{T}} = U \cup \left\{\frac{3}{4}\right\}.$$

Exercise 1.8. Find

$$[a, b]_{\mathbb{T}}, \quad [a, b)_{\mathbb{T}}, \quad (a, b]_{\mathbb{T}}, \quad (a, b)_{\mathbb{T}},$$

where
1. $\mathbb{T} = 4\mathbb{Z} + 1$, $a = -3$, $b = 9$.
2. $\mathbb{T} = (-2\mathbb{N}_0) \cup 3^{\mathbb{N}_0}$, $a = -4$, $b = 9$.
3. $\mathbb{T} = P_{3,7} \cup [4, 6]$, $a = 0$, $b = 12$.
4. $\mathbb{T} = \{0\} \cup 11^{\mathbb{N}_0}$, $a = 0$, $b = 121$.
5. $\mathbb{T} = [1, 2] \cup [3, 4] \cup [7, 8] \cup 9^{\mathbb{N}}$, $a = 2$, $b = 9$.

Definition 1.8. Define the sets

$$\mathbb{T}^{\kappa} = \begin{cases} \mathbb{T} \backslash (\rho(\sup \mathbb{T}), \sup \mathbb{T}]_{\mathbb{T}} & \text{if } \sup \mathbb{T} < \infty, \\ \mathbb{T} & \text{if } \sup \mathbb{T} = \infty \end{cases}$$

and

$$\mathbb{T}_{\kappa} = \begin{cases} \mathbb{T} \backslash [\inf \mathbb{T}, \sigma(\inf \mathbb{T}))_{\mathbb{T}} & \text{if } \inf \mathbb{T} > -\infty, \\ \mathbb{T} & \text{if } \inf \mathbb{T} > -\infty. \end{cases}$$

Example 1.71. Let $\mathbb{T} = h\mathbb{Z}$, $h > 0$. Then

$$\inf \mathbb{T} = -\infty,$$

$$\sup \mathbb{T} = \infty$$

and

$$\begin{aligned} \mathbb{T}^\kappa &= \mathbb{T}, \\ \mathbb{T}_\kappa &= \mathbb{T}. \end{aligned}$$

Example 1.72. Let $\mathbb{T} = 3^{\mathbb{N}_0}$. Then

$$\begin{aligned} \sup \mathbb{T} &= \infty, \\ \inf \mathbb{T} &= 1, \\ \sigma(\inf \mathbb{T}) &= \sigma(1) \\ &= 3. \end{aligned}$$

Hence

$$\mathbb{T}^\kappa = \mathbb{T},$$

and

$$\begin{aligned} \mathbb{T}_\kappa &= \mathbb{T} \backslash [\inf \mathbb{T}, \sigma(\inf \mathbb{T}))_{\mathbb{T}} \\ &= \mathbb{T} \backslash [1, 3)_{\mathbb{T}} \\ &= \mathbb{T} \backslash \{1\} \\ &= 3^{\mathbb{N}}. \end{aligned}$$

Example 1.73. Let $\mathbb{T} = \mathbb{N}_0^k$, $k \in \mathbb{N}$. Then

$$\begin{aligned} \sup \mathbb{T} &= \infty, \\ \inf \mathbb{T} &= 0, \\ \sigma(\inf \mathbb{T}) &= \sigma(0) \\ &= 1. \end{aligned}$$

Hence

$$\mathbb{T}^\kappa = \mathbb{T},$$

and

$$\begin{aligned} \mathbb{T}_\kappa &= \mathbb{T} \backslash [\inf \mathbb{T}, \sigma(\inf \mathbb{T}))_{\mathbb{T}} \\ &= \mathbb{T} \backslash [0, 1)_{\mathbb{T}} \\ &= \mathbb{T} \backslash \{0\} \\ &= \mathbb{N}^k. \end{aligned}$$

Example 1.74. Let $\mathbb{T} = P_{1,3}$. Then

$$\begin{aligned}\sup \mathbb{T} &= \infty,\\ \inf \mathbb{T} &= 0,\\ \sigma(\inf \mathbb{T}) &= \sigma(0)\\ &= 0.\end{aligned}$$

Hence

$$\mathbb{T}^{\kappa} = \mathbb{T},$$

and

$$\begin{aligned}\mathbb{T}_{\kappa} &= \mathbb{T}\backslash[\inf \mathbb{T}, \sigma(\inf \mathbb{T}))_{\mathbb{T}}\\ &= \mathbb{T}\backslash[0,0)_{\mathbb{T}}\\ &= \mathbb{T}\backslash\emptyset\\ &= \mathbb{T}.\end{aligned}$$

Example 1.75. Let $\mathbb{T} = C$, where C is the Cantor set. Then

$$\begin{aligned}\sup \mathbb{T} &= 1,\\ \rho(\sup \mathbb{T}) &= \rho(1)\\ &= 1,\\ \inf \mathbb{T} &= 0,\\ \sigma(\inf \mathbb{T}) &= \sigma(0)\\ &= 0.\end{aligned}$$

Hence

$$\begin{aligned}\mathbb{T}^{\kappa} &= \mathbb{T}\backslash(\rho(\sup \mathbb{T}), \sup \mathbb{T}]\\ &= \mathbb{T}\backslash(1,1]\\ &= \mathbb{T}\backslash\emptyset\\ &= \mathbb{T},\end{aligned}$$

and

$$\begin{aligned}\mathbb{T}_{\kappa} &= \mathbb{T}\backslash[\inf \mathbb{T}, \sigma(\inf \mathbb{T}))_{\mathbb{T}}\\ &= \mathbb{T}\backslash[0,0)_{\mathbb{T}}\\ &= \mathbb{T}\backslash\emptyset\\ &= \mathbb{T}.\end{aligned}$$

Example 1.76. Let $\mathbb{T} = \{\sum_{k=0}^{n-1} a_k : n \in \mathbb{N},\ a_k > 0,\ k \in \mathbb{N}_0\}$. Then

$$\begin{aligned}\sup \mathbb{T} &= \infty,\\ \inf \mathbb{T} &= a_0,\\ \sigma(\inf \mathbb{T}) &= \sigma(a_0)\\ &= a_0 + a_1.\end{aligned}$$

Hence

$$\mathbb{T}^\kappa = \mathbb{T},$$

and

$$\begin{aligned}\mathbb{T}_\kappa &= \mathbb{T}\backslash[\inf \mathbb{T}, \sigma(\inf \mathbb{T}))_{\mathbb{T}}\\ &= \mathbb{T}\backslash[a_0, a_0 + a_1)_{\mathbb{T}}\\ &= \mathbb{T}\backslash\{a_0\}\\ &= \left\{\sum_{k=0}^{n-1} a_k : n \in \mathbb{N},\ n \geq 2,\ a_k > 0,\ k \in \mathbb{N}_0\right\}.\end{aligned}$$

Example 1.77. Let $\mathbb{T} = \{\frac{1}{n} : n \in \mathbb{N}\} \cup \mathbb{N}_0$. Then

$$\begin{aligned}\sup \mathbb{T} &= \infty,\\ \inf \mathbb{T} &= 0,\\ \sigma(\inf \mathbb{T}) &= \sigma(0)\\ &= 0.\end{aligned}$$

Hence

$$\mathbb{T}^\kappa = \mathbb{T},$$

and

$$\begin{aligned}\mathbb{T}_\kappa &= \mathbb{T}\backslash[\inf \mathbb{T}, \sigma(\inf \mathbb{T}))_{\mathbb{T}}\\ &= \mathbb{T}\backslash[0, 0)_{\mathbb{T}}\\ &= \mathbb{T}\backslash\emptyset\\ &= \mathbb{T}.\end{aligned}$$

Example 1.78. Let $\mathbb{T} = \{(\frac{1}{2})^{2^n} : n \in \mathbb{N}_0\} \cup \{0, 1\}$. Then

$$\begin{aligned}\sup \mathbb{T} &= 1,\\ \rho(\sup \mathbb{T}) &= \rho(1)\end{aligned}$$

$$\begin{aligned}&= \frac{1}{2},\\ \inf \mathbb{T} &= 0,\\ \sigma(\inf \mathbb{T}) &= \sigma(0)\\ &= 0.\end{aligned}$$

Hence

$$\begin{aligned}\mathbb{T}^\kappa &= \mathbb{T}\backslash(\rho(\sup \mathbb{T}), \sup \mathbb{T}]\\ &= \mathbb{T}\backslash\left(\frac{1}{2}, 1\right]\\ &= \mathbb{T}\backslash\{1\}\\ &= \left\{\left(\frac{1}{2}\right)^{2^n} : n \in \mathbb{N}_0\right\} \cup \{0\},\end{aligned}$$

and

$$\begin{aligned}\mathbb{T}_\kappa &= \mathbb{T}\backslash[\inf \mathbb{T}, \sigma(\inf \mathbb{T}))_{\mathbb{T}}\\ &= \mathbb{T}\backslash[0, 0)_{\mathbb{T}}\\ &= \mathbb{T}\backslash\emptyset\\ &= \mathbb{T}.\end{aligned}$$

Example 1.79. Let $U = \{\frac{1}{2^n} : n \in \mathbb{N}\}$, and let

$$\mathbb{T} = \{0\} \cup U \cup (1 - U) \cup (1 + U) \cup (2 - U) \cup (2 + U) \cup \{1, 2\}.$$

Then

$$\begin{aligned}\sup \mathbb{T} &= \frac{5}{2},\\ \rho(\sup \mathbb{T}) &= \rho\left(\frac{5}{2}\right)\\ &= \frac{9}{4},\\ \inf \mathbb{T} &= 0,\\ \sigma(\inf \mathbb{T}) &= \sigma(0)\\ &= 0.\end{aligned}$$

Hence

$$\begin{aligned}\mathbb{T}^\kappa &= \mathbb{T}\backslash(\rho(\sup \mathbb{T}), \sup \mathbb{T}]\\ &= \mathbb{T}\backslash\left(\frac{9}{4}, \frac{5}{2}\right]\end{aligned}$$

$$= \mathbb{T}\backslash\left\{\frac{5}{2}\right\},$$

and

$$\begin{aligned}\mathbb{T}_\kappa &= \mathbb{T}\backslash[\inf\mathbb{T}, \sigma(\inf\mathbb{T}))_\mathbb{T}\\ &= \mathbb{T}\backslash[0,0)_\mathbb{T}\\ &= \mathbb{T}\backslash\emptyset\\ &= \mathbb{T}.\end{aligned}$$

Exercise 1.9. Find $\mathbb{T}^\kappa$ and $\mathbb{T}_\kappa$, where $\mathbb{T}$ are the time scales in Exercise 1.3.

1.5 Functions and jump operators

In this section, we consider compositions of real-valued functions with jump operators.

Definition 1.9. For a function $f : \mathbb{T} \to \mathbb{R}$ and $k \in \mathbb{N}_0$, define

$$f^{\sigma^k}(t) = f(\sigma^k(t)), \quad t \in \mathbb{T}.$$

Example 1.80. Let $\mathbb{T} = h\mathbb{Z}$, $h > 0$, and

$$f(t) = 1 + 2t + 3t^2, \quad t \in \mathbb{T}.$$

We will find $f^\sigma(t)$ and $f^{\sigma^2}(t)$ for $t \in \mathbb{T}$. By Example 1.10 we have

$$\sigma(t) = t + h, \quad t \in \mathbb{T},$$

and

$$\begin{aligned}f^\sigma(t) &= f(\sigma(t))\\ &= 1 + 2\sigma(t) + 3(\sigma(t))^2\\ &= 1 + 2(t+h) + 3(t+h)^2\\ &= 1 + 2t + 2h + 3t^2 + 6ht + 3h^2\\ &= 1 + 2h + 3h^2 + 2(1+3h)t + 3t^2, \quad t \in \mathbb{T}.\end{aligned}$$

Hence

$$\begin{aligned}f^{\sigma^2}(t) &= f(\sigma^2(t))\\ &= f(\sigma(\sigma(t)))\\ &= f^\sigma(\sigma(t))\\ &= 1 + 2h + 3h^2 + 2(1+3h)\sigma(t) + 3(\sigma(t))^2\end{aligned}$$

$$\begin{aligned}
&= 1 + 2h + 3h^2 + 2(1 + 3h)(t + h) + 3(t + h)^2 \\
&= 1 + 2h + 3h^2 + 2h(1 + 3h) + 2(1 + 3h)t + 3t^2 + 6ht + 3h^2 \\
&= 1 + 2h + 6h^2 + 2h + 6h^2 + 2(1 + 3h + 3h)t + 3t^2 \\
&= 1 + 4h + 12h^2 + 2(1 + 6h)t + 3t^2, \quad t \in \mathbb{T}.
\end{aligned}$$

Example 1.81. Let $\mathbb{T} = 3^{\mathbb{N}_0}$, and let

$$f(t) = \frac{1+t}{2+t}, \quad t \in \mathbb{T}.$$

We will find $f^{\sigma}(t)$ and $f^{\sigma^2}(t)$ for $t \in \mathbb{T}$. By Example 1.11 we have

$$\sigma(t) = 3t, \quad t \in \mathbb{T},$$

and

$$\begin{aligned}
f^{\sigma}(t) &= f(\sigma(t)) \\
&= \frac{1+\sigma(t)}{2+\sigma(t)} \\
&= \frac{1+3t}{2+3t}, \quad t \in \mathbb{T}.
\end{aligned}$$

Hence

$$\begin{aligned}
f^{\sigma^2}(t) &= f(\sigma^2(t)) \\
&= f(\sigma(\sigma(t))) \\
&= f^{\sigma}(\sigma(t)) \\
&= \frac{1+3\sigma(t)}{2+3\sigma(t)} \\
&= \frac{1+9t}{2+9t}, \quad t \in \mathbb{T}.
\end{aligned}$$

Example 1.82. Let $\mathbb{T} = \mathbb{N}_0^2$, and let

$$f(t) = \frac{1+2t}{1+t^4}, \quad t \in \mathbb{T}.$$

We will find $f^{\sigma}(t)$ and $f^{\sigma^2}(t)$ for $t \in \mathbb{T}$. By Example 1.12 we have

$$\sigma(t) = (\sqrt{t}+1)^2, \quad t \in \mathbb{T}.$$

Then

$$f^{\sigma}(t) = f(\sigma(t))$$

$$
\begin{aligned}
&= \frac{1+2\sigma(t)}{1+(\sigma(t))^2} \\
&= \frac{1+2(\sqrt{t}+1)^2}{1+(\sqrt{t}+1)^4}, \quad t \in \mathbb{T}.
\end{aligned}
$$

Hence

$$
\begin{aligned}
f^{\sigma^2}(t) &= f(\sigma^2(t)) \\
&= f(\sigma(\sigma(t))) \\
&= f^{\sigma}(\sigma(t)) \\
&= \frac{1+2(\sqrt{\sigma(t)}+1)^2}{1+(\sqrt{\sigma(t)}+1)^4} \\
&= \frac{1+(\sqrt{(\sqrt{t}+1)^2}+1)^2}{1+(\sqrt{(\sqrt{t}+1)^4}+1)^4} \\
&= \frac{1+2(\sqrt{t}+1+1)^2}{1+(\sqrt{t}+1+1)^4} \\
&= \frac{1+2(\sqrt{t}+2)^2}{1+(\sqrt{t}+2)^4}, \quad t \in \mathbb{T}.
\end{aligned}
$$

Example 1.83. Let $\mathbb{T} = \{H_n : n \in \mathbb{N}_0\}$, where H_n, $n \in \mathbb{N}_0$, are the harmonic numbers, and let

$$
f(t) = \frac{1-t}{1+4t}, \quad t \in \mathbb{T}.
$$

We will find $f^{\sigma}(t)$ and $f^{\sigma^2}(t)$ for $t \in \mathbb{T}$. By Example 1.13 we have

$$
\sigma(H_n) = H_{n+1}, \quad n \in \mathbb{N}_0.
$$

Then

$$
\begin{aligned}
f^{\sigma}(H_n) &= f(\sigma(H_n)) \\
&= \frac{1-\sigma(H_n)}{1+4\sigma(H_n)} \\
&= \frac{1-H_{n+1}}{1+4H_{n+1}}, \quad n \in \mathbb{N}_0.
\end{aligned}
$$

Hence

$$
\begin{aligned}
f^{\sigma^2}(H_n) &= f(\sigma^2(H_n)) \\
&= f(\sigma(\sigma(H_n))) \\
&= f^{\sigma}(\sigma(H_n))
\end{aligned}
$$

$$= \frac{1-\sigma(H_{n+1})}{1+4\sigma(H_{n+1})}$$
$$= \frac{1-H_{n+2}}{1+4H_{n+2}}, \quad n \in \mathbb{N}_0.$$

Example 1.84. Let $\mathbb{T} = P_{1,3}$, and let

$$f(t) = 1 - 2t - t^2, \quad t \in \mathbb{T}.$$

We will find $f^{\sigma}(t)$ and $f^{\sigma^2}(t)$ for $t \in \mathbb{T}$. By Example 1.14 we have

$$\sigma(t) = \begin{cases} t & \text{if } t \in \bigcup_{k=0}^{\infty}[4k, 4k+1), \\ t+3 & \text{if } t \in \bigcup_{k=0}^{\infty}\{4k+1\}. \end{cases}$$

Then we consider the following cases.

1. $t \in \bigcup_{k=0}^{\infty}[4k, 4k+1)$. Then $\sigma = I$, where I is the identity operator, and then

$$f^{\sigma}(t) = f(t),$$
$$f^{\sigma^2}(t) = f(t), \quad t \in \mathbb{T}.$$

2. $t \in \bigcup_{k=0}^{\infty}\{4k+1\}$. Then

$$\begin{aligned} f^{\sigma}(t) &= f(\sigma(t)) \\ &= 1 - 2\sigma(t) - (\sigma(t))^2 \\ &= 1 - 2(t+3) - (t+3)^2 \\ &= 1 - 2t - 6 - t^2 - 6t - 9 \\ &= -14 - 8t - t^2, \quad t \in \mathbb{T}, \end{aligned}$$

and

$$\begin{aligned} f^{\sigma^2}(t) &= f(\sigma^2(t)) \\ &= f(\sigma(\sigma(t))) \\ &= f^{\sigma}(\sigma(t)) \\ &= -14 - 8\sigma(t) - (\sigma(t))^2 \\ &= -14 - 8(t+3) - (t+3)^2 \\ &= -14 - 8t - 24 - t^2 - 6t - 9 \\ &= -47 - 14t - t^2, \quad t \in \mathbb{T}. \end{aligned}$$

Example 1.85. Let $\mathbb{T} = C$, where C is the Cantor set, and let

$$f(t) = t^3, \quad t \in \mathbb{T}.$$

We will find $f^{\sigma}(t)$ and $f^{\sigma^2}(t)$ for $t \in \mathbb{T}$. By Example 1.15 we have

$$\sigma(t) = \begin{cases} t + \frac{1}{3^{m+1}}, \ t \in C_1, \ t = \sum_{k=1}^{m} \frac{a_k}{3^k} + \frac{1}{3^{m+1}}, \\ t, \ t \in \mathbb{T} \backslash C_1. \end{cases}$$

Then we have the following cases.

1. $t \in C_1$, and

$$t = \sum_{k=1}^{m} \frac{a_k}{3^k} + \frac{1}{3^{m+1}}.$$

   Then

$$\begin{aligned} f^{\sigma}(t) &= f(\sigma(t)) \\ &= (\sigma(t))^3 \\ &= \left(t + \frac{1}{3^{m+1}}\right)^3, \end{aligned}$$

   and

$$\begin{aligned} f^{\sigma^2}(t) &= f(\sigma^2(t)) \\ &= f(\sigma(\sigma(t))) \\ &= f^{\sigma}(\sigma(t)) \\ &= \left(\sigma(t) + \frac{1}{3^{m+1}}\right)^3 \\ &= \left(t + \frac{1}{3^{m+1}} + \frac{1}{3^{m+1}}\right)^3 \\ &= \left(t + \frac{2}{3^{m+1}}\right)^3. \end{aligned}$$

2. Let $t \in \mathbb{T} \backslash C_1$. Then

$$\begin{aligned} f^{\sigma}(t) &= f(t), \\ f^{\sigma^2}(t) &= f(t). \end{aligned}$$

Example 1.86. Let

$$\mathbb{T} = \left\{ \sum_{k=0}^{n-1} a_k : n \in \mathbb{N}, \ a_k > 0, \ k \in \mathbb{N}_0 \right\},$$

and let

$$f(t) = \frac{1+t^3}{3+t^4}, \quad t \in \mathbb{T}.$$

By Example 1.16 we have

$$\sigma(t_n) = t_{n+1}, \quad n \in \mathbb{N}.$$

Then

$$\begin{aligned} f^\sigma(t_n) &= f(\sigma(t_n)) \\ &= \frac{1+(\sigma(t_n))^3}{3+(\sigma(t_n))^4} \\ &= \frac{1+t_{n+1}^3}{3+t_{n+1}^4}, \quad n \in \mathbb{N}, \end{aligned}$$

and

$$\begin{aligned} f^{\sigma^2}(t_n) &= f(\sigma^2(t_n)) \\ &= f(\sigma(\sigma(t_n))) \\ &= f^\sigma(\sigma(t_n)) \\ &= \frac{1+(\sigma(t_{n+1}))^3}{3+(\sigma(t_{n+1}))^4} \\ &= \frac{1+t_{n+2}^3}{3+t_{n+2}^4}, \quad n \in \mathbb{N}. \end{aligned}$$

Example 1.87. Let $\mathbb{T} = \{-\frac{1}{n} : n \in \mathbb{N}\} \cup \mathbb{N}_0$, and let

$$f(t) = 1 + t - t^2, \quad t \in \mathbb{T}.$$

By Example 1.17 we have

$$\sigma(t) = \begin{cases} -\frac{t}{t-1} & \text{if } t \in \{-\frac{1}{n} : n \in \mathbb{N}\}, \\ t+1 & \text{if } t \in \mathbb{N}_0. \end{cases}$$

Then we consider the following cases.

1. $t \in \{-\frac{1}{n} : n \in \mathbb{N}\}$. Then

$$\begin{aligned} f^\sigma(t) &= f(\sigma(t)) \\ &= 1 + \sigma(t) - (\sigma(t))^2 \\ &= 1 - \frac{t}{t-1} - \frac{t^2}{(t-1)^2} \\ &= \frac{(t-1)^2 - t(t-1) - t^2}{(t-1)^2} \end{aligned}$$

$$
\begin{aligned}
&= \frac{t^2 - 2t + 1 - t^2 + t - t^2}{(t-1)^2} \\
&= \frac{-t^2 - t + 1}{(t-1)^2},
\end{aligned}
$$

and

$$
\begin{aligned}
f^{\sigma^2}(t) &= f(\sigma^2(t)) \\
&= f(\sigma(\sigma(t))) \\
&= f^{\sigma}(\sigma(t)) \\
&= \frac{-(\sigma(t))^2 - \sigma(t) + 1}{(\sigma(t) - 1)^2} \\
&= \frac{-\frac{t^2}{(t-1)^2} + \frac{t}{t-1} + 1}{(-\frac{t}{t-1} - 1)^2} \\
&= \frac{\frac{-t^2 + t(t-1) + (t-1)^2}{(t-1)^2}}{(\frac{t+t-1}{t-1})^2} \\
&= \frac{-t^2 + t^2 - t + t^2 - 2t + 1}{(2t-1)^2} \\
&= \frac{t^2 - 3t + 1}{(2t-1)^2}.
\end{aligned}
$$

2. $t \in \mathbb{N}_0$. Then

$$
\begin{aligned}
f^{\sigma}(t) &= f(\sigma(t)) \\
&= 1 + \sigma(t) - (\sigma(t))^2 \\
&= 1 + t + 1 - (t+1)^2 \\
&= 2 + t - t^2 - 2t - 1 \\
&= -t^2 - t + 1,
\end{aligned}
$$

and

$$
\begin{aligned}
f^{\sigma^2}(t) &= f(\sigma^2(t)) \\
&= f(\sigma(\sigma(t))) \\
&= f^{\sigma}(\sigma(t)) \\
&= -(\sigma(t))^2 - \sigma(t) + 1 \\
&= -(t+1)^2 - (t+1) + 1 \\
&= -t^2 - 2t - 1 - t - 1 + 1 \\
&= -t^2 - 3t - 1.
\end{aligned}
$$

Example 1.88. Let $\mathbb{T} = \{(\frac{1}{2})^{2^n} : n \in \mathbb{N}_0\} \cup \{0,1\}$, and let

$$f(t) = t - t^3, \quad t \in \mathbb{T}.$$

Then

$$\begin{aligned} f^{\sigma}(t) &= f(\sigma(t)) \\ &= \sigma(t) - (\sigma(t))^3, \quad t \in \mathbb{T}, \end{aligned}$$

and

$$\begin{aligned} f^{\sigma^2}(t) &= f(\sigma^2(t)) \\ &= \sigma^2(t) - (\sigma^2(t))^3, \quad t \in \mathbb{T}. \end{aligned}$$

By Example 1.18 we have the following cases.

1. $t \in \{(\frac{1}{2})^{2^n} : n \in \mathbb{N}\}$. Then

$$\begin{aligned} \sigma(t) &= \sqrt{t}, \\ \sigma^2(t) &= \sqrt{\sigma(t)} \\ &= \sqrt{\sqrt{t}} \\ &= \sqrt[4]{t}, \quad t \in \mathbb{T}. \end{aligned}$$

   Hence

$$f^{\sigma}(t) = \sqrt{t} - (\sqrt{t})^3,$$

   and

$$f^{\sigma^2}(t) = \sqrt[4]{t} - (\sqrt[4]{t})^3.$$

2. $t = \frac{1}{2}$. Then

$$\sigma\left(\frac{1}{2}\right) = 1,$$

   and

$$\begin{aligned} \sigma^2\left(\frac{1}{2}\right) &= \sigma\left(\sigma\left(\frac{1}{2}\right)\right) \\ &= \sigma(1) \\ &= 1. \end{aligned}$$

   Hence

$$\begin{aligned} f^{\sigma}\left(\frac{1}{2}\right) &= \sigma\left(\frac{1}{2}\right) - \left(\sigma\left(\frac{1}{2}\right)\right)^3 \\ &= 1 - 1^3 \\ &= 0, \end{aligned}$$

and

$$\begin{aligned} f^{\sigma^2}\left(\frac{1}{2}\right) &= \sigma^2\left(\frac{1}{2}\right) - \left(\sigma^2\left(\frac{1}{2}\right)\right)^3 \\ &= 1 - 1^3 \\ &= 0. \end{aligned}$$

3. $t = 0$. Then

$$\sigma(0) = 0,$$

and

$$\begin{aligned} \sigma^2(0) &= \sigma(\sigma(0)) \\ &= \sigma(0) \\ &= 0. \end{aligned}$$

Hence

$$\begin{aligned} f^{\sigma}(0) &= \sigma(0) - (\sigma(0))^3 \\ &= 0 - 0^3 \\ &= 0, \end{aligned}$$

and

$$\begin{aligned} f^{\sigma^2}(0) &= \sigma^2(0) - (\sigma^2(0))^3 \\ &= 0 - 0^3 \\ &= 0. \end{aligned}$$

4. $t = 1$. Then

$$\sigma(1) = 1,$$

and

$$\begin{aligned} \sigma^2(1) &= \sigma(\sigma(1)) \\ &= \sigma(1) \\ &= 1. \end{aligned}$$

Hence

$$\begin{aligned} f^{\sigma}(1) &= \sigma(1) - (\sigma(1))^3 \\ &= 1 - 1^3 \\ &= 0, \end{aligned}$$

and

$$\begin{aligned} f^{\sigma^2}(1) &= \sigma^2(1) - (\sigma^2(1))^3 \\ &= 1 - 1^3 \\ &= 0. \end{aligned}$$

Example 1.89. Let $U = \{\frac{1}{2^n} : n \in \mathbb{N}\}$, let

$$\mathbb{T} = \{0\} \cup U \cup (1-U) \cup (1+U) \cup (2-U) \cup (2+U) \cup \{1,2\},$$

and let

$$f(t) = \frac{1+t}{1+4t}, \quad t \in \mathbb{T}.$$

We have

$$\begin{aligned} f^{\sigma}(t) &= f(\sigma(t)) \\ &= \frac{1+\sigma(t)}{1+4\sigma(t)}, \quad t \in \mathbb{T}, \end{aligned}$$

and

$$\begin{aligned} f^{\sigma^2}(t) &= f(\sigma^2(t)) \\ &= \frac{1+\sigma^2(t)}{1+4\sigma^2(t)}, \quad t \in \mathbb{T}. \end{aligned}$$

By Example 1.19 we have the following cases.

1. $t \in U \backslash \{\frac{1}{2}\}$. Then

$$\sigma(t) = 2t,$$

and

$$\begin{aligned} \sigma^2(t) &= \sigma(\sigma(t)) \\ &= 2\sigma(t) \\ &= 2(2t) \\ &= 4t. \end{aligned}$$

Hence

$$f^{\sigma}(t) = \frac{1+2t}{1+4(2t)} = \frac{1+2t}{1+8t},$$

and

$$f^{\sigma^2}(t) = \frac{1+4t}{1+4(4t)} = \frac{1+4t}{1+16t}.$$

2. $t \in (1-U)\backslash\{\frac{1}{2}\}$. Then

$$\sigma(t) = \frac{1+t}{2},$$

and

$$\begin{aligned}\sigma^2(t) &= \sigma(\sigma(t))\\ &= \frac{1+\sigma(t)}{2}\\ &= \frac{1+\frac{1+t}{2}}{2}\\ &= \frac{2+1+t}{4}\\ &= \frac{3+t}{4}.\end{aligned}$$

Hence

$$f^{\sigma}(t) = \frac{1+\frac{1+t}{2}}{1+4\frac{1+t}{2}} = \frac{2+1+t}{2+4+4t} = \frac{3+t}{6+4t},$$

and

$$f^{\sigma^2}(t) = \frac{1+\frac{3+t}{4}}{1+4\frac{3+t}{4}} = \frac{4+3+t}{4+12+4t} = \frac{7+t}{16+4t}.$$

3. $t \in (1+U)\backslash\{\frac{3}{2}\}$. Then

$$\sigma(t) = 2t - 1,$$

and

$$\begin{aligned}\sigma^2(t) &= \sigma(\sigma(t))\\ &= 2\sigma(t) - 1\\ &= 2(2t-1) - 1\\ &= 4t - 2 - 1\\ &= 4t - 3.\end{aligned}$$

Hence

$$\begin{aligned}f^{\sigma}(t) &= \frac{1+2t-1}{1+4(2t-1)}\\ &= \frac{2t}{1+8t-4}\\ &= \frac{2t}{8t-3},\end{aligned}$$

and

$$\begin{aligned}f^{\sigma^2}(t) &= \frac{1+4t-3}{1+4(4t-3)}\\ &= \frac{4t-2}{1+16t-12}\\ &= \frac{4t-2}{16t-11}.\end{aligned}$$

4. $t \in (2-U)\backslash\{\frac{3}{2}\}$. Then

$$\sigma(t) = \frac{t+2}{2},$$

and

$$\begin{aligned}\sigma^2(t) &= \sigma(\sigma(t))\\ &= \frac{\sigma(t)+2}{2}\\ &= \frac{\frac{t+2}{2}+2}{2}\\ &= \frac{t+2+4}{4}\\ &= \frac{t+6}{4}.\end{aligned}$$

Hence

$$\begin{aligned} f^{\sigma}(t) &= \frac{1+\frac{t+2}{2}}{1+4\frac{t+2}{2}} \\ &= \frac{2+t+2}{2(1+2t+4)} \\ &= \frac{t+4}{2(2t+5)}, \end{aligned}$$

and

$$\begin{aligned} f^{\sigma^2}(t) &= \frac{1+\frac{t+6}{4}}{1+4\frac{t+6}{4}} \\ &= \frac{4+t+6}{4(t+6+1)} \\ &= \frac{t+10}{4(t+7)}. \end{aligned}$$

5. $t \in (2+U)\backslash\{\frac{5}{2}\}$. Then

$$\sigma(t) = 2(t-1),$$

and

$$\begin{aligned} \sigma^2(t) &= \sigma(\sigma(t)) \\ &= 2(\sigma(t)-1) \\ &= 2(2(t-1)-1) \\ &= 2(2t-2-1) \\ &= 2(2t-3). \end{aligned}$$

Hence

$$\begin{aligned} f^{\sigma}(t) &= \frac{1+2(t-1)}{1+8(t-1)} \\ &= \frac{1+2t-2}{1+8t-8} \\ &= \frac{2t-1}{8t-7}, \end{aligned}$$

and

$$\begin{aligned} f^{\sigma^2}(t) &= \frac{1+2(2t-3)}{1+4(2(2t-3))} \\ &= \frac{1+4t-6}{1+16t-24} \end{aligned}$$

$$= \frac{4t-5}{16t-23}.$$

6. $t = 0$. Then

$$\sigma(0) = 0,$$

and

$$\begin{aligned}\sigma^2(0) &= \sigma(\sigma(0))\\ &= \sigma(0)\\ &= 0.\end{aligned}$$

Hence

$$\begin{aligned}f^{\sigma}(0) &= \frac{1+0}{1+4\cdot 0}\\ &= 1,\end{aligned}$$

and

$$\begin{aligned}f^{\sigma^2}(0) &= \frac{1+0}{1+4\cdot 0}\\ &= 1.\end{aligned}$$

7. $t = \frac{1}{2}$. Then

$$\sigma\left(\frac{1}{2}\right) = \frac{3}{4},$$

and

$$\begin{aligned}\sigma^2\left(\frac{1}{2}\right) &= \sigma\left(\sigma\left(\frac{1}{2}\right)\right)\\ &= \sigma\left(\frac{3}{4}\right)\\ &= \frac{7}{8}.\end{aligned}$$

Hence

$$\begin{aligned}f^{\sigma}\left(\frac{1}{2}\right) &= \frac{1+\frac{3}{4}}{1+4\cdot\frac{3}{4}}\\ &= \frac{4+3}{4(1+3)}\\ &= \frac{7}{16},\end{aligned}$$

and

$$\begin{aligned} f^{\sigma^2}\left(\frac{1}{2}\right) &= \frac{1+\frac{7}{8}}{1+4\cdot\frac{7}{8}} \\ &= \frac{8+7}{8+28} \\ &= \frac{15}{36} \\ &= \frac{5}{12}. \end{aligned}$$

8. $t = 1$. Then

$$\sigma(1) = 1,$$

and

$$\begin{aligned} \sigma^2(1) &= \sigma(\sigma(1)) \\ &= \sigma(1) \\ &= 1. \end{aligned}$$

Hence

$$\begin{aligned} f^{\sigma}(1) &= \frac{1+1}{1+4\cdot 1} \\ &= \frac{2}{5}, \end{aligned}$$

and

$$\begin{aligned} f^{\sigma^2}(1) &= \frac{1+1}{1+4\cdot 1} \\ &= \frac{2}{5}. \end{aligned}$$

9. $t = \frac{3}{2}$. Then

$$\sigma\left(\frac{3}{2}\right) = \frac{7}{4},$$

and

$$\begin{aligned} \sigma^2\left(\frac{3}{2}\right) &= \sigma\left(\sigma\left(\frac{3}{2}\right)\right) \\ &= \sigma\left(\frac{7}{4}\right) \end{aligned}$$

$$= \frac{15}{8}.$$

Hence

$$\begin{aligned} f^{\sigma}\left(\frac{3}{2}\right) &= \frac{1+\frac{7}{4}}{1+4\cdot\frac{7}{4}} \\ &= \frac{4+7}{4(1+7)} \\ &= \frac{11}{32}, \end{aligned}$$

and

$$\begin{aligned} f^{\sigma^2}\left(\frac{3}{2}\right) &= \frac{1+\frac{15}{8}}{1+4\cdot\frac{15}{8}} \\ &= \frac{8+15}{8+60} \\ &= \frac{23}{68}. \end{aligned}$$

10. $t = 2$. Then

$$\sigma(2) = 2,$$

and

$$\begin{aligned} \sigma^2(2) &= \sigma(\sigma(2)) \\ &= \sigma(2) \\ &= 2. \end{aligned}$$

Hence

$$\begin{aligned} f^{\sigma}(2) &= \frac{1+2}{1+4\cdot 2} \\ &= \frac{3}{9} \\ &= \frac{1}{3}, \end{aligned}$$

and

$$\begin{aligned} f^{\sigma^2}(2) &= \frac{1+2}{1+4\cdot 2} \\ &= \frac{1}{3}. \end{aligned}$$

11. $t = \frac{5}{2}$. Then

$$\sigma\left(\frac{5}{2}\right) = \frac{5}{2},$$

and

$$\begin{aligned}\sigma^2\left(\frac{5}{2}\right) &= \sigma\left(\sigma\left(\frac{5}{2}\right)\right) \\ &= \sigma\left(\frac{5}{2}\right) \\ &= \frac{5}{2}.\end{aligned}$$

Hence

$$\begin{aligned}f^\sigma\left(\frac{5}{2}\right) &= \frac{1+\frac{5}{2}}{1+4\cdot\frac{5}{2}} \\ &= \frac{2+5}{2(1+10)} \\ &= \frac{7}{22},\end{aligned}$$

and

$$\begin{aligned}f^{\sigma^2}\left(\frac{5}{2}\right) &= \frac{1+\frac{5}{2}}{1+4\cdot\frac{5}{2}} \\ &= \frac{7}{22}.\end{aligned}$$

Exercise 1.10. Let

$$f(t) = 2 - 4t + t^2, \quad t \in \mathbb{T},$$

where $\mathbb{T}$ are the time scales from Exercise 1.3. Find

$$f^\sigma(t) \quad \text{and} \quad f^{\sigma^2}(t) \quad \text{for } t \in \mathbb{T}.$$

Definition 1.10. For a function $f : \mathbb{T} \to \mathbb{R}$ and $k, l \in \mathbb{N}_0$, define

$$\begin{aligned}f^{\rho^k}(t) &= f(\rho^k(t)), \quad t \in \mathbb{T}, \\ f^{\rho^k\sigma^l}(t) &= f(\sigma^l(\rho^k(t))), \quad t \in \mathbb{T},\end{aligned}$$

and

$$f^{\sigma^l\rho^k}(t) = f(\rho^k(\sigma^l(t))), \quad t \in \mathbb{T}.$$

Remark 1.1. Note that in the general case, we have

$$f^{\rho^k\sigma^l}(t) \neq f^{\sigma^l\rho^k}(t), \quad t \in \mathbb{T}.$$

Indeed, let $\mathbb{T} = \{-1, 0\} \cup \{\frac{1}{n}\}_{n\in\mathbb{N}}$, and let

$$f(t) = t, \quad t \in \mathbb{T}.$$

Then

$$\begin{aligned} \rho(\sigma(0)) &= \rho(0) \\ &= -1, \end{aligned}$$

and

$$\begin{aligned} \sigma(\rho(0)) &= \sigma(-1) \\ &= 0. \end{aligned}$$

Thus

$$\rho(\sigma(0)) \neq \sigma(\rho(0)).$$

Example 1.90. Let $\mathbb{T} = \{-1, 0\} \cup \{\frac{1}{n}\}_{n\in\mathbb{N}}$, and let

$$f(t) = 1 + 4t^2, \quad t \in \mathbb{T}.$$

We will find

$$f^{\rho^2}(t), \quad f^{\sigma\rho\sigma}(t), \quad \text{and} \quad f^{\sigma^2\rho^3}(t) \quad \text{for } t \in \mathbb{T}.$$

Firstly, we will determine the forward and backward jump operators for the time scale $\mathbb{T}$. We have the following cases.

1. $t = \frac{1}{n}$, $n \in \mathbb{N}$, $n \geq 2$. Then

$$\begin{aligned} \sigma(t) &= \frac{1}{n-1} \\ &= \frac{1}{\frac{1}{t} - 1} \\ &= \frac{t}{1-t}, \end{aligned}$$

and

$$\rho(t) = \frac{1}{n+1}$$

$$= \frac{1}{\frac{1}{t}+1}$$
$$= \frac{t}{1+t}.$$

2. $t = 1$. Then

$$\sigma(1) = 1,$$
$$\rho(1) = \frac{1}{2}.$$

3. $t = 0$. Then

$$\sigma(0) = 0,$$
$$\rho(0) = -1.$$

4. $t = -1$. Then

$$\sigma(-1) = 0,$$
$$\rho(-1) = -1.$$

Therefore

$$\sigma(t) = \begin{cases} \frac{t}{1-t} & \text{if } t \in \{\frac{1}{n}\}_{n\in\mathbb{N}, n\geq 2}, \\ 1 & \text{if } t = 1, \\ 0 & \text{if } t = 0, \\ 0 & \text{if } t = -1, \end{cases}$$

and

$$\rho(t) = \begin{cases} \frac{t}{1+t} & \text{if } t \in \{\frac{1}{n}\}_{n\in\mathbb{N}, n\geq 2}, \\ \frac{1}{2} & \text{if } t = 1, \\ -1 & \text{if } t = 0, \\ -1 & \text{if } t = -1. \end{cases}$$

Hence we have the following cases.

1. $t \in \{\frac{1}{n}\}_{n\in\mathbb{N}, n\geq 2}$. Then

$$\rho^2(t) = \rho(\rho(t))$$
$$= \rho\left(\frac{t}{1+t}\right)$$
$$= \frac{\rho(t)}{1+\rho(t)}$$

$$= \frac{\frac{t}{1+t}}{1+\frac{t}{1+t}} = \frac{t}{1+2t},$$

and

$$\begin{aligned} f^{\rho^2}(t) &= f(\rho^2(t)) \\ &= f\left(\frac{t}{1+2t}\right) \\ &= 1+4\frac{t^2}{(1+2t)^2} \\ &= \frac{(1+2t)^2+4t^2}{(1+2t)^2} \\ &= \frac{1+4t+4t^2+4t^2}{(1+2t)^2} \\ &= \frac{1+4t+8t^2}{(1+2t)^2}. \end{aligned}$$

Next,

$$\begin{aligned} \sigma(\rho(\sigma(t))) &= \sigma\left(\rho\left(\frac{t}{1-t}\right)\right) \\ &= \sigma\left(\frac{\rho(t)}{1-\rho(t)}\right) \\ &= \sigma\left(\frac{\frac{t}{1+t}}{1-\frac{t}{1+t}}\right) \\ &= \sigma(t) \\ &= \frac{t}{1-t}, \end{aligned}$$

and

$$\begin{aligned} f^{\sigma\rho\sigma}(t) &= f(\sigma(\rho(\sigma(t)))) \\ &= f\left(\frac{t}{1-t}\right) \\ &= 1+\frac{4t^2}{(1-t)^2} \\ &= \frac{(1-t)^2+4t^2}{(1-t)^2} \\ &= \frac{1-2t+t^2+4t^2}{(1-t)^2} \end{aligned}$$

$$= \frac{1 - 2t + 5t^2}{(1 - t)^2}.$$

Moreover,

$$\begin{aligned}
\rho^3(\sigma^2(t)) &= \rho^3(\sigma(\sigma(t))) \\
&= \rho^3\left(\sigma\left(\frac{t}{1-t}\right)\right) \\
&= \rho^3\left(\frac{\sigma(t)}{1-\sigma(t)}\right) \\
&= \rho^3\left(\frac{\frac{t}{1-t}}{1-\frac{t}{1-t}}\right) \\
&= \rho^3\left(\frac{t}{1-2t}\right) \\
&= \rho^2\left(\rho\left(\frac{t}{1-2t}\right)\right) \\
&= \rho^2\left(\frac{\rho(t)}{1-2\rho(t)}\right) \\
&= \rho^2\left(\frac{\frac{t}{1+t}}{1-\frac{2t}{1+t}}\right) \\
&= \rho^2\left(\frac{t}{1-t}\right) \\
&= \rho\left(\rho\left(\frac{t}{1-t}\right)\right) \\
&= \rho\left(\frac{\rho(t)}{1-\rho(t)}\right) \\
&= \rho\left(\frac{\frac{t}{1+t}}{1-\frac{t}{1+t}}\right) \\
&= \rho(t) \\
&= \frac{t}{t+1},
\end{aligned}$$

and

$$\begin{aligned}
f^{\sigma^2\rho^3}(t) &= f(\rho^3(\sigma^2(t))) \\
&= f\left(\frac{t}{1+t}\right) \\
&= 1 + \frac{4t^2}{(1+t)^2} \\
&= \frac{(1+t)^2 + 4t^2}{(1+t)^2}
\end{aligned}$$

$$
\begin{aligned}
&= \frac{1+2t+t^2+4t^2}{(1+t)^2} \\
&= \frac{1+2t+5t^2}{(1+t)^2}.
\end{aligned}
$$

2. $t = 1$. Then

$$
\begin{aligned}
\rho^2(1) &= \rho(\rho(1)) \\
&= \rho\left(\frac{1}{2}\right) \\
&= \frac{1}{3},
\end{aligned}
$$

and

$$
\begin{aligned}
f^{\rho^2}(1) &= f(\rho^2(1)) \\
&= f\left(\frac{1}{3}\right) \\
&= 1 + \frac{4}{9} \\
&= \frac{13}{9}.
\end{aligned}
$$

Next,

$$
\begin{aligned}
\sigma\rho\sigma(1) &= \sigma(\rho(\sigma(1))) \\
&= \sigma(\rho(1)) \\
&= \sigma\left(\frac{1}{2}\right) \\
&= 1,
\end{aligned}
$$

and

$$
\begin{aligned}
f^{\sigma\rho\sigma}(1) &= f(\sigma(\rho(\sigma(1)))) \\
&= f(1) \\
&= 1 + 4 \\
&= 5.
\end{aligned}
$$

Moreover,

$$
\begin{aligned}
\rho^3\sigma^2(1) &= \rho^3(\sigma(\sigma(1))) \\
&= \rho^3(\sigma(1)) \\
&= \rho^3(1)
\end{aligned}
$$

$$
\begin{aligned}
&= \rho(\rho(\rho(1))) \\
&= \rho\left(\rho\left(\frac{1}{2}\right)\right) \\
&= \rho\left(\frac{1}{3}\right) \\
&= \frac{1}{4},
\end{aligned}
$$

and

$$
\begin{aligned}
f^{\sigma^2\rho^3}(1) &= f(\rho^3\sigma^2(1)) \\
&= f\left(\frac{1}{4}\right) \\
&= 1 + 4 \cdot \frac{1}{16} \\
&= 1 + \frac{1}{4} \\
&= \frac{5}{4}.
\end{aligned}
$$

3. $t = 0$. Then

$$
\begin{aligned}
\sigma\rho\sigma(0) &= \sigma\rho(\sigma(0)) \\
&= \sigma(\rho(0)) \\
&= \sigma(-1) \\
&= 0,
\end{aligned}
$$

and

$$
\begin{aligned}
f^{\sigma\rho\sigma}(0) &= f(\sigma\rho\sigma(0)) \\
&= f(0) \\
&= 1.
\end{aligned}
$$

Next,

$$
\begin{aligned}
\rho^2(0) &= \rho(\rho(0)) \\
&= \rho(-1) \\
&= -1,
\end{aligned}
$$

and

$$
\begin{aligned}
f^{\rho^2}(0) &= f(\rho^2(0)) \\
&= f(-1)
\end{aligned}
$$

$$= 1 + 4$$
$$= 5.$$

Moreover,

$$\begin{aligned} \rho^3\sigma^2(0) &= \rho^3\sigma(\sigma(0)) \\ &= \rho^3(\sigma(0)) \\ &= \rho^3(0) \\ &= \rho(\rho(\rho(0))) \\ &= \rho(\rho(-1)) \\ &= \rho(-1) \\ &= -1, \end{aligned}$$

and

$$\begin{aligned} f^{\sigma^2\rho^3}(0) &= f(\rho^3\sigma^2(0)) \\ &= f(-1) \\ &= 1 + 4 \\ &= 5. \end{aligned}$$

4. $t = -1$. Then

$$\begin{aligned} \rho^2(-1) &= \rho(\rho(-1)) \\ &= \rho(-1) \\ &= -1, \end{aligned}$$

and

$$\begin{aligned} f^{\rho^2}(-1) &= f(\rho^2(-1)) \\ &= f(-1) \\ &= 1 + 4 \\ &= 5. \end{aligned}$$

Next,

$$\begin{aligned} \sigma\rho\sigma(-1) &= \sigma(\rho(\sigma(-1))) \\ &= \sigma(\rho(0)) \\ &= \sigma(-1) \\ &= 0, \end{aligned}$$

and

$$\begin{aligned} f^{\sigma\rho\sigma}(-1) &= f(\sigma\rho\sigma(-1)) \\ &= f(0) \\ &= 1. \end{aligned}$$

Moreover,

$$\begin{aligned} \rho^3\sigma^2(-1) &= \rho^3\sigma(\sigma(-1)) \\ &= \rho^3(\sigma(0)) \\ &= \rho^3(0) \\ &= \rho^2(\rho(0)) \\ &= \rho^2(-1) \\ &= \rho(\rho(-1)) \\ &= \rho(-1) \\ &= -1, \end{aligned}$$

and

$$\begin{aligned} f^{\sigma^2\rho^3}(-1) &= f(\rho^3\sigma^2(-1)) \\ &= f(-1) \\ &= 1+4 \\ &= 5. \end{aligned}$$

Example 1.91. Let $\mathbb{T} = 3^{\mathbb{N}_0}$, and let

$$f(t) = \frac{2-t}{4+3t}, \quad t \in \mathbb{T}.$$

We will find $f^{\sigma\rho^2\sigma^4}(1)$ and $f^{\rho\sigma\rho}(27)$. By Example 1.11 we have

$$\sigma(t) = 3t, \quad t \in \mathbb{T},$$

and by Example 1.31 we find

$$\rho(t) = \begin{cases} 1 & \text{if } t = 1, \\ \frac{t}{3} & \text{if } t \in 3^{\mathbb{N}}. \end{cases}$$

Then

$$\sigma^4\rho^2\sigma(1) = \sigma^4\rho^2(\sigma(1))$$

$$
\begin{aligned}
&= \sigma^4\rho^2(3)\\
&= \sigma^4\rho(\rho(3))\\
&= \sigma^4(\rho(1))\\
&= \sigma^4(1)\\
&= \sigma^3(\sigma(1))\\
&= \sigma^3(3)\\
&= \sigma^2(\sigma(3))\\
&= \sigma^2(9)\\
&= \sigma(\sigma(9))\\
&= \sigma(27)\\
&= 81,
\end{aligned}
$$

and

$$
\begin{aligned}
\rho\sigma\rho(27) &= \rho(\sigma(\rho(27)))\\
&= \rho(\sigma(9))\\
&= \rho(27)\\
&= 9.
\end{aligned}
$$

Hence

$$
\begin{aligned}
f^{\sigma\rho^2\sigma^4}(1) &= f(\sigma^4\rho^2\sigma(1))\\
&= f(81)\\
&= \frac{2-81}{4+3\cdot 81}\\
&= \frac{-79}{4+243}\\
&= -\frac{79}{247},
\end{aligned}
$$

and

$$
\begin{aligned}
f^{\rho\sigma\rho}(27) &= f(\rho\sigma\rho(27))\\
&= f(9)\\
&= \frac{2-9}{4+3\cdot 9}\\
&= \frac{-7}{4+27}\\
&= -\frac{7}{31}.
\end{aligned}
$$

Example 1.92. Let $\mathbb{T} = \{(\frac{1}{2})^{2^n} : n \in \mathbb{N}_0\} \cup \{0, 1\}$, and let

$$f(t) = \sqrt{1 + t + t^2}, \quad t \in \mathbb{T}.$$

We will find

$$f^{\rho^2}(t), \quad f^{\sigma^3\rho^2}(t), \quad f^{\rho^2\sigma^3}(t), \quad t \in \mathbb{T}.$$

By Example 1.18 we have

$$\sigma(t) = \begin{cases} \sqrt{t} & \text{if } t \in \{(\frac{1}{2})^{2^n} : n \in \mathbb{N}\}, \\ 1 & \text{if } t = \frac{1}{2}, \\ 0 & \text{if } t = 0, \\ 1 & \text{if } t = 1, \end{cases}$$

and by Example 1.38 we have

$$\rho(t) = \begin{cases} t^2 & \text{if } t \in \{(\frac{1}{2})^{2^n} : n \in \mathbb{N}_0\}, \\ 0 & \text{if } t = 0, \\ \frac{1}{2} & \text{if } t = 1. \end{cases}$$

Thus we have the following cases.

1. $t \in \{(\frac{1}{2})^{2^n} : n \in \mathbb{N}\}$. Then

$$\begin{aligned} \rho^2(t) &= \rho(\rho(t)) \\ &= \rho(t^2) \\ &= (\rho(t))^2 \\ &= t^4, \end{aligned}$$

and

$$\begin{aligned} \sigma^3\rho^2(t) &= \sigma^3(\rho^2(t)) \\ &= \sigma^3(t^4) \\ &= \sigma^2(\sigma(t^4)) \\ &= \sigma^2(t^2) \\ &= \sigma(\sigma(t^2)) \\ &= \sigma(t) \\ &= \sqrt{t}, \end{aligned}$$

and

$$\begin{aligned}\rho^2\sigma^3(t) &= \rho^2\sigma^2(\sigma(t))\\ &= \rho^2\sigma^2(\sqrt{t})\\ &= \rho^2\sigma(\sigma(\sqrt{t}))\\ &= \rho^2(\sigma(\sqrt[4]{t}))\\ &= \rho^2(\sqrt[8]{t})\\ &= \rho(\rho(\sqrt[8]{t}))\\ &= \rho(\sqrt[4]{t})\\ &= \sqrt{t}.\end{aligned}$$

Hence

$$\begin{aligned}f^{\rho^2}(t) &= f(\rho^2(t))\\ &= f(t^2)\\ &= \sqrt{1+t^4+t^8},\\ f^{\sigma^3\rho^2}(t) &= f(\rho^2(\sigma^3(t)))\\ &= f(\sqrt{t})\\ &= \sqrt{1+\sqrt{t}+t},\end{aligned}$$

and

$$\begin{aligned}f^{\rho^2\sigma^3}(t) &= f(\sigma^3\rho^2(t))\\ &= f(\sqrt{t})\\ &= \sqrt{1+\sqrt{t}+t}.\end{aligned}$$

2. $t = 0$. Then

$$\begin{aligned}\rho^2(0) &= \rho(\rho(0))\\ &= \rho(0)\\ &= 0,\\ \sigma^3\rho^2(0) &= \sigma^3(\rho^2(0))\\ &= \sigma^3(0)\\ &= \sigma^2(\sigma(0))\\ &= \sigma^2(0)\\ &= \sigma(\sigma(0))\\ &= \sigma(0)\\ &= 0,\end{aligned}$$

and

$$\begin{aligned}\rho^2\sigma^3(0) &= \rho^2\sigma^2(\sigma(0))\\ &= \rho^2\sigma(\sigma(0))\\ &= \rho^2(\sigma(0))\\ &= \rho^2(0)\\ &= \rho(\rho(0))\\ &= \rho(0)\\ &= 0.\end{aligned}$$

Hence

$$\begin{aligned}f^{\rho^2}(0) &= f(\rho^2(0))\\ &= f(0)\\ &= 1,\\ f^{\sigma^3\rho^2}(0) &= f(\rho^2\sigma^3(0))\\ &= f(0)\\ &= 1,\end{aligned}$$

and

$$\begin{aligned}f^{\rho^2\sigma^3}(0) &= f(\sigma^3\rho^2(0))\\ &= f(0)\\ &= 1.\end{aligned}$$

3. $t = \frac{1}{2}$. Then

$$\begin{aligned}\rho^2\left(\frac{1}{2}\right) &= \rho\left(\rho\left(\frac{1}{2}\right)\right)\\ &= \rho\left(\frac{1}{4}\right)\\ &= \frac{1}{16},\\ \sigma^3\rho^2\left(\frac{1}{2}\right) &= \sigma^3\left(\rho^2\left(\frac{1}{2}\right)\right)\\ &= \sigma^3\left(\frac{1}{16}\right)\\ &= \sigma^2\left(\sigma\left(\frac{1}{16}\right)\right)\end{aligned}$$

$$
\begin{aligned}
&= \sigma^2\left(\frac{1}{4}\right)\\
&= \sigma\left(\sigma\left(\frac{1}{4}\right)\right)\\
&= \sigma\left(\frac{1}{2}\right)\\
&= 1,
\end{aligned}
$$

and

$$
\begin{aligned}
\rho^2\sigma^3\left(\frac{1}{2}\right) &= \rho^2\sigma^2\left(\sigma\left(\frac{1}{2}\right)\right)\\
&= \rho^2\sigma(\sigma(1))\\
&= \rho^2\sigma(1)\\
&= \rho^2(\sigma(1))\\
&= \rho^2(1)\\
&= \rho(\rho(1))\\
&= \rho\left(\frac{1}{2}\right)\\
&= \frac{1}{4}.
\end{aligned}
$$

Hence

$$
\begin{aligned}
f^{\rho^2}\left(\frac{1}{2}\right) &= f\left(\rho^2\left(\frac{1}{2}\right)\right)\\
&= f\left(\frac{1}{16}\right)\\
&= \sqrt{1 + \frac{1}{16} + \frac{1}{256}}\\
&= \sqrt{\frac{256 + 16 + 1}{256}}\\
&= \frac{\sqrt{273}}{16},\\
f^{\sigma^3\rho^2}\left(\frac{1}{2}\right) &= f\left(\rho^2\sigma^3\left(\frac{1}{2}\right)\right)\\
&= f\left(\frac{1}{4}\right)\\
&= \sqrt{1 + \frac{1}{4} + \frac{1}{16}}\\
&= \sqrt{\frac{16 + 4 + 1}{16}}
\end{aligned}
$$

$$= \frac{\sqrt{21}}{4},$$

and

$$\begin{aligned} f^{\rho^2\sigma^3}\left(\frac{1}{2}\right) &= f\left(\sigma^3\rho^2\left(\frac{1}{2}\right)\right) \\ &= f(1) \\ &= \sqrt{1+1+1} \\ &= \sqrt{3}. \end{aligned}$$

4. $t = 1$. Then

$$\begin{aligned} \rho^2(1) &= \rho(\rho(1)) \\ &= \rho\left(\frac{1}{2}\right) \\ &= \frac{1}{4}, \\ \sigma^3\rho^2(1) &= \sigma^3(\rho^2(1)) \\ &= \sigma^3\left(\frac{1}{4}\right) \\ &= \sigma^2\left(\sigma\left(\frac{1}{4}\right)\right) \\ &= \sigma^2\left(\frac{1}{2}\right) \\ &= \sigma\left(\sigma\left(\frac{1}{2}\right)\right) \\ &= \sigma(1) \\ &= 1, \end{aligned}$$

and

$$\begin{aligned} \rho^2\sigma^3(1) &= \rho^2\sigma^2(\sigma(1)) \\ &= \rho^2\sigma(\sigma(1)) \\ &= \rho^2(\sigma(1)) \\ &= \rho^2(1) \\ &= \frac{1}{4}. \end{aligned}$$

Hence

$$f^{\rho^2}(1) = f(\rho^2(1))$$

$$
\begin{aligned}
&= f\left(\frac{1}{4}\right) \\
&= \sqrt{1+\frac{1}{4}+\frac{1}{16}} \\
&= \frac{\sqrt{21}}{4}, \\
f^{\sigma^3\rho^2}(1) &= f(\rho^2\sigma^3(1)) \\
&= f\left(\frac{1}{4}\right) \\
&= \frac{\sqrt{21}}{4},
\end{aligned}
$$

and

$$
\begin{aligned}
f^{\rho^2\sigma^3}(1) &= f(\sigma^3\rho^2(1)) \\
&= f(1) \\
&= \sqrt{1+1+1} \\
&= \sqrt{3}.
\end{aligned}
$$

Exercise 1.11. Let

$$f(t) = \sqrt[3]{\frac{1+t}{2+7t}}, \quad t \in \mathbb{T},$$

where $\mathbb{T}$ are the time scales in Exercise 1.3. Find

$$f^{\rho^2}(t), \quad f^{\sigma^3\rho^2}(t), \quad f^{\rho^2\sigma^3}(t), \quad t \in \mathbb{T}.$$

Example 1.93. We will simplify

$$A(t) = \left(\frac{5\sigma(t) - 2(\sigma(t))^2}{(\sigma(t)-2)^2} + \sigma(t)\right) : \left(\sigma(t) - \frac{3\sigma(t)-8}{(\sigma(t)-2)^2} - 2\right), \quad t \in \mathbb{T},\ \sigma(t) \neq 2.$$

We have

$$
\begin{aligned}
A(t) &= \left(\frac{5\sigma(t) - 2(\sigma(t))^2 + \sigma(t)(\sigma(t)-2)^2}{(\sigma(t)-2)^2}\right) : \left(\frac{(\sigma(t)-2)^3 - 3\sigma(t) + 8}{(\sigma(t)-2)^2}\right) \\
&= \left(\frac{5\sigma(t) - 2(\sigma(t))^2 + (\sigma(t))^3 - 4(\sigma(t))^2 + 4\sigma(t)}{(\sigma(t)-2)^2}\right) \\
&\quad \times \left(\frac{(\sigma(t)-2)^2}{(\sigma(t))^3 - 6(\sigma(t))^2 + 12\sigma(t) - 8 - 3\sigma(t) + 8}\right) \\
&= \frac{9\sigma(t) - 6(\sigma(t))^2 + (\sigma(t))^3}{(\sigma(t))^3 - 6(\sigma(t))^2 + 9\sigma(t)}
\end{aligned}
$$

$$
\begin{aligned}
&= \frac{\sigma(t)(9 - 6\sigma(t) + (\sigma(t))^2)}{\sigma(t)(9 - 6\sigma(t) + (\sigma(t))^2)} \\
&= 1, \quad t \in \mathbb{T},\ \sigma(t) \neq 2.
\end{aligned}
$$

Example 1.94. We will simplify

$$
\begin{aligned}
A(t) = {} & \left(\frac{2\rho(t)}{\sigma(t) - 4\rho(t)} - \frac{10\sigma(t)(\rho(t))^2 - 32(\rho(t))^3}{(4\rho(t) - \sigma(t))^3} \right) \\
& \times \left(\frac{21\sigma(t)(\rho(t))^2 - 64(\rho(t))^3}{(\sigma(t))^2 - 3\sigma(t)\rho(t)} + \sigma(t) - 9\rho(t) \right), \quad t \in \mathbb{T},\ \sigma(t) \neq 0, 3\rho(t), 4\rho(t).
\end{aligned}
$$

We have

$$
\begin{aligned}
A(t) = {} & \left(\frac{2\rho(t)}{\sigma(t) - 4\rho(t)} + \frac{10\sigma(t)(\rho(t))^2 - 32(\rho(t))^3}{(\sigma(t) - 4\rho(t))^3} \right) \\
& \times \left(\frac{21\sigma(t)(\rho(t))^2 - 64(\rho(t))^3}{(\sigma(t))^2 - 3\sigma(t)\rho(t)} + \sigma(t) - 9\rho(t) \right) \\
= {} & \left(\frac{2\rho(t)(\sigma(t) - 4\rho(t))^2 + 10\sigma(t)(\rho(t))^2 - 32(\rho(t))^3}{(\sigma(t) - 4\rho(t))^3} \right) \\
& \times \left(\frac{21\sigma(t)(\rho(t))^2 - 64(\rho(t))^3 + (\sigma(t) - 9\rho(t))((\sigma(t))^2 - 3\sigma(t)\rho(t))}{\sigma(t)(\sigma(t) - 3\rho(t))} \right) \\
= {} & \left(\frac{2\rho(t)((\sigma(t))^2 - 8\sigma(t)\rho(t) + 16(\rho(t))^2) + 10\sigma(t)(\rho(t))^2 - 32(\rho(t))^3}{(\sigma(t) - 4\rho(t))^3} \right) \\
& \times \left(\frac{21\sigma(t)(\rho(t))^2 - 64(\rho(t))^3 + (\sigma(t))^3 - 3(\sigma(t))^2\rho(t) - 9(\sigma(t))^2\rho(t) + 27\sigma(t)(\rho(t))^2}{\sigma(t)(\sigma(t) - 3\rho(t))} \right) \\
= {} & \left(\frac{2\rho(t)(\sigma(t))^2 - 16\sigma(t)(\rho(t))^2 + 32(\rho(t))^3 + 10\sigma(t)(\rho(t))^2 - 32(\rho(t))^3}{(\sigma(t) - 4\rho(t))^3} \right) \\
& \times \left(\frac{48\sigma(t)(\rho(t))^2 - 12(\sigma(t))^2\rho(t) - 64(\rho(t))^3 + (\sigma(t))^3}{\sigma(t)(\sigma(t) - 3\rho(t))} \right) \\
= {} & \left(\frac{2\rho(t)(\sigma(t))^2 + 6\sigma(t)(\rho(t))^2}{(\sigma(t) - 4\rho(t))^3} \right) \left(\frac{(\sigma(t) - 4\rho(t))^3}{\sigma(t)(\sigma(t) - 3\rho(t))} \right) \\
= {} & \frac{2\rho(t)\sigma(t)(\sigma(t) - 3\rho(t))}{\sigma(t)(\sigma(t) - 3\rho(t))} \\
= {} & 2\rho(t), \quad t \in \mathbb{T},\ \sigma(t) \neq 0, 3\rho(t), 4\rho(t).
\end{aligned}
$$

Example 1.95. We will simplify

$$
A(t) = \frac{3}{\sigma(t) + \rho(t)} - \frac{3\sigma(t) - 3\rho(t)}{2\sigma(t) - 3\rho(t)} \left(\frac{2\sigma(t) - 3\rho(t)}{(\sigma(t))^2 - (\rho(t))^2} - 2\sigma(t) + 3\rho(t) \right),
$$

$t \in \mathbb{T}, \sigma(t) \neq \pm\rho(t), \frac{3}{2}\rho(t)$.

We have

$$\begin{aligned}
A(t) &= \frac{3}{\sigma(t)+\rho(t)} - \frac{3(\sigma(t)-\rho(t))}{2\sigma(t)-3\rho(t)}(2\sigma(t)-3\rho(t))\left(\frac{1}{(\sigma(t))^2-(\rho(t))^2}-1\right) \\
&= \frac{3}{\sigma(t)+\rho(t)} - 3(\sigma(t)-\rho(t))\left(\frac{1-(\sigma(t))^2+(\rho(t))^2}{(\sigma(t))^2-(\rho(t))^2}\right) \\
&= 3\frac{(\sigma(t))^2-(\rho(t))^2-(\sigma(t)+\rho(t))(\sigma(t)-\rho(t))(1-(\sigma(t))^2+(\rho(t))^2)}{(\sigma(t)+\rho(t))((\sigma(t))^2-(\rho(t))^2)} \\
&= 3\frac{(\sigma(t))^2-(\rho(t))^2-((\sigma(t))^2-(\rho(t))^2)(1-(\sigma(t))^2+(\rho(t))^2)}{(\sigma(t)+\rho(t))((\sigma(t))^2-(\rho(t))^2)} \\
&= 3\frac{(\sigma(t))^2-(\rho(t))^2-(\sigma(t))^2+(\rho(t))^2+((\sigma(t))^2-(\rho(t))^2)^2}{(\sigma(t)+\rho(t))((\sigma(t))^2-(\rho(t))^2)} \\
&= 3\frac{(\sigma(t)-\rho(t))^2(\sigma(t)+\rho(t))^2}{(\sigma(t)+\rho(t))^2(\sigma(t)-\rho(t))} \\
&= 3(\sigma(t)-\rho(t)), \quad t\in\mathbb{T},\ \sigma(t)\neq\pm\rho(t), \frac{3}{2}\rho(t).
\end{aligned}$$

Exercise 1.12. Simplify

1.
$$\left(\rho(t)+1-\frac{5(\rho(t))^2+2\rho(t)+1}{(\rho(t)+1)^2}\right)\left(\frac{1}{\rho(t)}+\frac{4}{(\rho(t)-1)^2}\right),$$
$t\in\mathbb{T}, \rho(t)\neq 0, \pm 1$.

2.
$$\left(\frac{2\sigma(t)+\rho(t)}{(\sigma(t))^2-4(\rho(t))^2}-2\sigma(t)-\rho(t)\right)\frac{4\rho(t)+2\sigma(t)}{2\sigma(t)+\rho(t)}-\frac{2}{\sigma(t)-2\rho(t)},$$
$t\in\mathbb{T}, \sigma(t)\neq \pm 2\rho(t), -\frac{\rho(t)}{2}$.

3.
$$\frac{\rho(t)}{\nu(t)}+\frac{\sigma(t)\rho(t)}{\mu(t)}\left(\frac{t+\sigma(t)}{t\sigma(t)-\sigma(t)\rho(t)}+\frac{t+\rho(t)}{(\rho(t))^2-t\rho(t)}+\frac{t}{\sigma(t)\rho(t)}\right),$$
$\sigma(t)\neq 0, t, \rho(t)\neq 0, t, t\in\mathbb{T}$.

4.
$$\left(\frac{\sigma(t)-\rho(t)}{\sigma(t)\rho(t)}-\frac{3\sigma(t)+\rho(t)}{\sigma(t)\rho(t)-(\sigma(t))^2}+\frac{3\rho(t)+\sigma(t)}{\sigma(t)\rho(t)-(\rho(t))^2}\right):\frac{2\sigma(t)+2\rho(t)}{\sigma(t)\rho(t)}+\frac{2\sigma(t)}{\rho(t)-\sigma(t)},$$
$t\in\mathbb{T}, \sigma(t)\neq 0, \pm\rho(t), \rho(t)\neq 0$.

5.
$$\frac{\sigma(t)-2}{(2\sigma(t)+4)^2}:\left(\frac{\sigma(t)}{2\sigma(t)-4}-\frac{(\sigma(t))^2+4}{2(\sigma(t))^2-8}-\frac{2}{(\sigma(t))^2+2\sigma(t)}\right),$$
$t\in\mathbb{T}, \sigma(t)\neq \pm 2, 0$.

1.6 The induction principle

The classical mathematical induction is a concept that helps to prove mathematical results and theorems for all natural numbers. The principle of the classical mathematical induction is a specific technique used to prove certain statements in algebra formulated in terms of n, where n are natural numbers. Any mathematical statement or expression is proved based on the premise that it is true for $n = 1$, $n = k$, and then it is proved for $n = k + 1$.

The time scale analogues of the classical mathematical induction read as follows.

Induction principle Let $t_0 \in \mathbb{T}$ and assume that

$$\{S(t) : t \in [t_0, \infty)_{\mathbb{T}}\}$$

is a family of statements satisfying

(i) $S(t_0)$ is true;

(ii) If $t \in [t_0, \infty)_{\mathbb{T}}$ is right-scattered and $S(t)$ is true, then $S(\sigma(t))$ is true;

(iii) If $t \in [t_0, \infty)_{\mathbb{T}}$ is right-dense and $S(t)$ is true, then there is a neighborhood U of t such that $S(s)$ is true for all $s \in U \cap (t, \infty)_{\mathbb{T}}$;

(iv) If $t \in (t_0, \infty)_{\mathbb{T}}$ is left-dense and $S(s)$ is true for $s \in [t_0, t)_{\mathbb{T}}$, then $S(t)$ is true.

Then $S(t)$ is true for all $t \in [t_0, \infty)_{\mathbb{T}}$.

Dual version of induction principle Let $t_0 \in \mathbb{T}$ and assume that

$$\{S(t) : t \in (-\infty, t_0]_{\mathbb{T}}\}$$

is a family of statements satisfying

(i) $S(t_0)$ is true;

(ii) If $t \in (-\infty, t_0]_{\mathbb{T}}$ is left-scattered and $S(t)$ is true, then $S(\rho(t))$ is true;

(iii) If $t \in (-\infty, t_0]_{\mathbb{T}}$ is left-dense and $S(t)$ is true, then there is neighborhood U of t such that $S(s)$ is true for all $s \in U \cap (-\infty, t)_{\mathbb{T}}$;

(iv) If $t \in (-\infty, t_0)_{\mathbb{T}}$ is right-dense and $S(s)$ is true for $s \in (t, t_0)_{\mathbb{T}}$, then $S(t)$ is true.

Then $S(t)$ is true for all $t \in (-\infty, t_0]_{\mathbb{T}}$.

Example 1.96. Let $\mathbb{T} = 3^{\mathbb{N}}$. We will prove the inequality

$$\frac{1}{4}(s^2 - r^2) \le \frac{1}{13}(s^3 - r^3) \quad \text{for all } r \le s,\ r, s \in \mathbb{T}. \tag{1.1}$$

Fix $r, s \in \mathbb{T}$ such that $r \le s$. Let

$$S(t) : \frac{1}{4}(t^2 - r^2) \le \frac{1}{13}(t^3 - r^3), \quad t \in [r, s]_{\mathbb{T}}. \tag{1.2}$$

1. The statement $S(r)$ is true.

2. Observe that any $t \in \mathbb{T}$ is right-scattered. Assume that $S(t)$ is true. Then

$$\sigma(t) = 3t,$$

and

$$t^2 \leq t^3.$$

Hence, using (1.2), we find

$$\frac{1}{4}(t^2 - r^2) + 2t^2 \leq \frac{1}{13}(t^3 - r^3) + 2t^3,$$

or

$$\frac{1}{4}(9t^2 - r^2) \leq \frac{1}{13}(27t^3 - r^3).$$

Thus

$$\frac{1}{4}((\sigma(t))^2 - r^2) \leq \frac{1}{3}((\sigma(t))^3 - r^3).$$

Consequently, $S(\sigma(t))$ is true.

Applying the induction principle, we conclude that (1.2) is true for all $t \in [r, s]_{\mathbb{T}}$, and thus (1.1) is true for all $r, s \in \mathbb{T}$, $r \leq s$.

Example 1.97. Let $\mathbb{T} = [-2, -1) \cup \{-\frac{1}{n}\}_{n\in\mathbb{N}} \cup \{0\} \cup \{\frac{1}{n}\}_{n\in\mathbb{N}} \cup (1, \frac{3}{2}] \cup [\frac{7}{4}, \frac{11}{6}]$. We will prove that

$$S(t) : \sqrt{t+2} > t \quad \text{for all } t \in \mathbb{T}.$$

We have the following cases.

1. For $t = -2$, we have

$$0 > -2,$$

i. e., the given inequality is true.

2. $t \in \mathbb{T}$ is right-scattered. Then we have the following subcases.
   a. $t = -\frac{1}{n}$ for some $n \in \mathbb{N}$. Then

$$\begin{aligned}\sqrt{-\frac{1}{n} + 2} &> 0 \\ &> -\frac{1}{n},\end{aligned}$$

i. e., $S(t)$ is true. Note that

$$\sigma(t) = -\frac{1}{n+1}$$

and

$$\sqrt{-\frac{1}{n+1} + 2} > 0$$
$$> -\frac{1}{n+1},$$

i. e., $S(\sigma(t))$ is true.

b. $t = \frac{1}{n}$ for some $n \in \mathbb{N}$, $n \geq 2$. Then

$$\frac{1}{n} > \frac{1}{n^2},$$

and

$$\frac{1}{n} + 2 > \frac{1}{n^2},$$

whereupon

$$\sqrt{\frac{1}{n} + 2} > \frac{1}{n},$$

i. e., $S(t)$ is true. Note that

$$\sigma(t) = \frac{1}{n-1}.$$

Then

$$\frac{1}{n-1} > \frac{1}{(n-1)^2},$$

and

$$\frac{1}{n-1} + 2 > \frac{1}{(n-1)^2},$$

whereupon

$$\sqrt{\frac{1}{n-1} + 2} > \frac{1}{n-1}.$$

Thus $S(\sigma(t))$ is true.

c. $t = \frac{3}{2}$. Then

$$\sqrt{\frac{3}{2} + 2} = \sqrt{\frac{7}{2}}$$

$$> \sqrt{\frac{9}{4}}$$
$$= \frac{3}{2},$$

i. e., $S(t)$ is true. Note that

$$\sigma(t) = \frac{7}{4}$$

and

$$\sqrt{\frac{7}{4} + 2} = \sqrt{\frac{15}{4}}$$
$$> \sqrt{\frac{49}{16}}$$
$$= \frac{7}{4},$$

i. e., $S(\sigma(t))$ is true.

3. t is right-dense. Then $t \in [-2, -1) \cup \{0\} \cup [1, \frac{3}{2}] \cup [\frac{7}{4}, \frac{11}{6}]$. We have the following subcases.
   a. $t \in [-2, -1) \cup [1, \frac{3}{2}] \cup [\frac{7}{4}, \frac{11}{6}]$. Then

$$t^2 - t - 2 < 0, \quad t \in \left[1, \frac{3}{2}\right] \cup \left[\frac{7}{4}, \frac{11}{6}\right],$$

   and hence

$$t^2 < t + 2, \quad t \in \left[1, \frac{3}{2}\right] \cup \left[\frac{7}{4}, \frac{11}{6}\right],$$

   whereupon

$$\sqrt{t+2} > t, \quad t \in \left[1, \frac{3}{2}\right] \cup \left[\frac{7}{4}, \frac{11}{6}\right].$$

   For $t \in [-2, -1)$, the inequality is true.
   b. $t = 0$. Then

$$\sqrt{2} > 0,$$

   and by case 2 it follows that there is a neighborhood U of 0 such that $S(s)$ is true for all $s \in U \cap \mathbb{T}$.
   c. t is left-dense. Then $t \in [-2, -1) \cup \{0\} \cup (1, \frac{3}{2}] \cup (\frac{7}{4}, \frac{11}{6}]$. By subcases a and b we get that if $s \in [t_0, t)_{\mathbb{T}}$ for some $t_0 \in \mathbb{T}$ and $S(s)$ is true, then $S(t)$ is true.

By the induction principle it follows that $S(t)$ is true for all $t \in \mathbb{T}$.

Example 1.98. Let $\mathbb{T} = (-\infty, -8) \cup \{-7 - \frac{1}{n}\}_{n\in\mathbb{N}} \cup \{-7\} \cup \{-7 + \frac{1}{n}\}_{n\in\mathbb{N}} \cup \{8\} \cup \{8 + \frac{1}{n}\}_{n\in\mathbb{N}} \cup 10^{\mathbb{N}}$. We will prove that

$$S(t) : \frac{2t - 15}{t + 5} \geq 0 \quad \text{for all } t \in \mathbb{T}. \tag{1.3}$$

1. We will prove that (1.3) holds for all $t \in (-\infty, -6]_{\mathbb{T}}$ using the dual induction principle.
   a. We have

$$\begin{aligned} \frac{2 \cdot (-6) - 15}{-6 + 5} &= \frac{-12 - 15}{-1} \\ &= 27 \\ &> 0, \end{aligned}$$

   i. e., $S(-6)$ is true.
   b. Let $t \in (-\infty, -6]_{\mathbb{T}}$ be left-scattered. Then we have the following cases.
      i. $t = -7 - \frac{1}{n}$ for some $n \in \mathbb{N}$, $n \geq 2$. Then

$$\rho(t) = -7 - \frac{1}{n - 1}.$$

      We have

$$\begin{aligned} \frac{2(-7 - \frac{1}{n}) - 15}{-7 - \frac{1}{n} + 5} &= \frac{-14 - \frac{2}{n} - 15}{-2 - \frac{1}{n}} \\ &= \frac{29 + \frac{2}{n}}{2 + \frac{1}{n}} \\ &> 0, \end{aligned}$$

      i. e., $S(t)$ is true. Moreover, we have

$$\begin{aligned} \frac{2(-7 - \frac{1}{n-1}) - 15}{-7 - \frac{1}{n-1} + 5} &= \frac{-14 - \frac{2}{n-1} - 15}{-2 - \frac{1}{n-1}} \\ &= \frac{29 + \frac{2}{n-1}}{2 + \frac{1}{n-1}} \\ &> 0. \end{aligned}$$

      Thus $S(\rho(t))$ is true.
      ii. $t = -7 + \frac{1}{n}$ for some $n \in \mathbb{N}$. Then

$$\rho(t) = -7 + \frac{1}{n + 1}.$$

We have

$$\frac{2(-7+\frac{1}{n})-15}{-7+\frac{1}{n}+5} = \frac{-14+\frac{2}{n}-15}{-2+\frac{1}{n}}$$
$$= \frac{29-\frac{2}{n}}{2-\frac{1}{n}}$$
$$> 0,$$

because $29-\frac{2}{n} > 0$ and $2-\frac{1}{n} > 0$. Thus $S(t)$ is true. Next,

$$\frac{2(-7+\frac{1}{n+1})-15}{-7+\frac{1}{n+1}+5} = \frac{-14+\frac{2}{n+1}-15}{-2+\frac{1}{n+1}}$$
$$= \frac{29-\frac{2}{n+1}}{2-\frac{1}{n+1}}$$
$$> 0,$$

$29-\frac{2}{n+1} > 0$, and $2-\frac{1}{n+1} > 0$. Therefore $S(\rho(t))$ is true.

c. $t \in (-\infty,-6]_{\mathbb{T}}$ is left-dense. Then we have the following cases.

i. $t \in (-\infty,-8]_{\mathbb{T}}$. Then

$$\frac{2t-15}{t+5} > 0,$$

and

$$\frac{2s-15}{s+5} > 0 \quad \text{for all } s \in (-\infty,t)_{\mathbb{T}}.$$

ii. $t = -7$. By 1. b. ii it follows that there is a neighborhood U of -7 such that $S(s)$ holds for all $s \in U \cap (-\infty,-7)_{\mathbb{T}}$.

d. $t \in (-\infty,-6]_{\mathbb{T}}$ is right-dense. Then we have the following cases.

i. $t \in (-\infty,-8)_{\mathbb{T}}$. By the previous cases we get that $S(s)$ is true for all $s \in (t,-6)_{\mathbb{T}}$ and $S(t)$ is true.

ii. $t = -7$. By 1. b. i it follows that there is a neighborhood U of -7 such that $S(s)$ holds for all $s \in U \cap (-\infty,-7)_{\mathbb{T}}$.

Now applying the dual induction principle, we conclude that $S(t)$ is true for all $t \in (-\infty,-6]_{\mathbb{T}}$.

2. Now we will prove that (1.3) holds for all $t \in [8,\infty)_{\mathbb{T}}$ using the induction principle.

a. We have

$$\frac{2\cdot 8-15}{8+5} = \frac{16-15}{13}$$

$$= \frac{1}{13}$$
$$> 0.$$

Thus $S(8)$ is true.

b. Let $t \in [8, \infty)_{\mathbb{T}}$ be right-scattered, and let $S(t)$ be true. Then we have the following cases.

i. $t = 8 + \frac{1}{n}$ for some $n \in \mathbb{N}$, $n \geq 2$. Then

$$\sigma(t) = 8 + \frac{1}{n-1}.$$

We have

$$\begin{aligned} \frac{2(8 + \frac{1}{n}) - 15}{8 + \frac{1}{n} + 5} &= \frac{16 + \frac{2}{n} - 15}{13 + \frac{1}{n}} \\ &= \frac{1 + \frac{2}{n}}{13 + \frac{1}{n}} \\ &> 0 \end{aligned}$$

and

$$\begin{aligned} \frac{2(8 + \frac{1}{n-1}) - 15}{8 + \frac{1}{n-1} + 5} &= \frac{16 + \frac{2}{n-1} - 15}{13 + \frac{1}{n-1}} \\ &= \frac{1 + \frac{2}{n-1}}{13 + \frac{1}{n-1}} \\ &> 0, \end{aligned}$$

i. e., $S(t)$ and $S(\sigma(t))$ are true.

ii. $t = 9$. Then

$$\sigma(9) = 10.$$

We have

$$\begin{aligned} \frac{2 \cdot 9 - 15}{9 + 5} &= \frac{18 - 15}{14} \\ &= \frac{3}{14} \\ &> 0 \end{aligned}$$

and

$$\frac{2 \cdot 10 - 15}{10 + 5} = \frac{20 - 15}{15}$$

$$= \frac{5}{15}$$
$$= \frac{1}{3}$$
$$> 0.$$

So $S(9)$ and $S(\sigma(9))$ are true.

iii. $t = 10^n$ for some $n \in \mathbb{N}$. Then

$$\sigma(t) = 10^{n+1}.$$

We have

$$\frac{2 \cdot 10^n - 15}{10^n + 5} > 0$$

and

$$\frac{2 \cdot 10^{n+1} - 15}{10^{n+1} + 5} > 0.$$

Thus $S(t)$ and $S(\sigma(t))$ are true.

c. $t = 8$. We have

$$\begin{aligned} \frac{2 \cdot 8 - 15}{8 + 5} &= \frac{16 - 15}{13} \\ &= \frac{1}{13} \\ &> 0, \end{aligned}$$

i. e., $S(8)$ is true. By 2. b. i it follows that there is a neighborhood U of 8 such that $S(t)$ is true for all $t \in U \cap [8, \infty)_{\mathbb{T}}$.

d. Note that there are no left-dense points in $(8, \infty)_{\mathbb{T}}$.

Now applying the induction principle, we conclude that $S(t)$ is true for all $t \in [8, \infty)_{\mathbb{T}}$.

Example 1.99. Let $\mathbb{T} = \{-1\} \cup \{-1 + \frac{1}{n^2}\}_{n \in \mathbb{N}} \cup \{\frac{1}{n^3}\}_{n \in \mathbb{N}} \cup \{0\} \cup (1, 2]$. We will prove that

$$S(t) : t^2 - t - 2 \leq 0 \quad \text{for all } t \in \mathbb{T}.$$

1. We have

$$\begin{aligned} (-1)^2 - (-1) - 2 &= 1 + 1 - 2 \\ &= 0, \end{aligned}$$

i. e., $S(-1)$ is true.

2. Let $t \in \mathbb{T}$ be right-scattered. We have the following cases.
   a. $t = -1 + \frac{1}{n^2}$ for some $n \in \mathbb{N}$, $n \geq 2$. Then

$$\sigma(t) = -1 + \frac{1}{(n-1)^2}.$$

We have

$$\begin{aligned}\left(-1+\frac{1}{n^2}\right)^2 - \left(-1+\frac{1}{n^2}\right) - 2 &= 1 - \frac{2}{n^2} + \frac{1}{n^4} + 1 - \frac{1}{n^2} - 2\\ &= \frac{1}{n^4} - \frac{3}{n^2}\\ &= \frac{1-3n^2}{n^4}\\ &\leq 0,\end{aligned}$$

i. e., $S(t)$ is true. Next,

$$\begin{aligned}\left(-1+\frac{1}{(n-1)^2}\right)^2 - \left(-1+\frac{1}{(n-1)^2}\right) - 2 &= 1 - \frac{2}{(n-1)^2} + \frac{1}{(n-1)^4} + 1 - \frac{1}{(n-1)^2} - 2\\ &= \frac{1}{(n-1)^4} - \frac{3}{(n-1)^2}\\ &= \frac{1-3(n-1)^2}{(n-1)^4}\\ &\leq 0.\end{aligned}$$

Thus $S(\sigma(t))$ is true.

   b. $t = \frac{1}{n^3}$ for some $n \in \mathbb{N}$, $n \geq 2$. Then

$$\sigma(t) = \frac{1}{(n-1)^3}.$$

We have

$$\begin{aligned}\left(-1+\frac{1}{n^3}\right)^2 - \left(-1+\frac{1}{n^3}\right) - 2 &= 1 - \frac{2}{n^3} + \frac{1}{n^6} + 1 - \frac{1}{n^3} - 2\\ &= \frac{1}{n^6} - \frac{3}{n^3}\\ &= \frac{1-3n^3}{n^6}\\ &\leq 0,\end{aligned}$$

i. e., $S(t)$ is true. Moreover,

$$
\begin{aligned}
\left(-1+\frac{1}{(n-1)^3}\right)^2-\left(-1+\frac{1}{(n-1)^3}\right)-2 &= 1-\frac{2}{(n-1)^3}+\frac{1}{(n-1)^6}+1-\frac{1}{(n-1)^3}-2\\
&= \frac{1}{(n-1)^6}-\frac{3}{(n-1)^3}\\
&= \frac{1-3(n-1)^3}{(n-1)^6}\\
&\le 0,
\end{aligned}
$$

and thus $S(\sigma(t))$ is true.

3. Let $t \in \mathbb{T}$ be right-dense, and let $S(t)$ be true. We have the following cases.
   a. $t = -1$. Then by 2.a it follows that there is a neighborhood U of -1 such that $S(s)$ is true for all $s \in U \cap \mathbb{T}$.
   b. $t = 0$. Then by 2.b it follows that there is a neighborhood U of 0 such that $S(s)$ is true for all $s \in U \cap \mathbb{T}$.
   c. $t \in (1, 2]$. Then

$$t^2 - t - 2 \le 0,$$

   and there is a neighborhood U of t such that $S(s)$ is true for all $s \in U \cap \mathbb{T}$.
   d. $t \in \mathbb{T}$ be left-dense, and $S(s)$ is true for all $s \in [-1, t)_{\mathbb{T}}$. We have that $t \subset (1, 2]_{\mathbb{T}}$. By the previous cases it follows that $S(t)$ is true.

Now applying the induction principle, we conclude that $S(t)$ is true for all $t \in \mathbb{T}$.

Example 1.100. Let $\mathbb{T} = \mathbb{N}$. We will prove that

$$S(n) : 1^4 + 2^4 + \cdots + n^4 = \frac{1}{30} n(n+1)(2n+1)(3n^2+3n-1) \quad \text{for all } n \in \mathbb{T}.$$

1. We have

$$
\begin{aligned}
\frac{1}{30} \cdot 1 \cdot (1+1) \cdot (2 \cdot 1 + 1)(3 \cdot 1^2 + 3 \cdot 1 - 1) &= \frac{1}{30} \cdot 2 \cdot 3 \cdot 5\\
&= 1\\
&= 1^4.
\end{aligned}
$$

   Thus $S(1)$ is true.
2. Firstly, note that all points of $\mathbb{T}$ are right-scattered. Assume that $S(n)$ is true for some $n \in \mathbb{T}$. We have

$$\sigma(n) = n + 1.$$

   Then

$$1^4 + 2^4 + \cdots + (n+1)^4 = 1^4 + 2^4 + \cdots + n^4 + (n+1)^4$$

$$
\begin{aligned}
&= \frac{1}{30} n(n+1)(2n+1)(3n^2+3n-1) + (n+1)^4 \\
&= (n+1)\left(\frac{1}{30} n(2n+1)(3n^2+3n-1) + (n+1)^3\right) \\
&= \frac{1}{30}(n+1)((2n^2+n)(3n^2+3n-1) + 30(n+1)^3) \\
&= \frac{1}{30}(n+1)(6n^4+6n^3-2n^2+3n^3+3n^2-n \\
&\quad + 30n^3+90n^2+90n+30) \\
&= \frac{1}{30}(n+1)(6n^4+39n^3+91n^2+89n+30) \\
&= \frac{1}{30}(n+1)(n+2)(6n^3+27n^2+37n+15) \\
&= \frac{1}{30}(n+1)(n+2)(2n+3)(3n^2+9n+5) \\
&= \frac{1}{30}(n+1)(n+2)(2n+3)(3n^2+6n+3+3n+3-1) \\
&= \frac{1}{30}(n+1)(n+2)(2n+3)(3(n+1)^2+3(n+1)-1).
\end{aligned}
$$

Thus $S(\sigma(n))$ is true.

Applying the induction principle, we conclude that $S(n)$ is true for all $n \in \mathbb{N}$.

Exercise 1.13. Let

$$
\mathbb{T} = -15^{\mathbb{N}} \cup \left\{-14 - \frac{1}{n}\right\}_{n\in\mathbb{N}} \cup \left\{-14 + \frac{1}{n}\right\}_{n\in\mathbb{N}} \cup \{-14\} \cup (-13,-10] \cup \{-9\} \cup \left\{-9 + \frac{1}{n}\right\}_{n\in\mathbb{N}}.
$$

Prove that

$$
\frac{2t-1}{3} < \frac{t-3}{2} \quad \text{for all } t \in \mathbb{T}.
$$

Exercise 1.14. Let

$$
\mathbb{T} = \{2\} \cup \left\{2 + \frac{1}{n}\right\}_{n\in\mathbb{N}} \cup \left\{3 - \frac{1}{n}\right\}_{n\in\mathbb{N}} \cup \left\{3 + \frac{1}{n}\right\}_{n\in\mathbb{N}} \cup \{3\} \cup 4^{\mathbb{N}}.
$$

Prove that

$$
\frac{2-t}{t+1} + \frac{t+2}{t-1} > \frac{t-3}{2} \quad \text{for all } t \in \mathbb{T}.
$$

Exercise 1.15. Let

$$
\mathbb{T} = (-\infty, 0] \cup \left\{\frac{1}{n}\right\}_{n\in\mathbb{N}} \cup \{2\} \cup \left\{2 - \frac{1}{n}\right\}_{n\in\mathbb{N}} \cup \{5\} \cup \left\{5 - \frac{1}{n}\right\}_{n\in\mathbb{N}} \cup \left\{5 + \frac{1}{n}\right\}_{n\in\mathbb{N}} \cup 7^{\mathbb{N}}.
$$

Prove that

$$5(t-3)(t-4) > 0 \quad \text{for all } t \in \mathbb{T}.$$

Exercise 1.16. Let

$$\mathbb{T} = (-\infty, -10] \cup \{-9\} \cup \left\{-9 - \frac{1}{n}\right\}_{n \in \mathbb{N}} \cup \left\{-9 + \frac{1}{n}\right\}_{n \in \mathbb{N}} \cup (-8, -1] \cup \{4\} \cup \left\{4 + \frac{1}{n}\right\}_{n \in \mathbb{N}}$$
$$\cup \{6\} \cup \left\{6 - \frac{1}{n}\right\}_{n \in \mathbb{N}} \cup \left\{6 + \frac{1}{n}\right\}_{n \in \mathbb{N}} \cup 9^{\mathbb{N}}.$$

Prove that

$$5t^2 - 2(t+3)(t-1) > 4t + 9 \quad \text{for all } t \in \mathbb{T}.$$

Exercise 1.17. Let $\mathbb{T} = \mathbb{N}$. Prove that

$$1^5 + 2^5 + \cdots + n^5 = \frac{1}{12} n^2 (n+1)^2 \left(2n^2 + 2n - 1\right) \quad \text{for all } n \in \mathbb{T}.$$

1.7 Advanced practical problems

Problem 1.1. Check if the following sets are time scales:
1. $3^{\mathbb{N}_0}$.
2. $[-1, 1] \cup (2, 7] \cup [8, 11] \cup [15, 27]$.
3. $(\frac{1}{4})^{\mathbb{N}_0}$.
4. $(3 - U) \cup (3 + U) \cup \{3\}$, $U = \{\frac{1}{12^n} : n \in \mathbb{N}_0\}$.
5. $(-1, 4] \cup 7^{\mathbb{N}_0}$.

Problem 1.2. Find $\sigma(t)$, $t \in \mathbb{T}$, where
1. $\mathbb{T} = 4^{\mathbb{N}_0} \cup (-5\mathbb{N}_0)$.
2. $\mathbb{T} = [-1, 3] \cup [7, 9] \cup [10, \infty)$.
3. $\mathbb{T} = \{0\} \cup \{1 - \frac{1}{4^n}\}_{n \in \mathbb{N}_0} \cup 2^{\mathbb{N}_0}$.
4. $\mathbb{T} = \{2\} \cup \{2 + \frac{1}{n}\}_{n \in \mathbb{N}} \cup [4, 9] \cup \{9 + \frac{1}{n^2}\}_{n \in \mathbb{N}}$.
5. $\mathbb{T} = [1, 3] \cup 5^{\mathbb{N}}$.

Problem 1.3. Find $\mu(t)$, $t \in \mathbb{T}$, where $\mathbb{T}$ are the time scales in Problem 1.2.

Problem 1.4. Let $\mathbb{T} = \{t_1 < t_2 < \cdots < t_p\}$. Find $\sum_{k=1}^{p} \mu(t_k)$.

Problem 1.5. Find $\rho(t)$, $t \in \mathbb{T}$, where $\mathbb{T}$ are the time scales in Problem 1.2.

Problem 1.6. Find $\nu(t)$, $t \in \mathbb{T}$, where T are the time scales in Problem 1.2.

Problem 1.7. Let $\mathbb{T} = \{t_1 < t_2 < \cdots < t_p\}$. Find $\sum_{k=1}^{p} \nu(t_k)$.

Problem 1.8. Classify the points of $\mathbb{T}$, where $\mathbb{T}$ are the time scales in Problem 1.2.

Problem 1.9. Find

$$[a,b]_{\mathbb{T}}, \quad [a,b)_{\mathbb{T}}, \quad (a,b]_{\mathbb{T}}, \quad (a,b)_{\mathbb{T}},$$

where
1. $\mathbb{T} = (-5\mathbb{N}_0) \cup 4^{\mathbb{N}_0}$, $a = -10$, $b = 16$.
2. $\mathbb{T} = [-1,3] \cup [7,9] \cup [10,\infty)$, $a = 0$, $b = 11$.
3. $\mathbb{T} = \{0\} \cup \{1 - \frac{1}{4^n}\}_{n\in\mathbb{N}_0} \cup 2^{\mathbb{N}_0}$, $a = \frac{63}{64}$, $b = 8$.
4. $\mathbb{T} = \{2\} \cup \{2 + \frac{1}{n}\}_{n\in\mathbb{N}} \cup [4,9] \cup \{9 + \frac{1}{n^2}\}_{n\in\mathbb{N}}$, $a = 3$, $b = 10$.
5. $\mathbb{T} = [1,3] \cup 5^{\mathbb{N}}$, $a = 3$, $b = 125$.

Problem 1.10. Find $\mathbb{T}_\kappa$ and $\mathbb{T}^k$, where $\mathbb{T}$ are the time scales in Problem 1.2.

Problem 1.11. Let

$$f(t) = \frac{1+t}{2+3t}, \quad t \in \mathbb{T},$$

where $\mathbb{T}$ are the time scales in Problem 1.2. Find

$$f^{\sigma}(t) \quad \text{and} \quad f^{\sigma^2}(t) \quad \text{for } t \in \mathbb{T}.$$

Problem 1.12. Let

$$f(t) = \sqrt{1+4t^2}, \quad t \in \mathbb{T},$$

where $\mathbb{T}$ are the time scales in Problem 1.2. Find

$$f^{\rho^2}(t) \quad \text{and} \quad f^{\sigma\rho\sigma}(t) \quad \text{for } t \in \mathbb{T}.$$

Problem 1.13. Simplify
1.
$$\left(\frac{(\sigma(t))^2+9}{27-3(\sigma(t))^2} + \frac{\sigma(t)}{3\sigma(t)+9} - \frac{3}{(\sigma(t))^2-3\sigma(t)}\right) : \frac{(3\sigma(t)+9)^2}{3(\sigma(t))^2-(\sigma(t))^3},$$
$t \in \mathbb{T}, \sigma(t) \neq 0, \pm 3$.
2.
$$\frac{2}{\sigma(t)\rho(t)} : \left(\frac{1}{\sigma(t)} - \frac{1}{\rho(t)}\right)^2 - \frac{(\sigma(t))^2+(\rho(t))^2}{(\sigma(t)-\rho(t))^2},$$
$t \in \mathbb{T}, \sigma(t) \neq 0, \rho(t) \neq 0, \sigma(t) \neq \rho(t)$.
3.
$$\left(\frac{\sigma(t)\rho(t)+(\sigma(t))^2}{5(\sigma(t))^2-5\sigma(t)\rho(t)} + \sigma(t)\rho(t) + (\rho(t))^2\right)\frac{5\sigma(t)}{\sigma(t)+\rho(t)} - \frac{\rho(t)}{\sigma(t)-\rho(t)},$$
$t \in \mathbb{T}, \sigma(t) \neq \pm\rho(t)$.

4.
$$\frac{(1+\sigma(t)\rho(t))(1+t\sigma(t))}{(\sigma(t)-\rho(t))\mu(t)}-\frac{(1+t\rho(t))(1+\sigma(t)\rho(t))}{\nu(t)(\rho(t)-\sigma(t))}-\frac{(1+t\sigma(t))(1+t\rho(t))}{\mu(t)\nu(t)},$$

$t\in\mathbb{T}, \sigma(t)\neq t, \rho(t), \rho(t)\neq t.$

5.
$$\frac{1}{\sigma(t)(\sigma(t)-\rho(t))\mu(t)}-\frac{1}{\rho(t)(\rho(t)-\sigma(t))\nu(t)}-\frac{1}{t\mu(t)\nu(t)},$$

$t\in\mathbb{T}, \mu(t)\neq 0, \sigma(t)\neq 0, \rho(t), \rho(t)\neq 0, \nu(t)\neq 0, t\in\mathbb{T}.$

Problem 1.14. Let

$$\mathbb{T}=\{-2\}\cup\left\{-2+\frac{1}{n}\right\}\cup\left\{-1-\frac{1}{n}\right\}_{n\in\mathbb{N}}\cup\{2\}\cup\left\{2-\frac{1}{n}\right\}_{n\in\mathbb{N}}\cup\left\{2+\frac{1}{n}\right\}_{n\in\mathbb{N}}\cup 4^{\mathbb{N}}.$$

Prove that

$$\frac{t}{t+3}-\frac{t-3}{t}>\frac{9-t}{t^2+3t}\quad\text{for all } t\in\mathbb{T}.$$

Problem 1.15. Let

$$\mathbb{T}=(-\infty,-5]\cup\{-3\}\cup\left\{-3-\frac{1}{n}\right\}_{n\in\mathbb{N}}\cup\left\{-3+\frac{1}{n}\right\}_{n\in\mathbb{N}}\cup\left\{-2-\frac{1}{n}\right\}_{n\in\mathbb{N}}$$
$$\cup\left\{-\frac{1}{2}\right\}\cup\left\{-\frac{1}{2}-\frac{1}{3n^2}\right\}_{n\in\mathbb{N}}\cup\left\{-\frac{1}{2}+\frac{1}{4n}\right\}_{n\in\mathbb{N}}.$$

Prove that

$$\frac{2-t}{t+1}+\frac{t+2}{t-1}<\frac{7t-t^3}{t^2-1}\quad\text{for all } t\in\mathbb{T}.$$

Problem 1.16. Let

$$\mathbb{T}=(-\infty,-3]\cup\{-2\}\cup\left\{-2-\frac{1}{2n}\right\}_{n\in\mathbb{N}}\cup\left\{-2+\frac{1}{4n^2}\right\}_{n\in\mathbb{N}}\cup[0,1]$$
$$\cup\left\{7-\frac{1}{n}\right\}_{n\in\mathbb{N}}\cup\{7\}\cup\left\{7+\frac{1}{n}\right\}_{n\in\mathbb{N}}\cup 9^{\mathbb{N}}.$$

Prove that

$$\frac{6-t}{5t-7}<0\quad\text{for all } t\in\mathbb{T}.$$

Problem 1.17. Let

$$\mathbb{T}=-10^{\mathbb{N}}\cup\{-7\}\cup\left\{-7-\frac{1}{n^2}\right\}_{n\in\mathbb{N}}\cup\left\{-7+\frac{1}{n^2}\right\}_{n\in\mathbb{N}}\cup\{3\}\cup\left\{3+\frac{1}{n^2}\right\}_{n\in\mathbb{N}}\cup 5^{\mathbb{N}}.$$

Prove that

$$\frac{6t-3}{t+4} \geq 2 \quad \text{for all } t \in \mathbb{T}.$$

Problem 1.18. Let $\mathbb{T} = \mathbb{N}$. Prove that

$$1^7 + 2^7 + \cdots + n^7 = \frac{1}{24} n^2 (n+1)^2 (3n^4 + 6n^3 - n^2 - 4n + 2) \quad \text{for all } n \in \mathbb{T}.$$

2 Delta differentiation

This chapter deals with the delta differential calculus for one-variable functions on time scales. The basic definition of delta differentiation is due to Hilger. We have included several examples on differentiation and the Leibniz formula for the nth delta derivative of a product of two functions. We present mean value results and sufficient conditions for convexity and concavity of one-variable functions and for complete differentiability of a one-variable function. Several versions of chain and L'Hôpital's rules, which do not appear in the usual form, are included.

2.1 Definition of delta derivative. Examples

Definition 2.1. Let $f : \mathbb{T} \to \mathbb{R}$, and let $t \in \mathbb{T}^\kappa$. We define $f^\Delta(t)$ as the number, provided that it exists, such that for any $\varepsilon > 0$, there is a neighborhood of t, $U = (t-\delta, t+\delta) \cap \mathbb{T}$ for some $\delta > 0$, such that

$$|f(\sigma(t)) - f(s) - f^\Delta(t)(\sigma(t) - s)| \le \varepsilon|\sigma(t) - s| \quad \text{for all } s \in U,\ s \neq \sigma(t).$$

We call $f^\Delta(t)$ the delta or Hilger derivative of f at t.

We say that f is delta or Hilger differentiable (shortly differentiable) in $\mathbb{T}^\kappa$ if $f^\Delta(t)$ exists for all $t \in \mathbb{T}^\kappa$. The function $f^\Delta : \mathbb{T} \to \mathbb{R}$ is called the delta derivative or Hilger derivative (shortly derivative) of f in $\mathbb{T}^\kappa$.

Remark 2.1. If $\mathbb{T} = \mathbb{R}$, then the delta derivative coincides with the classical derivative.

Remark 2.2. The delta derivative is well defined. Indeed, let $t \in \mathbb{T}^\kappa$, and let $f_1^\Delta(t)$ and $f_2^\Delta(t)$ be such that

$$\begin{aligned}
|f(\sigma(t)) - f(s) - f_1^\Delta(t)(\sigma(t) - s)| &\le \frac{\varepsilon}{2}|\sigma(t) - s|, \\
|f(\sigma(t)) - f(s) - f_2^\Delta(t)(\sigma(t) - s)| &\le \frac{\varepsilon}{2}|\sigma(t) - s|
\end{aligned}$$

for all $\varepsilon > 0$ and all s belonging to a neighborhood U of t, $U = (t-\delta, t+\delta) \cap \mathbb{T}$ for some $\delta > 0$, $s \neq \sigma(t)$. Hence

$$\begin{aligned}
|f_1^\Delta(t) - f_2^\Delta(t)| &= \left| f_1^\Delta(t) - \frac{f(\sigma(t)) - f(s)}{\sigma(t) - s} + \frac{f(\sigma(t)) - f(s)}{\sigma(t) - s} - f_2^\Delta(t) \right| \\
&\le \left| f_1^\Delta(t) - \frac{f(\sigma(t)) - f(s)}{\sigma(t) - s} \right| + \left| \frac{f(\sigma(t)) - f(s)}{\sigma(t) - s} - f_2^\Delta(t) \right| \\
&= \frac{|f(\sigma(t)) - f(s) - f_1^\Delta(t)(\sigma(t) - s)|}{|\sigma(t) - s|} + \frac{|f(\sigma(t)) - f(s) - f_2^\Delta(t)(\sigma(t) - s)|}{|\sigma(t) - s|} \\
&\le \frac{\varepsilon}{2} + \frac{\varepsilon}{2} \\
&= \varepsilon.
\end{aligned}$$

https://doi.org/10.1515/9783112232088-002

Since $\varepsilon > 0$ was arbitrarily chosen, we conclude that

$$f_1^{\Delta}(t) = f_2^{\Delta}(t).$$

Remark 2.3. Let us assume that $\sup \mathbb{T} < \infty$ and $\sup \mathbb{T}$ is left-scattered, and $f^{\Delta}(t)$ is defined at a point $t \in \mathbb{T} \backslash \mathbb{T}^{\kappa}$ as in Definition 2.1. Then the unique point $t \in \mathbb{T} \backslash \mathbb{T}^{\kappa}$ is $\sup \mathbb{T}$. Hence, for any $\varepsilon > 0$, there is a neighborhood $U = (t - \delta, t + \delta) \cap (\mathbb{T} \backslash \mathbb{T}^{\kappa})$ for some $\delta > 0$ such that

$$f(\sigma(t)) = f(s) = f(\sigma(\sup \mathbb{T})) = f(\sup \mathbb{T}), \quad s \in U,\ s \neq \sigma(t).$$

Therefore for all $\alpha \in \mathbb{R}$ and $s \in U$, we have

$$\begin{aligned} |f(\sigma(t)) - f(s) - \alpha(\sigma(t) - s)| &= |f(\sup \mathbb{T}) - f(\sup \mathbb{T}) - \alpha(\sup \mathbb{T} - \sup \mathbb{T})| \\ &\leq \varepsilon |\sigma(t) - s|, \end{aligned}$$

i. e., any $\alpha \in \mathbb{R}$ is the delta derivative of f at the point $t \in \mathbb{T} \backslash \mathbb{T}^{\kappa}$.

Example 2.1. Let $f(t) = t^3$, $t \in \mathbb{T}$. We will prove that

$$f^{\Delta}(t) = (\sigma(t))^2 + t\sigma(t) + t^2, \quad t \in \mathbb{T}^{\kappa}.$$

Fix $t \in \mathbb{T}^{\kappa}$. Take arbitrary $\varepsilon > 0$ and

$$\delta = \frac{-|\sigma(t) + 2t| + \sqrt{(\sigma(t) + 2t)^2 + 4\varepsilon}}{2}.$$

Then $\delta > 0$, and

$$\delta^2 + |\sigma(t) + 2t|\delta - \varepsilon = 0.$$

Let

$$U = \{s \in \mathbb{T} : |t - s| \leq \delta\}.$$

Then for all $s \in U$, we get

$$\begin{aligned} &|f(\sigma(t)) - f(s) - ((\sigma(t))^2 + t\sigma(t) + t^2)(\sigma(t) - s)| \\ &\quad = |(\sigma(t))^3 - s^3 - ((\sigma(t))^2 + t\sigma(t) + t^2)(\sigma(t) - s)| \\ &\quad = |(\sigma(t) - s)((\sigma(t))^2 + s\sigma(t) + s^2) - ((\sigma(t))^2 + t\sigma(t) + t^2)(\sigma(t) - s)| \\ &\quad = |(\sigma(t))^2 + s\sigma(t) + s^2 - (\sigma(t))^2 - t\sigma(t) - t^2||\sigma(t) - s| \\ &\quad = |(s - t)\sigma(t) + (s - t)(s + t)||\sigma(t) - s| \\ &\quad = |s - t||\sigma(t) + s + t||\sigma(t) - s| \end{aligned}$$

$$
\begin{aligned}
&= |t-s|\big|\sigma(t)+2t+s-t\big|\big|\sigma(t)-s\big| \\
&\le |t-s|\big(\big|\sigma(t)+2t\big|+|s-t|\big)\big|\sigma(t)-s\big| \\
&\le \delta\big(\delta+\big|\sigma(t)+2t\big|\big)\big|\sigma(t)-s\big| \\
&= \varepsilon\big|\sigma(t)-s\big|.
\end{aligned}
$$

Example 2.2. Let $f(t) = \sqrt{t}$, $t \in \mathbb{T}$, $t > 0$. We will prove that

$$f^{\Delta}(t) = \frac{1}{\sqrt{t}+\sqrt{\sigma(t)}}, \qquad t \in \mathbb{T}^{\kappa},\ t > 0.$$

Fix arbitrary $t \in \mathbb{T}^{\kappa}$, $t > 0$. Take arbitrary $\varepsilon > 0$ and

$$\delta = \sqrt{t}\sqrt{\sigma(t)}\big(\sqrt{t}+\sqrt{\sigma(t)}\big)\varepsilon.$$

Let

$$U = \big\{s \in \mathbb{T} : s > 0, \quad |t-s| \le \delta\big\}.$$

Then for $s \in U$, we get

$$
\begin{aligned}
&\left|f(\sigma(t)) - f(s) - \frac{1}{\sqrt{t}+\sqrt{\sigma(t)}}(\sigma(t)-s)\right| \\
&\quad= \left|\sqrt{\sigma(t)} - \sqrt{s} - \frac{1}{\sqrt{t}+\sqrt{\sigma(t)}}(\sigma(t)-s)\right| \\
&\quad= \left|\frac{\sigma(t)-s}{\sqrt{\sigma(t)}+\sqrt{s}} - \frac{1}{\sqrt{t}+\sqrt{\sigma(t)}}(\sigma(t)-s)\right| \\
&\quad= \left|\frac{1}{\sqrt{\sigma(t)}+\sqrt{s}} - \frac{1}{\sqrt{t}+\sqrt{\sigma(t)}}\right|\big|\sigma(t)-s\big| \\
&\quad= \left|\frac{\sqrt{t}+\sqrt{\sigma(t)}-\sqrt{\sigma(t)}-\sqrt{s}}{\big(\sqrt{\sigma(t)}+\sqrt{s}\big)\big(\sqrt{t}+\sqrt{\sigma(t)}\big)}\right|\big|\sigma(t)-s\big| \\
&\quad= \left|\frac{\sqrt{t}-\sqrt{s}}{\big(\sqrt{\sigma(t)}+\sqrt{s}\big)\big(\sqrt{t}+\sqrt{\sigma(t)}\big)}\right|\big|\sigma(t)-s\big| \\
&\quad= \frac{|t-s||\sigma(t)-s|}{\big(\sqrt{\sigma(t)}+\sqrt{s}\big)\big(\sqrt{t}+\sqrt{\sigma(t)}\big)\big(\sqrt{t}+\sqrt{s}\big)} \\
&\quad\le \frac{|t-s||\sigma(t)-s|}{\sqrt{t}\sqrt{\sigma(t)}\big(\sqrt{t}+\sqrt{\sigma(t)}\big)} \\
&\quad\le \frac{\delta|\sigma(t)-s|}{\sqrt{t}\sqrt{\sigma(t)}\big(\sqrt{t}+\sqrt{\sigma(t)}\big)} \\
&\quad= \varepsilon\big|\sigma(t)-s\big|.
\end{aligned}
$$

Example 2.3. Let $\mathbb{T} = h\mathbb{Z}$, $h > 0$, and

$$f(t) = t^2 - t, \quad t \in \mathbb{T}.$$

We will prove that

$$f^{\Delta}(t) = 2t + h - 1, \quad t \in \mathbb{T}.$$

Take arbitrary $t \in \mathbb{T}$. Let $\varepsilon > 0$ and $\delta = \varepsilon$. Let also,

$$U = \{s \in \mathbb{T} : |t - s| < \delta\}.$$

Then for $s \in U$, we get

$$\begin{aligned}
&|f(\sigma(t)) - f(s) - (2t + h - 1)(\sigma(t) - s)| \\
&\quad = |(\sigma(t))^2 - \sigma(t) - (s^2 - s) - (2t + h - 1)(\sigma(t) - s)| \\
&\quad = |(\sigma(t))^2 - s^2 - (\sigma(t) - s) - (2t + h - 1)(\sigma(t) - s)| \\
&\quad = |\sigma(t) + s - 1 - (2t + h - 1)||\sigma(t) - s| \\
&\quad = |t + h + s - 1 - 2t - h + 1||\sigma(t) - s| \\
&\quad = |t - s||\sigma(t) - s| \\
&\quad \le \delta|\sigma(t) - s| \\
&\quad = \varepsilon|\sigma(t) - s|.
\end{aligned}$$

Example 2.4. Let $T = 3^{\mathbb{N}_0}$, and let

$$f(t) = t^3 - 2t^2 + 4t + 1, \quad t \in \mathbb{T}.$$

We will prove that

$$f^{\Delta}(t) = 13t^2 - 8t + 4, \quad t \in \mathbb{T}.$$

Here

$$\sigma(t) = 3t, \quad t \in \mathbb{T}.$$

Fix $t \in \mathbb{T}$. Take arbitrary $\varepsilon > 0$. Let

$$\delta = \frac{-(5t + 2) + \sqrt{(5t + 2)^2 + 4\varepsilon}}{2}.$$

Then

$$\delta(\delta + 5t + 2) = \varepsilon.$$

Let

$$U = \{s \in \mathbb{T} : |t - s| \le \delta\}.$$

Then for $s \in U$, we get

$$\begin{aligned}
&|f(\sigma(t)) - f(s) - (13t^2 - 8t + 4)(\sigma(t) - s)| \\
&\quad = |f(3t) - f(s) - (13t^2 - 8t + 4)(3t - s)| \\
&\quad = |(3t)^3 - 2(3t)^2 + 4(3t) + 1 - s^3 + 2s^2 - 4s - 1 - (13t^2 - 8t + 4)(3t - s)| \\
&\quad = |(3t - s)(9t^2 + 3st + s^2) - 2(3t - s)(3t + s) + 4(3t - s) - (13t^2 - 8t + 4)(3t - s)| \\
&\quad = |9t^2 + 3st + s^2 - 6t - 2s + 4 - 13t^2 + 8t - 4||3t - s| \\
&\quad = |-4t^2 + 3st + s^2 + 2t - 2s||3t - s| \\
&\quad = |-3t^2 + 3st - t^2 + s^2 + 2(t - s)||3t - s| \\
&\quad = |-3t(t - s) - (t - s)(t + s) + 2(t - s)||3t - s| \\
&\quad = |-3t - t - s + 2||t - s||3t - s| \\
&\quad = |-4t - s + 2||t - s||3t - s| \\
&\quad = |-5t + t - s + 2||t - s||3t - s| \\
&\quad \le (5t + 2 + |t - s|)|t - s||3t - s| \\
&\quad \le (5t + 2 + \delta)\delta|3t - s| \\
&\quad = \varepsilon|3t - s|.
\end{aligned}$$

Example 2.5. Let $\mathbb{T} = \mathbb{N}_0^2$, and let

$$f(t) = \frac{1 + t}{2 + t}, \quad t \in \mathbb{T}.$$

We will prove that

$$f^{\Delta}(t) = \frac{1}{(2 + t)(t + 2\sqrt{t} + 3)}, \quad t \in \mathbb{T}.$$

Here

$$\begin{aligned}
\sigma(t) &= (\sqrt{t} + 1)^2 \\
&= t + 2\sqrt{t} + 1, \quad t \in \mathbb{T}.
\end{aligned}$$

Fix $t \in \mathbb{T}$. Take arbitrary $\varepsilon > 0$ and

$$\delta = 4(t + 2\sqrt{t} + 3)\varepsilon.$$

Let

$$U = \{s \in \mathbb{T} : |t - s| \le \delta\}.$$

Then for $s \in U$, we get

$$\begin{aligned}
&\left| f(\sigma(t)) - f(s) - \frac{1}{(2+t)(t+2\sqrt{t}+3)}(\sigma(t) - s) \right| \\
&\quad = \left| \frac{t+2\sqrt{t}+2}{t+2\sqrt{t}+3} - \frac{1+s}{2+s} - \frac{1}{(2+t)(t+2\sqrt{t}+3)}(t+2\sqrt{t}+1-s) \right| \\
&\quad = \left| \frac{(t+2\sqrt{t}+2)(2+s) - (1+s)(t+2\sqrt{t}+3)}{(t+2\sqrt{t}+3)(2+s)} - \frac{t+2\sqrt{t}+1-s}{(2+t)(t+2\sqrt{t}+3)} \right| \\
&\quad = \left| \frac{t+2\sqrt{t}+2+(1+s)(t+2\sqrt{t}+2-t-2\sqrt{t}-3)}{(t+2\sqrt{t}+3)(2+s)} - \frac{t+2\sqrt{t}+1-s}{(2+t)(t+2\sqrt{t}+3)} \right| \\
&\quad = \left| \frac{t+2\sqrt{t}+1-s}{(t+2\sqrt{t}+3)(2+s)} - \frac{t+2\sqrt{t}+1-s}{(2+t)(t+2\sqrt{t}+3)} \right| \\
&\quad = \left| \frac{1}{2+s} - \frac{1}{2+t} \right| \frac{|t+2\sqrt{t}+1-s|}{t+2\sqrt{t}+3} \\
&\quad = \frac{|2+t-2-s||t+2\sqrt{t}+1-s|}{(2+s)(2+t)(t+2\sqrt{t}+3)} \\
&\quad = \frac{|t-s||t+2\sqrt{t}+1-s|}{(2+s)(2+t)(t+2\sqrt{t}+3)} \\
&\quad \le \frac{|t-s||t+2\sqrt{t}+1-s|}{4(t+2\sqrt{t}+3)} \\
&\quad \le \frac{\delta|t+2\sqrt{t}+1-s|}{4(t+2\sqrt{t}+3)} \\
&\quad = \varepsilon|t+2\sqrt{t}+1-s|.
\end{aligned}$$

Example 2.6. Let $\mathbb{T} = \{H_n : n \in \mathbb{N}_0\}$, and let

$$f(t) = \frac{1-t}{1+2t}, \quad t \in \mathbb{T}.$$

We will prove that

$$f^{\Delta}(H_n) = -\frac{3}{(1+2H_n)(1+2H_{n+1})}, \quad n \in \mathbb{N}_0.$$

Fix arbitrary $n \in \mathbb{N}_0$. Take arbitrary $\varepsilon > 0$ and $\delta = \frac{\varepsilon}{6}$. Let

$$U = \{H_s \in \mathbb{T} : |H_s - H_n| \le \delta\}.$$

Then for $H_s \in U$, we get

$$\left| f(\sigma(H_n)) - f(H_s) + \frac{3}{(1+2H_n)(1+2H_{n+1})}(\sigma(H_n) - H_s) \right|$$

$$
\begin{aligned}
&= \left| f(H_{n+1}) - f(H_s) + \frac{3(H_{n+1} - H_s)}{(1 + 2H_n)(1 + 2H_{n+1})} \right| \\
&= \left| \frac{1 - H_{n+1}}{1 + 2H_{n+1}} - \frac{1 - H_s}{1 + 2H_s} + \frac{3(H_{n+1} - H_s)}{(1 + 2H_n)(1 + 2H_{n+1})} \right| \\
&= \left| \frac{(1 - H_{n+1})(1 + 2H_s) - (1 - H_s)(1 + 2H_{n+1})}{(1 + 2H_{n+1})(1 + 2H_s)} + \frac{3(H_{n+1} - H_s)}{(1 + 2H_n)(1 + 2H_{n+1})} \right| \\
&= \left| \frac{1 + 2H_s - H_{n+1} - 2H_{n+1}H_s - 1 - 2H_{n+1} + H_s + 2H_sH_{n+1}}{(1 + 2H_{n+1})(1 + 2H_s)} + \frac{3(H_{n+1} - H_s)}{(1 + 2H_n)(1 + 2H_{n+1})} \right| \\
&= \left| \frac{3}{(1 + 2H_{n+1})(1 + 2H_s)}(H_s - H_{n+1}) - \frac{3(H_s - H_{n+1})}{(1 + 2H_n)(1 + 2H_{n+1})} \right| \\
&= 3\left| \frac{1}{(1 + 2H_{n+1})(1 + 2H_s)} - \frac{1}{(1 + 2H_n)(1 + 2H_{n+1})} \right| |H_{n+1} - H_s| \\
&= 3\frac{|1 + 2H_n - (1 + 2H_s)|}{(1 + 2H_n)(1 + 2H_s)(1 + 2H_{n+1})} |H_{n+1} - H_s| \\
&= \frac{6|H_n - H_s|}{(1 + 2H_n)(1 + 2H_s)(1 + 2H_{n+1})} |H_{n+1} - H_s| \\
&\le 6\delta |H_{n+1} - H_s| \\
&= \varepsilon |H_{n+1} - H_s|.
\end{aligned}
$$

Example 2.7. Let $\mathbb{T} = P_{1,3}$, and let

$$f(t) = 1 + t + t^2, \quad t \in \mathbb{T}.$$

We will prove that

$$
f^{\Delta}(t) = \begin{cases} 2t + 1 & \text{if } t \in \bigcup_{k=0}^{\infty} [4k, 4k + 1), \\ 2t + 4 & \text{if } t \in \bigcup_{k=0}^{\infty} \{4k + 1\}. \end{cases}
$$

We will consider the following cases.

1. $t \in \bigcup_{k=0}^{\infty} [4k, 4k + 1)$. Take arbitrary $\varepsilon > 0$ and $\delta = \varepsilon$. Let

   $$U = \left\{ s \in \bigcup_{k=0}^{\infty} [4k, 4k + 1) : |t - s| \le \delta \right\}.$$

   Then for $s \in U$, we get

   $$
   \begin{aligned}
   &|f(\sigma(t)) - f(s) - (2t + 1)(\sigma(t) - s)| \\
   &\quad = |f(t) - f(s) - (2t + 1)(t - s)| \\
   &\quad = |1 + t + t^2 - 1 - s - s^2 - (2t + 1)(t - s)| \\
   &\quad = |(t - s) + (t - s)(t + s) - (2t + 1)(t - s)| \\
   &\quad = |1 + t + s - 2t - 1||t - s|
   \end{aligned}
   $$

$$\begin{aligned}&= |t-s|^2\\&\le \delta|t-s|\\&= \varepsilon|t-s|.\end{aligned}$$

2. $t \in \bigcup_{k=0}^{\infty}\{4k+1\}$. Take arbitrary $\varepsilon > 0$ and $\delta = \varepsilon$. Let

$$U = \left\{ s \in \bigcup_{k=0}^{\infty}\{4k+1\} : |t-s| \le \delta \right\}.$$

   Then for $s \in U$, we get

$$\begin{aligned}&\big|f(\sigma(t)) - f(s) - (2t+4)(\sigma(t)-s)\big|\\&\quad= \big|f(t+3) - f(s) - (2t+4)(t+3-s)\big|\\&\quad= \big|1+t+3+(t+3)^2-1-s-s^2-(2t+4)(t+3-s)\big|\\&\quad= \big|1+t+3+t^2+6t+9-1-s-s^2-(2t+4)(t+3-s)\big|\\&\quad= \big|(t+3-s)+(t+3)^2-s^2-(2t+4)(t+3-s)\big|\\&\quad= \big|(t+3-s)+(t+3-s)(t+3+s)-(2t+4)(t+3-s)\big|\\&\quad= |1+t+3+s-2t-4||t+3-s|\\&\quad= |t-s||t+3-s|\\&\quad\le \delta|t+3-s|\\&\quad= \varepsilon|t+3-s|.\end{aligned}$$

Example 2.8. Let $\mathbb{T} = C$, where C is the Cantor set, and let

$$f(t) = \frac{2+t}{1+4t}, \quad t \in \mathbb{T}.$$

We will prove that

$$f(t) = \begin{cases} -\frac{7}{(1+4t)(1+4t+\frac{4}{3^{m+1}})} & \text{if } t \in C_1, t = \sum_{k=1}^{m}\frac{a_k}{3^k} + \frac{1}{3^{m+1}},\\ -\frac{7}{(1+4t)^2} & \text{if } t \in \mathbb{T}\backslash C_1.\end{cases}$$

We will consider the following cases.

1. $t \in C_1$, $t = \sum_{k=1}^{m}\frac{a_k}{3^k} + \frac{1}{3^{m+1}}$ for some $m \in \mathbb{N}$. Take arbitrary $\varepsilon > 0$ and $\delta = \frac{\varepsilon}{28}$. Let

$$U = \left\{ s \in C_1 : s = \sum_{k=1}^{p}\frac{a_k}{3^k} + \frac{1}{3^{p+1}}, p \in \mathbb{N} : |t-s| \le \delta \right\}.$$

   Then for $s \in U$, we get

$$\begin{aligned}
&\left|f(\sigma(t)) - f(s) + \frac{7}{(1+4t)(1+4t+\frac{4}{3^{m+1}})}(\sigma(t)-s)\right| \\
&= \left|f\left(t+\frac{1}{3^{m+1}}\right) - f(s) + \frac{7}{(1+4t)(1+4t+\frac{4}{3^{m+1}})}\left(t+\frac{1}{3^{m+1}}-s\right)\right| \\
&= \left|\frac{2+t+\frac{1}{3^{m+1}}}{1+4t+\frac{4}{3^{m+1}}} - \frac{2+s}{1+4s} + \frac{7(t+\frac{1}{3^{m+1}}-s)}{(1+4t)(1+4t+\frac{4}{3^{m+1}})}\right| \\
&= \left|\frac{(2+t+\frac{1}{3^{m+1}})(1+4s) - (2+s)(1+4t+\frac{4}{3^{m+1}})}{(1+4t+\frac{4}{3^{m+1}})(1+4s)} + \frac{7(t+\frac{1}{3^{m+1}}-s)}{(1+4t)(1+4t+\frac{4}{3^{m+1}})}\right| \\
&= \left|\frac{2+8s+t+4st+\frac{1}{3^{m+1}}+\frac{4s}{3^{m+1}}-2-8t-\frac{8}{3^{m+1}}-s-4st-\frac{4s}{3^{m+1}}}{(1+4t+\frac{4}{3^{m+1}})(1+4s)}\right. \\
&\qquad \left. + \frac{7(t+\frac{1}{3^{m+1}}-s)}{(1+4t)(1+4t+\frac{4}{3^{m+1}})}\right| \\
&= \left|\frac{7s-7t-\frac{7}{3^{m+1}}}{(1+4t+\frac{4}{3^{m+1}})(1+4s)} - \frac{7(s-t-\frac{1}{3^{m+1}})}{(1+4t)(1+4t+\frac{4}{3^{m+1}})}\right| \\
&= 7\left|\frac{1}{(1+4t+\frac{4}{3^{m+1}})(1+4s)} - \frac{1}{(1+4t+\frac{4}{3^{m+1}})(1+4t)}\right|\left|s-t-\frac{1}{3^{m+1}}\right| \\
&= 7\frac{|1+4t-1-4s|}{(1+4t+\frac{4}{3^{m+1}})(1+4s)(1+4t)}\left|s-t-\frac{1}{3^{m+1}}\right| \\
&= \frac{28|t-s|}{(1+4t+\frac{4}{3^{m+1}})(1+4s)(1+4t)}\left|s-t-\frac{1}{3^{m+1}}\right| \\
&\le 28\delta\left|s-t-\frac{1}{3^{m+1}}\right| \\
&= \varepsilon\left|s-t-\frac{1}{3^{m+1}}\right|.
\end{aligned}$$

2. $t \in \mathbb{T}\backslash C_1$. Take arbitrary $\varepsilon > 0$ and $\delta = \frac{\varepsilon}{28}$. Let

$$U = \{s \in \mathbb{T}\backslash C_1 : |t-s| \le \delta\}.$$

Then for $s \in U$, we get

$$\begin{aligned}
&\left|f(\sigma(t)) - f(s) + \frac{7}{(1+4t)^2}(\sigma(t)-s)\right| \\
&= \left|f(t) - f(s) + \frac{7}{(1+4t)^2}(t-s)\right| \\
&= \left|\frac{2+t}{1+4t} - \frac{2+s}{1+4s} + \frac{7}{(1+4t)^2}(t-s)\right|
\end{aligned}$$

$$
\begin{aligned}
&= \left|\frac{(2+t)(1+4s)-(2+s)(1+4t)}{(1+4t)(1+4s)} + \frac{7}{(1+4t)^2}(t-s)\right| \\
&= \left|\frac{2+8s+t+4ts-2-8t-s-4ts}{(1+4t)(1+4s)} + \frac{7}{(1+4t)^2}(t-s)\right| \\
&= \left|\frac{7(s-t)}{(1+4s)(1+4t)} - \frac{7}{(1+4t)^2}(s-t)\right| \\
&= 7\left|\frac{1}{(1+4t)(1+4s)} - \frac{1}{(1+4t)^2}\right||t-s| \\
&= 7\left|\frac{1+4t-1-4s}{(1+4t)^2(1+4s)}\right||t-s| \\
&= 28\frac{|t-s|^2}{(1+4t)^2(1+4s)} \\
&\le 28\delta|t-s| \\
&= \varepsilon|t-s|.
\end{aligned}
$$

Example 2.9. Let

$$\mathbb{T} = \left\{t_n = \sum_{k=0}^{n-1} a_k,\ n \in \mathbb{N},\ a_k > 0,\ k \in \mathbb{N}_0\right\}$$

and

$$f(t) = t^4 + t, \quad t \in \mathbb{T}.$$

We will prove that

$$f^\Delta(t_n) = t_{n+1}^3 + t_n t_{n+1}^2 + t_n^2 t_{n+1} + t_n^3 + 1, \quad n \in \mathbb{N}.$$

Fix $n \in \mathbb{N}$. Take arbitrary $\varepsilon > 0$ and $\delta > 0$ such that

$$(t_{n+1}^2 + (2t_n + \delta)t_{n+1} + (t_n + \delta)^2 + (t_n + \delta)t_n + t_n^2)\delta = \varepsilon.$$

Let

$$U = \{t_s \in \mathbb{T} : |t_s - t_n| \le \delta\}.$$

Then for $t_s \in \mathbb{T}$, we get

$$t_s \le t_n + \delta$$

and

$$|f(\sigma(t_n)) - f(t_s) - (t_{n+1}^3 + t_n t_{n+1}^2 + t_n^2 t_{n+1} + t_n^3 + 1)(\sigma(t_n) - t_s)|$$

$$
\begin{aligned}
&= \left|f(t_{n+1}) - f(t_s) - (t_{n+1}^3 + t_n t_{n+1}^2 + t_n^2 t_{n+1} + t_n^3 + 1)(t_{n+1} - t_s)\right| \\
&= \left|t_{n+1}^4 + t_{n+1} - t_s^4 - t_s - (t_{n+1}^3 + t_n t_{n+1}^2 + t_n^2 t_{n+1} + t_n^3 + 1)(t_{n+1} - t_s)\right| \\
&= \left|(t_{n+1} - t_s)(t_{n+1}^3 + t_s t_{n+1}^2 + t_s^2 t_{n+1} + t_s^3) + (t_{n+1} - t_s)\right. \\
&\quad \left.- (t_{n+1}^3 + t_n t_{n+1}^2 + t_n^2 t_{n+1} + t_n^3 + 1)(t_{n+1} - s)\right| \\
&= \left|t_{n+1}^3 + t_s t_{n+1}^2 + t_s^2 t_{n+1} + t_s^3 + 1 - t_{n+1}^3 - t_n t_{n+1}^2 - t_n^2 t_{n+1} - t_n^3 - 1\right||t_{n+1} - s| \\
&= \left|(t_s - t_n)t_{n+1}^2 + (t_s - t_n)(t_s + t_n)t_{n+1} + (t_s - t_n)(t_s^2 + t_s t_n + t_n^2)\right||t_{n+1} - t_s| \\
&= (t_{n+1}^2 + (t_s + t_n)t_{n+1} + t_s^2 + t_s t_n + t_n^2)|t_n - t_s||t_{n+1} - t_s| \\
&\le (t_{n+1}^2 + (2t_n + \delta)t_{n+1} + (t_n + \delta)^2 + (t_n + \delta)t_n + t_n^2)|t_s - t_n||t_{n+1} - t_s| \\
&\le (t_{n+1}^2 + (2t_n + \delta)t_{n+1} + (t_n + \delta)^2 + (t_n + \delta)t_n + t_n^2)\delta|t_{n+1} - t_s| \\
&= \varepsilon|t_{n+1} - t_s|.
\end{aligned}
$$

Example 2.10. Let $\mathbb{T} = \{-\frac{1}{n} : n \in \mathbb{N}\} \cup \mathbb{N}_0$, and let

$$f(t) = 2t^2 - 3t + 1, \quad t \in \mathbb{T}.$$

We will prove that

$$
f^\Delta(t) = \begin{cases} \frac{2t^2-7t+3}{t-1} & \text{if } t \in \{-\frac{1}{n} : n \in \mathbb{N}\}, \\ 4t - 1 & \text{if } t \in \mathbb{N}_0. \end{cases}
$$

We will consider the following cases.

1. $t \in \{-\frac{1}{n} : n \in \mathbb{N}\}$. Take arbitrary $\varepsilon > 0$ and $\delta = \frac{\varepsilon}{2}$. Let

   $$U = \{s \in \mathbb{T} : |t - s| \le \delta\}.$$

   We have

   $$\sigma(t) = -\frac{t}{t-1}.$$

   For $s \in U$, we get

$$
\begin{aligned}
&\left|f(\sigma(t)) - f(s) - \frac{2t^2 - 7t + 3}{t - 1}(\sigma(t) - s)\right| \\
&\quad = \left|f\left(-\frac{t}{t-1}\right) - f(s) - \frac{2t^2 - 7t + 3}{t-1}\left(-\frac{t}{t-1} - s\right)\right| \\
&\quad = \left|2\frac{t^2}{(t-1)^2} + \frac{3t}{t-1} + 1 - 2s^2 + 3s - 1 + \frac{2t^2 - 7t + 3}{t-1}\left(\frac{t}{t-1} + s\right)\right| \\
&\quad = \left|2\left(\frac{t^2}{(t-1)^2} - s^2\right) + 3\left(\frac{t}{t-1} + s\right) + \frac{2t^2 - 7t + 3}{t-1}\left(\frac{t}{t-1} + s\right)\right|
\end{aligned}
$$

$$\begin{aligned}
&= \left|2\left(\frac{t}{t-1} - s\right)\left(\frac{t}{t-1} + s\right) + 3\left(\frac{t}{t-1} + s\right) + \frac{2t^2 - 7t + 3}{t-1}\left(\frac{t}{t-1} + s\right)\right| \\
&= \left|\frac{2t}{t-1} - 2s + 3 + \frac{2t^2 - 7t + 3}{t-1}\right|\left|\frac{t}{t-1} + s\right| \\
&= \left|\frac{2t - (2s-3)(t-1) + 2t^2 - 7t + 3}{t-1}\right|\left|\frac{t}{t-1} + s\right| \\
&= \left|\frac{2t - 2st + 2s + 3t - 3 + 2t^2 - 7t + 3}{t-1}\right|\left|\frac{t}{t-1} + s\right| \\
&= \left|\frac{2t(t-s) - 2(t-s)}{t-1}\right|\left|\frac{t}{t-1} + s\right| \\
&= 2|t-s|\left|\frac{t}{t-1} + s\right| \\
&\le 2\delta\left|\frac{t}{t-1} + s\right| \\
&= \varepsilon\left|\frac{t}{t-1} + s\right|.
\end{aligned}$$

2. $t \in \mathbb{N}_0$. Take arbitrary $\varepsilon > 0$ and $\delta = \frac{\varepsilon}{2}$. Let

$$U = \{s \in \mathbb{N}_0 : |t-s| \le \delta\}.$$

We have

$$\sigma(t) = t + 1.$$

Then for $s \in U$, we get

$$\begin{aligned}
&|f(\sigma(t)) - f(s) - (4t-1)(\sigma(t) - s)| \\
&\quad = |f(t+1) - f(s) - (4t-1)(t+1-s)| \\
&\quad = |2(t+1)^2 - 3(t+1) + 1 - 2s^2 + 3s - 1 - (4t-1)(t+1-s)| \\
&\quad = |2(t+1-s)(t+1+s) - 3(t+1-s) - (4t-1)(t+1-s)| \\
&\quad = |2t + 2 + 2s - 3 - 4t + 1||t+1-s| \\
&\quad = |2s - 2t||t+1-s| \\
&\quad = 2|t-s||t+1-s| \\
&\quad \le 2\delta|t+1-s| \\
&\quad = \varepsilon|t+1-s|.
\end{aligned}$$

Example 2.11. Let $\mathbb{T} = \{(\frac{1}{2})^{2^n} : n \in \mathbb{N}_0\} \cup \{0, 1\}$, and let

$$f(t) = \frac{1+3t}{3-2t}, \quad t \in \mathbb{T}.$$

We will prove that

$$f^{\Delta}(t)=\begin{cases}\frac{11}{(3-2t)(3-2\sqrt{t})} & \text{if } t\in\{(\frac{1}{2})^{2^n}:n\in\mathbb{N}\},\\ \frac{11}{2} & \text{if } t=\frac{1}{2},\\ \frac{11}{9} & \text{if } t=0.\end{cases}$$

We will consider the following cases.

1. $t\in\{(\frac{1}{2})^{2^n}:n\in\mathbb{N}\}$. Take arbitrary $\varepsilon>0$ and

$$\delta=\frac{(3-2\sqrt{t})(3-2t)}{22}\varepsilon.$$

Let

$$U=\left\{s\in\left\{\left(\frac{1}{2}\right)^{2^n}:n\in\mathbb{N}\right\}:|t-s|\le\delta\right\}.$$

We have

$$\sigma(t)=\sqrt{t}.$$

Then for $s\in U$, we get

$$\begin{aligned}&\left|f(\sigma(t))-f(s)-\frac{11}{(3-2t)(3-2\sqrt{t})}(\sigma(t)-s)\right|\\ &=\left|f(\sqrt{t})-f(s)-\frac{11}{(3-2t)(3-2\sqrt{t})}(\sqrt{t}-s)\right|\\ &=\left|\frac{1+3\sqrt{t}}{3-2\sqrt{t}}-\frac{1+3s}{3-2s}-\frac{11}{(3-2t)(3-2\sqrt{t})}(\sqrt{t}-s)\right|\\ &=\left|\frac{(1+3\sqrt{t})(3-2s)-(1+3s)(3-2\sqrt{t})}{(3-2\sqrt{t})(3-2s)}-\frac{11}{(3-2t)(3-2\sqrt{t})}(\sqrt{t}-s)\right|\\ &=\left|\frac{3-2s+9\sqrt{t}-6s\sqrt{t}-3+2\sqrt{t}-9s+6s\sqrt{t}}{(3-2\sqrt{t})(3-2s)}-\frac{11}{(3-2t)(3-2\sqrt{t})}(\sqrt{t}-s)\right|\\ &=\left|\frac{11(\sqrt{t}-s)}{(3-2\sqrt{t})(3-2s)}-\frac{11}{(3-2t)(3-2\sqrt{t})}(\sqrt{t}-s)\right|\\ &=\left|\frac{11}{(3-2\sqrt{t})(3-2s)}-\frac{11}{(3-2t)(3-2\sqrt{t})}\right||\sqrt{t}-s|\\ &=\frac{|11(3-2t)-11(3-2s)|}{(3-2\sqrt{t})(3-2t)(3-2s)}|\sqrt{t}-s|\\ &=\frac{|33-22t-33+22s|}{(3-2\sqrt{t})(3-2t)(3-2s)}|\sqrt{t}-s|\\ &=\frac{22|t-s|}{(3-2\sqrt{t})(3-2t)(3-2s)}|\sqrt{t}-s|\end{aligned}$$

$$
\begin{aligned}
&\leq \frac{22\delta}{(3-2\sqrt{t})(3-2t)}|\sqrt{t}-s| \\
&= \varepsilon|\sqrt{t}-s|.
\end{aligned}
$$

2. $t = \frac{1}{2}$. Take arbitrary $\varepsilon > 0$ and $\delta = \frac{1}{11}\varepsilon$. Let

$$U = \left\{s \in \mathbb{T} : \left|\frac{1}{2} - s\right| \leq \delta\right\}.$$

We have

$$\sigma\left(\frac{1}{2}\right) = 1.$$

Then for $s \in U$, we get

$$
\begin{aligned}
&\left|f\left(\sigma\left(\frac{1}{2}\right)\right) - f(s) - \frac{11}{2}\left(\sigma\left(\frac{1}{2}\right) - s\right)\right| \\
&= \left|f(1) - f(s) - \frac{11}{2}(1-s)\right| \\
&= \left|4 - \frac{1+3s}{3-2s} - \frac{11}{2}(1-s)\right| \\
&= \left|\frac{4(3-2s)-1-3s}{3-2s} - \frac{11}{2}(1-s)\right| \\
&= \left|\frac{12-8s-1-3s}{3-2s} - \frac{11}{2}(1-s)\right| \\
&= \left|\frac{11(1-s)}{3-2s} - \frac{11}{2}(1-s)\right| \\
&= 11\left|\frac{1}{3-2s} - \frac{1}{2}\right|(1-s) \\
&= 11\frac{|2-3+2s|}{2(3-2s)}(1-s) \\
&= \frac{11|\frac{1}{2}-s|}{(3-2s)}(1-s) \\
&\leq 11\delta(1-s) \\
&= \varepsilon(1-s).
\end{aligned}
$$

3. $t = 0$. Take arbitrary $\varepsilon > 0$ and $\delta = \frac{\varepsilon}{22}$. Let

$$U = \{s \in \mathbb{T} : s \leq \delta\}.$$

We have

$$\sigma(0) = 0.$$

Then for $s \in U$, we get

$$
\begin{aligned}
&\left|f(\sigma(0)) - f(s) - \frac{11}{9}(\sigma(0) - s)\right| \\
&= \left|f(0) - f(s) + \frac{11}{9}s\right| \\
&= \left|\frac{1}{3} - \frac{1+3s}{3-2s} + \frac{11}{9}s\right| \\
&= \left|\frac{3-2s-3-9s}{3(3-2s)} + \frac{11}{9}s\right| \\
&= \left|-\frac{11s}{3(3-2s)} + \frac{11}{9}s\right| \\
&= 11\left|-\frac{1}{3(3-2s)} + \frac{1}{9}\right|s \\
&= 11\left|\frac{3-2s-3}{9(3-2s)}\right|s \\
&= 11\left|\frac{2s}{9(3-2s)}\right|s \\
&= 22\frac{s^2}{3-2s} \\
&< 22\delta s \\
&= \varepsilon s.
\end{aligned}
$$

Example 2.12. Let $U = \{\frac{1}{2^n} : n \in \mathbb{N}\}$, let

$$\mathbb{T} = \{0\} \cup U \cup (1-U) \cup (1+U) \cup (2-U) \cup (2+U) \cup \{1,2\},$$

and let

$$f(t) = 3t^2 + 7t - 2, \quad t \in \mathbb{T}.$$

We will prove that

$$
f^\Delta(t) = \begin{cases}
7 & \text{if } t = 0, \\
\frac{43}{4} & \text{if } t = \frac{1}{2}, \\
13 & \text{if } t = 1, \\
\frac{67}{4} & \text{if } t = \frac{3}{2}, \\
19 & \text{if } t = 2, \\
9t + 7 & \text{if } t \in U\backslash\{\frac{1}{2}\}, \\
\frac{17+9t}{2} & \text{if } t \in (1-U)\backslash\{\frac{1}{2}\}, \\
9t + 4 & \text{if } t \in (1+U)\backslash\{\frac{3}{2}\}, \\
\frac{9t+20}{2} & \text{if } t \in (2-U)\backslash\{\frac{3}{2}\}, \\
9t + 1 & \text{if } t \in (2+U)\backslash\{\frac{5}{2}\}.
\end{cases}
$$

We will consider the following cases.

1. $t = 0$. Take arbitrary $\varepsilon > 0$ and $\delta = \frac{\varepsilon}{3}$. Let

$$V = \{s \in \mathbb{T} : s \leq \delta\}.$$

We have

$$\sigma(0) = 0.$$

Then for $s \in V$, we get

$$\begin{aligned}
&|f(\sigma(0)) - f(s) - 7(\sigma(0) - s)| \\
&= |f(0) - f(s) + 7s| \\
&= |-2 - 3s^2 - 7s + 2 + 7s| \\
&= 3s^2 \\
&\leq 3\delta s \\
&= \varepsilon s.
\end{aligned}$$

2. $t = \frac{1}{2}$. Take arbitrary $\varepsilon > 0$ and $\delta = \frac{\varepsilon}{3}$. Let

$$V = \left\{s \in \mathbb{T} : \left|s - \frac{1}{2}\right| \leq \delta\right\}.$$

We have

$$\sigma\left(\frac{1}{2}\right) = \frac{3}{4}.$$

Then for $s \in V$, we get

$$\begin{aligned}
&\left|f\left(\sigma\left(\frac{1}{2}\right)\right) - f(s) - \frac{43}{4}\left(\sigma\left(\frac{1}{2}\right) - s\right)\right| \\
&= \left|f\left(\frac{3}{4}\right) - f(s) - \frac{43}{4}\left(\frac{3}{4} - s\right)\right| \\
&= \left|\frac{27}{16} + \frac{21}{4} - 2 - 3s^2 - 7s + 2 - \frac{43}{4}\left(\frac{3}{4} - s\right)\right| \\
&= \left|3\left(\frac{9}{16} - s^2\right) + 7\left(\frac{3}{4} - s\right) - \frac{43}{4}\left(\frac{3}{4} - s\right)\right| \\
&= \left|3\left(\frac{3}{4} - s\right)\left(\frac{3}{4} + s\right) + 7\left(\frac{3}{4} - s\right) - \frac{43}{4}\left(\frac{3}{4} - s\right)\right| \\
&= \left|\frac{9}{4} + 3s + 7 - \frac{43}{4}\right|\left|\frac{3}{4} - s\right|
\end{aligned}$$

$$
\begin{aligned}
&= \left|3s - \frac{3}{2}\right|\left|\frac{3}{4} - s\right| \\
&= 3\left|s - \frac{1}{2}\right|\left|\frac{3}{4} - s\right| \\
&\leq 3\delta\left|\frac{3}{4} - s\right| \\
&= \varepsilon\left|\frac{3}{4} - s\right|.
\end{aligned}
$$

3. $t = 1$. Take arbitrary $\varepsilon > 0$ and $\delta = \frac{\varepsilon}{3}$. Let

$$V = \{s \in \mathbb{T} : |s - 1| \leq \delta\}.$$

We have

$$\sigma(1) = 1.$$

Then for $s \in V$, we get

$$
\begin{aligned}
&\left|f(\sigma(1)) - f(s) - 13(\sigma(1) - s)\right| \\
&\quad = \left|f(1) - f(s) - 13(1 - s)\right| \\
&\quad = \left|8 - 3s^2 - 7s + 2 - 13(1 - s)\right| \\
&\quad = \left|3(1 - s^2) + 7(1 - s) - 13(1 - s)\right| \\
&\quad = \left|3(1 - s)(1 + s) + 7(1 - s) - 13(1 - s)\right| \\
&\quad = |3 + 3s + 7 - 13||1 - s| \\
&\quad = |3s - 3||1 - s| \\
&\quad = 3(1 - s)^2 \\
&\quad \leq 3\delta|1 - s| \\
&\quad = \varepsilon|1 - s|.
\end{aligned}
$$

4. $t = \frac{3}{2}$. Take arbitrary $\varepsilon > 0$ and $\delta = \frac{\varepsilon}{3}$. Let

$$V = \left\{s \in \mathbb{T} : \left|\frac{3}{2} - s\right| \leq \delta\right\}.$$

We have

$$\sigma\left(\frac{3}{2}\right) = \frac{7}{4}.$$

Then for $s \in V$, we get

$$\begin{aligned}
&\left|f\left(\sigma\left(\frac{3}{2}\right)\right)-f(s)-\frac{67}{4}\left(\sigma\left(\frac{3}{2}\right)-s\right)\right| \\
&\quad=\left|f\left(\frac{7}{4}\right)-f(s)-\frac{67}{4}\left(\frac{7}{4}-s\right)\right| \\
&\quad=\left|3\left(\frac{7}{4}\right)^2+7\left(\frac{7}{4}\right)-2-3s^2-7s+2-\frac{67}{4}\left(\frac{7}{4}-s\right)\right| \\
&\quad=\left|3\left(\frac{7}{4}-s\right)\left(\frac{7}{4}+s\right)+7\left(\frac{7}{4}-s\right)-\frac{67}{4}\left(\frac{7}{4}-s\right)\right| \\
&\quad=\left|\frac{21}{4}+3s+7-\frac{67}{4}\right|\left|\frac{7}{4}-s\right| \\
&\quad=\left|3s-\frac{9}{2}\right|\left|\frac{7}{4}-s\right| \\
&\quad=3\left|s-\frac{3}{2}\right|\left|\frac{7}{4}-s\right| \\
&\quad\leq 3\delta\left|\frac{7}{4}-s\right| \\
&\quad=\varepsilon\left|\frac{7}{4}-s\right|.
\end{aligned}$$

5. $t=2$. Take arbitrary $\varepsilon>0$ and $\delta=\frac{\varepsilon}{3}$. Let

$$V=\{s\in\mathbb{T}:|s-2|\leq\delta\}.$$

We have

$$\sigma(2)=2.$$

Then for $s\in V$, we get

$$\begin{aligned}
&|f(\sigma(2))-f(s)-19(\sigma(2)-s)| \\
&\quad=|f(2)-f(s)-19(2-s)| \\
&\quad=|3\cdot 2^2+7\cdot 2-2-3s^2-7s+2-19(2-s)| \\
&\quad=|3(2^2-s^2)+7(2-s)-19(2-s)| \\
&\quad=|3(2-s)(2+s)+7(2-s)-19(2-s)| \\
&\quad=|6+3s+7-19||2-s| \\
&\quad=|3s-6||2-s| \\
&\quad=3(s-2)^2 \\
&\quad\leq 3\delta|2-s| \\
&\quad=\varepsilon|2-s|.
\end{aligned}$$

6. $t \in U \backslash \{\frac{1}{2}\}$. Take arbitrary $\varepsilon > 0$ and $\delta = \frac{\varepsilon}{3}$. Let

$$V = \{s \in \mathbb{T} : |s - t| \le \delta\}.$$

We have

$$\sigma(t) = 2t.$$

Then for $s \in V$, we get

$$\begin{aligned}
&|f(\sigma(t)) - f(s) - (9t + 7)(\sigma(t) - s)| \\
&\quad = |f(2t) - f(s) - (9t + 7)(2t - s)| \\
&\quad = |3(2t)^2 + 7(2t) - 2 - 3s^2 - 7s + 2 - (9t + 7)(2t - s)| \\
&\quad = |3(2t - s)(2t + s) + 7(2t - s) - (9t + 7)(2t - s)| \\
&\quad = |6t + 3s + 7 - 9t - 7||2t - s| \\
&\quad = 3|t - s||2t - s| \\
&\quad \le 3\delta|2t - s| \\
&\quad = \varepsilon|2t - s|.
\end{aligned}$$

7. $t \in (1 - U) \backslash \{\frac{1}{2}\}$. Take arbitrary $\varepsilon > 0$ and $\delta = \frac{\varepsilon}{3}$. Let

$$V = \{s \in \mathbb{T} : |s - t| \le \delta\}.$$

We have

$$\sigma(t) = \frac{1 + t}{2}.$$

Then for $s \in V$, we get

$$\begin{aligned}
&\left|f(\sigma(t)) - f(s) - \frac{17 + 9t}{2}(\sigma(t) - s)\right| \\
&\quad = \left|f\left(\frac{1 + t}{2}\right) - f(s) - \frac{17 + 9t}{2}\left(\frac{1 + t}{2} - s\right)\right| \\
&\quad = \left|3\left(\frac{1 + t}{2}\right)^2 + 7\left(\frac{1 + t}{2}\right) - 2 - 3s^2 - 7s + 2 - \frac{17 + 9t}{2}\left(\frac{1 + t}{2} - s\right)\right| \\
&\quad = \left|3\left(\left(\frac{1 + t}{2}\right)^2 - s^2\right) + 7\left(\frac{1 + t}{2} - s\right) - \frac{17 + 9t}{2}\left(\frac{1 + t}{2} - s\right)\right| \\
&\quad = \left|3\left(\frac{1 + t}{2} - s\right)\left(\frac{1 + t}{2} + s\right) + 7\left(\frac{1 + t}{2} - s\right) - \frac{17 + 9t}{2}\left(\frac{1 + t}{2} - s\right)\right| \\
&\quad = \left|\frac{3 + 3t}{2} + 3s + 7 - \frac{17 + 9t}{2}\right|\left|\frac{1 + t}{2} - s\right|
\end{aligned}$$

$$
\begin{aligned}
&= |3s - 3t|\left|\frac{1+t}{2} - s\right| \\
&= 3|s - t|\left|\frac{1+t}{2} - s\right| \\
&\le 3\delta\left|\frac{1+t}{2} - s\right| \\
&= \varepsilon\left|\frac{1+t}{2} - s\right|.
\end{aligned}
$$

8. $t \in (1 + U)\backslash\{\frac{3}{2}\}$. Take arbitrary $\varepsilon > 0$ and $\delta = \frac{\varepsilon}{3}$. Let

$$V = \{s \in \mathbb{T} : |t - s| \le \delta\}.$$

We have

$$\sigma(t) = 2t - 1.$$

Then for $s \in V$, we get

$$
\begin{aligned}
&|f(\sigma(t)) - f(s) - (9t + 4)(\sigma(t) - s)| \\
&\quad = |f(2t - 1) - f(s) - (9t + 4)(2t - 1 - s)| \\
&\quad = |3(2t - 1)^2 + 7(2t - 1) - 2 - 3s^2 - 7s + 2 - (9t + 4)(2t - 1 - s)| \\
&\quad = ||3((2t - 1)^2 - s^2) + 7(2t - 1 - s) - (9t + 4)(2t - 1 - s)| \\
&\quad = |3(2t - 1 - s)(2t - 1 + s) + 7(2t - 1 - s) - (9t + 4)(2t - 1 - s)| \\
&\quad = |6t - 3 + 3s + 7 - 9t - 4||2t - 1 - s| \\
&\quad = |3s - 3t||2t - 1 - s| \\
&\quad = 3|s - t||2t - 1 - s| \\
&\quad \le 3\delta|2t - 1 - s| \\
&\quad = \varepsilon|2t - 1 - s|.
\end{aligned}
$$

9. $t \in (2 - U)\backslash\{\frac{3}{2}\}$. Take arbitrary $\varepsilon > 0$ and $\delta = \frac{\varepsilon}{3}$. Let

$$V = \{s \in \mathbb{T} : |t - s| \le \delta\}.$$

We have

$$\sigma(t) = \frac{t + 2}{2}.$$

Then for $s \in V$, we get

$$\left|f(\sigma(t)) - f(s) - \frac{9t + 20}{2}(\sigma(t) - s)\right|$$

$$
\begin{aligned}
&= \left|f\left(\frac{t+2}{2}\right) - f(s) - \frac{9t+20}{2}\left(\frac{t+2}{2} - s\right)\right| \\
&= \left|3\left(\frac{t+2}{2}\right)^2 + 7\left(\frac{t+2}{2}\right) - 2 - 3s^2 - 7s + 2 - \frac{9t+20}{2}\left(\frac{t+2}{2} - s\right)\right| \\
&= \left|3\left(\left(\frac{t+2}{2}\right)^2 - s^2\right) + 7\left(\frac{t+2}{2} - s\right) - \frac{9t+20}{2}\left(\frac{t+2}{2} - s\right)\right| \\
&= \left|3\left(\frac{t+2}{2} - s\right)\left(\frac{t+2}{2} + s\right) + 7\left(\frac{t+2}{2} - s\right) - \frac{9t+20}{2}\left(\frac{t+2}{2} - s\right)\right| \\
&= \left|\frac{3t+6}{2} + 3s + 7 - \frac{9t+20}{2}\right|\left|\frac{t+2}{2} - s\right| \\
&= |3s - 3t|\left|\frac{t+2}{2} - s\right| \\
&= 3|t-s|\left|\frac{t+2}{2} - s\right| \\
&\le 3\delta\left|\frac{t+2}{2} - s\right| \\
&= \varepsilon\left|\frac{t+2}{2} - s\right|.
\end{aligned}
$$

10. $t \in (2+U)\backslash\{\frac{5}{2}\}$. Take arbitrary $\varepsilon > 0$ and $\delta = \frac{\varepsilon}{3}$. Let

$$V = \{s \in \mathbb{T} : |s-t| \le \delta\}.$$

We have

$$\sigma(t) = 2(t-1).$$

Then for $s \in \mathbb{T}$, we get

$$
\begin{aligned}
&|f(\sigma(t)) - f(s) - (9t+1)(\sigma(t) - s)| \\
&\quad = |f(2t-2) - f(s) - (9t+1)(2t-2-s)| \\
&\quad = |3(2t-2)^2 + 7(2t-2) - 2 - 3s^2 - 7s + 2 - (9t+1)(2t-2-s)| \\
&\quad = |3((2t-2)^2 - s^2) + 7(2t-2-s) - (9t+1)(2t-2-s)| \\
&\quad = |3(2t-2-s)(2t-2+s) + 7(2t-2-s) - (9t+1)(2t-2-s)| \\
&\quad = |6t - 6 + 3s + 7 - 9t - 1||2t-2-s| \\
&\quad = |3s - 3t||2t-2-s| \\
&\quad = 3|t-s||2t-2-s| \\
&\quad \le 3\delta|2t-2-s| \\
&\quad = \varepsilon|2t-2-s|.
\end{aligned}
$$

Exercise 2.1. Let $\mathbb{T} = h\mathbb{Z} + k$, $h > 0$, $k \in \mathbb{R}$, and let

$$f(t) = 1 + 4t - 8t^2, \quad t \in \mathbb{T}.$$

Prove that

$$f^{\Delta}(t) = 4 - 8h - 16t, \quad t \in \mathbb{T}.$$

Exercise 2.2. Let $\mathbb{T} = (-2\mathbb{N}_0) \cup 3^{\mathbb{N}_0}$, and let

$$f(t) = t^3 - t, \quad t \in \mathbb{T}.$$

Prove that

$$f^{\Delta}(t) = \begin{cases} 3t^2 + 6t + 3 & \text{if } t \in (-2\mathbb{N}), \\ 0 & \text{if } t = 0, \\ 13t^2 - 1 & \text{if } t \in 3^{\mathbb{N}_0}. \end{cases}$$

Exercise 2.3. Let $\mathbb{T} = P_{3,7} \cup [4, 6]$, and let

$$f(t) = \frac{t - 5}{1 + 2t^2}.$$

Prove that

$$f^{\Delta}(t) = \begin{cases} \frac{1+20t-2t^2}{(1+2t^2)^2} & \text{if } t \in (\bigcup_{k=0}^{\infty}[10k, 10k + 3)) \cup [4, 6), \\ \frac{71+6t-2t^2}{(1+2t^2)(99+28t+2t^2)} & \text{if } t = 10k + 3, k \neq 0. \end{cases}$$

Exercise 2.4. Let $\mathbb{T} = \{0\} \cup 11^{\mathbb{N}_0}$, and let

$$f(t) = \frac{1 + 7t}{5 - 4t}, \quad t \in \mathbb{T}.$$

Prove that

$$f^{\Delta}(t) = \begin{cases} \frac{39}{5} & \text{if } t = 0, \\ \frac{39}{(5-4t)(5-44t)} & \text{if } t \in 11^{\mathbb{N}_0}. \end{cases}$$

Exercise 2.5. Let $\mathbb{T} = [1, 2] \cup [3, 4] \cup [7, 8] \cup 9^{\mathbb{N}}$, and let

$$f(t) = 1 - 8t - 2t^2, \quad t \in \mathbb{T}.$$

Prove that

$$f^{\Delta}(t) = \begin{cases} -8 - 4t & \text{if } t \in [1, 2) \cup [3, 4) \cup [7, 8), \\ -8 - 20t & \text{if } t \in 9^{\mathbb{N}}, \\ -18 & \text{if } t = 2, \\ -30 & \text{if } t = 4, \\ -42 & \text{if } t = 8. \end{cases}$$

2.2 Basic rules of delta differentiation

The basic rules of delta differentiation are as follows. Let $f, g : \mathbb{T} \to \mathbb{R}$.

1. If f is differentiable at t, then f is continuous at t.
2. If f is continuous at t and t is right-scattered, then f is differentiable at t with

$$f^{\Delta}(t) = \frac{f(\sigma(t)) - f(t)}{\mu(t)}.$$

3. If t is right-dense, then f is differentiable iff the finite limit

$$\lim_{s\to t} \frac{f(t) - f(s)}{t - s}$$

exists. In this case,

$$f^{\Delta}(t) = \lim_{s\to t} \frac{f(t) - f(s)}{t - s}.$$

4. If f is differentiable at t, then

$$f(\sigma(t)) = f(t) + \mu(t) f^{\Delta}(t).$$

We further suppose that f and g are differentiable at $t \in \mathbb{T}^{\kappa}$. Then we have the following.

5. The sum $f + g : \mathbb{T} \to \mathbb{R}$ is differentiable at t with

$$(f + g)^{\Delta}(t) = f^{\Delta}(t) + g^{\Delta}(t).$$

6. For any constant α, the function $\alpha f : \mathbb{T} \to \mathbb{R}$ is differentiable at t with

$$(\alpha f)^{\Delta}(t) = \alpha f^{\Delta}(t).$$

7. If $g(t)g(\sigma(t)) \neq 0$, then $\frac{f}{g} : \mathbb{T} \to \mathbb{R}$ is differentiable at t with

$$\left(\frac{f}{g}\right)^{\Delta}(t) = \frac{f^{\Delta}(t)g(t) - f(t)g^{\Delta}(t)}{g(t)g(\sigma(t))}.$$

8. The product $fg : \mathbb{T} \to \mathbb{R}$ is differentiable at t with

$$\begin{aligned}(fg)^{\Delta}(t) &= f^{\Delta}(t)g(t) + f(\sigma(t))g^{\Delta}(t)\\ &= f(t)g^{\Delta}(t) + f^{\Delta}(t)g(\sigma(t))\\ &= f(t)g^{\Delta}(t) + f^{\Delta}(t)g(t) + \mu(t)f^{\Delta}(t)g^{\Delta}(t).\end{aligned}$$

Example 2.13. Let $\mathbb{T} = h\mathbb{Z}$, $h > 0$, and let

$$f(t) = (t+1)(t+2)(t+3)(t+4), \quad t \in \mathbb{T}.$$

We will find $f^{\Delta}(t)$, $t \in \mathbb{T}^{\kappa}$. For this aim, we will use the rule of delta differentiation of the product of two functions. Set

$$\begin{aligned} f_1(t) &= t+1, \\ f_2(t) &= (t+2)(t+3)(t+4), \\ f_3(t) &= t+2, \\ f_4(t) &= (t+3)(t+4), \\ f_5(t) &= t+3, \\ f_6(t) &= t+4, \quad t \in \mathbb{T}. \end{aligned}$$

Then

$$\begin{aligned} f(t) &= f_1(t) f_2(t), \\ f_2(t) &= f_3(t) f_4(t), \\ f_4(t) &= f_5(t) f_6(t), \quad t \in \mathbb{T}. \end{aligned}$$

Hence

$$\begin{aligned} f^{\Delta}(t) &= (f_1 f_2)^{\Delta}(t) \\ &= f_1^{\Delta}(t) f_2(t) + f_1(\sigma(t)) f_2^{\Delta}(t), \quad t \in \mathbb{T}^{\kappa}, \\ f_2^{\Delta}(t) &= (f_3 f_4)^{\Delta}(t) \\ &= f_3^{\Delta}(t) f_4(t) + f_3(\sigma(t)) f_4^{\Delta}(t), \quad t \in \mathbb{T}^{\kappa}, \end{aligned}$$

and

$$\begin{aligned} f_4^{\Delta}(t) &= (f_5 f_6)^{\Delta}(t) \\ &= f_5^{\Delta}(t) f_6(t) + f_5(\sigma(t)) f_6^{\Delta}(t), \quad t \in \mathbb{T}^{\kappa}. \end{aligned}$$

Note that

$$\begin{aligned} f_1^{\Delta}(t) &= 1, \\ f_3^{\Delta}(t) &= 1, \\ f_5^{\Delta}(t) &= 1, \\ f_6^{\Delta}(t) &= 1, \quad t \in \mathbb{T}^{\kappa}, \end{aligned}$$

and

$$\begin{aligned} f_1(\sigma(t)) &= \sigma(t) + 1 \\ &= t + h + 1, \\ f_3(\sigma(t)) &= \sigma(t) + 2 \end{aligned}$$

$$
\begin{aligned}
&= t + h + 2,\\
f_5(\sigma(t)) &= \sigma(t) + 3\\
&= t + h + 3, \quad t \in \mathbb{T}^\kappa.
\end{aligned}
$$

Therefore

$$
\begin{aligned}
f_4^\Delta(t) &= t + 4 + t + h + 3\\
&= 2t + h + 7,\\
f_2^\Delta(t) &= (t+3)(t+4) + (t+h+2)(2t+h+7)\\
&= t^2 + 7t + 12 + 2t^2 + (h+7)t + 2(h+2)t + (h+2)(h+7)\\
&= 3t^2 + (3h+21)t + 12 + h^2 + 9h + 14\\
&= 3t^2 + (3h+21)t + h^2 + 9h + 36, \quad t \in \mathbb{T}^\kappa,
\end{aligned}
$$

and

$$
\begin{aligned}
f^\Delta(t) &= (t+2)(t+3)(t+4) + (t+h+1)(3t^2 + (3h+21)t + h^2 + 9h + 36)\\
&= (t+2)(t^2 + 7t + 12) + 3t^3 + (3h+21)t^2 + (h^2 + 9h + 36)t\\
&\quad +3(h+1)t^2 + (h+1)(3h+21)t + (h+1)(h^2 + 9h + 36)\\
&= t^3 + 7t^2 + 12t + 2t^2 + 14t + 24 + 3t^3 + (6h+24)t^2\\
&\quad +(h^2 + 9h + 36 + 3h^2 + 21h + 3h + 21)t\\
&\quad +h^3 + 9h^2 + 36h + h^2 + 9h + 36\\
&= 4t^3 + 3(2h+11)t^2 + (4h^2 + 33h + 57)t + h^3 + 10h^2 + 45h + 60, \quad t \in \mathbb{T}^\kappa.
\end{aligned}
$$

Example 2.14. Let $\mathbb{T} = 3^{\mathbb{N}_0}$, and let

$$f(t) = t^3 + t^2 + t + 1, \quad t \in \mathbb{T}.$$

We will find $f^\Delta(t)$, $t \in \mathbb{T}$. Note that all points $t \in \mathbb{T}$ are right-scattered. Then

$$f^\Delta(t) = \frac{f(\sigma(t)) - f(t)}{\sigma(t) - t}, \quad t \in \mathbb{T}.$$

We have

$$\sigma(t) = 3t, \quad t \in \mathbb{T},$$

and

$$
\begin{aligned}
f(\sigma(t)) - f(t) &= (\sigma(t))^3 + (\sigma(t))^2 + \sigma(t) + 1 - (t^3 + t^2 + t + 1)\\
&= ((\sigma(t))^3 - t^3) + ((\sigma(t))^2 - t^2) + (\sigma(t) - t)\\
&= (\sigma(t) - t)((\sigma(t))^2 + t\sigma(t) + t^2) + (\sigma(t) - t)(\sigma(t) + t) + (\sigma(t) - t)\\
&= (\sigma(t) - t)((\sigma(t))^2 + t\sigma(t) + t^2 + \sigma(t) + t + 1), \quad t \in \mathbb{T}.
\end{aligned}
$$

Hence

$$
\begin{aligned}
f^{\Delta}(t) &= \frac{(\sigma(t)-t)((\sigma(t))^2 + t\sigma(t) + t^2 + \sigma(t) + t + 1)}{\sigma(t)-t} \\
&= (\sigma(t))^2 + t\sigma(t) + t^2 + \sigma(t) + t + 1 \\
&= (3t)^2 + t(3t) + t^2 + 3t + t + 1 \\
&= 9t^2 + 3t^2 + t^2 + 4t + 1 \\
&= 13t^2 + 4t + 1, \quad t \in \mathbb{T}.
\end{aligned}
$$

Example 2.15. Let $\mathbb{T} = \mathbb{N}_0^3$, and let

$$
f(t) = \frac{t^2 - 5t + 6}{t^2 + t + 7}, \quad t \in \mathbb{T}.
$$

We will find $f^{\Delta}(t)$, $t \in \mathbb{T}$. For this aim, we will use the rule of delta derivative of the quotient of two functions. Here

$$
\sigma(t) = (\sqrt[3]{t} + 1)^3, \quad t \in \mathbb{T},
$$

and all points $t \in \mathbb{T}$ are right-scattered. Let

$$
\begin{aligned}
g(t) &= t^2 - 5t + 6, \\
h(t) &= t^2 + t + 7, \quad t \in \mathbb{T}.
\end{aligned}
$$

Then

$$
\begin{aligned}
g^{\Delta}(t) &= \frac{g(\sigma(t)) - g(t)}{\sigma(t)-t} \\
&= \frac{(\sigma(t))^2 - 5\sigma(t) + 6 - (t^2 - 5t + 6)}{\sigma(t)-t} \\
&= \frac{((\sigma(t))^2 - t^2) - 5(\sigma(t)-t)}{\sigma(t)-t} \\
&= \frac{(\sigma(t)-t)(\sigma(t)+t) - 5(\sigma(t)-t)}{\sigma(t)-t} \\
&= \frac{(\sigma(t)-t)(\sigma(t)+t-5)}{\sigma(t)-t} \\
&= \sigma(t) + t - 5 \\
&= (\sqrt[3]{t} + 1)^3 + t - 5 \\
&= t + 3\sqrt[3]{t^2} + 3\sqrt[3]{t} + 1 + t - 5 \\
&= 2t + 3\sqrt[3]{t^2} + 3\sqrt[3]{t} - 4, \quad t \in \mathbb{T}, \\
h^{\Delta}(t) &= \frac{h(\sigma(t)) - h(t)}{\sigma(t)-t}
\end{aligned}
$$

$$
\begin{aligned}
&= \frac{(\sigma(t))^2 + \sigma(t) + 7 - (t^2 + t + 7)}{\sigma(t) - t} \\
&= \frac{((\sigma(t))^2 - t^2) + (\sigma(t) - t)}{\sigma(t) - t} \\
&= \frac{(\sigma(t) - t)(\sigma(t) + t) + (\sigma(t) - t)}{\sigma(t) - t} \\
&= \sigma(t) + t + 1 \\
&= \left(\sqrt[3]{t} + 1\right)^3 + t + 1 \\
&= t + 3\sqrt[3]{t^2} + 3\sqrt[3]{t} + 1 + t + 1 \\
&= 2t + 3\sqrt[3]{t^2} + 3\sqrt[3]{t} + t + 2, \quad t \in \mathbb{T},
\end{aligned}
$$

and

$$
\begin{aligned}
h(\sigma(t)) &= \left(\sigma(t)\right)^2 + \sigma(t) + 7 \\
&= \left(\sqrt[3]{t} + 1\right)^6 + (\sqrt[3]{t} + 1)^3 + 7 \\
&= (4 + 3\sqrt[3]{t^2} + 3\sqrt[3]{t} + 1)^2 + t + 3\sqrt[3]{t^2} + 3\sqrt[3]{t} + 1 + 7 \\
&= t^2 + 15t\sqrt[3]{t} + 6t\sqrt[3]{t^2} + 18\sqrt[3]{t^2} + 9\sqrt[3]{t} + 21t + 9, \quad t \in \mathbb{T}.
\end{aligned}
$$

Hence

$$
\begin{aligned}
f^{\Delta}(t) &= \left(\frac{g}{h}\right)^{\Delta}(t) \\
&= \frac{g^{\Delta}(t)h(t) - g(t)h^{\Delta}(t)}{h(t)h(\sigma(t))} \\
&= \frac{(2t + 3\sqrt[3]{t^2} + 3\sqrt[3]{t} - 4)(t^2 + t + 7) - (t^2 - 5t + 6)(2t + 3\sqrt[3]{t^2} + 3\sqrt[3]{t} + t + 2)}{(t^2 + t + 7)(t^2 + 15t\sqrt[3]{t} + 6t\sqrt[3]{t^2} + 18\sqrt[3]{t^2} + 9\sqrt[3]{t} + 21t + 9)} \\
&= \frac{1}{(t^2 + t + 7)(t^2 + 15t\sqrt[3]{t} + 6t\sqrt[3]{t^2} + 18\sqrt[3]{t^2} + 9\sqrt[3]{t} + 21t + 9)}(2t^3 + 2t^2 + 14t \\
&\quad +3t^2\sqrt[3]{t^2} + 3t\sqrt[3]{t^2} + 21\sqrt[3]{t^2} \\
&\quad + 3t^2\sqrt[3]{t} + 3t\sqrt[3]{t} + 21\sqrt[3]{t} - 4t^2 - 4t - 28 - 2t^3 - 3t^2\sqrt[3]{t^2} - 3t^2\sqrt[3]{t} - t^3 - 2t^2 \\
&\quad + 10t^2 + 15t\sqrt[3]{t^2} + 15t\sqrt[3]{t} + 5t^2 + 10t - 12t - 18\sqrt[3]{t^2} - 18\sqrt[3]{t} - 6t - 12) \\
&= \frac{-t^3 + 11t^2 + 2t + 18t\sqrt[3]{t^2} + 3\sqrt[3]{t^2} + 21\sqrt[3]{t} - 40}{(t^2 + t + 7)(t^2 + 15t\sqrt[3]{t} + 6t\sqrt[3]{t^2} + 18\sqrt[3]{t^2} + 9\sqrt[3]{t} + 21t + 9)}, \quad t \in \mathbb{T}.
\end{aligned}
$$

Example 2.16. Let $\mathbb{T} = \{H_n : n \in \mathbb{N}_0\}$, where H_n, $n \in \mathbb{N}_0$, are the harmonic numbers, and let

$$
f(t) = \frac{t}{t^2 - 4t - 12}, \quad t \in \mathbb{T}.
$$

We will find $f^{\Delta}(t)$, $t \in \mathbb{T}$. For this aim, we will use the rule of delta derivative of the quotient of two functions. Here

$$\sigma(H_n) = H_{n+1}, \quad n \in \mathbb{N}_0.$$

Note that all points $t \in \mathbb{T}$ are right-scattered. Let

$$\begin{aligned} g(t) &= t, \\ h(t) &= t^2 - 4t - 12, \quad t \in \mathbb{T}. \end{aligned}$$

We have

$$\begin{aligned} g^{\Delta}(t) &= 1, \\ h^{\Delta}(t) &= \frac{h(\sigma(t)) - h(t)}{\sigma(t) - t} \\ &= \frac{(\sigma(t))^2 - 4\sigma(t) - 12 - t^2 + 4t + 12}{\sigma(t) - t} \\ &= \frac{((\sigma(t))^2 - t^2) - 4(\sigma(t) - t)}{\sigma(t) - t} \\ &= \frac{(\sigma(t) - t)(\sigma(t) + t) - 4(\sigma(t) - t)}{\sigma(t) - t} \\ &= \sigma(t) + t - 4, \quad t \in \mathbb{T}, \end{aligned}$$

and

$$h(\sigma(t)) = (\sigma(t))^2 - 4\sigma(t) - 12, \quad t \in \mathbb{T}.$$

Hence

$$\begin{aligned} g^{\Delta}(H_n) &= 1, \\ h^{\Delta}(H_n) &= H_{n+1} + H_n - 4, \\ h(\sigma(H_n)) &= H_{n+1}^2 - 4H_{n+1} - 12, \quad n \in \mathbb{N}, \end{aligned}$$

and

$$\begin{aligned} f^{\Delta}(H_n) &= \left(\frac{g}{h}\right)^{\Delta}(H_n) \\ &= \frac{g^{\Delta}(H_n)h(H_n) - g(H_n)h^{\Delta}(H_n)}{h(H_n)h(\sigma(H_n))} \\ &= \frac{H_n^2 - 4H_n - 12 - H_n(H_{n+1} + H_n - 4)}{(H_n^2 - 4H_n - 12)(H_{n+1}^2 - 4H_{n+1} - 12)} \end{aligned}$$

$$
\begin{aligned}
&= \frac{H_n^2 - 4H_n - 12 - H_nH_{n+1} - H_n^2 + 4H_n}{(H_n^2 - 4H_n - 12)(H_{n+1}^2 - 4H_{n+1} - 12)} \\
&= -\frac{H_nH_{n+1} + 12}{(H_n^2 - 4H_n - 12)(H_{n+1}^2 - 4H_{n+1} - 12)}, \quad n \in \mathbb{N}_0.
\end{aligned}
$$

Example 2.17. Let $\mathbb{T} = P_{1,3}$, and let

$$f(t) = \frac{1-t}{1+t}, \quad t \in \mathbb{T}.$$

We will find $f^\Delta(t)$, $t \in \mathbb{T}$. We have the following two cases.

1. $t \in \mathbb{T}$ is right-scattered. Set

$$
\begin{aligned}
g(t) &= 1 - t, \\
h(t) &= 1 + t, \quad t \in \mathbb{T}.
\end{aligned}
$$

   We have

$$\sigma(t) = t + 3.$$

   Then

$$
\begin{aligned}
g^\Delta(t) &= -1, \\
h^\Delta(t) &= 1.
\end{aligned}
$$

   Using the rule of delta derivative of the quotient of two functions, we arrive at

$$
\begin{aligned}
f^\Delta(t) &= \frac{g^\Delta(t)h(t) - g(t)h^\Delta(t)}{h(t)h(\sigma(t))} \\
&= \frac{-(1+t) - (1-t)}{(1+t)(1+t+3)} \\
&= \frac{-1-t-1+t}{(1+t)(4+t)} \\
&= -\frac{2}{(1+t)(4+t)}.
\end{aligned}
$$

2. $t \in \mathbb{T}$ is right-dense. Then

$$
\begin{aligned}
f^\Delta(t) &= \lim_{s\to t} \frac{f(t) - f(s)}{t - s} \\
&= \lim_{s\to t} \frac{\frac{1-t}{1+t} - \frac{1-s}{1+s}}{t - s} \\
&= \lim_{s\to t} \frac{(1-t)(1+s) - (1-s)(1+t)}{(1+t)(1+s)(t-s)}
\end{aligned}
$$

$$\begin{aligned}
&= \lim_{s\to t} \frac{1+s-t-st-1-t+s+st}{(1+t)(1+s)(t-s)}\\
&= \lim_{s\to t} \frac{2(s-t)}{(1+t)(1+s)(t-s)}\\
&= -\lim_{s\to t} \frac{2}{(1+t)(1+s)}\\
&= -\frac{2}{(1+t)^2}.
\end{aligned}$$

Consequently,

$$f^{\Delta}(t) = \begin{cases} -\frac{2}{(1+t)(4+t)} & \text{if } t \text{ is right-scattered,}\\ -\frac{2}{(1+t)^2} & \text{if } t \text{ is right-dense.}\end{cases}$$

Example 2.18. Let $\mathbb{T} = C$, where C is the Cantor set, and let

$$f(t) = \frac{1-t^2}{1+t^2}, \quad t \in \mathbb{T}.$$

We will find $f^{\Delta}(t)$, $t \in \mathbb{T}$. We have the following cases.

1. $t \in \mathbb{C}_1$. Then t is right-scattered, and

$$\sigma(t) = t + \frac{1}{3^{m+1}}, \quad t = \sum_{k=1}^{m} \frac{a_k}{3^k} + \frac{1}{3^{m+1}}.$$

Set

$$\begin{aligned} g(t) &= 1-t^2,\\ h(t) &= 1+t^2. \end{aligned}$$

Then

$$\begin{aligned}
g^{\Delta}(t) &= \frac{g(\sigma(t)) - g(t)}{\sigma(t)-t}\\
&= \frac{1-(\sigma(t))^2-(1-t^2)}{\sigma(t)-t}\\
&= -\frac{(\sigma(t))^2-t^2}{\sigma(t)-t}\\
&= -\frac{(\sigma(t)-t)(\sigma(t)+t)}{\sigma(t)-t}\\
&= -\sigma(t)-t,\\
h^{\Delta}(t) &= \frac{h(\sigma(t))-h(t)}{\sigma(t)-t}\\
&= \frac{1+(\sigma(t))^2-(1+t^2)}{\sigma(t)-t}
\end{aligned}$$

$$\begin{aligned}&= \frac{(\sigma(t))^2 - t^2}{\sigma(t) - t}\\&= \frac{(\sigma(t) - t)(\sigma(t) + t)}{\sigma(t) - t}\\&= \sigma(t) + t,\end{aligned}$$

and

$$\begin{aligned}h(\sigma(t)) &= 1 + (\sigma(t))^2\\&= 1 + \left(t + \frac{1}{3^{m+1}}\right)^2\\&= 1 + \frac{1}{3^{2m+2}} + \frac{2}{3^{m+1}}t + t^2.\end{aligned}$$

Now applying the rule of delta derivative of the quotient of two functions, we arrive at

$$\begin{aligned}f^{\Delta}(t) &= \frac{g^{\Delta}(t)h(t) - g(t)h^{\Delta}(t)}{h(t)h(\sigma(t))}\\&= \frac{(-2t - \frac{1}{3^{m+1}})(1 + t^2) - (2t + \frac{1}{3^{m+1}})(1 - t^2)}{(1 + t^2)(1 + \frac{1}{3^{2m+2}} + \frac{2}{3^{m+1}}t + t^2)}\\&= \frac{-2t - 2t^3 - \frac{1}{3^{m+1}} - \frac{t^2}{3^{m+1}} - 2t + 2t^3 - \frac{1}{3^{m+1}} + \frac{t^2}{3^{m+1}}}{(1 + t^2)(1 + \frac{1}{3^{2m+2}} + \frac{2}{3^{m+1}}t + t^2)}\\&= \frac{-4t - \frac{2}{3^{m+1}}}{(1 + t^2)(1 + \frac{1}{3^{2m+2}} + \frac{2}{3^{m+1}}t + t^2)}.\end{aligned}$$

2. $t \in \mathbb{T}\backslash C_1$. Then t is right-dense. Hence

$$\begin{aligned}f^{\Delta}(t) &= \lim_{s\to t} \frac{f(t) - f(s)}{t - s}\\&= \lim_{s\to t} \frac{\frac{1-t^2}{1+t^2} - \frac{1-s^2}{1+s^2}}{t - s}\\&= \lim_{s\to t} \frac{(1 - t^2)(1 + s^2) - (1 - s^2)(1 + t^2)}{(1 + s^2)(1 + t^2)(t - s)}\\&= \lim_{s\to t} \frac{1 + s^2 - t^2 - s^2t^2 - 1 - t^2 + s^2 + s^2t^2}{(1 + t^2)(1 + s^2)(t - s)}\\&= \lim_{s\to t} \frac{2(s^2 - t^2)}{(1 + t^2)(1 + s^2)(t - s)}\\&= -2\lim_{s\to t} \frac{(t - s)(t + s)}{(1 + s^2)(1 + t^2)(t - s)}\\&= -2\lim_{s\to t} \frac{t + s}{(1 + s^2)(1 + t^2)}\end{aligned}$$

$$= -\frac{4t}{(1+t^2)^2}.$$

Consequently,

$$f^{\Delta}(t) = \begin{cases} \frac{-4t-\frac{2}{3^{m+1}}}{(1+t^2)(1+\frac{1}{3^{2m+2}}+\frac{2}{3^{m+1}}t+t^2)} & \text{if } t \in C_1, \\ -\frac{4t}{(1+t^2)^2} & \text{if } t \in \mathbb{T}\backslash C_1. \end{cases}$$

Example 2.19. Let $\mathbb{T} = \{t_n = \sum_{k=0}^{n-1} \alpha_k,\ n \in \mathbb{N},\ \alpha_k > 0,\ k \in \mathbb{N}_0\}$, and let

$$f(t) = \frac{1+2t}{3+4t}, \quad t \in \mathbb{T}.$$

We will find $f^{\Delta}(t_n)$, $n \in \mathbb{N}$. Note that any point $t_n \in \mathbb{T}$, $n \in \mathbb{N}$, is right-scattered and

$$\sigma(t_n) = t_{n+1}, \quad n \in \mathbb{N}.$$

Let

$$\begin{aligned} g(t) &= 1 + 2t, \\ h(t) &= 3 + 4t, \quad t \in \mathbb{T}. \end{aligned}$$

Then

$$\begin{aligned} g^{\Delta}(t_n) &= \frac{g(\sigma(t_n)) - g(t_n)}{\sigma(t_n) - t_n} \\ &= \frac{g(t_{n+1}) - g(t_n)}{t_{n+1} - t_n} \\ &= \frac{1 + 2t_{n+1} - 1 - 2t_n}{t_{n+1} - t_n} \\ &= \frac{2(t_{n+1} - t_n)}{t_{n+1} - t_n} \\ &= 2, \quad n \in \mathbb{N}, \\ h^{\Delta}(t_n) &= \frac{h(\sigma(t_n)) - h(t_n)}{\sigma(t_n) - t_n} \\ &= \frac{h(t_{n+1}) - h(t_n)}{t_{n+1} - t_n} \\ &= \frac{3 + 4t_{n+1} - 3 - 4t_n}{t_{n+1} - t_n} \\ &= \frac{4(t_{n+1} - t_n)}{t_{n+1} - t_n} \\ &= 4, \quad n \in \mathbb{N}, \end{aligned}$$

and

$$\begin{aligned} h(\sigma(t_n)) &= 3 + 4\sigma(t_n) \\ &= 3 + 4t_{n+1}, \quad n \in \mathbb{N}. \end{aligned}$$

Hence, applying the rule of delta derivative of the quotient of two functions, we arrive at

$$\begin{aligned} f^{\Delta}(t_n) &= \frac{g^{\Delta}(t_n)h(t_n) - g(t_n)h^{\Delta}(t_n)}{h(t_n)h(\sigma(t_n))} \\ &= \frac{2(3 + 4t_n) - 4(1 + 2t_n)}{(3 + 4t_n)(3 + 4t_{n+1})} \\ &= \frac{6 + 8t_n - 4 - 8t_n}{(3 + 4t_n)(3 + 4t_{n+1})} \\ &= \frac{2}{(3 + 4t_n)(3 + 4t_{n+1})}, \quad n \in \mathbb{N}. \end{aligned}$$

Example 2.20. Let $\mathbb{T} = \{t_n = -\frac{1}{n} : n \in \mathbb{N}\} \cup \mathbb{N}_0$, and let

$$f(t) = \frac{1 + t^2}{1 + 7t}, \quad t \in \mathbb{T}.$$

We will find $f^{\Delta}(t)$, $t \in \mathbb{T}$. Note that all points $t \in \mathbb{T}$ are right-scattered. Let

$$\begin{aligned} g(t) &= 1 + t^2, \\ h(t) &= 1 + 7t, \quad t \in \mathbb{T}. \end{aligned}$$

Then

$$\begin{aligned} g^{\Delta}(t) &= \frac{g(\sigma(t)) - g(t)}{\sigma(t) - t} \\ &= \frac{1 + (\sigma(t))^2 - 1 - t^2}{\sigma(t) - t} \\ &= \frac{(\sigma(t))^2 - t^2}{\sigma(t) - t} \\ &= \frac{(\sigma(t) - t)(\sigma(t) + t)}{\sigma(t) - t} \\ &= \sigma(t) + t, \quad t \in \mathbb{T}, \\ h^{\Delta}(t) &= \frac{h(\sigma(t)) - h(t)}{\sigma(t) - t} \\ &= \frac{1 + 7\sigma(t) - 1 - 7t}{\sigma(t) - t} \\ &= \frac{7(\sigma(t) - t)}{\sigma(t) - t} \\ &= 7, \quad t \in \mathbb{T}, \end{aligned}$$

and

$$h(\sigma(t)) = 1 + 7\sigma(t), \quad t \in \mathbb{T}.$$

Now applying the rule of delta derivative of the quotient of two functions, we arrive at

$$\begin{aligned} f^{\Delta}(t) &= \left(\frac{g}{h}\right)^{\Delta}(t) \\ &= \frac{g^{\Delta}(t)h(t) - g(t)h^{\Delta}(t)}{h(t)h(\sigma(t))} \\ &= \frac{(\sigma(t) + t)(1 + 7t) - (1 + t^2)7}{(1 + 7t)(1 + 7\sigma(t))} \\ &= \frac{\sigma(t) + 7t\sigma(t) + t + 7t^2 - 7 - 7t^2}{(1 + 7t)(1 + 7\sigma(t))} \\ &= \frac{-7 + t + \sigma(t) + 7t\sigma(t)}{(1 + 7t)(1 + 7\sigma(t))}, \quad t \in \mathbb{T}. \end{aligned}$$

Then we have the following cases.

1. $t \in \{-\frac{1}{n} : n \in \mathbb{N}\}$. Then

$$\sigma(t) = -\frac{t}{t-1},$$

and

$$\begin{aligned} f^{\Delta}(t) &= \frac{-7 + t - \frac{t}{t-1} - \frac{7t^2}{t-1}}{(1 + 7t)(1 - \frac{7t}{t-1})} \\ &= \frac{(-7 + t)(t - 1) - t - 7t^2}{(1 + 7t)(t - 1 - 7t)} \\ &= \frac{-7t + 7 + t^2 - t - t - 7t^2}{-(1 + 7t)(1 + 6t)} \\ &= \frac{-6t^2 - 9t + 7}{-(1 + 7t)(1 + 6t)} \\ &= \frac{6t^2 + 9t - 7}{(1 + 7t)(1 + 6t)}. \end{aligned}$$

2. $t \in \mathbb{N}_0$. Then

$$\sigma(t) = t + 1,$$

and

$$f^{\Delta}(t) = \frac{-7 + t + t + 1 + 7(t + 1)t}{(1 + 7t)(1 + 7t + 7)}$$

$$
\begin{aligned}
&= \frac{-6 + 2t + 7t^2 + 7t}{(1 + 7t)(8 + 7t)} \\
&= \frac{-6 + 9t + 7t^2}{(1 + 7t)(8 + 7t)}.
\end{aligned}
$$

Consequently,

$$
f^{\Delta}(t) = \begin{cases} \frac{6t^2+9t-7}{(1+7t)(1+6t)} & \text{if } t \in \{-\frac{1}{n} : n \in \mathbb{N}\}, \\ \frac{7t^2+9t-6}{(1+7t)(8+7t)} & \text{if } t \in \mathbb{N}_0. \end{cases}
$$

Example 2.21. Let $\mathbb{T} = \{(\frac{1}{2})^{2^n} : n \in \mathbb{N}_0\} \cup \{0, 1\}$, and let

$$
f(t) = \frac{1 + 7t}{4 + 5t}, \quad t \in \mathbb{T}.
$$

We will find $f^{\Delta}(t)$, $t \in \mathbb{T}^{\kappa}$. Here

$$
\mathbb{T}^{\kappa} = \left\{ \left(\frac{1}{2}\right)^{2^n} : n \in \mathbb{N}_0 \right\} \cup \{0\}.
$$

Set

$$
\begin{aligned}
g(t) &= 1 + 7t, \\
h(t) &= 4 + 5t, \quad t \in \mathbb{T}^{\kappa}.
\end{aligned}
$$

We have the following cases.

1. $t \in \mathbb{T}$ is right-scattered. Then

$$
\begin{aligned}
g^{\Delta}(t) &= \frac{g(\sigma(t)) - g(t)}{\sigma(t) - t} \\
&= \frac{1 + 7\sigma(t) - 1 - 7t}{\sigma(t) - t} \\
&= \frac{7(\sigma(t) - t)}{\sigma(t) - t} \\
&= 7, \\
h^{\Delta}(t) &= \frac{h(\sigma(t)) - h(t)}{\sigma(t) - t} \\
&= \frac{4 + 5\sigma(t) - 4 - 5t}{\sigma(t) - t} \\
&= \frac{5(\sigma(t) - t)}{\sigma(t) - t} \\
&= 5,
\end{aligned}
$$

and

$$h(\sigma(t)) = 4 + 5\sigma(t).$$

Now applying the rule of delta derivative of the quotient of two functions, we arrive at

$$\begin{aligned} f^{\Delta}(t) &= \left(\frac{g}{h}\right)^{\Delta}(t) \\ &= \frac{g^{\Delta}(t)h(t) - g(t)h^{\Delta}(t)}{h(t)h(\sigma(t))} \\ &= \frac{7(4+5t) - 5(1+7t)}{(4+5t)(4+5\sigma(t))} \\ &= \frac{28 + 35t - 5 - 35t}{(4+5t)(4+5\sigma(t))} \\ &= \frac{23}{(4+5t)(4+5\sigma(t))}. \end{aligned}$$

We have the following subcases.

a. $t \in \{(\frac{1}{2})^{2^n} : n \in \mathbb{N}\}$. Then

$$\sigma(t) = \sqrt{t},$$

and

$$f^{\Delta}(t) = \frac{23}{(4+5t)(4+5\sqrt{t})}.$$

b. $t = \frac{1}{2}$. Then

$$\sigma\left(\frac{1}{2}\right) = 1,$$

and

$$\begin{aligned} f^{\Delta}\left(\frac{1}{2}\right) &= \frac{23}{(4+\frac{5}{2})(4+5)} \\ &= \frac{23}{\frac{13}{2}\cdot 9} \\ &= \frac{46}{117}. \end{aligned}$$

2. $t \in \mathbb{T}$ is right-dense. Then $t = 0$. Hence

$$\begin{aligned} f^{\Delta}(0) &= \lim_{s\to 0}\frac{f(s) - f(0)}{s} \\ &= \lim_{s\to 0}\frac{\frac{1+7s}{4+5s} - \frac{1}{4}}{s} \end{aligned}$$

$$
\begin{aligned}
&= \lim_{s\to 0} \frac{4+28s-4-5s}{4s(4+5s)}\\
&= \lim_{s\to 0} \frac{23s}{4s(4+5s)}\\
&= \lim_{s\to 0} \frac{23}{4(4+5s)}\\
&= \frac{23}{16}.
\end{aligned}
$$

Consequently,

$$
f^{\Delta}(t) = \begin{cases} \frac{23}{(4+5t)(4+5\sqrt{t})} & \text{if } t \in \{(\frac{1}{2})^{2^n} : n \in \mathbb{N}\},\\ \frac{46}{117} & \text{if } t = \frac{1}{2},\\ \frac{23}{16} & \text{if } t = 0. \end{cases}
$$

Example 2.22. Let $U = \{\frac{1}{2^n} : n \in \mathbb{N}\}$, let

$$
\mathbb{T} = \{0\} \cup U \cup (1-U) \cup (1+U) \cup (2-U) \cup (2+U) \cup \{1,2\},
$$

and let

$$
f(t) = (2+t^2)(1-t^2), \quad t \in \mathbb{T}.
$$

We will find $f^{\Delta}(t)$, $t \in \mathbb{T}^{\kappa}$. Here

$$
\mathbb{T}^{\kappa} = \mathbb{T}\backslash\left\{\frac{5}{2}\right\}.
$$

Set

$$
\begin{aligned}
g(t) &= 2+t^2,\\
h(t) &= 1-t^2, \quad t \in \mathbb{T}^{\kappa}.
\end{aligned}
$$

We have the following cases.

1. $t \in \mathbb{T}^{\kappa}$ is right-scattered. Then

$$
\begin{aligned}
g^{\Delta}(t) &= \frac{g(\sigma(t)) - g(t)}{\sigma(t) - t}\\
&= \frac{2 + (\sigma(t))^2 - 2 - t^2}{\sigma(t) - t}\\
&= \frac{(\sigma(t))^2 - t^2}{\sigma(t) - t}\\
&= \frac{(\sigma(t) - t)(\sigma(t) + t)}{\sigma(t) - t}\\
&= \sigma(t) + t,
\end{aligned}
$$

$$\begin{aligned}
h^{\Delta}(t) &= \frac{h(\sigma(t)) - h(t)}{\sigma(t) - t} \\
&= \frac{1 - (\sigma(t))^2 - 1 + t^2}{\sigma(t) - t} \\
&= -\frac{(\sigma(t))^2 - t^2}{\sigma(t) - t} \\
&= -\frac{(\sigma(t) - t)(\sigma(t) + t)}{\sigma(t) - t} \\
&= -\sigma(t) - t,
\end{aligned}$$

and

$$h(\sigma(t)) = 1 - (\sigma(t))^2.$$

Now applying the rule of delta derivative of the product of two functions, we find

$$\begin{aligned}
f^{\Delta}(t) &= (gh)^{\Delta}(t) \\
&= g^{\Delta}(t)h(\sigma(t)) + g(t)h^{\Delta}(t) \\
&= (\sigma(t)+t)(1 - (\sigma(t))^2) + (2 + t^2)(-\sigma(t) - t) \\
&= \sigma(t) - (\sigma(t))^3 + t - t(\sigma(t))^2 - 2\sigma(t) - 2t - t^2\sigma(t) - t^3 \\
&= -(\sigma(t))^3 - \sigma(t) - t - t^3 - t(\sigma(t))^2 - t^2\sigma(t), \quad t \in \mathbb{T}^{\kappa}.
\end{aligned}$$

Now we consider the following subcases.

a. $t = \frac{1}{2}$. Then

$$\sigma\left(\frac{1}{2}\right) = \frac{3}{4},$$

and

$$\begin{aligned}
f^{\Delta}\left(\frac{1}{2}\right) &= -\left(\sigma\left(\frac{1}{2}\right)\right)^3 - \sigma\left(\frac{1}{2}\right) - \frac{1}{2} - \left(\frac{1}{2}\right)^3 - \frac{1}{2}\left(\sigma\left(\frac{1}{2}\right)\right)^2 - \left(\frac{1}{2}\right)^2\sigma\left(\frac{1}{2}\right) \\
&= -\left(\frac{3}{4}\right)^3 - \frac{3}{4} - \frac{1}{2} - \frac{1}{8} - \frac{1}{2}\cdot\left(\frac{3}{4}\right)^2 - \frac{1}{4}\cdot\frac{3}{4} \\
&= -\frac{27}{64} - \frac{3}{4} - \frac{1}{2} - \frac{1}{8} - \frac{9}{32} - \frac{3}{16} \\
&= -\frac{27}{64} - \frac{3}{4} - \frac{1}{2} - \frac{13}{32} - \frac{3}{16} \\
&= -\frac{27}{64} - \frac{3}{4} - \frac{29}{32} - \frac{3}{16} \\
&= \frac{-27 - 48 - 58 - 12}{64} \\
&= -\frac{145}{64}.
\end{aligned}$$

b. $t = \frac{3}{2}$. Then

$$\sigma\left(\frac{3}{2}\right) = \frac{7}{4},$$

and

$$\begin{aligned} f^{\Delta}\left(\frac{3}{2}\right) &= -\left(\sigma\left(\frac{3}{2}\right)\right)^3 - \sigma\left(\frac{3}{2}\right) - \frac{3}{2} - \left(\frac{3}{2}\right)^3 - \frac{3}{2}\left(\sigma\left(\frac{3}{2}\right)\right)^2 - \left(\frac{3}{2}\right)^2\sigma\left(\frac{3}{2}\right) \\ &= -\left(\frac{7}{4}\right)^3 - \frac{7}{4} - \frac{3}{2} - \frac{27}{8} - \frac{3}{2}\cdot\left(\frac{7}{4}\right)^2 - \frac{9}{4}\cdot\frac{7}{4} \\ &= -\frac{343}{64} - \frac{7}{4} - \frac{3}{2} - \frac{27}{8} - \frac{98}{32} - \frac{63}{16} \\ &= -\frac{343}{64} - \frac{7}{4} - \frac{3}{2} - \frac{206}{32} - \frac{63}{16} \\ &= \frac{-343 - 112 - 96 - 412 - 252}{64} \\ &= -\frac{1215}{64}. \end{aligned}$$

c. $t \in U\backslash\{\frac{1}{2}\}$. Then

$$\sigma(t) = 2t,$$

and

$$\begin{aligned} f^{\Delta}(t) &= -(2t)^3 - 2t - t - t^3 - t(2t)^2 - t^2(2t) \\ &= -8t^3 - 3t - t^3 - 4t^3 - 2t^3 \\ &= -15t^3 - 3t. \end{aligned}$$

d. $t \in (1 - U)\backslash\{\frac{1}{2}\}$. Then

$$\sigma(t) = \frac{1+t}{2},$$

and

$$\begin{aligned} f^{\Delta}(t) &= -\left(\frac{1+t}{2}\right)^3 - \frac{1+t}{2} - t - t^3 - t\left(\frac{1+t}{2}\right)^2 - t^2\frac{1+t}{2} \\ &= -\frac{(1+t)^3}{8} - \frac{1+t}{2} - t - t^3 - t\frac{(1+t)^2}{4} - t^2\frac{1+t}{2} \\ &= -\frac{(1+t)^3 + 4(1+t) + 8t + 8t^3 + 2t(1+t)^2 + 4t^2(1+t)}{8} \\ &= -\frac{t^3 + 3t^2 + 3t + 1 + 4t + 4 + 8t + 8t^3 + 2t(t^2 + 2t + 1) + 4t^2 + 4t^3}{8} \end{aligned}$$

$$\begin{aligned}
&= -\frac{13t^3 + 7t^2 + 15t + 5 + 2t^3 + 4t^2 + 2t}{8} \\
&= -\frac{15t^3 + 11t^2 + 17t + 5}{8}.
\end{aligned}$$

e. $t \in (1 + U) \backslash \{\frac{3}{2}\}$. Then

$$\sigma(t) = 2t - 1,$$

and

$$\begin{aligned}
f^{\Delta}(t) &= -(2t-1)^3 - (2t-1) - t - t^3 - t(2t-1)^2 - t^2(2t-1) \\
&= -8t^3 + 12t^2 - 6t + 1 - 2t + 1 - t - t^3 - t(4t^2 - 4t + 1) - 2t^3 + t^2 \\
&= -11t^3 + 13t^2 - 9t + 2 - 4t^3 + 4t^2 - t \\
&= -15t^3 + 17t^2 - 10t + 2.
\end{aligned}$$

f. $t \in (2 - U) \backslash \{\frac{3}{2}\}$. Then

$$\sigma(t) = \frac{t+2}{2},$$

and

$$\begin{aligned}
f^{\Delta}(t) &= -\left(\frac{t+2}{2}\right)^3 - \frac{t+2}{2} - t - t^3 - t\left(\frac{t+2}{2}\right)^2 - t^2\left(\frac{t+2}{2}\right) \\
&= -\frac{(t+2)^3}{8} - \frac{t+2}{2} - t - t^3 - t\frac{(t+2)^2}{4} - t^2\frac{t+2}{2} \\
&= -\frac{(t+2)^3 + 4(t+2) + 8t + 8t^3 + 2t(t+2)^2 + 4t^2(t+2)}{8} \\
&= -\frac{t^3 + 6t^2 + 12t + 8 + 4t + 8 + 8t + 8t^3 + 2t(t^2 + 4t + 4) + 4t^3 + 8t^2}{8} \\
&= -\frac{13t^3 + 14t^2 + 24t + 8 + 2t^3 + 8t^2 + 8t}{8} \\
&= -\frac{15t^3 + 22t^2 + 32t + 8}{8}.
\end{aligned}$$

g. $t \in (2 + U) \backslash \{\frac{5}{2}\}$. Then

$$\sigma(t) = 2(t-1),$$

and

$$\begin{aligned}
f^{\Delta}(t) &= -(2(t-1))^3 - 2(t-1) - t - t^3 - t(2(t-1))^2 - t^2(2(t-1)) \\
&= -8(t-1)^3 - 2(t-1) - t - t^3 - 4t(t-1)^2 - 2t^2(t-1)
\end{aligned}$$

$$= -8t^3 + 24t^2 - 24t + 8 - 2t + 2 - t - t^3 - 4t(t^2 - 2t + 1) - 2t^3 + 2t^2$$
$$= -11t^3 + 26t^2 - 27t + 10 - 4t^3 + 8t^2 - 4t$$
$$= -15t^3 + 34t^2 - 31t + 10.$$

2. $t \in \mathbb{T}^\kappa$ is right-dense. Then $t = 0$, or $t = 1$, or $t = 2$.
   a. $t = 0$. Then

$$\begin{aligned} f^\Delta(0) &= \lim_{s\to 0} \frac{f(s) - f(0)}{s} \\ &= \lim_{s\to 0} \frac{(2 + s^2)(1 - s^2) - 2}{s} \\ &= \lim_{s\to 0} \frac{2 - 2s^2 + s^2 - s^4 - 2}{s} \\ &= -\lim_{s\to 0} \frac{s(s + s^3)}{s} \\ &= -\lim_{s\to 0}(s + s^3) \\ &= 0. \end{aligned}$$

   b. $t = 1$. Then

$$\begin{aligned} f^\Delta(1) &= \lim_{s\to 1} \frac{f(s) - f(1)}{s - 1} \\ &= \lim_{s\to 1} \frac{(2 + s^2)(1 - s^2) - (2 + 1)(1 - 1)}{s - 1} \\ &= -\lim_{s\to 1} \frac{(2 + s^2)(s - 1)(s + 1)}{s - 1} \\ &= -\lim_{s\to 1}(2 + s^2)(s + 1) \\ &= -(2 + 1)(1 + 1) \\ &= -6. \end{aligned}$$

   c. $t = 2$. Then

$$\begin{aligned} f^\Delta(2) &= \lim_{s\to 2} \frac{f(s) - f(2)}{s - 2} \\ &= \lim_{s\to 2} \frac{(2 + s^2)(1 - s^2) - (2 + 4)(1 - 4)}{s - 2} \\ &= \lim_{s\to 2} \frac{2 - 2s^2 + -s^2 - s^4 + 18}{s - 2} \\ &= -\lim_{s\to 2} \frac{s^4 + s^2 - 20}{s - 2} \\ &= -\lim_{s\to 2} \frac{(s - 2)(s^3 + 2s^2 + 5s + 10)}{s - 2} \end{aligned}$$

$$
\begin{aligned}
&= -\lim_{s\to 2}(s^3+2s^2+5s+10)\\
&= -(2^3+2\cdot 2^2+5\cdot 2+10)\\
&= -(8+8+10+10)\\
&= -36.
\end{aligned}
$$

Consequently,

$$
f^{\Delta}(t) = \begin{cases}
-\frac{145}{64} & \text{if } t = \frac{1}{2},\\
-\frac{1215}{64} & \text{if } t = \frac{3}{2},\\
-15t^3 - 3t & \text{if } t \in U\backslash\{\frac{1}{2}\},\\
-\frac{15t^3+11t^2+17t+5}{8} & \text{if } t \in (1-U)\backslash\{\frac{1}{2}\},\\
-15t^3 + 17t^2 - 10t + 2 & \text{if } t \in (1+U)\backslash\{\frac{3}{2}\},\\
-\frac{15t^3+22t^2+32t+8}{8} & \text{if } t \in (2-U)\backslash\{\frac{3}{2}\},\\
-15t^3 + 34t^2 - 31t + 10 & \text{if } t \in (2+U)\backslash\{\frac{5}{2}\},\\
0 & \text{if } t = 0,\\
-6 & \text{if } t = 1,\\
-36 & \text{if } t = 2.
\end{cases}
$$

Example 2.23. Let

$$
f(t) = t^n, \quad t \in \mathbb{T},\ n \in \mathbb{N}.
$$

We will prove that

$$
f^{\Delta}(t) = (\sigma(t))^{n-1} + t(\sigma(t))^{n-2} + \cdots + t^{n-2}\sigma(t) + t^{n-1}
$$

if $t \in \mathbb{T}$ is right-scattered. Indeed, let $t \in \mathbb{T}$ be right-scattered. Then

$$
\begin{aligned}
f^{\Delta}(t) &= \frac{f(\sigma(t)) - f(t)}{\sigma(t) - t}\\
&= \frac{(\sigma(t))^n - t^n}{\sigma(t) - t}\\
&= \frac{(\sigma(t) - t)((\sigma(t))^{n-1} + t(\sigma(t))^{n-2} + \cdots + t^{n-2}\sigma(t) + t^{n-1})}{\sigma(t) - t}\\
&= (\sigma(t))^{n-1} + t(\sigma(t))^{n-2} + \cdots + t^{n-2}\sigma(t) + t^{n-1}.
\end{aligned}
$$

Example 2.24. Let

$$
\mathbb{T} = \left\{\frac{1}{2n+1} : n \in \mathbb{N}_0\right\} \cup \{0\},
$$

and let $f(t) = \sigma(t)$, $t \in \mathbb{T}$. We will find $f^{\Delta}(t)$, $t \in \mathbb{T}$. For

$$t \in \mathbb{T}, \quad t = \frac{1}{2n+1}, \quad n = \frac{1-t}{2t}, \quad n \geq 1,$$

we have

$$\begin{aligned}\sigma(t) &= \inf\left\{\frac{1}{2l+1}, 0 : \frac{1}{2l+1}, 0 > \frac{1}{2n+1}, l \in \mathbb{N}_0\right\}\\ &= \frac{1}{2n-1}\\ &= \frac{1}{2\frac{1-t}{2t} - 1}\\ &= \frac{t}{1-2t}\\ &> t,\end{aligned}$$

i. e., any point $t = \frac{1}{2n+1}$, $n \geq 1$, is right-scattered. At these points,

$$\begin{aligned}f^{\Delta}(t) &= \frac{f(\sigma(t)) - f(t)}{\sigma(t) - t}\\ &= \frac{\sigma(\sigma(t)) - \sigma(t)}{\sigma(t) - t}\\ &= 2\frac{(\sigma(t))^2}{(1-2\sigma(t))(\sigma(t)-t)}\\ &= 2\frac{(\frac{t}{1-2t})^2}{(1-\frac{2t}{1-2t})(\frac{t}{1-2t} - t)}\\ &= 2\frac{\frac{t^2}{(1-2t)^2}}{\frac{1-4t}{1-2t}\frac{2t^2}{1-2t}}\\ &= 2\frac{t^2}{2t^2(1-4t)}\\ &= \frac{1}{1-4t}.\end{aligned}$$

Let $n = 0$, i. e., $t = 1$. Then

$$\begin{aligned}\sigma(1) &= \inf\left\{\frac{1}{2l+1}, 0 : \frac{1}{2l+1}, 0 > 1, l \in \mathbb{N}_0\right\}\\ &= \inf\emptyset\\ &= \sup\mathbb{T}\\ &= 1,\end{aligned}$$

i. e., $t = 1$ is a right-dense point. Also,

$$\begin{aligned}\lim_{h\to 0}\frac{f(1+h)-f(h)}{h} &= \lim_{h\to 0}\frac{\sigma(1+h)-\sigma(h)}{h}\\ &= \lim_{h\to 0}\frac{\frac{1+h}{1-2(1+h)}-\frac{h}{1-2h}}{h}\\ &= -\lim_{h\to 0}\frac{\frac{1+h}{1+2h}+\frac{h}{1-2h}}{h}\\ &= -\lim_{h\to 0}\frac{(1+h)(1-2h)+h(1+2h)}{(1+2h)(1-2h)h}\\ &= -\lim_{h\to 0}\frac{1}{(1+2h)(1-2h)h}\\ &= -\infty.\end{aligned}$$

Therefore $\sigma^{\Delta}(1)$ does not exist.

Let now $t = 0$. Then

$$\begin{aligned}\sigma(0) &= \inf\left\{\frac{1}{2l+1}, 0 : \frac{1}{2l+1}, 0 > 0, l \in \mathbb{N}_0\right\}\\ &= 0.\end{aligned}$$

Consequently, $t = 0$ is right-dense. Also,

$$\begin{aligned}\lim_{h\to 0}\frac{\sigma(h)-\sigma(0)}{h} &= \lim_{h\to 0}\frac{\frac{h}{1-2h}-0}{h}\\ &= \lim_{h\to 0}\frac{1}{1-2h}\\ &= 1.\end{aligned}$$

Therefore $\sigma^{\Delta}(0) = 1$.

Example 2.25. Let $f, g, h : \mathbb{T} \to \mathbb{R}$ be differentiable at $t \in \mathbb{T}^{\kappa}$. Then

$$\begin{aligned}(fgh)^{\Delta}(t) &= ((fg)h)^{\Delta}(t)\\ &= (fg)^{\Delta}(t)h(t) + (fg)(\sigma(t))h^{\Delta}(t)\\ &= (f^{\Delta}(t)g(t) + f(\sigma(t))g^{\Delta}(t))h(t) + f^{\sigma}(t)g^{\sigma}(t)h^{\Delta}(t)\\ &= f^{\Delta}(t)g(t)h(t) + f^{\sigma}(t)g^{\Delta}(t)h(t) + f^{\sigma}(t)g^{\sigma}(t)h^{\Delta}(t).\end{aligned}$$

Example 2.26. Let $f : \mathbb{T} \to \mathbb{R}$ be differentiable at $t \in \mathbb{T}^{\kappa}$. Then

$$\begin{aligned}(f^2)^{\Delta}(t) &= (ff)^{\Delta}(t)\\ &= f^{\Delta}(t)f(t) + f(\sigma(t))f^{\Delta}(t)\\ &= f^{\Delta}(t)(f^{\sigma}(t) + f(t)).\end{aligned}$$

Also,

$$
\begin{aligned}
(f^3)^\Delta(t) &= (ff^2)^\Delta(t)\\
&= f^\Delta(t)(f(t))^2 + f(\sigma(t))(f^2)^\Delta(t)\\
&= f^\Delta(t)(f(t))^2 + f^\sigma(t)f^\Delta(t)(f^\sigma(t) + f(t))\\
&= f^\Delta(t)((f(t))^2 + f(t)f^\sigma(t) + (f^\sigma(t))^2).
\end{aligned}
$$

Assume that

$$
(f^n)^\Delta(t) = f^\Delta(t)\sum_{k=0}^{n-1}(f(t))^k(f^\sigma(t))^{n-1-k}
$$

for some $n \in \mathbb{N}$. We will prove that

$$
(f^{n+1})^\Delta(t) = f^\Delta(t)\sum_{k=0}^{n}(f(t))^k(f^\sigma(t))^{n-k}.
$$

Indeed,

$$
\begin{aligned}
(f^{n+1})^\Delta(t) &= (ff^n)^\Delta(t)\\
&= f^\Delta(t)(f(t))^n + f^\sigma(t)(f^n)^\Delta(t)\\
&= f^\Delta(t)(f(t))^n + f^\Delta(t)f^\sigma(t)((f(t))^{n-1} + (f(t))^{n-2}f^\sigma(t)\\
&\quad + \cdots + f(t)(f^\sigma(t))^{n-2} + (f^\sigma(t))^{n-1}f^\sigma(t)\\
&= f^\Delta(t)((f(t))^n + (f(t))^{n-1}f^\sigma(t) + (f(t))^{n-2}(f^\sigma(t))^2 + \cdots + (f^\sigma(t))^n)\\
&= f^\Delta(t)\sum_{k=0}^{n}(f(t))^k(f^\sigma(t))^{n-k}.
\end{aligned}
$$

Example 2.27. Now consider $f(t) = (t-a)^m$, $t \in \mathbb{T}$, for $a \in \mathbb{R}$ and $m \in \mathbb{N}$. We set

$$
h(t) = t - a, \quad t \in \mathbb{T}.
$$

Then

$$
h^\Delta(t) = 1, \quad t \in \mathbb{T}^\kappa.
$$

By example 2.26 we get

$$
\begin{aligned}
(h^m)^\Delta(t) &= h^\Delta(t)\sum_{k=0}^{m-1}(h(t))^k(h^\sigma(t))^{m-1-k}\\
&= \sum_{k=0}^{m-1}(t-a)^k(\sigma(t)-a)^{m-1-k}, \quad t \in \mathbb{T}^\kappa.
\end{aligned}
$$

Let now $g(t) = \frac{1}{f(t)}$, $t \in \mathbb{T}$. Then

$$g^{\Delta}(t) = -\frac{f^{\Delta}(t)}{f(\sigma(t))f(t)}, \quad t \in \mathbb{T}^{\kappa},$$

whereupon

$$\begin{aligned}\left(\frac{1}{h^m}\right)^{\Delta}(t) &= -\frac{1}{(\sigma(t)-a)^m(t-a)^m}\sum_{k=0}^{m-1}(t-a)^k(\sigma(t)-a)^{m-1-k} \\ &= -\sum_{k=0}^{m-1}\frac{1}{(t-a)^{m-k}}\frac{1}{(\sigma(t)-a)^{k+1}}, \quad t \in \mathbb{T}^{\kappa}.\end{aligned}$$

Exercise 2.6. Let $\mathbb{T} = k\mathbb{Z}$, $k > 0$. Prove that

$$\begin{aligned} t &= \frac{1}{2}f_2^{\Delta}(t) - \frac{k}{2}, \\ t^2 &= \frac{1}{3}f_3^{\Delta}(t) - \frac{k}{2}f_2^{\Delta}(t) + \frac{1}{6}k^2, \\ t^3 &= \frac{1}{4}f_4^{\Delta}(t) - \frac{1}{2}kf_3^{\Delta}(t) + \frac{1}{4}k^2f_2^{\Delta}(t), \\ t^4 &= \frac{1}{5}f_5^{\Delta}(t) - \frac{1}{2}kf_4^{\Delta}(t) + \frac{1}{3}k^2f_3^{\Delta}(t) - \frac{1}{30}k^4, \\ t^5 &= \frac{1}{6}f_6^{\Delta}(t) - \frac{1}{2}kf_5^{\Delta}(t) + \frac{5}{12}k^2f_4^{\Delta}(t) - \frac{1}{12}k^4f_2^{\Delta}(t), \end{aligned}$$

where $f_i(t) = t^i$, $i = 1, \ldots, 6$, $t \in \mathbb{T}$.

Exercise 2.7. Let $t \in \mathbb{T}$ be right-scattered. Find $f^{\Delta}(t)$, $t \in \mathbb{T}^{\kappa}$, where

$$f(t) = 2\sin t + t^2 - 3t^3, \quad t \in \mathbb{T}.$$

Exercise 2.8. Let $t \in 2^{\mathbb{N}_0}$. Find $f^{\Delta}(t)$, $t \in \mathbb{T}$, where

$$f(t) = \frac{t^3 + t^2 - 2t}{2t^2 + 3t + 1}, \quad t \in \mathbb{T}.$$

Exercise 2.9. Let $\mathbb{T} = 3^{\mathbb{N}_0}$. Find $f^{\Delta}(t)$, $t \in \mathbb{T}^{\kappa}$, where

$$f(t) = \frac{1}{4}t^4 + \frac{1}{3}t^3 - t^2 + 1, \quad t \in \mathbb{T}.$$

Exercise 2.10. Let $\mathbb{T} = \mathbb{Z}$. Find $f^{\Delta}(t)$, $t \in \mathbb{T}^{\kappa}$, where

$$f(t) = \frac{1}{3}t^3 - \frac{5}{2}t^2 - 4t, \quad t \in \mathbb{T}.$$

Exercise 2.11. Let $\mathbb{T} = \{\sqrt[4]{n} : n \in \mathbb{N}\}$. Find $f^{\Delta}(t)$, $t \in \mathbb{T}^{\kappa}$, where

$$f(t) = \frac{1}{3}t^3 - 2t^2 - 3t + 2, \quad t \in \mathbb{T}.$$

Exercise 2.12. Let $\mathbb{T} = [1,2] \cup [3,4] \cup [7,8] \cup 9^{\mathbb{N}}$. Find $f^{\Delta}(t)$, $t \in \mathbb{T}^{\kappa}$, where

$$f(t) = \frac{2-t}{3+7t}, \quad t \in \mathbb{T}.$$

2.3 Higher-order delta differentiation

Definition 2.2. Let $f : \mathbb{T} \to \mathbb{R}$ and $t \in (\mathbb{T}^{\kappa})^{\kappa} = \mathbb{T}^{\kappa^2}$. We define the second derivative of f at t, provided that it exists, as

$$f^{\Delta^2} = (f^{\Delta})^{\Delta} : \mathbb{T}^{\kappa^2} \to \mathbb{R}.$$

Similarly, we define higher-order derivatives $f^{\Delta^n} : \mathbb{T}^{\kappa^n} \to \mathbb{R}$ for $n \in \mathbb{N}$.

Example 2.28. Let $\mathbb{T} = h\mathbb{Z}$, $h > 0$, and let

$$f(t) = (t+1)(t+2)(t+3)(t+4), \quad t \in \mathbb{T}.$$

We will find $f^{\Delta^2}(t)$, $t \in \mathbb{T}$. By Example 2.13 we have

$$f^{\Delta}(t) = 4t^3 + 3(2h+11)t^2 + (4h^2 + 33h + 57)t + h^3 + 10h^2 + 45h + 60, \quad t \in \mathbb{T}.$$

Now applying Example 2.23 and using that

$$\sigma(t) = t + h, \quad t \in \mathbb{T},$$

we obtain

$$\begin{aligned} f^{\Delta^2}(t) &= 4\left((\sigma(t))^2 + t\sigma(t) + t^2\right) + 3(2h+11)(\sigma(t) + t) + 4h^2 + 33h + 57 \\ &= 4\left((t+h)^2 + t(t+h) + t^2\right) + 3(2h+11)(t+h+t) + 4h^2 + 33h + 57 \\ &= 4(t^2 + 2ht + h^2 + t^2 + ht + t^2) + 3(2h+11)(2t+h) + 4h^2 + 33h + 57 \\ &= 4(3t^2 + 3ht + h^2) + 6(2h+11)t + 3h(2h+11) + 4h^2 + 33h + 57 \\ &= 12t^2 + 12ht + 4h^2 + 6(2h+11)t + 6h^2 + 33h + 4h^2 + 33h + 57 \\ &= 12t^2 + 6(4h+11)t + 14h^2 + 66h + 57, \quad t \in \mathbb{T}. \end{aligned}$$

Example 2.29. Let $\mathbb{T} = 3^{\mathbb{N}_0}$, and let

$$f(t) = t^3 + t^2 + t + 1, \quad t \in \mathbb{T}.$$

We will find $f^{\Delta^2}(t)$, $t \in \mathbb{T}$. By Example 2.14 we have

$$f^{\Delta}(t) = 13t^2 + 4t + 1, \quad t \in \mathbb{T}.$$

Now applying Example 2.23 and using that

$$\sigma(t) = 3t, \quad t \in \mathbb{T},$$

we obtain

$$\begin{aligned} f^{\Delta^2}(t) &= 13(\sigma(t) + t) + 4 \\ &= 13(3t + t) + 4 \\ &= 52t + 4, \quad t \in \mathbb{T}. \end{aligned}$$

Example 2.30. Let $\mathbb{T} = \mathbb{N}_0^2$, and let

$$f(t) = \frac{1-t}{5+t}, \quad t \in \mathbb{T}.$$

We will find $f^{\Delta^2}(t)$, $t \in \mathbb{T}$. Here

$$\sigma(t) = (\sqrt{t} + 1)^2, \quad t \in \mathbb{T}.$$

Note that all points of $\mathbb{T}$ are right-scattered. Then

$$\begin{aligned} f^{\Delta}(t) &= \frac{f(\sigma(t)) - f(t)}{\sigma(t) - t} \\ &= \frac{1}{\sigma(t) - t}\left(\frac{1-\sigma(t)}{5+\sigma(t)} - \frac{1-t}{5+t}\right) \\ &= \frac{1}{\sigma(t) - t}\left(\frac{(1-\sigma(t))(5+t) - (1-t)(5+\sigma(t))}{(5+t)(5+\sigma(t))}\right) \\ &= \frac{5 + t - 5\sigma(t) - t\sigma(t) - 5 - \sigma(t) + 5t + t\sigma(t)}{(\sigma(t) - t)(5+t)(5+\sigma(t))} \\ &= -\frac{6(\sigma(t) - t)}{(\sigma(t) - t)(5+t)(5+\sigma(t))} \\ &= -\frac{6}{(5+t)(5+\sigma(t))} \\ &= -\frac{6}{(5+t)(6 + 2\sqrt{t} + t)}, \quad t \in \mathbb{T}. \end{aligned}$$

Hence

$$f^{\Delta^2}(t) = \frac{f^{\Delta}(\sigma(t)) - f^{\Delta}(t)}{\sigma(t) - t}$$

$$
\begin{aligned}
&= \frac{1}{\sigma(t)-t}\left(-\frac{6}{(5+\sigma(t))(6+2\sqrt{\sigma(t)}+\sigma(t))} + \frac{6}{(5+t)(6+2\sqrt{t}+t)}\right) \\
&= \frac{6}{\sigma(t)-t}\left(\frac{(5+\sigma(t))(6+2\sqrt{\sigma(t)}+\sigma(t)) - (5+t)(6+2\sqrt{t}+t)}{(5+\sigma(t))(5+t)(6+2\sqrt{t}+t)(6+2\sqrt{\sigma(t)}+\sigma(t))}\right) \\
&= \frac{6}{(\sigma(t)-t)(5+t)(5+\sigma(t))(6+2\sqrt{t}+t)(6+2\sqrt{\sigma(t)}+\sigma(t))} \\
&\quad \times (30 + 10\sqrt{\sigma(t)} + 5\sigma(t) + 6\sigma(t) + 2\sigma(t)\sqrt{\sigma(t)} + (\sigma(t))^2 \\
&\quad - 30 - 10\sqrt{t} - 5t - 6t - 2t\sqrt{t} - t^2) \\
&= \frac{6}{(\sigma(t)-t)(5+t)(5+\sigma(t))(6+2\sqrt{t}+t)(6+2\sqrt{\sigma(t)}+\sigma(t))} \\
&\quad \times (10(\sqrt{\sigma(t)} - \sqrt{t}) + 11(\sigma(t)-t) + 2(\sigma(t)\sqrt{\sigma(t)} - t\sqrt{t}) + (\sigma(t))^2 - t^2)) \\
&= \frac{6}{(\sigma(t)-t)(5+t)(5+\sigma(t))(6+2\sqrt{t}+t)(6+2\sqrt{\sigma(t)}+\sigma(t))} \\
&\quad \times \left(10\frac{\sigma(t)-t}{\sqrt{\sigma(t)}+\sqrt{t}} + 11(\sigma(t)-t) + 2\frac{(\sigma(t))^3 - t^3}{\sigma(t)\sqrt{\sigma(t)} + t\sqrt{t}} + (\sigma(t)-t)(\sigma(t)+t)\right) \\
&= \frac{6}{(\sigma(t)-t)(5+t)(5+\sigma(t))(6+2\sqrt{t}+t)(6+2\sqrt{\sigma(t)}+\sigma(t))} \\
&\quad \times \left(10\frac{\sigma(t)-t}{\sqrt{\sigma(t)}+\sqrt{t}} + 11(\sigma(t)-t) + 2\frac{(\sigma(t)-t)((\sigma(t))^2 + t\sigma(t) + t^2)}{\sigma(t)\sqrt{\sigma(t)} + t\sqrt{t}} + \sigma(t) + t\right) \\
&= \frac{6}{(5+t)(5+\sigma(t))(6+2\sqrt{t}+t)(6+2\sqrt{\sigma(t)}+\sigma(t))} \\
&\quad \times \left(\frac{10}{\sqrt{\sigma(t)}+\sqrt{t}} + 11 + 2\frac{(\sigma(t))^2 + t\sigma(t) + t^2}{\sigma(t)\sqrt{\sigma(t)} + t\sqrt{t}} + \sigma(t) + t\right) \\
&= \frac{6}{(5+t)(5+(\sqrt{t}+1)^2)(6+2\sqrt{t}+t)(6+2(\sqrt{t}+1)+(\sqrt{t}+1)^2)} \\
&\quad \times \left(\frac{10}{\sqrt{t}+1+\sqrt{t}} + 11 + 2\frac{(\sqrt{t}+1)^4 + t(\sqrt{t}+1)^2 + t^2}{(\sqrt{t}+1)^2(\sqrt{t}+1) + t\sqrt{t}} + (\sqrt{t}+1)^2 + t\right) \\
&= \frac{6}{(5+t)(6+2\sqrt{t}+t)^2(8+2\sqrt{t}+t+2\sqrt{t}+1)} \\
&\quad \times \left(\frac{10}{2\sqrt{t}+1} + 11\right. \\
&\quad \left. + 2\frac{t^2 + 4t + 1 + 4t\sqrt{t} + 2t + 4\sqrt{t} + t^2 + 2t\sqrt{t} + t + t^2}{(t+2\sqrt{t}+1)(\sqrt{t}+1) + t\sqrt{t}} + 2t + 2\sqrt{t} + 1\right) \\
&= \frac{6}{(5+t)(6+2\sqrt{t}+t)^2(9+4\sqrt{t}+t)} \\
&\quad \times \left(\frac{10}{2\sqrt{t}+1} + 11 + 2\frac{3t^2 + 6t\sqrt{t} + 7t + 4\sqrt{t} + 1}{t\sqrt{t} + t + 2t + 2\sqrt{t} + \sqrt{t} + 1 + t\sqrt{t}} + 2t + 2\sqrt{t} + 1\right) \\
&= \frac{6}{(5+t)(6+2\sqrt{t}+t)^2(9+4\sqrt{t}+t)}
\end{aligned}
$$

$$\times\left(\frac{10}{2\sqrt{t}+1}+11+2\frac{3t^2+6t\sqrt{t}+7t+4\sqrt{t}+1}{2t\sqrt{t}+3t+3\sqrt{t}+1}+2t+2\sqrt{t}+1\right)$$
$$=\frac{6}{(5+t)(6+2\sqrt{t}+t)^2(9+4\sqrt{t}+t)(2\sqrt{t}+1)(2t\sqrt{t}+3t+3\sqrt{t}+1)}$$
$$\times\,(20t\sqrt{t}+30t+30\sqrt{t}+10+11(2\sqrt{t}+1)(2t\sqrt{t}+3t+3\sqrt{t}+1)$$
$$+\,2(3t^2+6t\sqrt{t}+7t+4\sqrt{t}+1)(2\sqrt{t}+1)$$
$$+\,(2\sqrt{t}+1)(2t\sqrt{t}+3t+3\sqrt{t}+1)(2t+2\sqrt{t}+1))$$
$$=\frac{6}{(5+t)(6+2\sqrt{t}+t)^2(9+4\sqrt{t}+t)(2\sqrt{t}+1)(2t\sqrt{t}+3t+3\sqrt{t}+1)}$$
$$\times(20\sqrt{t}+30t+30\sqrt{t}+10+44t^2+66t\sqrt{t}+66t+22\sqrt{t}+22t\sqrt{t}$$
$$+33t+33\sqrt{t}+11+12t^2\sqrt{t}+6t^2+24t^2+12t\sqrt{t}+28t\sqrt{t}$$
$$+14t+16t+8\sqrt{t}+4\sqrt{t}+2$$
$$+(2\sqrt{t}+1)(4t^2\sqrt{t}+4t^2+2t\sqrt{t}+6t^2+6t\sqrt{t}+3t+6t\sqrt{t}+6t+3\sqrt{t}$$
$$+2t+2\sqrt{t}+1))$$
$$=\frac{6}{(5+t)(6+2\sqrt{t}+t)^2(9+4\sqrt{t}+t)(2\sqrt{t}+1)(2t\sqrt{t}+3t+3\sqrt{t}+1)}$$
$$\times(12t^2\sqrt{t}+74t^2+128t\sqrt{t}+159t+97\sqrt{t}+23$$
$$+(2\sqrt{t}+1)(4t^2\sqrt{t}+10t^2+14t\sqrt{t}+11t+5\sqrt{t}+1))$$
$$=\frac{6}{(5+t)(6+2\sqrt{t}+t)^2(9+4\sqrt{t}+t)(2\sqrt{t}+1)(2t\sqrt{t}+3t+3\sqrt{t}+1)}$$
$$\times(12t^2\sqrt{t}+74t^2+128t\sqrt{t}+159t+97\sqrt{t}+23$$
$$+8t^3+20t^2\sqrt{t}+28t^2+22t\sqrt{t}+10t+2\sqrt{t}+4t^2\sqrt{t}+10t^2+14t\sqrt{t}$$
$$+11t+5\sqrt{t}+1)$$
$$=\frac{6}{(5+t)(6+2\sqrt{t}+t)^2(9+4\sqrt{t}+t)(2\sqrt{t}+1)(2t\sqrt{t}+3t+3\sqrt{t}+1)}$$
$$\times(8t^3+36t^2\sqrt{t}+112t^2+164t\sqrt{t}+180t+104\sqrt{t}+24),$$

$t \in \mathbb{T}^{\kappa^2}$.

Example 2.31. Let $\mathbb{T} = \{H_n : n \in \mathbb{N}_0\}$, where H_n, $n \in \mathbb{N}_0$, are the harmonic numbers, and let

$$f(t) = \frac{t}{t^2 - 4t - 12}, \qquad t \in \mathbb{T}.$$

We will find $f^{\Delta^2}(H_n)$, $n \in \mathbb{N}_0$. By Example 2.16 we have

$$f^{\Delta}(H_n) = -\frac{H_n H_{n+1} + 12}{(H_n^2 - 4H_n - 12)(H_{n+1}^2 - 4H_{n+1} - 12)}, \qquad n \in \mathbb{N}_0.$$

Hence, using that

$$\sigma(H_n) = H_{n+1},$$
$$\sigma(H_{n+1}) = H_{n+2}, \quad n \in \mathbb{N}_0,$$

we find

$$\begin{aligned}
f^{\Delta^2}(H_n) &= \frac{f^\Delta(\sigma(H_n)) - f^\Delta(H_n)}{\sigma(H_n) - H_n} \\
&= \frac{1}{H_{n+1} - H_n}\left(-\frac{\sigma(H_n)\sigma(H_{n+1}) + 12}{((\sigma(H_n))^2 - 4\sigma(H_n) - 12)((\sigma(H_{n+1}))^2 - 4\sigma(H_{n+1}) - 12)}\right. \\
&\quad \left. + \frac{H_nH_{n+1} + 12}{(H_n^2 - 4H_n - 12)(H_{n+1}^2 - 4H_{n+1} - 12)}\right) \\
&= \frac{1}{H_{n+1} - H_n}\left(-\frac{H_{n+1}H_{n+2} + 12}{(H_{n+1}^2 - 4H_{n+1} - 12)(H_{n+2}^2 - 4H_{n+2} - 12)}\right. \\
&\quad \left. + \frac{H_nH_{n+1} + 12}{(H_n^2 - 4H_n - 12)(H_{n+1}^2 - 4H_{n+1} - 12)}\right) \\
&= \frac{1}{(H_{n+1} - H_n)(H_{n+1}^2 - 4H_{n+1} - 12)} \\
&\quad \times \left(\frac{H_nH_{n+1} + 12}{H_n^2 - 4H_n - 12} - \frac{H_{n+1}H_{n+2} + 12}{H_{n+2}^2 - 4H_{n+2} - 12}\right) \\
&= \frac{1}{(H_{n+1} - H_n)(H_n^2 - 4H_n - 12)(H_{n+1}^2 - 4H_{n+1} - 12)(H_{n+2}^2 - 4H_{n+2} - 12)} \\
&\quad \times \left((H_nH_{n+1} + 12)(H_{n+2}^2 - 4H_{n+2} - 12) - (H_{n+1}H_{n+2} + 12)(H_n^2 - 4H_n - 12)\right) \\
&= \frac{1}{(H_{n+1} - H_n)(H_n^2 - 4H_n - 12)(H_{n+1}^2 - 4H_{n+1} - 12)(H_{n+2}^2 - 4H_{n+2} - 12)} \\
&\quad \times (H_nH_{n+1}H_{n+2}^2 - 4H_nH_{n+1}H_{n+2} - 12H_nH_{n+1} + 12H_{n+2}^2 - 48H_{n+2} - 144 \\
&\quad - H_n^2H_{n+1}H_{n+2} + 4H_nH_{n+1}H_{n+2} + 12H_nH_{n+2} - 12H_n^2 + 48H_n + 144) \\
&= \frac{1}{(H_{n+1} - H_n)(H_n^2 - 4H_n - 12)(H_{n+1}^2 - 4H_{n+1} - 12)(H_{n+2}^2 - 4H_{n+2} - 12)} \\
&\quad \times (H_nH_{n+1}H_{n+2}(H_{n+2} - H_n) + 12H_{n+1}(H_{n+2} - H_n) \\
&\quad + 12(H_{n+2}^2 - H_n^2) - 48(H_{n+2} - H_n)) \\
&= \frac{1}{(H_{n+1} - H_n)(H_n^2 - 4H_n - 12)(H_{n+1}^2 - 4H_{n+1} - 12)(H_{n+2}^2 - 4H_{n+2} - 12)} \\
&\quad \times (H_nH_{n+1}H_{n+2}(H_{n+2} - H_n) + 12H_{n+1}(H_{n+2} - H_n) \\
&\quad + 12(H_{n+2} - H_n)(H_{n+2} + H_n) - 48(H_{n+2} - H_n)) \\
&= \frac{H_{n+2} - H_n}{(H_{n+1} - H_n)(H_n^2 - 4H_n - 12)(H_{n+1}^2 - 4H_{n+1} - 12)(H_{n+2}^2 - 4H_{n+2} - 12)} \\
&\quad \times (H_nH_{n+1}H_{n+2} + 12H_{n+1} + 12H_{n+2} + 12H_n - 48), \quad n \in \mathbb{N}_0.
\end{aligned}$$

Example 2.32. Let $\mathbb{T} = P_{1,3}$, and let

$$f(t) = \frac{1-t}{1+t}, \quad t \in \mathbb{T}.$$

We will find $f^{\Delta^2}(t)$, $t \in \mathbb{T}^{\kappa^2}$. We will consider the following cases.

1. $t \in \mathbb{T}^{\kappa^2}$ is right-scattered. Then

$$\sigma(t) = t + 3.$$

By Example 2.17 we have

$$f^{\Delta}(t) = -\frac{2}{(1+t)(4+t)}.$$

Then

$$\begin{aligned} f^{\Delta^2}(t) &= \frac{f^{\Delta}(\sigma(t)) - f^{\Delta}(t)}{\sigma(t) - t} \\ &= \frac{1}{t+3-t}\left(-\frac{2}{(1+\sigma(t))(4+\sigma(t))} + \frac{2}{(1+t)(4+t)}\right) \\ &= \frac{2}{3}\left(-\frac{1}{(t+4)(t+7)} + \frac{1}{(t+1)(t+4)}\right) \\ &= \frac{2}{3(t+4)}\left(\frac{1}{t+1} - \frac{1}{t+7}\right) \\ &= \frac{2}{3(t+4)}\left(\frac{t+7-t-1}{(t+1)(t+7)}\right) \\ &= \frac{4}{(t+1)(t+4)(t+7)}. \end{aligned}$$

2. $t \in \mathbb{T}^{\kappa^2}$ is right-dense. By Example 2.17 we have

$$f^{\Delta}(t) = -\frac{2}{(1+t)^2}.$$

Then

$$\begin{aligned} f^{\Delta^2}(t) &= \lim_{s \to t} \frac{f^{\Delta}(t) - f^{\Delta}(s)}{t-s} \\ &= \lim_{s \to t} \frac{-\frac{2}{(1+t)^2} + \frac{2}{(1+s)^2}}{t-s} \\ &= 2\lim_{s \to t} \frac{(1+t)^2 - (1+s)^2}{(t-s)(1+t)^2(1+s)^2} \\ &= 2\lim_{s \to t} \frac{(1+t-1-s)(1+t+1+s)}{(t-s)(1+s)^2(1+t)^2} \end{aligned}$$

$$= 2\lim_{s\to t}\frac{t+s+2}{(1+s)^2(1+t)^2}$$
$$= \frac{4(1+t)}{(1+t)^4}$$
$$= \frac{4}{(1+t)^3}.$$

Consequently,

$$f^{\Delta^2}(t) = \begin{cases} \frac{4}{(t+1)(t+4)(t+7)} & \text{if } t \text{ is right-scattered,} \\ \frac{4}{(t+1)^3} & \text{if } t \text{ is right-dense.} \end{cases}$$

Example 2.33. Let $\mathbb{T} = C$, where C is the Cantor set, and let

$$f(t) = \frac{1-t^2}{1+t^2}, \quad t \in \mathbb{T}.$$

We will find $f^{\Delta^2}(t)$, $t \in \mathbb{T}^{\kappa^2}$. For this aim, we will consider the following cases.

1. $t \in C_1$,

$$t = \sum_{k=1}^{m} \frac{a_k}{3^k} + \frac{1}{3^{m+1}},$$

and

$$a = \frac{1}{3^{m+1}}.$$

Then t is right-scattered, and

$$\sigma(t) = t + a.$$

By Example 2.18 we find

$$f^{\Delta}(t) = \frac{-4t-2a}{(1+t^2)(1+a^2+2at+t^2)}$$
$$= \frac{-4t-2a}{(1+t^2)(1+(t+a)^2)}.$$

Hence

$$f^{\Delta^2}(t) = \frac{f^{\Delta}(\sigma(t)) - f^{\Delta}(t)}{\sigma(t)-t}$$
$$= \frac{1}{a}\left(\frac{-4\sigma(t)-2a}{(1+(\sigma(t))^2)(1+(\sigma(t)+a)^2)} + \frac{4t+2a}{(1+t^2)(1+(t+a)^2)}\right)$$

$$
\begin{aligned}
&= \frac{1}{a}\left(\frac{-4t-6a}{(1+(t+a)^2)(1+(t+2a)^2)} + \frac{4t+2a}{(1+t^2)(1+(t+a)^2)}\right)\\
&= \frac{2}{a(1+(t+a)^2)}\left(\frac{2t+a}{1+t^2} - \frac{2t+3a}{(1+(t+2a)^2)}\right)\\
&= \frac{2}{a(1+t^2)(1+a^2+2at+t^2)(1+4a^2+4at+t^2)}\\
&\quad \times ((2t+a)(1+(t+a)^2) - (1+t^2)(2t+3a))\\
&= \frac{2}{a(1+t^2)(1+a^2+2at+t^2)(1+4a^2+4at+t^2)}\\
&\quad \times (2t+a+(2t+a)(t^2+4at+4a^2) - 2t - 3a - 2t^3 - 3t^2a)\\
&= \frac{2}{a(1+t^2)(1+a^2+2at+t^2)(1+4a^2+4at+t^2)}\\
&\quad \times (2t+a+2t^3+8at^2+8a^2t+at^2+4a^2t+4a^3-2t-3a-2t^3-3t^2a)\\
&= \frac{2}{a(1+t^2)(1+a^2+2at+t^2)(1+4a^2+4at+t^2)}(6at^2+12a^2t+4a^3-2a)\\
&= \frac{4(3t^2+6at+2a^2-a)}{a(1+t^2)(1+a^2+2at+t^2)(1+4a^2+4at+t^2)}\\
&= \frac{4(3t^2+\frac{6}{3^{m+1}}t+\frac{2}{3^{2m+2}}-\frac{1}{3^{m+1}})}{(1+t^2)(1+\frac{1}{3^{2m+2}}+\frac{2}{3^{m+1}}t+t^2)(1+\frac{4}{3^{2m+2}}+\frac{4}{3^{m+1}}t+t^2)}.
\end{aligned}
$$

2. $t \in \mathbb{T}\backslash C_1$. Then t is right-dense. By Example 2.18 we have

$$
f^{\Delta}(t) = -\frac{4t}{(1+t^2)^2}.
$$

Hence

$$
\begin{aligned}
f^{\Delta^2}(t) &= \lim_{s\to t}\frac{f^{\Delta}(t)-f^{\Delta}(s)}{t-s}\\
&= \lim_{s\to t}\frac{-\frac{4t}{(1+t^2)^2}+\frac{4s}{(1+s^2)^2}}{t-s}\\
&= 4\lim_{s\to t}\frac{s(1+t^2)^2-t(1+s^2)^2}{(t-s)(1+s^2)^2(1+t^2)^2}\\
&= 4\lim_{s\to t}\frac{s+2st^2+st^4-t-2ts^2-ts^4}{(t-s)(1+s^2)^2(1+t^2)^2}\\
&= 4\lim_{s\to t}\frac{-(t-s)+2st(t-s)+st(t^3-s^3)}{(t-s)(1+s^2)^2(1+t^2)^2}\\
&= 4\lim_{s\to t}\frac{-(t-s)+2st(t-s)+st(t-s)(t^2st+s^2)}{(t-s)(1+s^2)^2(1+t^2)^2}\\
&= 4\lim_{s\to t}\frac{-1+2st+st(t^2+st+s^2)}{(1+s^2)^2(1+t^2)^2}
\end{aligned}
$$

$$
\begin{aligned}
&= 4\frac{-1 + 2t^2 + 3t^4}{(1 + t^2)^4} \\
&= \frac{4(1 + t^2)(3t^2 - 1)}{(1 + t^2)^4} \\
&= \frac{4(3t^2 - 1)}{(1 + t^2)^3}.
\end{aligned}
$$

Consequently,

$$
f^{\Delta^2}(t) = \begin{cases} \frac{4(3t^2 + \frac{6}{3^{m+1}}t + \frac{2}{3^{2m+2}} - \frac{1}{3^{m+1}})}{(1+t^2)(1+\frac{1}{3^{2m+2}} + \frac{2}{3^{m+1}}t + t^2)(1+\frac{4}{3^{2m+2}} + \frac{4}{3^{m+1}}t + t^2)} & \text{if } t \in C_1,\ t = \sum_{k=1}^{m} \frac{a_k}{3^k} + \frac{1}{3^{m+1}}, \\ \frac{4(3t^2-1)}{(1+t^2)^3} & \text{if } t \in \mathbb{T} \backslash C_1. \end{cases}
$$

Example 2.34. Let

$$
\mathbb{T} = \left\{ t_n = \sum_{k=0}^{n-1} a_k : n \in \mathbb{N},\ a_k > 0, \quad k \in \mathbb{N}_0 \right\},
$$

and let

$$
f(t) = \frac{1 + 2t}{3 + 4t}, \quad t \in \mathbb{T}.
$$

We will find $f^{\Delta^2}(t_n)$, $n \in \mathbb{N}$. Note that all points of $\mathbb{T}$ are right-scattered. By Example 2.19 we have

$$
f^{\Delta}(t_n) = \frac{2}{(3 + 4t_n)(3 + 4t_{n+1})}, \quad n \in \mathbb{N}.
$$

Hence, using that

$$
\begin{aligned}
\sigma(t_n) &= t_{n+1}, \\
\sigma(t_{n+1}) &= t_{n+2}, \quad n \in \mathbb{N},
\end{aligned}
$$

we get

$$
\begin{aligned}
f^{\Delta^2}(t_n) &= \frac{f^{\Delta}(\sigma(t_n)) - f^{\Delta}(t_n)}{\sigma(t_n) - t_n} \\
&= \frac{1}{t_{n+1} - t_n}\left(\frac{2}{(3 + 4\sigma(t_n))(3 + 4\sigma(t_{n+1}))} - \frac{2}{(3 + 4t_n)(3 + 4t_{n+1})} \right) \\
&= \frac{1}{t_{n+1} - t_n}\left(\frac{2}{(3 + 4t_{n+1})(3 + 4t_{n+2})} - \frac{2}{(3 + 4t_n)(3 + 4t_{n+1})} \right)
\end{aligned}
$$

$$= \frac{2(3+4t_n-3-4t_{n+2})}{(t_{n+1}-t_n)(3+4t_n)(3+4t_{n+1})(3+4t_{n+2})}$$
$$= \frac{8(t_n-t_{n+2})}{(t_{n+1}-t_n)(3+4t_n)(3+4t_{n+1})(3+4t_{n+2})}, \quad n \in \mathbb{N}_0.$$

Example 2.35. Let $\mathbb{T} = \{-\frac{1}{n} : n \in \mathbb{N}\} \cup \mathbb{N}_0$, and let

$$f(t) = \frac{1+t^2}{1+7t}, \quad t \in \mathbb{T}.$$

We will find $f^{\Delta^2}(t)$, $t \in \mathbb{T}$. We have the following cases.

1. $t \in \{-\frac{1}{n} : n \in \mathbb{N}\}$. Then

$$\sigma(t) = -\frac{t}{t-1}.$$

By Example 2.20 we have

$$f^{\Delta}(t) = \frac{6t^2+9t-7}{(1+7t)(1+6t)}.$$

Then

$$\begin{aligned} f^{\Delta}(\sigma(t)) &= \frac{6(\sigma(t))^2+9\sigma(t)-7}{(1+7\sigma(t))(1+6\sigma(t))} \\ &= \frac{6\frac{t^2}{(t-1)^2} - \frac{9t}{t-1} - 7}{(1-\frac{7t}{t-1})(1-\frac{6t}{t-1})} \\ &= \frac{6t^2-9t(t-1)-7(t-1)^2}{(t-1-7t)(t-1-6t)} \\ &= \frac{6t^2-9t^2+9t-7t^2+14t-7}{(1+6t)(1+5t)} \\ &= \frac{-10t^2+23t-7}{(1+6t)(1+5t)}, \end{aligned}$$

and

$$\begin{aligned} f^{\Delta^2}(t) &= \frac{f^{\Delta}(\sigma(t))-f^{\Delta}(t)}{\sigma(t)-t} \\ &= \frac{1}{-\frac{t}{t-1}-t}\left(\frac{-10t^2+23t-7}{(1+6t)(1+5t)} - \frac{6t^2+9t-7}{(1+6t)(1+7t)}\right) \\ &= -\frac{(t-1)((-10t^2+23t-7)(1+7t)-(6t^2+9t-7)(1+5t))}{t^2(1+5t)(1+6t)(1+7t)} \\ &= -\frac{(t-1)(-10t^2-70t^3+23t+161t^2-7-49t-6t^2-30t^3-9t-45t^2+7+35t)}{t^2(1+5t)(1+6t)(1+7t)} \end{aligned}$$

$$= -\frac{(t-1)(-100t^3+100t^2)}{t^2(1+5t)(1+6t)(1+7t)}$$
$$= \frac{100(t-1)^2}{(1+5t)(1+6t)(1+7t)}.$$

2. $t \in \mathbb{N}_0$. Then

$$\sigma(t) = t+1.$$

By Example 2.20 we have

$$f^{\Delta}(t) = \frac{7t^2+9t-6}{(1+7t)(8+7t)}.$$

Hence

$$\begin{aligned} f^{\Delta}(\sigma(t)) &= \frac{7(\sigma(t))^2+9\sigma(t)-6}{(1+7\sigma(t))(8+7\sigma(t))} \\ &= \frac{7(t+1)^2+9(t+1)-6}{(1+7t+7)(8+7t+7)} \\ &= \frac{7t^2+14t+7+9t+9-6}{(8+7t)(15+7t)} \\ &= \frac{7t^2+23t+10}{(8+7t)(15+7t)}, \end{aligned}$$

and

$$\begin{aligned} f^{\Delta^2}(t) &= \frac{f^{\Delta}(\sigma(t))-f^{\Delta}(t)}{\sigma(t)-t} \\ &= \frac{7t^2+23t+10}{(8+7t)(15+7t)} - \frac{7t^2+9t-6}{(1+7t)(8+7t)} \\ &= \frac{(7t^2+23t+10)(1+7t)-(7t^2+9t-6)(15+7t)}{(1+7t)(8+7t)(15+7t)} \\ &= \frac{7t^2+23t+10+49t^3+161t^2+70t-105t^2-49t^3-135t-63t^2+90+42t}{(1+7t)(8+7t)(15+7t)} \\ &= \frac{100}{(1+7t)(8+7t)(15+7t)}. \end{aligned}$$

Consequently,

$$f^{\Delta^2}(t) = \begin{cases} \frac{100(t-1)^2}{(1+5t)(1+6t)(1+7t)} & \text{if } t \in \{-\frac{1}{n} : n \in \mathbb{N}\}, \\ \frac{100}{(1+7t)(8+7t)(15+7t)} & \text{if } t \in \mathbb{N}_0. \end{cases}$$

Example 2.36. Let $\mathbb{T} = \{(\frac{1}{2})^{2^n}; n \in \mathbb{N}_0\} \cup \{0, 1\}$, and let

$$f(t) = \frac{1+7t}{4+5t}, \quad t \in \mathbb{T}.$$

We will find $f^{\Delta^2}(t)$, $t \in \mathbb{T}^{\kappa^2}$. Here

$$\mathbb{T}^{\kappa^2} = \left\{\left(\frac{1}{2}\right)^{2^n} : n \in \mathbb{N}\right\} \cup \{0\}.$$

We have the following cases.

1. $t \in \{(\frac{1}{2})^{2^n} : n \in \mathbb{N}\}$. Then

$$\sigma(t) = \sqrt{t}.$$

By Example 2.21 we have

$$f^{\Delta}(t) = \frac{23}{(4+5t)(4+5\sqrt{t})}.$$

Then

$$\begin{aligned} f^{\Delta}(\sigma(t)) &= \frac{23}{(4+5\sigma(t))(4+5\sqrt{\sigma(t)})} \\ &= \frac{23}{(4+5\sqrt{t})(4+5\sqrt[4]{t})}, \end{aligned}$$

and

$$\begin{aligned} f^{\Delta^2}(t) &= \frac{f^{\Delta}(\sigma(t)) - f^{\Delta}(t)}{\sigma(t) - t} \\ &= \frac{1}{\sqrt{t} - t}\left(\frac{23}{(4+5\sqrt{t})(4+5\sqrt[4]{t})} - \frac{23}{(4+5t)(4+5\sqrt{t})}\right) \\ &= \frac{23(4+5t-4-5\sqrt[4]{t})}{\sqrt{t}(1-\sqrt{t})(4+5t)(4+5\sqrt{t})(4+5\sqrt[4]{t})} \\ &= \frac{115(t-\sqrt[4]{t})}{\sqrt{t}(1-\sqrt{t})(4+5t)(4+5\sqrt{t})(4+5\sqrt[4]{t})}. \end{aligned}$$

2. $t = 0$. Then

$$\sigma(0) = 0.$$

By Example 2.21 we have

$$f^{\Delta}(0) = \frac{23}{16}.$$

Hence

$$
\begin{aligned}
f^{\Delta^2}(0) &= \lim_{s\to 0}\frac{f^{\Delta}(s)-f^{\Delta}(0)}{s}\\
&= \lim_{s\to 0}\frac{\frac{23}{(4+5s)(4+5\sqrt{s})}-\frac{23}{16}}{s}\\
&= \lim_{s\to 0}\frac{23(16-(4+5s)(4+5\sqrt{s}))}{s(4+5s)(4+5\sqrt{s})}\\
&= \lim_{s\to 0}\frac{23(16-16-20\sqrt{s}-20s-25s^{\frac{3}{2}})}{s(4+5s)(4+5\sqrt{s})}\\
&= -\lim_{s\to 0}\frac{115(4\sqrt{s}+4s+5s^{\frac{3}{2}})}{s(4+5s)(4+5\sqrt{s})}\\
&= -\lim_{s\to 0}\frac{115(4+4\sqrt{s}+5s)}{\sqrt{s}(4+5s)(4+5\sqrt{s})}\\
&= -\infty.
\end{aligned}
$$

Thus $f^{\Delta^2}(0)$ does not exist.

Consequently,

$$
f^{\Delta^2}(t) = \begin{cases} \frac{115(t-\sqrt[4]{t})}{\sqrt{t}(1-\sqrt{t})(4+5t)(4+5\sqrt{t})(4+5\sqrt[4]{t})} & \text{if } t \in \{(\frac{1}{2})^{2^n} : n \in \mathbb{N}\},\\ \text{does not exist} & \text{if } t = 0. \end{cases}
$$

Example 2.37. Let $U = \{\frac{1}{2^n} : n \in \mathbb{N}\}$, let

$$
\mathbb{T} = U \cup (1-U) \cup (1+U) \cup (2-U) \cup (2+U) \cup \{0,1,2\},
$$

and let

$$
f(t) = (2+t^2)(1-t^2), \quad t \in \mathbb{T}.
$$

We will find $f^{\Delta^2}(t)$, $t \in \mathbb{T}^{\kappa^2}$. Here $\mathbb{T}^{\kappa^2} = \mathbb{T}\backslash\{\frac{5}{2}, \frac{9}{4}\}$. We have the following cases.

1. $t \in U\backslash\{\frac{1}{2}\}$. Then

$$
\sigma(t) = 2t.
$$

By Example 2.22 we have

$$
f^{\Delta}(t) = -15t^3 - 3t.
$$

Then

$$
f^{\Delta}(\sigma(t)) = -15(\sigma(t))^3 - 3\sigma(t)
$$

$$= -15(8t^3) - 3(2t)$$
$$= -120t^3 - 6t,$$

and

$$\begin{aligned} f^{\Delta^2}(t) &= \frac{f^{\Delta}(\sigma(t)) - f^{\Delta}(t)}{\sigma(t) - t} \\ &= \frac{-120t^3 - 6t + 15t^3 + 3t}{2t - t} \\ &= \frac{-105t^3 - 3t}{t} \\ &= -105t^2 - 3. \end{aligned}$$

2. $t = \frac{1}{2}$. Then

$$\sigma\left(\frac{1}{2}\right) = \frac{3}{4}.$$

By Example 2.22 we have

$$f^{\Delta}\left(\frac{1}{2}\right) = -\frac{145}{64}$$

and

$$\begin{aligned} f^{\Delta}\left(\sigma\left(\frac{1}{2}\right)\right) &= f^{\Delta}\left(\frac{3}{4}\right) \\ &= -\frac{15(\frac{3}{4})^3 + 11(\frac{3}{4})^2 + 17(\frac{3}{4}) + 5}{8} \\ &= -\frac{15(\frac{27}{64}) + 11(\frac{9}{16}) + \frac{51}{4} + 5}{8} \\ &= -\frac{15 \cdot 27 + 11 \cdot 36 + 51 \cdot 16 + 5 \cdot 64}{512} \\ &= -\frac{405 + 396 + 816 + 320}{512} \\ &= -\frac{1937}{512}. \end{aligned}$$

Hence

$$\begin{aligned} f^{\Delta^2}\left(\frac{1}{2}\right) &= \frac{f^{\Delta}(\sigma(\frac{1}{2})) - f^{\Delta}(\frac{1}{2})}{\sigma(\frac{1}{2}) - \frac{1}{2}} \\ &= \frac{f^{\Delta}(\frac{3}{4}) - f^{\Delta}(\frac{1}{2})}{\frac{3}{4} - \frac{1}{2}} \end{aligned}$$

$$= \frac{-\frac{1937}{512} + \frac{145}{64}}{\frac{1}{4}}$$
$$= \frac{-1937 + 1160}{128}$$
$$= -\frac{777}{128}.$$

3. $t \in (1 - U)\backslash\{\frac{1}{2}\}$. Then

$$\sigma(t) = \frac{1+t}{2}.$$

By Example 2.22 we have

$$f^{\Delta}(t) = -\frac{15t^3 + 11t^2 + 17t + 5}{8}.$$

Then

$$\begin{aligned}
f^{\Delta}(\sigma(t)) &= -\frac{15(\frac{1+t}{2})^3 + 11(\frac{1+t}{2})^2 + 17(\frac{1+t}{2}) + 5}{8} \\
&= -\frac{15(1+t)^3 + 22(1+t)^2 + 68(1+t) + 40}{64} \\
&= -\frac{15 + 45t^2 + 45t + 15t^3 + 22 + 44t + 22t^2 + 68 + 68t + 40}{64} \\
&= -\frac{15t^3 + 67t^2 + 157t + 77}{64},
\end{aligned}$$

and

$$\begin{aligned}
f^{\Delta^2}(t) &= \frac{f^{\Delta}(\sigma(t)) - f^{\Delta}(t)}{\sigma(t) - t} \\
&= \frac{1}{\frac{1+t}{2} - t}\left(-\frac{15t^3 + 67t^2 + 157t + 77}{64} + \frac{15t^3 + 11t^2 + 17t + 5}{8}\right) \\
&= \frac{2}{1-t}\left(\frac{-15t^3 - 67t^2 - 157t - 77 + 120t^3 + 88t^2 + 136t + 40}{64}\right) \\
&= \frac{105t^3 + 21t^2 - 21t - 37}{32(1-t)}.
\end{aligned}$$

4. $t \in (1 + U)\backslash\{\frac{3}{2}\}$. Then

$$\sigma(t) = 2t - 1.$$

By Example 2.22 we have

$$f^{\Delta}(t) = -15t^3 + 17t^2 - 10t + 2.$$

Then

$$\begin{aligned} f^{\Delta}(\sigma(t)) &= -15(\sigma(t))^3 + 17(\sigma(t))^2 - 10\sigma(t) + 2 \\ &= -15(2t-1)^3 + 17(2t-1)^2 - 10(2t-1) + 2 \\ &= -15(8t^3 - 12t^2 + 6t - 1) + 17(4t^2 - 4t + 1) - 20t + 10 + 2 \\ &= -120t^3 + 180t^2 - 90t + 15 + 68t^2 - 68t + 17 - 20t + 12 \\ &= -120t^3 + 248t^2 - 178t + 44, \end{aligned}$$

and

$$\begin{aligned} f^{\Delta^2}(t) &= \frac{f^{\Delta}(\sigma(t)) - f^{\Delta}(t)}{\sigma(t) - t} \\ &= \frac{-120t^3 + 248t^2 - 178t + 44 + 15t^3 - 17t^2 + 10t - 2}{2t - 1 - t} \\ &= \frac{-105t^3 + 231t^2 - 168t + 42}{t - 1}. \end{aligned}$$

5. $t = \frac{3}{2}$. Then

$$\sigma\left(\frac{3}{2}\right) = \frac{7}{4}.$$

By Example 2.22 we have

$$f^{\Delta}\left(\frac{3}{2}\right) = -\frac{1215}{64}$$

and

$$\begin{aligned} f^{\Delta}\left(\sigma\left(\frac{3}{2}\right)\right) &= f^{\Delta}\left(\frac{7}{4}\right) \\ &= -\frac{15(\frac{7}{4})^3 + 22(\frac{7}{4})^2 + 32(\frac{7}{4}) + 8}{8} \\ &= -\frac{15(\frac{343}{64}) + 22(\frac{49}{16}) + 32(\frac{7}{4}) + 8}{8} \\ &= -\frac{15 \cdot 343 + 88 \cdot 49 + 32 \cdot 112 + 8 \cdot 64}{512} \\ &= -\frac{5145 + 4312 + 3584 + 512}{512} \\ &= -\frac{13553}{512}. \end{aligned}$$

Hence

$$f^{\Delta^2}\left(\frac{3}{2}\right) = \frac{f^{\Delta}(\sigma(\frac{3}{2})) - f^{\Delta}(\frac{3}{2})}{\sigma(\frac{3}{2}) - \frac{3}{2}}$$
$$= \frac{-\frac{13553}{512} + \frac{1215}{64}}{\frac{7}{4} - \frac{3}{2}}$$
$$= \frac{-13553 + 9720}{128}$$
$$= -\frac{3833}{128}.$$

6. $t \in (2 - U)\backslash\{\frac{3}{2}\}$. Then

$$\sigma(t) = \frac{t+2}{2}.$$

By Example 2.22 we have

$$f^{\Delta}(t) = -\frac{15t^3 + 22t^2 + 32t + 8}{8}.$$

Then

$$f^{\Delta^2}(\sigma(t)) = -\frac{15(\sigma(t))^3 + 22(\sigma(t))^2 + 32\sigma(t) + 8}{8}$$
$$= -\frac{15(\frac{t+2}{2})^3 + 22(\frac{t+2}{2})^2 + 32(\frac{t+2}{2}) + 8}{8}$$
$$= -\frac{15(t+2)^3 + 44(t+2)^2 + 128(t+2) + 64}{64}$$
$$= -\frac{15t^3 + 90t^2 + 180t + 120 + 44t^2 + 176t + 176 + 128t + 256 + 64}{64}$$
$$= -\frac{15t^3 + 134t^2 + 484t + 616}{64},$$

and

$$f^{\Delta^2}(t) = \frac{f^{\Delta}(\sigma(t)) - f^{\Delta}(t)}{\sigma(t) - t}$$
$$= \frac{1}{\frac{t+2}{2} - t}\left(-\frac{15t^3 + 134t^2 + 484t + 616}{64} + \frac{15t^3 + 22t^2 + 32t + 8}{8}\right)$$
$$= \frac{-15t^3 - 134t^2 - 484t - 616 + 120t^3 + 176t^2 + 256t + 64}{32(2-t)}$$
$$= \frac{105t^3 + 42t^2 - 228t - 552}{32(2-t)}.$$

7. $t \in (2+U)\backslash\{\frac{5}{2}\}$. Then

$$\sigma(t) = 2(t-1).$$

By Example 2.22 we have

$$f^{\Delta}(t) = -15t^3 + 34t^2 - 31t + 10.$$

Then

$$\begin{aligned} f^{\Delta}(\sigma(t)) &= -15(\sigma(t))^3 + 34(\sigma(t))^2 - 31\sigma(t) + 10 \\ &= -120(t-1)^3 + 136(t-1)^2 - 62(t-1) + 10 \\ &= -120t^3 + 360t^2 - 360t + 120 + 136t^2 - 272t + 136 - 62t + 62 + 10 \\ &= -120t^3 + 496t^2 - 694t + 192, \end{aligned}$$

and

$$\begin{aligned} f^{\Delta^2}(t) &= \frac{f^{\Delta}(\sigma(t)) - f^{\Delta}(t)}{\sigma(t) - t} \\ &= \frac{-120t^3 + 496t^2 - 694t + 192 + 15t^3 - 34t^2 + 31t - 10}{t-2} \\ &= \frac{-105t^3 + 462t^2 - 663t + 182}{t-2}. \end{aligned}$$

8. $t = 0$. Then

$$\sigma(0) = 0.$$

By Example 2.22 we have

$$f^{\Delta}(0) = 0.$$

Hence

$$\begin{aligned} f^{\Delta^2}(0) &= \lim_{s\to 0} \frac{f^{\Delta}(s) - f^{\Delta}(0)}{s} \\ &= \lim_{s\to 0} \frac{-15s^3 - 3s}{s} \\ &= \lim_{s\to 0}(-15s^2 - 3) \\ &= -3. \end{aligned}$$

9. $t = 1$. Then

$$\sigma(1) = 1.$$

By Example 2.22 we have

$$f^{\Delta}(1) = -6.$$

Hence

$$\begin{aligned}\lim_{s\to 1, s>1} \frac{f^{\Delta}(s) - f^{\Delta}(1)}{s-1} &= \lim_{s\to 1, s>1} \frac{-15s^3 + 17s^2 - 10s + 2 + 6}{s-1}\\ &= \lim_{s\to 1, s>1} \frac{-15s^3 + 17s^2 - 10s + 8}{s-1}\\ &= \lim_{s\to 1, s>1} \frac{(s-1)(-15s^2 + 2s - 8)}{s-1}\\ &= \lim_{s\to 1, s>1} \left(-15s^2 + 2s - 8\right)\\ &= -21,\end{aligned}$$

and

$$\begin{aligned}\lim_{s\to 1, s<1} \frac{f^{\Delta}(s) - f^{\Delta}(1)}{s-1} &= \lim_{s\to 1, s<1} \frac{-\frac{15s^3+11s^2+17s+5}{8} + 6}{s-1}\\ &= \lim_{s\to 1, s<1} \frac{-15s^3 - 11s^2 - 17s - 5 + 48}{8(s-1)}\\ &= \lim_{s\to 1, s<1} \frac{-15s^3 - 11s^2 - 17s + 43}{8(s-1)}\\ &= \lim_{s\to 1, s<1} \frac{(s-1)(-15s^2 - 26s - 43)}{8(s-1)}\\ &= \lim_{s\to 1, s<1} \frac{-15s^2 - 26s - 43}{8}\\ &= -\frac{21}{2}.\end{aligned}$$

Since

$$\lim_{s\to 1, s<1} \frac{f^{\Delta}(s) - f^{\Delta}(1)}{s-1} \neq \lim_{s\to 1, s>1} \frac{f^{\Delta}(s) - f^{\Delta}(1)}{s-1},$$

we conclude that $f^{\Delta^2}(1)$ does not exist.

10. $t = 2$. Then

$$\sigma(2) = 2.$$

By Example 2.22 we have

$$f^{\Delta}(2) = -36.$$

Then

$$\begin{aligned}
\lim_{s\to 2, s<2} \frac{f^{\Delta}(s) - f^{\Delta}(1)}{s-2} &= \lim_{s\to 2, s<2} \frac{1}{s-2}\left(-\frac{15s^3 + 22s^2 + 32s + 8}{8} + 36\right) \\
&= \lim_{s\to 2, s<2} \frac{1}{s-2}\left(\frac{-15s^3 - 22s^2 - 32s - 8 + 288}{8}\right) \\
&= \lim_{s\to 2, s<2} \frac{-15s^3 - 22s^2 - 32s + 280}{8(s-2)} \\
&= -\infty.
\end{aligned}$$

Therefore $f^{\Delta^2}(2)$ does not exist.

Consequently,

$$f^{\Delta^2}(t) = \begin{cases}
-105t^2 - 3 & \text{if } t \in U\backslash\{\frac{1}{2}\}, \\
-\frac{777}{128} & \text{if } t = \frac{1}{2}, \\
\frac{105t^3+21t^2-21t-37}{32(1-t)} & \text{if } t \in (1-U)\backslash\{\frac{1}{2}\}, \\
\frac{-105t^3+231t^2-168t+42}{t-1} & \text{if } t \in (1+U)\backslash\{\frac{3}{2}\}, \\
-\frac{3833}{128} & \text{if } t = \frac{3}{2}, \\
\frac{105t^3+42t^2-229t-552}{32(2-t)} & \text{if } t \in (2-U)\backslash\{\frac{3}{2}\}, \\
\frac{-105t^3+461t^2-663t+78}{t-2} & \text{if } t \in (2+U)\backslash\{\frac{5}{2}\}, \\
-3 & \text{if } t = 0, \\
\text{does not exist} & \text{if } t = 1, \\
\text{does not exist} & \text{if } t = 2.
\end{cases}$$

Example 2.38 (Leibniz formula). Let $S_k^{(n)}$ be the set consisting of all possible strings of length n, containing σ and Δ exactly k and $n-k$ times, respectively. Suppose also that

$$f^{\Lambda} \quad \text{exists for all } \Lambda \in S_k^{(n)}.$$

We will prove that

$$(fg)^{\Delta^n} = \sum_{k=0}^{n}\left(\sum_{\Lambda\in S_k^{(n)}} f^{\Lambda}\right) g^{\Delta^k}. \tag{2.1}$$

For this aim, we will use the classical induction principle.

1. $n = 1$. Then

$$S_0^{(1)} = \Delta, \quad S_1^{(1)} = \sigma.$$

Hence

$$\sum_{k=0}^{1}\Big(\sum_{\Lambda\in S_k^{(1)}} f^{\Lambda}\Big)g^{\Delta^k} = \sum_{\Lambda\in S_0^{(1)}} f^{\Lambda}g + \sum_{\Lambda\in S_1^{(1)}} f^{\Lambda}g^{\Delta}$$
$$= f^{\Delta}g + f^{\sigma}g^{\Delta}$$
$$= (fg)^{\Delta},$$

i. e., the statement holds for $n = 1$.

2. Assume that the statement is valid for some $n \in \mathbb{N}$.
3. We will prove that

$$(fg)^{\Delta^{n+1}} = \Big(\sum_{\Lambda\in S_k^{(n+1)}} f^{\Lambda}\Big)g^{\Delta^k}.$$

Using (2.1), we have

$$\begin{aligned}
(fg)^{\Delta^{n+1}} &= ((fg)^{\Delta^n})^{\Delta}\\
&= \Big(\sum_{k=0}^{n}\Big(\sum_{\Lambda\in S_k^{(n)}} f^{\Lambda}\Big)g^{\Delta^k}\Big)^{\Delta}\\
&= \sum_{k=0}^{n}\Big(\Big(\sum_{\Lambda\in S_k^{(n)}} f^{\Lambda}\Big)g^{\Delta^k}\Big)^{\Delta}\\
&= \sum_{k=0}^{n}\Big(\Big(\sum_{\Lambda\in S_k^{(n)}} f^{\Lambda\sigma}\Big)g^{\Delta^{k+1}} + \Big(\sum_{\Lambda\in S_k^{(n)}} f^{\Lambda\Delta}\Big)g^{\Delta^k}\Big)\\
&= \sum_{k=0}^{n}\Big(\sum_{\Lambda\in S_k^{(n)}} f^{\Lambda\sigma}\Big)g^{\Delta^{k+1}} + \sum_{k=0}^{n}\Big(\sum_{\Lambda\in S_k^{(n)}} f^{\Lambda\Delta}\Big)g^{\Delta^k}\\
&= \sum_{k=1}^{n+1}\Big(\sum_{\Lambda\in S_{k-1}^{(n)}} f^{\Lambda\sigma}\Big)g^{\Delta^k} + \sum_{k=0}^{n}\Big(\sum_{\Lambda\in S_k^{(n)}} f^{\Lambda\Delta}\Big)g^{\Delta^k}\\
&= \sum_{k=1}^{n}\Big(\sum_{\Lambda\in S_{k-1}^{(n)}} f^{\Lambda\sigma}\Big)g^{\Delta^k} + \Big(\sum_{\Lambda\in S_n^{(n)}} f^{\Lambda\sigma}\Big)g^{\Delta^{n+1}}\\
&\quad + \sum_{k=1}^{n}\Big(\sum_{\Lambda\in S_k^{(n)}} f^{\Lambda\Delta}\Big)g^{\Delta^k} + \Big(\sum_{\Lambda\in S_0^{(n)}} f^{\Lambda\Delta}\Big)g\\
&= \sum_{k=1}^{n}\Big(\sum_{\Lambda\in S_{k-1}^{(n)}} f^{\Lambda\sigma} + \sum_{\Lambda\in S_k^{(n)}} f^{\Lambda\Delta}\Big)g^{\Delta^k}\\
&\quad + \Big(\sum_{\Lambda\in S_n^{(n)}} f^{\Lambda\sigma}\Big)g^{\Delta^{n+1}} + \Big(\sum_{\Lambda\in S_0^{(n)}} f^{\Lambda\Delta}\Big)g
\end{aligned}$$

$$= \sum_{k=1}^{n} \Big(\sum_{\Lambda \in S_k^{(n+1)}} f^{\Lambda} \Big) g^{\Delta^k} + \Big(\sum_{\Lambda \in S_{n+1}^{(n+1)}} f^{\Lambda} \Big) g^{\Delta^{n+1}} + \Big(\sum_{\Lambda \in S_0^{(n+1)}} f^{\Lambda} \Big) g$$

$$= \sum_{k=0}^{n+1} \Big(\sum_{\Lambda \in S_k^{(n+1)}} f^{\Lambda} \Big) g^{\Delta^k}.$$

Example 2.39. Let f and μ be differentiable in $\mathbb{T}^\kappa$ and all points of $\mathbb{T}^\kappa$ be isolated. Then

$$f^{\Delta\sigma} = \frac{f^{\sigma^2} - f^{\sigma}}{\mu^{\sigma}} \quad \text{and} \quad f^{\sigma\Delta} = \frac{f^{\sigma^2} - f^{\sigma}}{\mu}.$$

Therefore

$$\begin{aligned} f^{\sigma\Delta} &= \frac{f^{\Delta\sigma}\mu^{\sigma}}{\mu} \\ &= \frac{f^{\Delta\sigma}\mu(1+\mu^{\Delta})}{\mu} \\ &= (1+\mu^{\Delta})f^{\Delta\sigma}. \end{aligned}$$

Also,

$$f^{\sigma^2\Delta} = \frac{f^{\sigma^3} - f^{\sigma^2}}{\mu} \quad \text{and} \quad f^{\sigma\Delta\sigma} = \frac{f^{\sigma^3} - f^{\sigma^2}}{\mu^{\sigma}}.$$

Therefore

$$\begin{aligned} f^{\sigma\Delta\sigma} &= \frac{f^{\sigma^2\Delta}\mu}{\mu^{\sigma}} \\ &= \frac{f^{\sigma^2\Delta}\mu}{\mu(1+\mu^{\Delta})} \\ &= \frac{f^{\sigma^2\Delta}}{1+\mu^{\Delta}}, \end{aligned}$$

whereupon

$$f^{\sigma^2\Delta} = (1+\mu^{\Delta})f^{\sigma\Delta\sigma}.$$

We have

$$\begin{aligned} f^{\Delta\sigma^2} &= \frac{f^{\sigma^3} - f^{\sigma^2}}{\mu^{\sigma^2}} \\ &= \frac{f^{\sigma^2\Delta}\mu}{\mu^{\sigma^2}} \end{aligned}$$

$$= \frac{\mu f^{\sigma^2 \Delta}}{\mu(1+\mu^{\Delta})(1+\mu^{\Delta\sigma})}$$
$$= \frac{f^{\sigma^2 \Delta}}{(1+\mu^{\Delta})(1+\mu^{\Delta\sigma})},$$

from which we get

$$f^{\sigma^2 \Delta} = (1+\mu^{\Delta})(1+\mu^{\Delta\sigma}) f^{\Delta\sigma^2}.$$

Exercise 2.13. Let $\mathbb{T} = \mathbb{Z}$. Find $f^{\Delta^2}(t)$, $t \in \mathbb{T}$, where

$$f(t) = t^2 - 3t + 1, \quad t \in \mathbb{T}.$$

Exercise 2.14. Let $\mathbb{T} = 3^{\mathbb{N}_0}$. Find $f^{\Delta^2}(t)$, $t \in \mathbb{T}$, where

$$f(t) = t^3 + 3t^2 + t + 1, \quad t \in \mathbb{T}.$$

Exercise 2.15. Let $\mathbb{T} = 2^{\mathbb{N}_0}$. Find $f^{\Delta^2}(t)$, $t \in \mathbb{T}$, where

$$f(t) = \frac{t^2+1}{t+1}, \quad t \in \mathbb{T}.$$

Exercise 2.16. Let $\mathbb{T} = 3^{\mathbb{N}_0}$. Find $f^{\Delta^2}(t)$, $t \in \mathbb{T}$, where

$$f(t) = \frac{t+2}{t+3}, \quad t \in \mathbb{T}.$$

Exercise 2.17. Let $\mathbb{T} = \mathbb{N}_0$. Find $f^{\Delta^2}(t)$, $t \in \mathbb{T}$, where

$$f(t) = \frac{1}{t+1}, \quad t \in \mathbb{T}.$$

Exercise 2.18. Let $\mathbb{T} = P_{3,7} \cup [4,6]$. Find $f^{\Delta^2}(t)$, $t \in \mathbb{T}^{\kappa^2}$, where

$$f(t) = \frac{t-2}{t-4}, \quad t \in \mathbb{T}.$$

Exercise 2.19. Let $\mathbb{T} = (-2\mathbb{N}_0) \cup 3^{\mathbb{N}_0}$. Find $f^{\Delta^2}(t)$, $t \in \mathbb{T}$, where

$$f(t) = t^3 - t^2 + t - 1, \quad t \in \mathbb{T}.$$

2.4 Nabla derivatives

Definition 2.3 (The nabla derivative). A function $f : \mathbb{T} \to \mathbb{R}$ is said to be nabla differentiable at $t \in \mathbb{T}_\kappa$ if

1. f is defined in a neighborhood U of t,
2. f is defined at $\rho(t)$, and
3. there exists a unique real number $f^\nabla(t)$, called the nabla derivative of f at t, such that for each $\varepsilon > 0$, there exists a neighborhood N of t with $N \subset U$, and

$$\left|f(\rho(t)) - f(s) - (\rho(t) - s)f^\nabla(t)\right| \le \varepsilon\left|\rho(t) - s\right| \quad \text{for all } s \in N.$$

The basic rules for nabla differentiation read as follows.

- Let $f, g : \mathbb{T} \to \mathbb{R}$, and let $t \in \mathbb{T}_\kappa$. Then we have the following.
  1. The nabla derivative is well defined.
  2. If f is nabla differentiable at t, then f is continuous at t.
  3. If f is continuous at t and t is left-scattered, then f is nabla differentiable at t with

$$f^\nabla(t) = \frac{f(t) - f(\rho(t))}{\nu(t)}.$$

  4. If t is left-dense, then f is nabla differentiable at t iff the finite limit

$$\lim_{s \to t} \frac{f(t) - f(s)}{t - s}$$

  exists. In this case,

$$f^\nabla(t) = \lim_{s \to t} \frac{f(t) - f(s)}{t - s}.$$

  5. If f is differentiable at t, then

$$f(\rho(t)) = f(t) + \nu(t) f^\nabla(t).$$

  6. If f and g are nabla differentiable at t, then
     a. The sum $f + g : \mathbb{T} \to \mathbb{R}$ is nabla differentiable at t with

$$(f + g)^\nabla(t) = f^\nabla(t) + g^\nabla(t).$$

     b. For any constant α, $\alpha f : \mathbb{T} \to \mathbb{R}$ is nabla differentiable at t with

$$(\alpha f)^\nabla(t) = \alpha f^\nabla(t).$$

     c. The product $fg : \mathbb{T} \to \mathbb{R}$ is nabla differentiable at t with

$$(fg)^\nabla(t) = f^\nabla(t)g(t) + f(\rho(t))g^\nabla(t) = f(t)g^\nabla(t) + f^\nabla(t)g(\rho(t)).$$

d. If $g(t)g(\rho(t)) \neq 0$, then $\frac{f}{g} : \mathbb{T} \to \mathbb{R}$ is nabla differentiable at t with

$$\left(\frac{f}{g}\right)^{\nabla}(t) = \frac{f^{\nabla}(t)g(t) - f(t)g^{\nabla}(t)}{g(t)g(\rho(t))}.$$

Definition 2.4. Let $f : \mathbb{T} \to \mathbb{R}$, and let $t \in (\mathbb{T}_\kappa)_\kappa = \mathbb{T}_{\kappa^2}$. We define the second nabla derivative of f at t, provided that it exists, as

$$f^{\nabla\nabla} = (f^{\nabla})^{\nabla} : \mathbb{T}_{\kappa^2} \to \mathbb{R}.$$

Similarly, we define the higher-order nabla derivatives $f^{\nabla^n} : \mathbb{T}_{\kappa^n} \to \mathbb{R}$, $n \in \mathbb{N}$. Let $t \in \mathbb{T}_\kappa^\kappa$. We define the second mixed derivatives of f at t, provided that they exist, as

$$f^{\nabla\Delta} = (f^{\nabla})^{\Delta} : \mathbb{T}_\kappa^\kappa \to \mathbb{R}$$

and

$$f^{\Delta\nabla} = (f^{\Delta})^{\nabla} : \mathbb{T}_\kappa^\kappa \to \mathbb{R}.$$

Example 2.40. Let $\mathbb{T} = \mathbb{Z}$, and let $f : \mathbb{T} \to \mathbb{R}$. Then

$$f^{\nabla}(t) = f(t) - f(t-1) \quad \text{for all } t \in \mathbb{T}.$$

Example 2.41. Let $\mathbb{T} = 2^{\mathbb{N}_0}$, and let

$$f(t) = t^3 + t^2 + t + 1, \quad t \in \mathbb{T}_\kappa.$$

We will find $f^{\nabla}(t)$ for $t \in \mathbb{T}_\kappa$. We have that $\mathbb{T}_\kappa = \mathbb{T} \setminus \{1\}$ and $\rho(t) = \frac{t}{2}$, $t \in \mathbb{T}_\kappa$. Hence

$$\begin{aligned} f^{\nabla}(t) &= (\rho(t))^2 + t\rho(t) + t^2 + \rho(t) + t + 1 \\ &= \frac{t^2}{4} + \frac{t^2}{2} + t^2 + \frac{t}{2} + t + 1 \\ &= \frac{7}{4}t^2 + \frac{3}{2}t + 1, \quad t \in \mathbb{T}_\kappa. \end{aligned}$$

Example 2.42. Let $\mathbb{T} = P_{1,3}$, and let

$$f(t) = \frac{1-t}{1+t}, \quad t \in \mathbb{T}.$$

We will find $f^{\nabla}(t)$ for $t \in \mathbb{T}_\kappa$ and $f^{\Delta\nabla}(t)$ for $t \in \mathbb{T}_\kappa^\kappa$. Firstly, we will find $f^{\nabla}(t)$ for $t \in \mathbb{T}_\kappa$. We have the following cases.

1. $t \in \mathbb{T}_\kappa$ is left-scattered. Then $\rho(t) = t - 3$, and

$$f^{\nabla}(t) = \frac{f(\rho(t)) - f(t)}{\rho(t) - t}$$

$$
\begin{aligned}
&= \frac{f(t-3)-f(t)}{t-3-t} \\
&= \frac{\frac{1-t+3}{1+t-3} - \frac{1-t}{1+t}}{-3} \\
&= \frac{1}{3}\left(\frac{1-t}{1+t} - \frac{4-t}{t-2}\right) \\
&= \frac{1}{3}\left(\frac{(1-t)(t-2)-(1+t)(4-t)}{(t+1)(t-2)}\right) \\
&= \frac{1}{3}\left(\frac{t-2-t^2+2t-4+t-4t+t^2}{(t+1)(t-2)}\right) \\
&= \frac{1}{3}\left(-\frac{6}{(t+1)(t-2)}\right) \\
&= -\frac{2}{(t+1)(t-2)}.
\end{aligned}
$$

2. $t \in \mathbb{T}_\kappa$ is left-dense. Then

$$
\begin{aligned}
f^\nabla(t) &= \lim_{s\to t} \frac{f(t)-f(s)}{t-s} \\
&= \lim_{s\to t} \frac{\frac{1-t}{1+t} - \frac{1-s}{1+s}}{t-s} \\
&= \lim_{s\to t} \frac{(1-t)(1+s)-(1-s)(1+t)}{(t-s)(1+t)(1+s)} \\
&= \lim_{s\to t} \frac{1+s-t-st-1-t+s+st}{(t-s)(1+t)(1+s)} \\
&= \lim_{s\to t} \frac{2s-2t}{(t-s)(1+t)(1+s)} \\
&= -2\lim_{s\to t} \frac{1}{(1+t)(1+s)} \\
&= -\frac{2}{(1+t)^2}.
\end{aligned}
$$

Now we will find $f^{\Delta\nabla}(t)$ for $t \in \mathbb{T}_\kappa^\kappa$. We have the following cases.

1. $t \in \mathbb{T}_\kappa^\kappa$ is right-scattered. We have

$$
f^\Delta(t) = -\frac{2}{(1+t)(4+t)}.
$$

and t is left-dense. Then

$$
\begin{aligned}
f^{\Delta\nabla}(t) &= \lim_{s\to t} \frac{f^\Delta(t)-f^\Delta(s)}{t-s} \\
&= \lim_{s\to t} \frac{-\frac{2}{(1+t)(4+t)} + \frac{2}{(1+s)(4+s)}}{t-s}
\end{aligned}
$$

$$
\begin{aligned}
&= -2\lim_{s\to t} \frac{\frac{1}{(1+t)(4+t)} - \frac{1}{(1+s)(4+s)}}{t-s}\\
&= -2\lim_{s\to t}\left(\frac{(1+s)(4+s)-(1+t)(4+t)}{(t-s)(1+t)(4+t)(1+s)(4+s)}\right)\\
&= -2\lim_{s\to t}\frac{4+s+4s+s^2-4-t-4t-t^2}{(t-s)(1+t)(4+t)(1+s)(4+s)}\\
&= -2\lim_{s\to t}\frac{5(s-t)+(s-t)(s+t)}{(t-s)(1+t)(4+t)(1+s)(4+s)}\\
&= 2\lim_{s\to t}\frac{5+s+t}{(1+t)(4+t)(1+s)(4+s)}\\
&= \frac{2(5+2t)}{(1+t)^2(4+t)^2}.
\end{aligned}
$$

2. $t \in \mathbb{T}^\kappa_\kappa$ is right-dense. We have

$$f^\Delta(t) = -\frac{2}{(1+t)^2}.$$

a. t is left-scattered. Then

$$
\begin{aligned}
f^{\Delta\nabla}(t) &= \frac{f^\Delta(\rho(t)) - f^\Delta(t)}{\rho(t)-t}\\
&= \frac{f^\Delta(t-3) - f^\Delta(t)}{t-3-t}\\
&= \frac{-\frac{2}{(1+t-3)^2} + \frac{2}{(1+t)^2}}{-3}\\
&= \frac{2}{3}\left(\frac{1}{(t-2)^2} - \frac{1}{(t+1)^2}\right)\\
&= \frac{2}{3}\left(\frac{(t+1)^2-(t-2)^2}{(t+1)^2(t-2)^2}\right)\\
&= \frac{2}{3}\left(\frac{t^2+2t+1-t^2+4t-4}{(t+1)^2(t-2)^2}\right)\\
&= \frac{2}{3}\left(\frac{6t-3}{(t+1)^2(t-2)^2}\right)\\
&= \frac{2(2t-1)}{(t+1)^2(t-2)^2}.
\end{aligned}
$$

b. t is left-dense. Then

$$
\begin{aligned}
f^{\Delta\nabla}(t) &= \lim_{s\to t}\frac{f^\Delta(t)-f^\Delta(s)}{t-s}\\
&= \lim_{s\to t}\frac{-\frac{2}{(1+t)^2} + \frac{2}{(1+s)^2}}{t-s}
\end{aligned}
$$

$$
\begin{aligned}
&= \lim_{s\to t} \frac{(1+t)^2-(1+s)^2}{(t-s)(1+t)^2(1+s)^2}\\
&= 2\lim_{s\to t} \frac{1+2t+t^2-1-2s-s^2}{(t-s)(1+t)^2(1+s)^2}\\
&= 2\lim_{s\to t} \frac{2(t-s)+(t-s)(t+s)}{(t-s)(1+t)^2(1+s)^2}\\
&= 2\lim_{s\to t} \frac{2+t+s}{(1+t)^2(1+s)^2}\\
&= \frac{4(1+t)}{(1+t)^4}\\
&= \frac{4}{(1+t)^3}.
\end{aligned}
$$

Example 2.43. Let $\mathbb{T} = \{-\frac{1}{n} : n \in \mathbb{N}\} \cup \mathbb{N}_0$, and let

$$
f(t) = \frac{1+t^2}{1+7t}, \qquad t \in \mathbb{T}.
$$

We will find $f^\nabla(t)$, $t \in \mathbb{T}_\kappa$. We have the following cases.

1. $t \in \{-\frac{1}{n} : n \in \mathbb{N}, n \geq 2\}$. Then

$$
\rho(t) = \frac{t}{t+1}.
$$

Hence

$$
\begin{aligned}
f^\nabla(t) &= \frac{f(\rho(t))-f(t)}{\rho(t)-t}\\
&= \frac{\frac{1+(\rho(t))^2}{1+7\rho(t)} - \frac{1+t^2}{1+7t}}{\frac{t}{t+1}-t}\\
&= \frac{\frac{1+\frac{t^2}{(t+1)^2}}{1+7\frac{t}{t+1}} - \frac{1+t^2}{1+7t}}{t(\frac{1}{t+1}-1)}\\
&= \frac{\frac{(t+1)^2+t^2}{(t+1)(t+1+7t)} - \frac{1+t^2}{1+7t}}{t(\frac{1-t-1}{t+1})}\\
&= \frac{\frac{2t^2+2t+1}{(t+1)(8t+1)} - \frac{1+t^2}{1+7t}}{-\frac{t^2}{t+1}}\\
&= -\frac{(2t^2+2t+1)(1+7t)-(1+t^2)(t+1)(8t+1)}{t^2(8t+1)(7t+1)}\\
&= -\frac{2t^2+14t^3+2t+14t^2+1+7t-(1+t^2)(8t^2+9t+1)}{t^2(8t+1)(7t+1)}
\end{aligned}
$$

$$
\begin{aligned}
&= -\frac{14t^3 + 16t^2 + 9t + 1 - 8t^2 - 9t - 1 - 8t^4 - 9t^3 - t^2}{t^2(8t+1)(7t+1)} \\
&= -\frac{-8t^4 + 5t^3 + 7t^2}{t^2(8t+1)(7t+1)} \\
&= \frac{8t^2 - 5t - 7}{(8t+1)(7t+1)}.
\end{aligned}
$$

2. $t = 0$. Then $\rho(0) = 0$, and

$$
\begin{aligned}
f^{\nabla}(0) &= \lim_{s\to 0} \frac{f(s) - f(0)}{s} \\
&= \lim_{s\to 0} \frac{\frac{1+s^2}{1+7s} - 1}{s} \\
&= \lim_{s\to 0} \frac{1 + s^2 - 1 - 7s}{s(1+7s)} \\
&= \lim_{s\to 0} \frac{s(s-7)}{s(1+7s)} \\
&= \lim_{s\to 0} \frac{s-7}{1+7s} \\
&= -7.
\end{aligned}
$$

3. $t \in \mathbb{N}$. Then $\rho(t) = t - 1$, and

$$
\begin{aligned}
f^{\nabla}(t) &= \frac{f(\rho(t)) - f(t)}{\rho(t) - t} \\
&= \frac{\frac{1+(t-1)^2}{1+7(t-1)} - \frac{1+t^2}{1+7t}}{t - 1 - t} \\
&= \frac{1+t^2}{1+7t} - \frac{1+(t-1)^2}{1+7(t-1)} \\
&= \frac{1+t^2}{1+7t} - \frac{t^2 - 2t + 2}{7t - 6} \\
&= \frac{(1+t^2)(7t-6) - (1+7t)(t^2-2t+2)}{(7t+1)(7t-6)} \\
&= \frac{7t - 6 + 7t^3 - 6t^2 - t^2 + 2t - 2 - 7t^3 + 14t^2 - 14t}{(7t+1)(7t-6)} \\
&= \frac{7t^2 - 5t - 8}{(7t+1)(7t-6)}.
\end{aligned}
$$

Example 2.44. Let $U = \{\frac{1}{2^n} : n \in \mathbb{N}\}$, let $\mathbb{T} = \{0\} \cup U \cup (1 - U) \cup \{1\}$, and let

$$f(t) = (2 + t^2)(1 - t^2), \quad t \in \mathbb{T}.$$

We will find $f^{\nabla}(t)$, $t \in \mathbb{T}_{\kappa}$, and $f^{\nabla^2}(t)$, $t \in \mathbb{T}_{\kappa^2}$. We have the following cases.

1. $t \in U$. By Example 1.39 we have

$$\rho(t) = \frac{1}{2}t.$$

Hence

$$\begin{aligned}
f^{\nabla}(t) &= \frac{f(\rho(t)) - f(t)}{\rho(t) - t} \\
&= \frac{(2 + \frac{t^2}{4})(1 - \frac{t^2}{4}) - (2 + t^2)(1 - t^2)}{\frac{t}{2} - t} \\
&= \frac{(2 + t^2)(1 - t^2) - \frac{1}{16}(8 + t^2)(4 - t^2)}{\frac{t}{2}} \\
&= \frac{16(2 + t^2)(1 - t^2) - (8 + t^2)(4 - t^2)}{8t} \\
&= \frac{32 - 32t^2 + 16t^2 - 16t^4 - 32 + 8t^2 - 4t^2 + t^4}{8t} \\
&= \frac{-15t^4 - 12t^2}{8t} \\
&= -\frac{15}{8}t^3 - \frac{3}{2}t.
\end{aligned}$$

Then

$$\begin{aligned}
f^{\nabla^2}(t) &= \frac{f^{\nabla}(\rho(t)) - f^{\nabla}(t)}{\rho(t) - t} \\
&= \frac{-\frac{15}{64}t^3 - \frac{3}{4}t}{\frac{t}{2} - t} \\
&= \frac{\frac{15}{64}t^3 + \frac{3}{4}t}{\frac{t}{2}} \\
&= \frac{15}{32}t^2 + \frac{3}{2}.
\end{aligned}$$

2. $t \in (1 - U)\backslash\{\frac{1}{2}\}$. By Example 1.39 we get

$$\rho(t) = 2t - 1,$$

and

$$\begin{aligned}
f^{\nabla}(t) &= \frac{f(\rho(t)) - f(t)}{\rho(t) - t} \\
&= \frac{(2 + (2t - 1)^2)(1 - (2t - 1)^2) - (2 + t^2)(1 - t^2)}{2t - 1 - t}
\end{aligned}$$

$$
\begin{aligned}
&= \frac{(4t^2-4t+3)(1-2t+1)(1+2t-1)-(2+t^2)(1-t)(1+t)}{t-1}\\
&= \frac{-4(4t^2-4t+3)t(t-1)+(2+t^2)(t-1)(t+1)}{t-1}\\
&= -4t(4t^2-4t+3)+(2+t^2)(t+1)\\
&= -16tr^3+16t^2-12t+2t+2+t^3+t^2\\
&= -15t^3+17t^2-10t+2.
\end{aligned}
$$

Then

$$
\begin{aligned}
f^{\nabla^2}(t) &= \frac{f^{\nabla}(\rho(t))-f^{\nabla}(t)}{\rho(t)-t}\\
&= \frac{-15(2t-1)^3+17(2t-1)^2-10(2t-1)+2+15t^3-17t^2+10t-2}{2t-1-t}\\
&= \frac{-15(t-1)(4t^2-4t+1+2t^2-t+t^2)+17(t-1)(2t-1+t)-10(t-1)}{t-1}\\
&= -15(7t^2-5t+1)+17(3t-1)-10\\
&= -105t^2+75t-15+51t-17-10\\
&= -105t^2+126t-42.
\end{aligned}
$$

3. $t=\frac{1}{2}$. By Example 1.39 we have

$$
\rho\left(\frac{1}{2}\right)=\frac{1}{4},
$$

and then

$$
\begin{aligned}
f^{\nabla}\left(\frac{1}{2}\right) &= \frac{f(\rho(\frac{1}{2}))-f(\frac{1}{2})}{\rho(\frac{1}{2})-\frac{1}{2}}\\
&= \frac{(2+\frac{1}{16})(1-\frac{1}{16})-(2+\frac{1}{4})(1-\frac{1}{4})}{\frac{1}{4}-\frac{1}{2}}\\
&= \frac{\frac{33}{16}\cdot\frac{15}{16}-\frac{9}{4}\cdot\frac{3}{4}}{-\frac{1}{4}}\\
&= \frac{27\cdot 16-33\cdot 15}{4}\\
&= \frac{432-495}{4}\\
&= -\frac{63}{4}.
\end{aligned}
$$

Note that

$$
\begin{aligned}
f^{\nabla}\left(\frac{1}{4}\right) &= -\frac{15}{8}\cdot\left(\frac{1}{4}\right)^3 - \frac{3}{2}\cdot\frac{1}{4}\\
&= -\frac{15}{8}\cdot\frac{1}{64} - \frac{3}{8}\\
&= -\frac{3}{8}\left(\frac{5}{64}+1\right)\\
&= -\frac{3\cdot 69}{8\cdot 64}\\
&= -\frac{207}{512}.
\end{aligned}
$$

Hence

$$
\begin{aligned}
f^{\nabla^2}\left(\frac{1}{2}\right) &= \frac{f^{\nabla}(\rho(\frac{1}{2})) - f^{\nabla}(\frac{1}{2})}{\rho(\frac{1}{2}) - \frac{1}{2}}\\
&= \frac{-\frac{207}{512} + \frac{63}{4}}{\frac{1}{4} - \frac{1}{2}}\\
&= \frac{207 - 63\cdot 128}{128}\\
&= \frac{207 - 8064}{128}\\
&= -\frac{7857}{128}.
\end{aligned}
$$

4. $t = 1$. By Example 1.39 we have $\rho(1) = 1$, and then

$$
\begin{aligned}
f^{\nabla}(1) &= \lim_{s\to 1}\frac{f(s) - f(1)}{s-1}\\
&= \lim_{s\to 1}\frac{(2+s^2)(1-s^2)}{s-1}\\
&= -\lim_{s\to 1}(2+s^2)(1+s)\\
&= -6.
\end{aligned}
$$

Hence

$$
\begin{aligned}
f^{\nabla^2}(1) &= \lim_{s\to 1}\frac{f^{\nabla}(s) - f^{\nabla}(1)}{s-1}\\
&= \lim_{s\to 1}\frac{-15s^3 + 17s^2 - 10s + 2 + 6}{s-1}\\
&= -\lim_{s\to 1}\frac{15s^3 - 17s^2 + 10s - 8}{s-1}\\
&= -\lim_{s\to 1}\frac{(15s^2 - 2s + 8)(s-1)}{s-1}
\end{aligned}
$$

$$\begin{aligned}&= -\lim_{s\to 1}(15s^2 - 2s + 8)\\&= -21.\end{aligned}$$

Example 2.45. Let $\mathbb{T} = (-\mathbb{N}_0) \cup [1,2] \cup 4^{\mathbb{N}}$, and let

$$f(t) = t^3 + t, \quad t \in \mathbb{T}.$$

We will find $f^{\nabla}(t), f^{\nabla^2}(t), f^{\nabla^3}(t)$, $t \in \mathbb{T}$. We have the following cases.

1. $t \in (-\mathbb{N}_0)$. Then

$$\rho(t) = t - 1,$$

and

$$\begin{aligned}f^{\nabla}(t) &= \frac{f(\rho(t)) - f(t)}{\rho(t) - t}\\&= \frac{f(t-1) - f(t)}{t-1-t}\\&= t^3 + t - (t-1)^3 - (t-1)\\&= t^3 - t^3 + 3t^2 - 3t + 1 + t - t + 1\\&= 3t^2 - 3t + 2,\\f^{\nabla^2}(t) &= \frac{f^{\nabla}(\rho(t)) - f^{\nabla}(t)}{\rho(t) - t}\\&= \frac{f^{\nabla}(t-1) - f^{\nabla}(t)}{t-1-t}\\&= 3t^2 - 3t + 2 - 3(t-1)^2 + 3(t-1) - 2\\&= 3t^2 - 3t^2 + 6t - 3 - 3t + 3t - 3\\&= 6t - 6,\end{aligned}$$

and

$$\begin{aligned}f^{\nabla^3}(t) &= \frac{f^{\nabla^2}(\rho(t)) - f^{\nabla^2}(t)}{\rho(t) - t}\\&= \frac{f^{\nabla^2}(t-1) - f^{\nabla^2}(t)}{t-1-t}\\&= 6t - 6 - 6(t-1) + 6\\&= 6t - 6t + 6\\&= 6.\end{aligned}$$

2. $t = 1$. Then $\rho(1) = 0$, and

$$\begin{aligned} f^{\nabla}(1) &= \frac{f(\rho(1)) - f(1)}{\rho(1) - 1} \\ &= \frac{f(0) - f(1)}{0 - 1} \\ &= f(1) - f(0) \\ &= 2. \end{aligned}$$

Note that

$$f^{\nabla}(0) = 2.$$

Then

$$\begin{aligned} f^{\nabla^2}(1) &= \frac{f^{\nabla}(\rho(1)) - f^{\nabla}(1)}{\rho(1) - 1} \\ &= \frac{f^{\nabla}(0) - f^{\nabla}(1)}{0 - 1} \\ &= 2 - 2 \\ &= 0. \end{aligned}$$

Next,

$$f^{\nabla^2}(0) = -6,$$

and

$$\begin{aligned} f^{\nabla^3}(1) &= \frac{f^{\nabla^2}(\rho(1)) - f^{\nabla^2}(1)}{\rho(1) - 1} \\ &= \frac{f^{\nabla^2}(0) - f^{\nabla^2}(1)}{0 - 1} \\ &= 0 - (-6) \\ &= 6. \end{aligned}$$

3. $t \in (1, 2]$. Then

$$\begin{aligned} f^{\nabla}(t) &= 3t^2 + 1, \\ f^{\nabla^2}(t) &= 6t, \\ f^{\nabla^3}(t) &= 6. \end{aligned}$$

4. $t = 4$. Then $\rho(4) = 2$, and

$$\begin{aligned} f^{\nabla}(4) &= \frac{f(\rho(4)) - f(4)}{\rho(4) - 4} \\ &= \frac{f(2) - f(4)}{2 - 4} \\ &= \frac{68 - 10}{2} \\ &= 29. \end{aligned}$$

Note that

$$\begin{aligned} f^{\nabla}(2) &= 3 \cdot 4 + 1 \\ &= 13, \end{aligned}$$

and then

$$\begin{aligned} f^{\nabla^2}(4) &= \frac{f^{\nabla}(\rho(4)) - f^{\nabla}(4)}{\rho(4) - 4} \\ &= \frac{f^{\nabla}(2) - f^{\nabla}(4)}{2 - 4} \\ &= \frac{29 - 13}{2} \\ &= 8. \end{aligned}$$

Next,

$$f^{\nabla^2}(2) = 12,$$

and

$$\begin{aligned} f^{\nabla^3}(4) &= \frac{f^{\nabla^2}(\rho(4)) - f^{\nabla^2}(4)}{\rho(4) - 4} \\ &= \frac{f^{\nabla^2}(2) - f^{\nabla^2}(4)}{2 - 4} \\ &= \frac{12 - 8}{-2} \\ &= -2. \end{aligned}$$

5. $t \in 4^{\mathbb{N}}$, $t \geq 16$. Then

$$\begin{aligned} \rho(t) &= \frac{t}{4}, \\ f^{\nabla}(t) &= \frac{f(\rho(t)) - f(t)}{\rho(t) - t} \end{aligned}$$

$$
\begin{aligned}
&= \frac{t^3 + t - \frac{t^3}{64} - \frac{t}{4}}{t - \frac{t}{4}} \\
&= \frac{\frac{63}{64}t^3 + \frac{3}{4}t}{\frac{3}{4}t} \\
&= \frac{21}{16}t^2 + 1, \\
f^{\nabla^2}(t) &= \frac{f^{\nabla}(\rho(t)) - f^{\nabla}(t)}{\rho(t) - t} \\
&= \frac{\frac{21}{16}t^2 + 1 - \frac{21}{256}t^2 - 1}{t - \frac{t}{4}} \\
&= \frac{\frac{315}{256}t^2}{\frac{3}{4}t} \\
&= \frac{105}{64}t,
\end{aligned}
$$

and

$$
\begin{aligned}
f^{\nabla^3}(t) &= \frac{f^{\nabla^2}(\rho(t)) - f^{\nabla^2}(t)}{\rho(t) - t} \\
&= \frac{\frac{105}{64}t - \frac{105}{256}t}{t - \frac{t}{4}} \\
&= \frac{\frac{315}{256}t}{\frac{3}{4}t} \\
&= \frac{105}{64}.
\end{aligned}
$$

Exercise 2.20. Let $\mathbb{T} = 2^{\mathbb{N}_0}$. Find $f^{\nabla}(t)$ for $t \in \mathbb{T}_\kappa$, where

$$f(t) = \frac{t^2 + 2t - 3}{t - 7}.$$

Exercise 2.21. Let $\mathbb{T} = (-2\mathbb{N}_0) \cup 3^{\mathbb{N}_0}$, and let

$$f(t) = t^3 - 3t^2 + 2t, \quad t \in \mathbb{T}.$$

Find $f^{\nabla}(t)$ and $f^{\nabla^2}(t)$ for $t \in \mathbb{T}$.

2.5 Mean value theorems

Let $a, b \in \mathbb{T}$, $a < b$, and let $f : \mathbb{T} \to \mathbb{R}$. The mean value theorems are stated as follows.

1. Suppose that f has a delta derivative at each point of $[a, b]_{\mathbb{T}}$. If

$$f(a) = f(b),$$

   then there exist points $\xi_1, \xi_2 \in [a, b]_{\mathbb{T}}$ such that

$$f^{\Delta}(\xi_2) \le 0 \le f^{\Delta}(\xi_1).$$

2. Suppose that f is continuous on $[a, b]_{\mathbb{T}}$ and has a delta derivative at each point of $[a, b)_{\mathbb{T}}$. Then there exist $\xi_1, \xi_2 \in (a, b)_{\mathbb{T}}$ such that

$$f^{\Delta}(\xi_1)(b - a) \le f(b) - f(a) \le f^{\Delta}(\xi_2)(b - a). \tag{2.2}$$

Example 2.46. Let $\mathbb{T} = \mathbb{Z}$, and let $f(t) = t^2$, $t \in \mathbb{T}$. We will find $\xi_1, \xi_2 \in (-3, 3)_{\mathbb{T}}$ such that

$$f^{\Delta}(\xi_2) \le 0 \le f^{\Delta}(\xi_1).$$

We have

$$\begin{aligned} \sigma(t) &= t + 1, \\ f(-3) &= f(3) \\ &= 9, \\ f^{\Delta}(t) &= \sigma(t) + t \\ &= 2t + 1, \quad t \in \mathbb{T}. \end{aligned}$$

Hence

$$\begin{aligned} f^{\Delta}(\xi_2) &= 2\xi_2 + 1 \\ &\le 0 \end{aligned}$$

for $\xi_2 \in \{-2, -1\}$, and

$$\begin{aligned} f^{\Delta}(\xi_1) &= 2\xi_1 + 1 \\ &\ge 0 \end{aligned}$$

for $\xi_1 \in \{0, 1, 2\}$.

Example 2.47. Let $\mathbb{T} = 2^{\mathbb{N}_0}$, and let $f(t) = t^3 - \frac{73}{9}t^2 + 2$, $t \in \mathbb{T}$. We will find $\xi_1, \xi_2 \in (1, 8)_{\mathbb{T}}$ such that

$$f^{\Delta}(\xi_2) \le 0 \le f^{\Delta}(\xi_1).$$

We have

$$\begin{aligned} f(1) &= 1 - \frac{73}{9} + 2 \\ &= 3 - \frac{73}{9} \\ &= -\frac{46}{9}, \\ f(8) &= 512 - \frac{73}{9}64 + 2 \\ &= \frac{72-73}{9}64 + 2 \\ &= -\frac{64}{9} + 2 \\ &= -\frac{46}{9}. \end{aligned}$$

Therefore $f(1) = f(8)$.

Also,

$$\begin{aligned} \sigma(t) &= 2t, \\ f^{\Delta}(t) &= (\sigma(t))^2 + t\sigma(t) + t^2 - \frac{73}{9}(\sigma(t) + t) \\ &= 4t^2 + 2t^2 + t^2 - \frac{73}{9}(3t) \\ &= 7t^2 - \frac{73}{3}t. \end{aligned}$$

Hence

$$\begin{aligned} f^{\Delta}(\xi_2) &= 7\xi_2^2 - \frac{73}{3}\xi_2 \leq 0 \quad \text{for } \xi_2 = 2, \\ f^{\Delta}(\xi_1) &= 7\xi_1^2 - \frac{73}{3}\xi_1 \geq 0 \quad \text{for } \xi_1 = 4. \end{aligned}$$

Example 2.48. Let $\mathbb{T} = 3^{\mathbb{N}_0} \cup \{0\}$, and let $f(t) = t^2 - 9t + 3$, $t \in \mathbb{T}$. We will find $\xi_1, \xi_2 \in (0,9)_{\mathbb{T}}$ such that

$$f^{\Delta}(\xi_2) \leq 0 \leq f^{\Delta}(\xi_1).$$

Here

$$\begin{aligned} \sigma(t) &= 3t, \quad t \neq 0, \\ \sigma(0) &= 1, \\ f(0) &= 3, \\ f(9) &= 3. \end{aligned}$$

Then

$$\begin{aligned} f^{\Delta}(t) &= \sigma(t) + t - 9 \\ &= 3t + t - 9 \\ &= 4t - 9. \end{aligned}$$

Hence

$$\begin{aligned} f^{\Delta}(\xi_2) &= 4\xi_2 - 9 \\ &\leq 0 \quad \text{for } \xi_2 = 1, \\ f^{\Delta}(\xi_1) &= 4\xi_1 - 9 \\ &\geq 0 \quad \text{for } \xi_1 = 3. \end{aligned}$$

Example 2.49. Let $\mathbb{T} = \mathbb{Z}, f(t) = t^2 + 2t, t \in \mathbb{T}$, and let $[a, b]_{\mathbb{T}} = [-2, 4]_{\mathbb{T}}$. Then

$$\begin{aligned} \sigma(t) &= t + 1, \\ f^{\Delta}(t) &= \sigma(t) + t + 2 \\ &= t + 1 + t + 2 \\ &= 2t + 3, \\ f(4) &= 24, \\ f(-2) &= 0. \end{aligned}$$

Hence

$$\begin{aligned} 2\xi_1 + 3 \leq \frac{24}{6} \leq 2\xi_2 + 3 &\iff \\ \begin{cases} 2\xi_1 + 3 \leq 4 \\ 2\xi_2 + 3 \geq 4 \end{cases} &\iff \\ \begin{cases} \xi_1 \leq \frac{1}{2} \\ \xi_2 \geq \frac{1}{2} \end{cases} &\iff \\ \begin{cases} \xi_1 = -1, 0 \\ \xi_2 = 1, 2, 3. \end{cases} & \end{aligned}$$

Example 2.50. Let $\mathbb{T} = P_{1,3}$, $a = 0$, $b = 5$, and let

$$f(t) = \frac{1 - t}{1 + t}, \quad t \in \mathbb{T}.$$

We have

$$f(b) = f(5)$$

$$\begin{aligned}&= \frac{1-5}{1+5}\\&= -\frac{2}{3}\end{aligned}$$

and

$$\begin{aligned}f(a) &= f(0)\\&= \frac{1-0}{1+0}\\&= 1.\end{aligned}$$

Hence

$$\begin{aligned}f(b) - f(a) &= -\frac{2}{3} - 1\\&= -\frac{5}{3}.\end{aligned}$$

We have the following cases.

1. t is right-scattered. By Example 2.17 we have

$$f^{\Delta}(t) = -\frac{2}{(1+t)(4+t)}.$$

Then

$$\begin{aligned}5f^{\Delta}(t) \le -\frac{5}{3} &\quad \text{if and only if}\\-\frac{2}{(1+t)(4+t)} \le -\frac{1}{3} &\quad \text{if and only if}\\\frac{2}{(1+t)(4+t)} \ge \frac{1}{3} &\quad \text{if and only if}\\6 \ge (t+1)(t+4) &\quad \text{if and only if}\\6 \ge t^2 + 5t + 4 &\quad \text{if and only if}\\t^2 + 5t - 2 \le 0 &\quad \text{if and only if}\\t \in \left(0, \frac{-5+\sqrt{33}}{2}\right]_{\mathbb{T}}.\end{aligned}$$

Moreover,

$$\begin{aligned}5f^{\Delta}(t) \ge -\frac{5}{3} &\quad \text{if and only if}\\t \in \left[\frac{-5+\sqrt{33}}{2}, 1\right]_{\mathbb{T}} \cup [4,5)_{\mathbb{T}}.\end{aligned}$$

2. t is right-dense. By Example 2.17 we have

$$f^{\Delta}(t) = -\frac{2}{(1+t)^2}.$$

Hence

$$\begin{aligned} 5f^{\Delta}(t) \leq -\frac{5}{3} \quad &\text{if and only if} \\ -\frac{5}{(1+t)^2} \leq -\frac{5}{3} \quad &\text{if and only if} \\ \frac{2}{(1+t)^2} \geq \frac{1}{3} \quad &\text{if and only if} \\ (1+t)^2 \leq 6 \quad &\text{if and only if} \\ t \in (0,1]_{\mathbb{T}}, \end{aligned}$$

and

$$\begin{aligned} 5f^{\Delta}(t) \geq -\frac{5}{3} \quad &\text{if and only if} \\ t \in [4,5)_{\mathbb{T}}. \end{aligned}$$

Example 2.51. Let $\mathbb{T} = \{-\frac{1}{n} : n \in \mathbb{N}\} \cup \mathbb{N}_0$, $a = -1$, $b = 1$, and let

$$f(t) = \frac{1+t^2}{1+7t}, \quad t \in \mathbb{T}.$$

We have

$$\begin{aligned} f(b) &= f(1) \\ &= \frac{1+1}{1+7} \\ &= \frac{1}{4} \end{aligned}$$

and

$$\begin{aligned} f(a) &= f(-1) \\ &= \frac{1+1}{1-7} \\ &= -\frac{1}{3}. \end{aligned}$$

Hence

$$f(b) - f(a) = \frac{1}{4} + \frac{1}{3} = \frac{7}{12}.$$

Consider the following cases.

1. $t \in \{-\frac{1}{n} : n \in \mathbb{N}\}$. By Example 2.20 we have

$$f^{\Delta}(t) = \frac{6t^2 + 9t - 7}{(1 + 7t)(1 + 6t)}.$$

   Then

$$\begin{aligned} f^{\Delta}(t) \le \frac{7}{12} \quad &\text{if and only if} \\ \frac{6t^2 + 9t - 7}{(1 + 7t)(1 + 6t)} \le \frac{7}{12} \quad &\text{if and only if} \\ \frac{6t^2 + 9t - 7}{(1 + 7t)(1 + 6t)} - \frac{7}{12} \le 0 \quad &\text{if and only if} \\ \frac{72t^2 + 108t - 84 - 7(42t^2 + 13t + 1)}{12(1 + 7t)(1 + 6t)} \le 0 \quad &\text{if and only if} \\ \frac{72t^2 + 108t - 84 - 294t^2 - 91t - 7}{12(1 + 7t)(1 + 6t)} \le 0 \quad &\text{if and only if} \\ \frac{-222t^2 + 17t - 91}{12(1 + 7t)(1 + 6t)} \le 0 \quad &\text{if and only if} \\ \frac{222t^2 - 17t + 91}{12(1 + 7t)(1 + 6t)} \ge 0 \quad &\text{if and only if} \end{aligned}$$

   $t \in \left[-1, -\frac{1}{6}\right)_{\mathbb{T}} \cup \left(-\frac{1}{7}, 0\right)_{\mathbb{T}}$, and $f^{\Delta}(t) \ge \frac{7}{12}$ is impossible.

2. $t = 0$ or $t = 1$. By Example 2.20 we have

$$f^{\Delta}(0) = -\frac{3}{4} \quad \text{and} \quad 2f^{\Delta}(0) \le \frac{7}{12},$$

   and

$$f^{\Delta}(1) = \frac{1}{12} \quad \text{and} \quad 2f^{\Delta}(1) \le \frac{7}{12}.$$

Exercise 2.22. Let $\mathbb{T} = \mathbb{Z}$, and let $f(t) = t^3 - 16t + 1$, $t \in \mathbb{T}$. Find $\xi_1, \xi_2 \in (-4, 4)$ such that

$$f^{\Delta}(\xi_2) \le 0 \le f^{\Delta}(\xi_1).$$

Exercise 2.23. Let $\mathbb{T} = 2^{\mathbb{N}_0} \cup \{0\}$. Prove that for every $t > 1$, there are $\xi_1, \xi_2 \in (0,t)$ such that

$$7\xi_1^2 + 3\xi_1 \le t^3 + t^2 \le 7\xi_2^2 + 3\xi_2.$$

Exercise 2.24. Let $\mathbb{T} = h\mathbb{Z} + k$, $h > 0$, $k \in \mathbb{R}$, $a = -h + k$, $b = 5h + k$, and let

$$f(t) = t^2 - 3t + 1, \quad t \in \mathbb{T}.$$

Prove that there are $\xi_1, \xi_2 \in (a,b)_{\mathbb{T}}$ such that

$$f^\Delta(\xi_1)(b-a) \le f(b) - f(a) \le f^\Delta(\xi_2)(b-a).$$

Exercise 2.25. Let $\mathbb{T} = (-2\mathbb{N}_0) \cup 3^{\mathbb{N}_0}$, $a = -10$, $b = 27$, and let

$$f(t) = 5 + 2t - 7t^2, \quad t \in \mathbb{T}.$$

Prove that there are $\xi_1, \xi_2 \in (a,b)_{\mathbb{T}}$ such that

$$f^\Delta(\xi_1)(b-a) \le f(b) - f(a) \le f^\Delta(\xi_2)(b-a).$$

Exercise 2.26. Let $\mathbb{T} = P_{3,7}$, $a = 0$, $b = 6$, and let

$$f(t) = t^3 + 2t^2 + 3t + 1, \quad t \in \mathbb{T}.$$

Prove that there are $\xi_1, \xi_2 \in (a,b)_{\mathbb{T}}$ such that

$$f^\Delta(\xi_1)(b-a) \le f(b) - f(a) \le f^\Delta(\xi_2)(b-a).$$

2.6 Increasing and decreasing functions

Using the first delta derivative of a function, we have the following criteria for increasing and decreasing of a function.

- Let f be a continuous function on $[a,b]_{\mathbb{T}}$ that has a delta derivative at each point of $[a,b)_{\mathbb{T}}$. Then f is strictly increasing, strictly decreasing, nondecreasing, or nonincreasing on $[a,b]_{\mathbb{T}}$ if $f^\Delta(t) > 0$, $f^\Delta(t) < 0$, $f^\Delta(t) \ge 0$, or $f^\Delta(t) \le 0$ for all $t \in [a,b)_{\mathbb{T}}$, respectively.

Example 2.52. Let $\mathbb{T} = \mathbb{Z}$, and let $f(t) = t^3 - 2t^2 - t$, $t \in \mathbb{T}$. Then

$$\begin{aligned}
\sigma(t) &= t + 1, \\
f^\Delta(t) &= (\sigma(t))^2 + t\sigma(t) + t^2 - 2(\sigma(t) + t) - 1 \\
&= (t+1)^2 + t(t+1) + t^2 - 2(t+1+t) - 1 \\
&= t^2 + 2t + 1 + t^2 + t + t^2 - 4t - 2 - 1
\end{aligned}$$

$$= 3t^2 - t - 2, \quad t \in \mathbb{T}.$$

Hence

$$f^{\Delta}(t) = 0 \quad \text{iff} \quad t \in \left\{-\frac{2}{3}, 1\right\}.$$

Therefore

$$f^{\Delta}(t) \geq 0 \quad \text{for } t \in \left(-\infty, -\frac{2}{3}]_{\mathbb{T}} \cup [1, \infty\right)_{\mathbb{T}},$$

and

$$f^{\Delta}(t) \leq 0 \quad \text{for } t \in \left[-\frac{2}{3}, 1\right]_{\mathbb{T}}.$$

Consequently, f is increasing in $(-\infty, -1]_{\mathbb{T}} \cup [1, +\infty)_{\mathbb{T}}$ and decreasing in $[-1, 1]_{\mathbb{T}}$.

Example 2.53. Let $\mathbb{T} = 2^{\mathbb{N}_0}$. We will investigate for monotonicity the function

$$f(t) = \frac{1-2t}{1+t}, \quad t \in \mathbb{T}.$$

Here $\sigma(t) = 2t$, $t \in \mathbb{T}$. Then

$$\begin{aligned} f^{\Delta}(t) &= \frac{-2(1+t) - (1-2t)}{(1+t)(1+2t)} \\ &= \frac{-2 - 2t - 1 + 2t}{(1+t)(1+2t)} \\ &= -\frac{3}{(1+t)(1+2t)}, \quad t \in \mathbb{T}. \end{aligned}$$

Therefore the function f is decreasing for all $t \in \mathbb{T}$.

Example 2.54. Let $\mathbb{T} = \mathbb{Z}$. We will investigate where the function

$$f(t) = 4t^3 - 21t^2 + 18t + 20, \quad t \in \mathbb{T},$$

is increasing or decreasing. Here $\sigma(t) = t + 1$, $t \in \mathbb{T}$, and

$$\begin{aligned} f^{\Delta}(t) &= 4((\sigma(t))^2 + t\sigma(t) + t^2) - 21(\sigma(t) + t) + 18 \\ &= 4((t+1)^2 + t(t+1) + t^2) - 21(t + 1 + t) + 18 \\ &= 4(t^2 + 2t + 1 + t^2 + t + t^2) - 21(2t + 1) + 18 \\ &= 12t^2 + 12t + 4 - 42t - 21 + 18 \\ &= 12t^2 - 30t + 1, \quad t \in \mathbb{T}. \end{aligned}$$

Hence

$$f^{\Delta}(t) = 0 \quad \text{and} \quad t \in \left\{ \frac{15+\sqrt{213}}{12}, \frac{15-\sqrt{213}}{12} \right\}.$$

Therefore $f^{\Delta}(t) \geq 0$ for $t \in (-\infty, 0]_{\mathbb{T}} \cup [3, +\infty)_{\mathbb{T}}$, and $f^{\Delta}(t) \leq 0$ for $t \in [1, 2]_{\mathbb{T}}$.

Example 2.55. Let $\mathbb{T} = h\mathbb{N}_0$, $h > 0$, and let

$$f(t) = (t+1)(t+2)(t+3)(t+4), \quad t \in \mathbb{T}.$$

By Example 2.13 we get

$$\begin{aligned} f^{\Delta}(t) &= 4t^3 + 3(2h+11)t^2 + (4h^2 + 33h + 57)t + h^2 + 10h + 60 \\ &\geq 0, \quad t \in \mathbb{T}. \end{aligned}$$

Thus f is increasing on $\mathbb{T}$.

Example 2.56. Let $\mathbb{T} = 3^{\mathbb{N}_0}$, and let

$$f(t) = t^3 + t^2 + t + 1, \quad t \in \mathbb{T}.$$

By Example 2.14 we get

$$\begin{aligned} f^{\Delta}(t) &= 13t^2 + 4t + 1 \\ &\geq 0, \quad t \in \mathbb{T}. \end{aligned}$$

Thus f is increasing on $\mathbb{T}$.

Example 2.57. Let $\mathbb{T} = \{H_n : n \in \mathbb{N}_0\}$, where H_n, $n \in \mathbb{N}$, are the harmonic numbers, and let

$$f(t) = \frac{t}{t^2 - 4t - 12}, \quad t \in \mathbb{T}.$$

By Example 2.16 we have

$$f^{\Delta}(H_n) = -\frac{H_n H_{n+1} + 12}{(H_n^2 - 4H_n - 12)(H_{n+1}^2 - 4H_{n+1} - 12)}, \quad n \in \mathbb{N}.$$

Thus, for

$$n \in \{s \in \mathbb{N} : H_{s+1} \leq 6\} \cup \{s \in \mathbb{N} : H_s \geq 6\},$$

the function f is decreasing, and for

$$n \in \{s \in \mathbb{N} : H_s \leq 6, H_{s+1} \geq 6\},$$

it is increasing.

Example 2.58. Let $\mathbb{T} = P_{1,3}$, and let

$$f(t) = \frac{1-t}{1+t}, \quad t \in \mathbb{T}.$$

By Example 2.17 we get

$$\begin{aligned} f^{\Delta}(t) &= \begin{cases} -\frac{2}{(1+t)(4+t)} & \text{if } t \text{ is right-scattered,} \\ -\frac{2}{(1+t)^2} & \text{if } t \text{ is right-dense} \end{cases} \\ &\leq 0. \end{aligned}$$

Thus f is decreasing on $\mathbb{T}$.

Example 2.59. Let $\mathbb{T} = C$, where C is the Cantor set, and let

$$f(t) = \frac{1-t^2}{1+t^2}, \quad t \in \mathbb{T}.$$

By Example 2.18 we get

$$\begin{aligned} f^{\Delta}(t) &= \begin{cases} \frac{-4t-2a}{(1+t^2)(1+a^2+2at+t^2)} & \text{if } t \in C_1, \\ -\frac{4t}{(1+t^2)^2} & \text{if } t \in \mathbb{T}\backslash C_1 \end{cases} \\ &\leq 0, \end{aligned}$$

where $a = \frac{1}{3^{m+1}}$, $m \in \mathbb{N}$. Thus f is decreasing on $\mathbb{T}$.

Example 2.60. Let $\mathbb{T} = \{t_n = \sum_{k=0}^{n-1} a_k,\ n \in \mathbb{N},\ a_k > 0,\ k \in \mathbb{N}_0\}$, and let

$$f(t) = \frac{1+2t}{3+4t}, \quad t \in \mathbb{T}.$$

By Example 2.19 we get

$$\begin{aligned} f^{\Delta}(t_n) &= \frac{2}{(3+4t_n)(3+4t_{n+1})} \\ &\geq 0, \quad n \in \mathbb{N}. \end{aligned}$$

Thus f is increasing on $\mathbb{T}$.

Example 2.61. Let $\mathbb{T} = \{-\frac{1}{n} : n \in \mathbb{N}\} \cup \mathbb{N}_0$, and let

$$f(t) = \frac{1+t^2}{1+7t}, \quad t \in \mathbb{T}.$$

We have the following cases.

1. $t \in \{-\frac{1}{n} : n \in \mathbb{N}\}$. By Example 2.20 we have

$$f^{\Delta}(t) = \frac{6t^2 + 9t - 7}{(1 + 7t)(1 + 6t)}$$

and

$$f^{\Delta}(0) = -6.$$

So for $t \in [-1, -\frac{1}{6}]_{\mathbb{T}} \cup [-\frac{1}{7}, 0]_{\mathbb{T}}$, the function f is decreasing, and for $t \in [-\frac{1}{6}, -\frac{1}{7}]_{\mathbb{T}}$, it is increasing.

2. $t \in \mathbb{N}$. By Example 2.20 we get

$$\begin{aligned} f^{\Delta}(t) &= \frac{7t^2 + 9t - 6}{(1 + 7t)(1 + 6t)} \\ &\geq 0. \end{aligned}$$

Thus f is increasing on $\mathbb{T}$.

Example 2.62. Let $\mathbb{T} = \{(\frac{1}{2})^{2^n} : n \in \mathbb{N}_0\} \cup \{0, 1\}$, and let

$$f(t) = \frac{1 + 7t}{4 + 5t}, \quad t \in \mathbb{T}.$$

By Example 2.21 we get

$$\begin{aligned} f^{\Delta}(t) &= \begin{cases} \frac{23}{(4+5t)(4+5\sqrt{t})} & \text{if } t \in \{(\frac{1}{2})^{2^n} : n \in \mathbb{N}\}, \\ \frac{46}{117} & \text{if } t = \frac{1}{2}, \\ \frac{23}{16} & \text{if } t = 0 \end{cases} \\ &\geq 0. \end{aligned}$$

Thus f is increasing on $\mathbb{T}$.

Exercise 2.27. Let $\mathbb{T}$ and f be as in Exercise 2.8. Investigate where f is increasing and decreasing.

Exercise 2.28. Let $\mathbb{T}$ and f be as in Exercise 2.10. Investigate where f is increasing and decreasing.

Exercise 2.29. Let $\mathbb{T}$ and f be as in Exercise 2.12. Investigate where f is increasing and decreasing.

2.7 Extreme values

Let f be defined on $D_f \subset \mathbb{T}$, and let $t_0 \in \mathbb{T}$.

Definition 2.5. We say that $f(t_0)$ is a local maximum value of f if there exists $\delta > 0$ such that

$$f(t) \le f(t_0) \quad \text{for all } t \in (t_0 - \delta, t_0 + \delta) \cap D_f$$

and $f(\rho(t_0)), f(\sigma(t_0)) \le f(t_0)$ or a local minimum value of f if there exists $\delta > 0$ such that

$$f(t) \ge f(t_0) \quad \text{for all } t \in (t_0 - \delta, t_0 + \delta) \cap D_f$$

and $f(\rho(t_0)), f(\sigma(t_0)) \ge f(t_0)$. The point t_0 is called a local extreme point of f or, more specifically, a local maximum or local minimum point of f.

Example 2.63. Let f be delta and nabla differentiable in a neighborhood $(t_0 - \delta, t_0 + \delta)$ of t_0. We will prove that if $f^{\Delta}(t) \le 0$ in $[t_0, t_0 + \delta)$ and $f^{\nabla}(t) \ge 0$ in $(t_0 - \delta, t_0]$, then t_0 is a local maximum point of f.

Indeed, we have the following cases.

1. t_0 is an isolated point. Then

$$\rho(t_0) < t_0 < \sigma(t_0),$$

and

$$\begin{aligned} f^{\Delta}(t_0) &= \frac{f(\sigma(t_0)) - f(t_0)}{\sigma(t_0) - t_0} \\ &\le 0, \\ f^{\nabla}(t_0) &= \frac{f(t_0) - f(\rho(t_0))}{t_0 - \rho(t_0)} \\ &\ge 0. \end{aligned}$$

Therefore $f(t_0) \ge f(\sigma(t_0))$ and $f(t_0) \ge f(\rho(t_0))$. Also, there exists $\delta_1 > 0$ such that $f(t) \le f(t_0)$ for all $t \in (t_0 - \delta_1, t_0 + \delta_1)$. Consequently, t_0 is a local maximum point.

2. t_0 is left-dense and right-scattered. As above, we have that $f(\sigma(t_0)) \le f(t_0)$. Since t_0 is left-dense, we have that $f^{\nabla}(t_0) = f'(t_0)$ and there exists $\delta_1 > 0$ such that $f^{\nabla}(t) = f'(t)$ for every $t \in (t_0 - \delta_1, t_0]$. For every $t_1 \in (t_0 - \delta_1, t_0]$, there exists $\xi_1 \in (t_1, t_0)$ such that

$$f'(\xi_1) = \frac{f(t_1) - f(t_0)}{t_1 - t_0}.$$

Because $f'(\xi_1) \ge 0$, we obtain that $f(t_1) \le f(t_0)$. Consequently, there exists $\delta_2 > 0$, $\delta_2 \le \delta_1$, such that for every $t \in (t_0 - \delta_2, t_0 + \delta_2)$, we have $f(t) \le f(t_0)$. Therefore t_0 is a local maximum point.

The cases where t_0 is left-scattered and right-dense and where t_0 is dense are left to the reader as an exercise.

Exercise 2.30. Let f be delta and nabla differentiable in a neighborhood $(t_0 - \delta, t_0 + \delta)$ of t_0. If $f^\Delta(t) \geq 0$ on $[t_0, t_0 + \delta)$ and $f^\nabla(t) \leq 0$ on $(t_0 - \delta, t_0]$, then prove that t_0 is a local minimum point of f.

Example 2.64. Let $\mathbb{T} = \mathbb{Z}$. Consider the function

$$f(t) = t^2 - 5t + 4, \quad t \in \mathbb{T}.$$

Then

$$\begin{aligned} f^\Delta(t) &= \sigma(t) + t - 5 \\ &= t + 1 + t - 5 \\ &= 2t - 4, \quad t \in \mathbb{T}, \end{aligned}$$

and

$$\begin{aligned} f^\nabla(t) &= \rho(t) + t - 5 \\ &= t - 1 + t - 5 \\ &= 2t - 6, \quad t \in \mathbb{T}. \end{aligned}$$

Hence

$$f^\Delta(t) \leq 0 \quad \text{and} \quad f^\nabla(t) \geq 0$$

iff

$$2t - 4 \leq 0 \quad \text{and} \quad 2t - 6 \geq 0$$

iff

$$t \leq 2 \quad \text{and} \quad t \geq 3.$$

Therefore f has no local maximum points. Also,

$$f^\Delta(t) \geq 0 \quad \text{and} \quad f^\nabla(t) \leq 0$$

iff

$$2t - 4 \geq 0 \quad \text{and} \quad 2t - 6 \leq 0$$

iff

$$t \geq 2 \quad \text{and} \quad t \leq 3.$$

Consequently, $t = 2$ and $t = 3$ are local minimum points. We have

$$\begin{aligned} f_{min} &= f(2) \\ &= f(3) \\ &= -2. \end{aligned}$$

Example 2.65. Let $\mathbb{T} = \mathbb{Z}$. We will find the local extreme values of the function

$$f(t) = \frac{t+1}{t^2+1}, \quad t \in \mathbb{T}.$$

Here

$$\sigma(t) = t+1, \quad t \in \mathbb{T},$$

and

$$\rho(t) = t-1, \quad t \in \mathbb{T}.$$

Also,

$$\begin{aligned} f^{\Delta}(t) &= \frac{t^2+1-(t+1)(\sigma(t)+t)}{(t^2+1)((t+1)^2+1)} \\ &= \frac{t^2+1-(t+1)(t+1+t)}{(t^2+1)(t^2+2t+2)} \\ &= \frac{t^2+1-(t+1)(2t+1)}{(t^2+1)(t^2+2t+2)} \\ &= \frac{t^2+1-(2t^2+t+2t+1)}{(t^2+1)(t^2+2t+2)} \\ &= \frac{t^2+1-2t^2-3t-1}{(t^2+1)(t^2+2t+2)} \\ &= \frac{-t^2-3t}{(t^2+1)(t^2+2t+2)} \\ &= -\frac{t(t+3)}{(t^2+1)(t^2+2t+2)}, \quad t \in \mathbb{T}, \end{aligned}$$

and

$$\begin{aligned} f^{\nabla}(t) &= \frac{t^2+1-(t+1)(\rho(t)+t)}{(t^2+1)((t-1)^2+1)} \\ &= \frac{t^2+1-(t+1)(2t-1)}{(t^2+1)(t^2-2t+2)} \end{aligned}$$

$$\begin{aligned}
&= \frac{t^2+1-(2t^2-t+2t-1)}{(t^2+1)(t^2-2t+2)}\\
&= \frac{t^2+1-2t^2-t+1}{(t^2+1)(t^2-2t+2)}\\
&= \frac{-t^2-t+2}{(t^2+1)(t^2-2t+2)}\\
&= -\frac{t^2+t-2}{(t^2+1)(t^2-2t+2)}\\
&= -\frac{(t+2)(t-1)}{(t^2+1)(t^2-2t+2)}, \quad t \in \mathbb{T}.
\end{aligned}$$

Hence

$$f^{\Delta}(t) \le 0 \quad \text{and} \quad f^{\nabla}(t) \ge 0$$

iff

$$-\frac{t(t+3)}{(t^2+1)(t^2+2t+2)} \le 0 \quad \text{and} \quad -\frac{(t+2)(t-1)}{(t^2+1)(t^2-2t+2)} \ge 0$$

iff

$$t(t+3) \ge 0 \quad \text{and} \quad (t-1)(t+2) \le 0,$$

so that

$$t = 0 \quad \text{and} \quad t = 1.$$

Therefore

$$\begin{aligned}
f_{\max} &= f(0)\\
&= f(1)\\
&= 1.
\end{aligned}$$

Also,

$$f^{\Delta}(t) \ge 0 \quad \text{and} \quad f^{\nabla}(t) \le 0$$

iff

$$-\frac{t(t+3)}{(t^2+1)(t^2+2t+2)} \ge 0 \quad \text{and} \quad -\frac{(t+2)(t-1)}{(t^2+1)(t^2-2t+2)} \le 0$$

iff

$$t(t+3) \le 0 \quad \text{and} \quad (t-1)(t+2) \ge 0,$$

so that

$$t = -2 \quad \text{and} \quad t = -1.$$

Consequently,

$$\begin{aligned} f_{\min} &= f(-2) \\ &= \frac{-2+1}{4+1} \\ &= -\frac{1}{5}, \end{aligned}$$

and

$$\begin{aligned} f_{\min} &= f(-1) \\ &= 0. \end{aligned}$$

Example 2.66. Let $\mathbb{T} = 2^{\mathbb{N}_0}$. We will find the extreme values of the function

$$f(t) = \frac{t^2+2}{t+2} \quad \text{for } t \geq 4.$$

Here

$$\sigma(t) = 2t,$$

and

$$\rho(t) = \frac{1}{2}t$$

for all $t \geq 4$ in $\mathbb{T}$. Then for $t \geq 4$, we have

$$\begin{aligned} f^{\Delta}(t) &= \frac{(\sigma(t)+t)(t+2) - (t^2+2)}{(t+2)(2t+2)} \\ &= \frac{3t(t+2) - (t^2+2)}{2(t+1)(t+2)} \\ &= \frac{3t^2 + 6t - t^2 - 2}{2(t+1)(t+2)} \\ &= \frac{t^2 + 3t - 1}{(t+1)(t+2)}, \quad t \in \mathbb{T}, \end{aligned}$$

and

$$f^{\nabla}(t) = \frac{(\rho(t)+t)(t+2) - (t^2+2)}{(t+2)(\rho(t)+2)}$$

$$
\begin{aligned}
&= \frac{\frac{3}{2}t(t+2) - t^2 - 2}{(t+2)(\frac{1}{2}t+2)}\\
&= \frac{\frac{3}{2}t^2 + 3t - t^2 - 2}{(t+2)(\frac{1}{2}t+2)}\\
&= \frac{\frac{1}{2}t^2 + 3t - 2}{(t+2)(\frac{1}{2}t+2)}\\
&= \frac{t^2 + 6t - 4}{(t+2)(t+4)}, \quad t \in \mathbb{T}.
\end{aligned}
$$

Note that $f^{\Delta}(t) \geq 0$ and $f^{\nabla}(t) \geq 0$ for all $t \geq 4$. Therefore the function f has no local extreme values.

Example 2.67. Let $\mathbb{T} = P_{1,3}$, and let

$$f(t) = \frac{1-t}{1+t}, \quad t \in \mathbb{T}.$$

We will find the local extreme points of f. Consider the following cases.

1. $t \subset \mathbb{T}$ is right-scattered. Then t is left-dense. By Example 2.17 we have

$$
\begin{aligned}
f^{\Delta}(t) &= -\frac{2}{(t+1)(t+4)}\\
&\leq 0.
\end{aligned}
$$

   By Example 2.42 we have

$$
\begin{aligned}
f^{\nabla}(t) &= -\frac{2}{(1+t)^2}\\
&\leq 0.
\end{aligned}
$$

   Thus f has no local extreme points.

2. $t \in \mathbb{T}$ is right-dense. By Example 2.17 we have

$$
\begin{aligned}
f^{\Delta}(t) &= -\frac{2}{(1+t)^2}\\
&\leq 0.
\end{aligned}
$$

   a. t is left-dense. By Example 2.42 we have

$$
\begin{aligned}
f^{\nabla}(t) &= -\frac{2}{(1+t)^2}\\
&\leq 0.
\end{aligned}
$$

   Thus f has no local extreme points.

b. t is left-scattered. By Example 2.42 we have

$$\begin{aligned} f^{\nabla}(t) &= -\frac{2}{(t+1)(t-2)} \\ &\le 0, \quad t \ge 2, \end{aligned}$$

and

$$f^{\nabla}(t) \ge 0, \quad t \le 2.$$

Therefore all points of $[0,1]$ are local maximum points.

Example 2.68. Let $\mathbb{T} = \{-\frac{1}{n} : n \in \mathbb{N}\} \cup \mathbb{N}_0$, and let

$$f(t) = \frac{1+t^2}{1+7t}, \quad t \in \mathbb{T}.$$

We will find the local extreme points of f. We have the following cases.

1. $t \in \{-\frac{1}{n} : n \in \mathbb{N}\}$. By Example 2.20 we have

$$\begin{aligned} f^{\Delta}(t) &= \frac{6t^2+9t-7}{(1+7t)(1+6t)} \\ &\le 0 \quad \text{for } t < -\frac{1}{6} \text{ and } t > -\frac{1}{7}. \end{aligned}$$

By Example 2.43 we have

$$\begin{aligned} f^{\nabla}(t) &= \frac{8t^2-5t-7}{(8t+1)(7t+1)} \\ &\ge 0 \quad \text{for } t \in \left[-1, -\frac{11}{16}\right]_{\mathbb{T}}. \end{aligned}$$

Since

$$\left[-1, -\frac{11}{16}\right]_{\mathbb{T}} \cap \left\{-\frac{1}{n} : n \in \mathbb{N}\right\} = \{-1\},$$

we conclude that $t = -1$ is a local maximum point and

$$f_{max} = f(-1) = \frac{1+1}{1-7} = -\frac{1}{3}.$$

2. $t = 0$. By Example 2.20 we have

$$f^{\Delta}(0) = -\frac{3}{4}.$$

By Example 2.43 we get

$$f^{\nabla}(0) = -7.$$

Thus 0 is not a local extreme point for the function f.

3. $t \in \mathbb{N}$. By Example 2.20 we have

$$\begin{aligned} f^{\Delta}(t) &= \frac{7t^2 + 9t - 6}{(1+7t)(1+6t)} \\ &\geq 0. \end{aligned}$$

By Example 2.43 we get

$$\begin{aligned} f^{\nabla}(t) &= \frac{7t^2 - 5t - 8}{(7t+1)(7t-6)} \\ &\leq 0 \quad \text{for } t = 1. \end{aligned}$$

Thus $t = 1$ is a local minimum point, and

$$\begin{aligned} f_{\min} &= f(1) \\ &= \frac{1+1}{1+7} \\ &= \frac{1}{4}. \end{aligned}$$

Exercise 2.31. Let $\mathbb{T} = \mathbb{Z}$. Find the local extreme values of the function

$$f(t) = t^3 - 3t^2 + 4, \quad t \in \mathbb{T}.$$

Exercise 2.32. Let $\mathbb{T} = \{0\} \cup \{\frac{1}{n^2} : n \in \mathbb{N}\} \cup [2, 3] \cup 4^{\mathbb{N}}$. Find the local extreme points of the function

$$f(t) = t^3 + t^2 + t + 1, \quad t \in \mathbb{T}.$$

Definition 2.6. Suppose that f is Δ-differentiable and ∇-differentiable at t_0. We say that t_0 is a critical point of f if

$$f^{\Delta}(t_0) \leq 0 \quad \text{and} \quad f^{\nabla}(t_0) \geq 0$$

or

$$f^{\Delta}(t_0) \geq 0 \quad \text{and} \quad f^{\nabla}(t_0) \leq 0.$$

The least (greatest) value of a continuous function f on a given interval $[a, b]_{\mathbb{T}}$ is attained at the critical points of f or at the endpoints of the interval $[a, b]_{\mathbb{T}}$.

Example 2.69. Let $\mathbb{T} = 2^{\mathbb{N}_0}$. We will prove that

$$\frac{t^2+2}{t+3} \geq \frac{3}{4} \quad \text{for all } t \in \mathbb{T}.$$

We have $\sigma(t) = 2t$ for all $t \in \mathbb{T}$ and

$$\begin{aligned} f^{\Delta}(t) &= \frac{(t+\sigma(t))(t+3)-(t^2+2)}{(t+3)(2t+3)} \\ &= \frac{3t(t+3)-t^2-2}{(t+3)(2t+3)} \\ &= \frac{3t^2+9t-t^2-2}{(t+3)(2t+3)} \\ &= \frac{2t^2+9t-2}{(t+3)(2t+3)} \geq 0 \end{aligned}$$

for all $t \in \mathbb{T}$. Consequently, f is increasing in $\mathbb{T}$. Hence

$$f(t) \geq f(1) = \frac{3}{4} \quad \text{for all } t \in \mathbb{T}.$$

Example 2.70. Let $\mathbb{T} = 3^{\mathbb{N}_0}$. We will find a positive constant a such that

$$1 + a \log t \leq t^2 \quad \text{for all } t \in \mathbb{T}.$$

Let

$$f(t) = t^2 - a \log t - 1, \quad t \in \mathbb{T}.$$

Here $\sigma(t) = 3t$ for all $t \in \mathbb{T}$, and

$$\begin{aligned} f^{\Delta}(t) &= \sigma(t) + t - a\frac{\log \sigma(t) - \log t}{\sigma(t) - t} \\ &= 3t + t - a\frac{\log(3t) - \log t}{3t - t} \\ &= 4t - a\frac{\log 3}{2t} \quad \text{for all } t \in \mathbb{T}. \end{aligned}$$

Since

$$\frac{\log 3}{2t} \leq \frac{\log 3}{2} \quad \text{for all } t \in \mathbb{T},$$

we conclude that

$$4t - a\frac{\log 3}{2t} \geq 4 - a\frac{\log 3}{2} \quad \text{for all } t \in \mathbb{T}.$$

Hence, if $0 < a < \frac{8}{\log 3}$, then f is increasing in $\mathbb{T}$, from which it follows that

$$f(t) \geq f(1) = 0 \quad \text{for all } t \in \mathbb{T} \text{ and for } 0 < a < \frac{8}{\log 3}.$$

2.8 Convex and concave functions

Let $f : \mathbb{T} \to \mathbb{R}$.

Definition 2.7. The function f is called convex if for all $t_1, t_2 \in \mathbb{T}$ and all $\lambda \in [0,1]$, we have the inequality

$$f(\lambda t_1 + (1-\lambda)t_2) \leq \lambda f(t_1) + (1-\lambda)f(t_2).$$

Definition 2.8. The function f is called strictly convex if for all $t_1, t_2 \in \mathbb{T}$ such that $t_1 \neq t_2$, and all $\lambda \in (0,1)$, we have the inequality

$$f(\lambda t_1 + (1-\lambda)t_2) < \lambda f(t_1) + (1-\lambda)f(t_2).$$

Definition 2.9. The function f is said to be (strictly) concave if $-f$ is (strictly) convex.

Criteria for convexity and concavity of a function f read as follows.

- Let f be twice delta differentiable on (a,b) and $f^{\Delta^2}(t) \geq 0$ for all $t \in (a,b)$. Then f is convex.
- Let f be twice delta differentiable on (a,b) and $f^{\Delta^2}(t) \leq 0$ for all $t \in (a,b)$. Then f is concave.

Example 2.71. Let $\mathbb{T} = \mathbb{Z}$. Consider

$$f(t) = t^3 - 7t^2 + t - 10, \quad t \in \mathbb{T}.$$

Here $\sigma(t) = t+1$, $t \in \mathbb{T}$, and

$$\begin{aligned} f^{\Delta}(t) &= (\sigma(t))^2 + t\sigma(t) + t^2 - 7(\sigma(t) + t) + 1 \\ &= (t+1)^2 + t(t+1) + t^2 - 7(t+1+t) + 1 \\ &= t^2 + 2t + 1 + t^2 + t + t^2 - 14t - 7 + 1 \\ &= 3t^2 - 11t - 5, \\ f^{\Delta^2}(t) &= 3(\sigma(t) + t) - 11 \\ &= 3(t+1+t) - 11 \\ &= 6t - 8, \quad t \in \mathbb{T}. \end{aligned}$$

Hence

$$f^{\Delta^2}(t) \geq 0 \text{ for } t \geq 2$$

and

$$f^{\Delta^2}(t) \leq 0 \text{ for } t \leq 1.$$

Therefore f is convex for $t \geq 2$ and concave for $t \leq 1$.

Example 2.72. Let $\mathbb{T} = 2^{\mathbb{N}_0}$. Consider

$$f(t) = t^4 - t^3 - t^2 - t, \quad t \in \mathbb{T}.$$

Then $\sigma(t) = 2t$, $t \in \mathbb{T}$, and

$$\begin{aligned} f^{\Delta}(t) &= (\sigma(t))^3 + t(\sigma(t))^2 + t^2\sigma(t) + t^3 \\ &\quad - ((\sigma(t))^2 + t\sigma(t) + t^2) - (\sigma(t) + t) - 1 \\ &= 8t^3 + 4t^3 + 2t^3 + t^3 - (4t^2 + 2t^2 + t^2) - (2t + t) - 1 \\ &= 15t^3 - 7t^2 - 3t - 1, \\ f^{\Delta^2}(t) &= 15((\sigma(t))^2 + t\sigma(t) + t^2) - 7(\sigma(t) + t) - 3 \\ &= 15(4t^2 + 2t^2 + t^2) - 7(2t + t) - 3 \\ &= 105t^2 - 21t - 3, \quad t \in \mathbb{T}. \end{aligned}$$

Hence $f^{\Delta^2}(t) \geq 0$ for all $t \in \mathbb{T}$. Therefore the function f is convex on $\mathbb{T}$.

Example 2.73. Let $\mathbb{T} = 3^{\mathbb{N}_0}$. Consider the function

$$f(t) = \frac{t-3}{t+2}, \quad t \in \mathbb{T}.$$

We have $\sigma(t) = 3t$, $t \in \mathbb{T}$, and

$$\begin{aligned} f^{\Delta}(t) &= \frac{t + 2 - (t - 3)}{(t + 2)(3t + 2)} \\ &= \frac{5}{3t^2 + 2t + 6t + 4} \\ &= \frac{5}{3t^2 + 8t + 4}, \\ f^{\Delta^2}(t) &= -5\frac{3(\sigma(t) + t) + 8}{(3t^2 + 8t + 4)(3(\sigma(t))^2 + 8\sigma(t) + 4)} \\ &= -5\frac{12t + 8}{(3t^2 + 8t + 4)(27t^2 + 24t + 4)} \\ &= -20\frac{3t + 2}{(3t^2 + 8t + 4)(27t^2 + 24t + 4)} \\ &\leq 0 \quad \text{for all } t \in \mathbb{T}. \end{aligned}$$

Therefore f is a concave function on $\mathbb{T}$.

Example 2.74. Let $\mathbb{T} = h\mathbb{N}_0$, $h > 0$, and let

$$f(t) = (t+1)(t+2)(t+3)(t+4), \quad t \in \mathbb{T}.$$

By Example 2.28 we get

$$\begin{aligned} f^{\Delta^2}(t) &= 12t^2 + 6(4h+11)t + 14h^2 + 66h + 57 \\ &\geq 0, \quad t \in \mathbb{T}. \end{aligned}$$

Thus f is convex on $\mathbb{T}$.

Example 2.75. Let $\mathbb{T} = 3^{\mathbb{N}_0}$, and let

$$f(t) = t^3 + t^2 + t + 1, \quad t \in \mathbb{T}.$$

By Example 2.29 we get

$$\begin{aligned} f^{\Delta^2}(t) &= 52t + 4 \\ &\geq 0, \quad t \in \mathbb{T}. \end{aligned}$$

Thus f is convex on $\mathbb{T}$.

Example 2.76. Let $\mathbb{T} = \mathbb{N}_0^2$, and let

$$f(t) = \frac{1-t}{5+t}, \quad t \in \mathbb{T}.$$

By Example 2.30 we get

$$\begin{aligned} f^{\Delta^2}(t) &= \frac{6(8t^3 + 36t^2\sqrt{t} + 112t^2 + 164t\sqrt{t} + 180t + 104\sqrt{t} + 24)}{(5+t)(6+2\sqrt{t}+t)^2(9+4\sqrt{t}+t)(2\sqrt{t}+1)(2t\sqrt{t}+3t+3\sqrt{t}+1)} \\ &\geq 0, \quad t \in \mathbb{T}. \end{aligned}$$

Thus f is convex on $\mathbb{T}$.

Example 2.77. Let $\mathbb{T} = P_{1,3}$, and let

$$f(t) = \frac{1-t}{1+t}, \quad t \in \mathbb{T}.$$

By Example 2.32 we get

$$\begin{aligned} f^{\Delta^2}(t) &= \begin{cases} \frac{4}{(t+1)(t+4)(t+7)} & \text{if } t \text{ is right-scattered,} \\ \frac{4}{(t+1)^3} & \text{if } t \text{ is right-dense} \end{cases} \\ &\geq 0, \quad t \in \mathbb{T}. \end{aligned}$$

Thus f is convex on $\mathbb{T}$.

Example 2.78. Let $\mathbb{T} = C$, where C is the Cantor set, and let

$$f(t) = \frac{1-t^2}{1+t^2}, \quad t \in \mathbb{T}.$$

By Example 2.33 we get

$$f^{\Delta^2}(t) = \begin{cases} \frac{4(3t^2+6at+2a^2-a)}{(1+t^2)(1+a^2+2at+t^2)(1+4a^2+4at+t^2)} & \text{if } t \in C_1, \\ \frac{4(3t^2-1)}{(1+t^2)^2} & \text{if } t \in \mathbb{T}\backslash C_1, \end{cases}$$

where $a = \frac{1}{3^{m+1}}$. We have the following cases.

1. $t \in C_1$. Then for $t \in [0, \frac{-3a+\sqrt{3a^2+3a}}{3}]_{\mathbb{T}}$, the function f is concave, and for $t \in [\frac{-3a+\sqrt{3a^2+3a}}{3}, 1]_{\mathbb{T}}$, it is convex.
2. Let $t \in \mathbb{T}\backslash C_1$. Then for $t \in [0, \frac{1}{\sqrt{3}}]_{\mathbb{T}}$, the function f is concave, and for $t \in [\frac{1}{\sqrt{3}}, 1]_{\mathbb{T}}$, it is convex.

Example 2.79. Let $\mathbb{T} = \{t_n = \sum_{k=0}^{n-1} a_k,\ n \in \mathbb{N},\ a_k > 0,\ k \in \mathbb{N}_0\}$, and let

$$f(t) = \frac{1+2t}{3+4t}, \quad t \in \mathbb{T}.$$

By Example 2.34 we get

$$\begin{aligned} f^{\Delta^2}(t_n) &= \frac{8(t_n - t_{n+2})}{(t_{n+1}-t_n)(3+4t_n)(3+4t_{n+1})(3+4t_{n+2})} \\ &\leq 0, \quad n \in \mathbb{N}. \end{aligned}$$

Thus f is concave on $\mathbb{T}$.

Example 2.80. Let $\mathbb{T} = \{-\frac{1}{n} : n \in \mathbb{N}\} \cup \mathbb{N}_0$, and let

$$f(t) = \frac{1+t^2}{1+7t}, \quad t \in \mathbb{T}.$$

By Example 2.35 we have

$$f^{\Delta^2}(t) = \begin{cases} \frac{100(t-1)^2}{(1+5t)(1+6t)(1+7t)} & \text{if } t \in \{-\frac{1}{n} : n \in \mathbb{N}\}, \\ \frac{100}{(1+7t)(8+7t)(15+7t)} & \text{if } t \in \mathbb{N}_0. \end{cases}$$

We have the following cases.

1. $t \in \{-\frac{1}{n} : n \in \mathbb{N}\}$. Then for $t \in [-1, -\frac{1}{5}]_{\mathbb{T}} \cup [-\frac{1}{6}, -\frac{1}{7}]_{\mathbb{T}}$, the function f is concave, and for $t \in [-\frac{1}{5}, -\frac{1}{6}]_{\mathbb{T}} \cup [-\frac{1}{7}, 0]_{\mathbb{T}}$, it is convex.
2. $t \in \mathbb{N}_0$. The function f is convex on $\mathbb{N}_0$.

Example 2.81. Let $\mathbb{T} = \{(\frac{1}{2})^{2^n} : n \in \mathbb{N}_0\} \cup \{0, 1\}$, and let

$$f(t) = \frac{1+7t}{4+5t}, \quad t \in \mathbb{T}.$$

By Example 2.36 we have that $f^{\Delta^2}(t)$ exists for $t \in \{(\frac{1}{2})^{2^n} : n \in \mathbb{N}\}$ and

$$\begin{aligned} f^{\Delta^2}(t) &= \frac{115(t - \sqrt[4]{t})}{\sqrt{t}(1-\sqrt{t})(4+5t)(4+5\sqrt{t})(4+5\sqrt[4]{t})} \\ &\leq 0, \quad t \in \left\{\left(\frac{1}{2}\right)^{2^n} : n \in \mathbb{N}\right\}. \end{aligned}$$

Thus f is concave on $\{(\frac{1}{2})^{2^n} : n \in \mathbb{N}\}$.

Exercise 2.33. Let $\mathbb{T} = \mathbb{Z}$. Find the intervals of convexity and concavity of the following functions:
1. $f(t) = t^3 - 6t^2 + 12t + 4$, $t \in \mathbb{T}$.
2. $f(t) = \frac{1}{t+3}$, $t \in \mathbb{T}$.

Exercise 2.34. Let $\mathbb{T}$ and f be as in Exercise 2.13. Determine where f is convex and concave.

Exercise 2.35. Let $\mathbb{T}$ and f be as in Exercise 2.14. Determine where f is convex and concave.

Exercise 2.36. Let $\mathbb{T}$ and f be as in Exercise 2.15. Determine where f is convex and concave.

Exercise 2.37. Let $\mathbb{T}$ and f be as in Exercise 2.16. Determine where f is convex and concave.

Exercise 2.38. Let $\mathbb{T}$ and f be as in Exercise 2.17. Determine where f is convex and concave.

Exercise 2.39. Let $\mathbb{T}$ and f be as in Exercise 2.18. Determine where f is convex and concave.

Exercise 2.40. Let $\mathbb{T}$ and f be as in Exercise 2.19. Determine where f is convex and concave.

Exercise 2.41. Let f and g be convex functions. Prove that so are

$$m(x) = \max\{f(x), g(x)\} \quad \text{and} \quad h(x) = f(x) + g(x).$$

2.9 Completely delta-differentiable functions

Definition 2.10. A function $f : \mathbb{T} \to \mathbb{R}$ is called completely delta differentiable at a point $t^0 \in \mathbb{T}^\kappa$ if there exist constants A_1 and A_2 such that

$$f(t^0) - f(t) = A_1(t^0 - t) + \alpha(t^0 - t) \quad \text{for all } t \in U_\delta(t^0) \tag{2.3}$$

and

$$f(\sigma(t^0)) - f(t) = A_2(\sigma(t^0) - t) + \beta(\sigma(t^0) - t) \quad \text{for all } t \in U_\delta(t^0), \tag{2.4}$$

where $\alpha = \alpha(t^0, t)$ and $\beta = \beta(t^0, t)$ are equal to zero at t^0, and

$$\lim_{t \to t^0} \alpha(t^0, t) = \lim_{t \to t^0} \beta(t^0, t) = 0.$$

A criterion for a function f to be completely delta differentiable reads as follows.

- Let $f : \mathbb{T} \to \mathbb{R}$ be a continuous function having the first-order delta derivative $f^\Delta(t)$ in some δ-neighborhood $U_\delta(t^0)$ of the point $t^0 \in \mathbb{T}^\kappa$. If this derivative is continuous at the point t^0, then f is completely delta differentiable at t^0.

Example 2.82. Let $\mathbb{T} = h\mathbb{Z}$, $h > 0$, and let

$$f(t) = (t + 1)(t + 2)(t + 3)(t + 4), \quad t \in \mathbb{T}.$$

By the computations of Example 2.13 it follows that f^Δ is continuous at all points of $\mathbb{T}$ and hence is completely delta differentiable at all points of $\mathbb{T}$.

Example 2.83. Let $\mathbb{T} = 3^{\mathbb{N}_0}$, and let

$$f(t) = t^3 + t^2 + t + 1, \quad t \in \mathbb{T}.$$

By the computations of Example 2.14 it follows that f^Δ is continuous at all points of $\mathbb{T}$ and hence is completely delta differentiable at all points of $\mathbb{T}$.

Example 2.84. Let $\mathbb{T} = \mathbb{N}_0^2$, and let

$$f(t) = \frac{t^2 - 5t + 6}{t^2 + t + 7}, \quad t \in \mathbb{T}.$$

By the computations of Example 2.15 it follows that f^Δ is continuous at all points of $\mathbb{T}$ and hence is completely delta differentiable at all points of $\mathbb{T}$.

Example 2.85. Let $\mathbb{T} = \{H_n : n \in \mathbb{N}_0\}$, where H_n, $n \in \mathbb{N}_0$, are the harmonic numbers, and let

$$f(t) = \frac{t}{t^2 - 4t - 12}, \quad t \in \mathbb{T}.$$

By the computations of Example 2.16 it follows that f^{Δ} is continuous at all points of $\mathbb{T}$ and hence is completely delta differentiable at all points of $\mathbb{T}$.

Example 2.86. Let $\mathbb{T} = P_{1,3}$, and let

$$f(t) = \frac{1-t}{1+t}, \quad t \in \mathbb{T}.$$

By the computations of Example 2.17 it follows that f^{Δ} is continuous at all points of $\mathbb{T}$ and hence is completely delta differentiable at all points of $\mathbb{T}$.

Example 2.87. Let $\mathbb{T} = C$, where C is the Cantor set, and let

$$f(t) = \frac{1-t^2}{1+t^2}, \quad t \in \mathbb{T}.$$

By the computations of Example 2.18 it follows that f^{Δ} is continuous at all points of $\mathbb{T}$ and hence is completely delta differentiable at all points of $\mathbb{T}$.

Example 2.88. Let $\mathbb{T} = \{t_n = \sum_{k=0}^{n-1} a_k,\ n \in \mathbb{N},\ a_k > 0,\ k \in \mathbb{N}_0\}$, and let

$$f(t) = \frac{1+2t}{3+4t}, \quad t \in \mathbb{T}.$$

By the computations of Example 2.19 it follows that f^{Δ} is continuous at all points of $\mathbb{T}$ and hence is completely delta differentiable at all points of $\mathbb{T}$.

Example 2.89. Let $\mathbb{T} = \{-\frac{1}{n} : n \in \mathbb{N}\} \cup \mathbb{N}_0$, and let

$$f(t) = \frac{1+t^2}{1+7t}, \quad t \in \mathbb{T}.$$

By the computations of Example 2.20 it follows that f^{Δ} is continuous at all points $t \in \mathbb{T}$, $t \neq 0$, and hence is completely delta differentiable at all points $t \in \mathbb{T}$, $t \neq 0$. Let $t = 0$. Then

$$f^{\Delta}(0) = -\frac{3}{4},$$

and

$$\begin{aligned}\lim_{t\to 0} f^{\Delta}(t) &= \lim_{t\to 0} \frac{6t^2+9t-7}{(1+7t)(1+6t)} \\ &= -7 \\ &\neq f^{\Delta}(0).\end{aligned}$$

Thus f^{Δ} is not continuous at $t = 0$ and thus not completely delta differentiable at $t = 0$.

Example 2.90. Let $\mathbb{T} = \{(\frac{1}{2})^{2^n} : n \in \mathbb{N}_0\} \cup \{0, 1\}$, and let

$$f(t) = \frac{1 + 7t}{4 + 3t}, \quad t \in \mathbb{T}.$$

By the computations of Example 2.21 it follows that f^Δ is continuous at all points of $\mathbb{T}$ and hence is completely delta differentiable at all points of $\mathbb{T}$.

Example 2.91. Let $U = \{\frac{1}{2^n} : n \in \mathbb{N}\}$, let

$$\mathbb{T} = U \cup (1 - U) \cup (1 + U) \cup (2 - U) \cup (2 + U) \cup \{0, 1, 2\},$$

and let

$$f(t) = (2 + t^2)(1 - t^2), \quad t \in \mathbb{T}.$$

By the computations of Example 2.22 it follows that f^Δ is continuous at all points t of $\mathbb{T}$, $t \neq 2$, and hence is completely delta differentiable at all points t of $\mathbb{T}$, $t \neq 2$.

Exercise 2.42. Let $\mathbb{T}$ and f be as in Exercise 2.8. Check if f is completely delta differentiable at all points of $\mathbb{T}$.

Exercise 2.43. Let $\mathbb{T}$ and f be as in Exercise 2.9. Check if f is completely delta differentiable at all points of $\mathbb{T}$.

Exercise 2.44. Let $\mathbb{T}$ and f be as in Exercise 2.10. Check if f is completely delta differentiable at all points of $\mathbb{T}$.

Exercise 2.45. Let $\mathbb{T}$ and f be as in Exercise 2.11. Check if f is completely delta differentiable at all points of $\mathbb{T}$.

Exercise 2.46. Let $\mathbb{T}$ and f be as in Exercise 2.12. Check if f is completely delta differentiable at all points of $\mathbb{T}$.

2.10 One-sided delta derivatives

Definition 2.11. Let f be defined on $[t_0, b) \subset \mathbb{T}$. Then the right-hand derivative of f at t_0 is defined as

$$f_+^\Delta(t_0) = \lim_{t \to t_0+} \frac{f(\sigma(t_0)) - f(t)}{\sigma(t_0) - t},$$

whereas if f is defined on $(a, t_0] \subset \mathbb{T}$, then the left-hand derivative of f at t_0 is defined as

$$f_-^\Delta(t_0) = \lim_{t \to t_0-} \frac{f(\sigma(t_0)) - f(t)}{\sigma(t_0) - t}.$$

Remark 2.4. f is differentiable at t_0 if and only if $f_+^\Delta(t_0)$ and $f_-^\Delta(t_0)$ exist and

$$f^\Delta(t_0) = f_-^\Delta(t_0) = f_+^\Delta(t_0).$$

Example 2.92. Consider

$$f(t) = \begin{cases} t^2 + t & \text{for } t \in \{1,2,3\}, \\ t+1 & \text{for } t \in [-1,1), \end{cases}$$

where $[-1,1)$ is a real-valued interval. Note that f is continuous on

$$[-1,1) \cup \{1,2,3\}.$$

Also,

$$\begin{aligned} f_-^\Delta(1) &= \lim_{t\to 1-} \frac{f(\sigma(1)) - f(t)}{\sigma(1) - t} \\ &= \lim_{t\to 1-} \frac{f(2) - f(t)}{2-t} \\ &= \lim_{t\to 1-} \frac{6-(t+1)}{2-t} \\ &= \lim_{t\to 1-} \frac{5-t}{2-t} \\ &= 4, \\ f_+^\Delta(1) &= \lim_{t\to 1+} \frac{f(\sigma(1)) - f(t)}{\sigma(1) - t} \\ &= \lim_{t\to 1+} \frac{f(2) - f(t)}{2-t} \\ &= \lim_{t\to 1+} \frac{6-t^2-t}{2-t} \\ &= \lim_{t\to 1+} \frac{(2-t)(t+3)}{2-t} \\ &= \lim_{t\to 1+} (t+3) \\ &= 4. \end{aligned}$$

Therefore

$$f_-^\Delta(1) = f_+^\Delta(1).$$

Hence f is differentiable at $t = 1$.

Example 2.93. Consider

$$f(t) = \begin{cases} t+3 & \text{for } t \in [-2,2), \\ t^2 + t & \text{for } t \in \{2,4,8\}. \end{cases}$$

We have

$$\begin{aligned} f_-^\Delta(2) &= \lim_{t\to 2-} \frac{f(\sigma(2)) - f(t)}{\sigma(2) - t} \\ &= \lim_{t\to 2-} \frac{f(4) - f(t)}{4 - t} \\ &= \lim_{t\to 2-} \frac{20 - (t+3)}{4 - t} \\ &= \lim_{t\to 2-} \frac{17 - t}{4 - t} \\ &= \frac{15}{2}, \\ f_+^\Delta(t) &= \lim_{t\to 2+} \frac{f(\sigma(2)) - f(t)}{\sigma(2) - t} \\ &= \lim_{t\to 2+} \frac{f(4) - f(t)}{4 - t} \\ &= \lim_{t\to 2+} \frac{20 - t^2 - t}{4 - t} \\ &= 7. \end{aligned}$$

Therefore

$$f_-^\Delta(2) \neq f_+^\Delta(2),$$

and thus f is not differentiable at $t = 2$.

Example 2.94. Consider

$$f(t) = \begin{cases} 1 & \text{for } t \in \{3, 5, 7, 9\}, \\ 3 & \text{for } t \in [0, 3), \end{cases}$$

where $[0, 3)$ is a real-valued interval.

We have

$$\begin{aligned} f_-^\Delta(3) &= \lim_{t\to 3-} \frac{f(\sigma(3)) - f(t)}{\sigma(3) - t} \\ &= \lim_{t\to 3-} \frac{f(5) - f(t)}{5 - t} \\ &= \lim_{t\to 3-} \frac{1 - 3}{5 - t} \\ &= \lim_{t\to 3-} \frac{-2}{5 - t} \\ &= -1, \\ f_+^\Delta(3) &= \lim_{t\to 3+} \frac{f(\sigma(3)) - f(t)}{\sigma(3) - t} \end{aligned}$$

$$\begin{aligned}
&= \lim_{t\to 3+} \frac{f(5)-f(t)}{5-t}\\
&= \lim_{t\to 3+} \frac{1-1}{5-t}\\
&= 0.
\end{aligned}$$

Consequently,

$$f_-^\Delta(3) \neq f_+^\Delta(3),$$

and thus f is not differentiable at 3.

Example 2.95. Let

$$f(t) = \begin{cases} t^2 + \alpha & \text{if } t \in [-1,1), \\ \frac{t+1}{t+2} & \text{if } t \in \{1,4,7,9\}, \end{cases}$$

where α is a parameter. We will find the values of α such that

$$f_-^\Delta(1) = f_+^\Delta(1).$$

We have

$$\begin{aligned}
\sigma(1) &= 4,\\
f_-^\Delta(1) &= \lim_{t\to 1-} \frac{f(\sigma(1))-f(t)}{\sigma(1)-t}\\
&= \lim_{t\to 1-} \frac{f(4)-f(t)}{4-t}\\
&= \lim_{t\to 1-} \frac{\frac{4+1}{4+2} - t^2 - \alpha}{4-t}\\
&= \frac{\frac{5}{6} - 1 - \alpha}{3}\\
&= \frac{-1-6\alpha}{18},
\end{aligned}$$

and

$$\begin{aligned}
f_+^\Delta(1) &= \lim_{t\to 1+} \frac{f(\sigma(1))-f(t)}{\sigma(1)-t}\\
&= \lim_{t\to 1+} \frac{f(4)-f(t)}{4-t}\\
&= \lim_{t\to 1+} \frac{\frac{5}{6} - \frac{t+1}{t+2}}{4-t}\\
&= \frac{\frac{5}{6} - \frac{2}{3}}{3}
\end{aligned}$$

$$= \frac{5-4}{18} = \frac{1}{18}.$$

Thus

$$\begin{aligned} f_-^\Delta(1) = f_+^\Delta(1) \quad &\text{if and only if} \\ \frac{-1-6\alpha}{18} = \frac{1}{18} \quad &\text{if and only if} \\ -1-6\alpha = 1 \quad &\text{if and only if} \\ -6\alpha = 2 \quad &\text{if and only if} \\ \alpha = -\frac{1}{3}. \end{aligned}$$

Example 2.96. Let

$$f(t) = \begin{cases} t^2 - 1 & \text{if } t \in [-2, 0), \\ t + \alpha & \text{if } t \in \{0, 2\} \cup 3^{\mathbb{N}}, \end{cases}$$

where α is a real parameter. We will find the values of α such that

$$f_-^\Delta(0) = f_+^\Delta(0).$$

We have

$$\begin{aligned} \sigma(0) &= 2, \\ f_-^\Delta(0) &= \lim_{t\to 0-} \frac{f(\sigma(0)) - f(t)}{\sigma(0) - t} \\ &= \lim_{t\to 0-} \frac{f(2) - f(t)}{2 - t} \\ &= \lim_{t\to 0-} \frac{2 + \alpha + (1 - t^2)}{2 - t} \\ &= \frac{3 + \alpha}{2}, \end{aligned}$$

and

$$\begin{aligned} f_+^\Delta(0) &= \lim_{t\to 0+} \frac{f(\sigma(0)) - f(t)}{\sigma(0) - t} \\ &= \lim_{t\to 0+} \frac{f(2) - f(t)}{2 - t} \\ &= \lim_{t\to 0+} \frac{2 + \alpha - t - \alpha}{2 - t} \\ &= 1. \end{aligned}$$

Thus

$$f_-^\Delta(0) = f_+^\Delta(0) \quad \text{if and only if}$$
$$\frac{3+a}{2} = 1 \quad \text{if and only if}$$
$$3 + a = 2 \quad \text{if and only if}$$
$$a = -1.$$

Exercise 2.47. Let

$$f(t) = \begin{cases} t^2 + 4t - 5 & \text{for } t \in \{0, 1, 2, 3, 4\}, \\ t^7 + 27 & \text{for } t \in [-3, 0). \end{cases}$$

Check if $f_-^\Delta(0) = f_+^\Delta(0)$.

Exercise 2.48. Let $\mathbb{T} = (-3\mathbb{N}_0) \cup 4^{\mathbb{N}_0}$, and let

$$f(t) = \frac{1 - 3t}{2 + t + t^2}, \quad t \in \mathbb{T}.$$

Check if $f_-^\Delta(t) = f_+^\Delta(t), t \in \mathbb{T}$.

Exercise 2.49. Let

$$f(t) = \begin{cases} \frac{2-3t}{5} & \text{if } t \in (-\infty, -2], \\ t^2 + t & \text{if } t \in (-2, 1] \cup 3^{\mathbb{N}}. \end{cases}$$

Check if $f_-^\Delta(-2) = f_+^\Delta(-2)$.

Exercise 2.50. Let

$$f(t) = \begin{cases} 1 + at & \text{if } t \in \{-1, 0, 1\}, \\ 1 - t^2 & \text{if } t \in (1, 2] \cup \{2 + \frac{1}{n}\}_{n \in \mathbb{N}}, \end{cases}$$

where a is a real parameter. Find the values of a such that $f_-^\Delta(1) = f_+^\Delta(1)$.

Exercise 2.51. Let

$$f(t) = \begin{cases} 1 - t + 2t^2 & \text{if } t \in [-3, 2), \\ at - 3 & \text{if } t \in \{2, 3\} \cup 4^{\mathbb{N}}, \end{cases}$$

where a is a real parameter. Find the values of a such that $f_-^\Delta(2) = f_+^\Delta(2)$.

2.11 Chain rules

The chain rules in the framework of time scale analysis read as follows.

- Let $g : \mathbb{R} \to \mathbb{R}$ be continuous, let $g : \mathbb{T} \to \mathbb{R}$ be delta differentiable on $\mathbb{T}^\kappa$, and let $f : \mathbb{R} \to \mathbb{R}$ be continuously differentiable. Then there exists $c \in [t, \sigma(t)]_\mathbb{T}$ such that

$$(f \circ g)^\Delta(t) = f'(g(c))g^\Delta(t), \quad t \in \mathbb{T}.$$

- Let $f : \mathbb{R} \to \mathbb{R}$ be continuously differentiable, and let $g : \mathbb{T} \to \mathbb{R}$ be delta differentiable. Then $f \circ g : \mathbb{T} \to \mathbb{R}$ is delta differentiable, and

$$(f \circ g)^\Delta(t) = \left\{ \int_0^1 f'(g(t) + h\mu(t)g^\Delta(t))dh \right\} g^\Delta(t), \quad t \in \mathbb{T}. \tag{2.5}$$

- Let $v : \mathbb{T} \to \mathbb{R}$ be strictly increasing, let $\tilde{\mathbb{T}} = v(\mathbb{T})$ be a time scale with delta differentiation operator $\tilde{\Delta}$, and let $w : \tilde{\mathbb{T}} \to \mathbb{R}$. If $v^\Delta(t)$ and $w^{\tilde{\Delta}}(v(t))$ exist for $t \in \mathbb{T}^\kappa$, then

$$(w \circ v)^\Delta = (w^{\tilde{\Delta}} \circ v)v^\Delta.$$

Example 2.97. Let $\mathbb{T} = \mathbb{Z}$, $f(t) = t^3 + 1$, and $g(t) = t^2$, $t \in \mathbb{T}$. Then $g : \mathbb{R} \to \mathbb{R}$ is continuous, $g : \mathbb{T} \to \mathbb{R}$ is delta differentiable on $\mathbb{T}^\kappa$, $f : \mathbb{R} \to \mathbb{R}$ is continuously differentiable, and $\sigma(t) = t + 1$, $t \in \mathbb{T}$. We have

$$\begin{aligned} g^\Delta(t) &= \sigma(t) + t, \quad t \in \mathbb{T}, \\ (f \circ g)^\Delta(1) &= f'(g(c))g^\Delta(1) \\ &= 3(g(c))^2(\sigma(1) + 1) \\ &= 9c^4. \end{aligned} \tag{2.6}$$

Here $c \in [1, \sigma[1]]_\mathbb{T} = [1, 2]_\mathbb{T}$. Also,

$$\begin{aligned} f \circ g(t) &= f(g(t)) \\ &= (g(t))^3 + 1 \\ &= t^6 + 1, \\ (f \circ g)^\Delta(t) &= (\sigma(t))^5 + t(\sigma(t))^4 + t^2(\sigma(t))^3 + t^3(\sigma(t))^2 + t^4\sigma(t) + t^5, \\ (f \circ g)^\Delta(1) &= (\sigma(1))^5 + (\sigma(1))^4 + (\sigma(1))^3 + (\sigma(1))^2 + \sigma(1) + 1 \\ &= 63, \quad t \in \mathbb{T}. \end{aligned}$$

By (2.6) we get

$$63 = 9c^4,$$

so

$$c^4 = 7,$$

and thus

$$\begin{aligned} c &= \sqrt[4]{7} \\ &\in [1, 2]. \end{aligned}$$

Example 2.98. Let $\mathbb{T} = \{2^n : n \in \mathbb{N}_0\}, f(t) = t + 2$, and $g(t) = t^2 - 1, t \in \mathbb{T}$. Then $g : \mathbb{T} \to \mathbb{R}$ is delta differentiable, $g : \mathbb{R} \to \mathbb{R}$ is continuous, and $f : \mathbb{R} \to \mathbb{R}$ is continuously differentiable. For $t = 2^n \in \mathbb{T}$ and $n = \log_2 t \in \mathbb{N}_0$, we have

$$\begin{aligned} \sigma(t) &= \inf\{2^l : 2^l > 2^n, l \in \mathbb{N}_0\} \\ &= 2^{n+1} \\ &= 2t \\ &> t. \end{aligned}$$

Therefore all points of $\mathbb{T}$ are right-scattered. Since $\sup \mathbb{T} = \infty$, we have that $\mathbb{T}^\kappa = \mathbb{T}$. Also, for $t \in \mathbb{T}$, we have

$$\begin{aligned} (f \circ g)(t) &= f(g(t)) \\ &= g(t) + 2 \\ &= t^2 - 1 + 2 = t^2 + 1, \\ (f \circ g)^\Delta(t) &= \sigma(t) + t \\ &= 2t + t \\ &= 3t, \quad t \in \mathbb{T}. \end{aligned}$$

Hence

$$(f \circ g)^\Delta(2) = 6. \tag{2.7}$$

Now we get that there is $c \in [2, \sigma(2)] = [2, 4]$ such that

$$\begin{aligned} (f \circ g)^\Delta(2) &= f'(g(c))g^\Delta(2) \\ &= g^\Delta(2) \\ &= \sigma(2) + 2 \\ &= 4 + 2 \\ &= 6. \end{aligned} \tag{2.8}$$

From (2.7) and (2.8) we find that for every $c \in [2, 4]$ one has

$$(f \circ g)^\Delta(2) = f'(g(c))g^\Delta(2).$$

Example 2.99. Let $\mathbb{T} = \{3^{n^2} : n \in \mathbb{N}_0\}$, $f(t) = t^2 + 1$, $g(t) = t^3$, $t \in \mathbb{T}$. Then $g : \mathbb{R} \to \mathbb{R}$ is continuous, $g : \mathbb{T} \to \mathbb{R}$ is delta differentiable, and $f : \mathbb{R} \to \mathbb{R}$ is continuously differentiable. For $t = 3^{n^2} \in \mathbb{T}$ and $n = (\log_3 t)^{\frac{1}{2}} \in \mathbb{N}_0$, we have

$$\begin{aligned}\sigma(t) &= \inf\{3^{l^2} : 3^{l^2} > 3^{n^2}, l \in \mathbb{N}_0\}\\ &= 3^{(n+1)^2}\\ &= 3 \cdot 3^{n^2} \cdot 3^{2n}\\ &= 3t3^{2(\log_3 t)^{\frac{1}{2}}} > t.\end{aligned}$$

Consequently, all points of $\mathbb{T}$ are right-scattered. Also, $\sup \mathbb{T} = \infty$. Then $\mathbb{T}^\kappa = \mathbb{T}$. Hence, for $t \in \mathbb{T}$, we have

$$\begin{aligned}(f \circ g)(t) &= f(g(t))\\ &= (g(t))^2 + 1\\ &= t^6 + 1,\\ (f \circ g)^\Delta(t) &= (t^6 + 1)^\Delta\\ &= (\sigma(t))^5 + t(\sigma(t))^4 + t^2(\sigma(t))^3 + t^3(\sigma(t))^2 + t^4\sigma(t) + t^5, \quad t \in \mathbb{T},\\ (f \circ g)^\Delta(1) &= (\sigma(1))^5 + (\sigma(1))^4 + (\sigma(1))^3 + (\sigma(1))^2 + \sigma(1) + 1\\ &= 3^5 + 3^4 + 3^3 + 3^2 + 3 + 1 \qquad (2.9)\\ &= 364.\end{aligned}$$

Then there is $c \in [1, \sigma(1)] = [1, 3]$ such that

$$\begin{aligned}(f \circ g)^\Delta(1) &= f'(g(c))g^\Delta(1)\\ &= 2g(c)g^\Delta(1) \qquad (2.10)\\ &= 2c^3 g^\Delta(1).\end{aligned}$$

Because all points of $\mathbb{T}$ are right-scattered, we have

$$\begin{aligned}g^\Delta(1) &= (\sigma(1))^2 + \sigma(1) + 1\\ &= 9 + 3 + 1\\ &= 13.\end{aligned}$$

By (2.10) we find

$$(f \circ g)^\Delta(1) = 26c^3.$$

From the last equation and from (2.9) we obtain

$$364 = 26c^3,$$

so

$$c^3 = \frac{364}{26} = 14,$$

and thus

$$c = \sqrt[3]{14}.$$

Example 2.100. Let $\mathbb{T} = \mathbb{Z}$, $f(t) = \frac{1}{1+t^2}$, and $g(t) = t+1$, $t \in \mathbb{T}$. Then $f : \mathbb{R} \to \mathbb{R}$ is continuously differentiable, and $g : \mathbb{T} \to \mathbb{R}$ is delta differentiable. We have

$$\begin{aligned} f'(t) &= -\frac{2t}{(1+t^2)^2}, \\ \mu(t) &= 1, \\ g^{\Delta}(t) &= 1, \\ g(t) + h\mu(t)g^{\Delta}(t) &= t+1+h, \quad t \in \mathbb{T}. \end{aligned}$$

Hence

$$\begin{aligned} f'(g(t) + h\mu(t))g^{\Delta}(t) &= f'(t+1+h) \\ &= -\frac{2(t+1+h)}{(1+(t+1+h)^2)^2}, \quad t \in \mathbb{T}. \end{aligned}$$

From here we conclude that $f \circ g : \mathbb{T} \to \mathbb{R}$ is delta differentiable and

$$\begin{aligned} (f \circ g)^{\Delta}(t) &= -\int_0^1 \frac{2(t+1+h)}{(1+(t+1+h)^2)^2} dh \\ &= -\int_0^1 \frac{d(t+1+h)^2}{(1+(t+1+h)^2)^2} \\ &= \frac{1}{1+(t+1+h)^2}\Big|_{h=0}^{h=1} \\ &= \frac{1}{1+(t+2)^2} - \frac{1}{1+(t+1)^2} \\ &= \frac{(t+1)^2 - (t+2)^2}{(t^2+4t+5)(t^2+2t+2)} \\ &= \frac{t^2+2t+1-t^2-4t-4}{(t^2+4t+5)(t^2+2t+2)} \\ &= \frac{-2t-3}{(t^2+4t+5)(t^2+2t+2)}, \quad t \in \mathbb{T}. \end{aligned}$$

Example 2.101. Let $\mathbb{T} = 2^{\mathbb{N}_0}$, $f(t) = \sin t$, and $g(t) = t^2 + 1$, $t \in \mathbb{T}$. Then $f : \mathbb{R} \to \mathbb{R}$ is continuously differentiable, $g : \mathbb{T} \to \mathbb{R}$ is delta differentiable, and

$$\begin{aligned}
\sigma(t) &= 2t,\\
\mu(t) &= t,\\
g^{\Delta}(t) &= \sigma(t) + t\\
&= 2t + t\\
&= 3t,\\
f'(t) &= \cos t,\\
g(t) + h\mu(t)g^{\Delta}(t) &= t^2 + 1 + ht(3t)\\
&= t^2 + 1 + 3t^2 h,\\
f'(g(t) + h\mu(t)g^{\Delta}(t)) &= \cos(t^2 + 1 + 3t^2 h), \quad t \in \mathbb{T}.
\end{aligned}$$

Then we conclude that $f \circ g : \mathbb{T} \to \mathbb{R}$ is delta differentiable and

$$\begin{aligned}
(f \circ g)^{\Delta}(t) &= \int_0^1 \cos(t^2 + 1 + 3t^2 h)dh(3t)\\
&= \frac{3t}{3t^2} \int_0^1 \cos(t^2 + 1 + 3t^2 h)d(t^2 + 1 + 3t^2 h)\\
&= \frac{1}{t} \sin(t^2 + 1 + 3t^2 h)\Big|_{h=0}^{h=1}\\
&= \frac{1}{t}(\sin(4t^2 + 1) - \sin(t^2 + 1))\\
&= \frac{2}{t} \sin \frac{3t^2}{2} \cos\left(\frac{5t^2}{2} + 1\right), \quad t \in \mathbb{T}.
\end{aligned}$$

Example 2.102. Let

$$\mathbb{T} = 3^{\mathbb{N}_0}, \quad f(t) = \log(1 + t^2), \quad \text{and} \quad g(t) = t^3 - 2t, \quad t \in \mathbb{T}.$$

Then $f : \mathbb{R} \to \mathbb{R}$ is continuously differentiable, $g : \mathbb{T} \to \mathbb{R}$ is delta differentiable, and

$$\begin{aligned}
\sigma(t) &= 3t,\\
\mu(t) &= 2t,\\
f'(t) &= \frac{2t}{1 + t^2},\\
g^{\Delta}(t) &= (\sigma(t))^2 + t\sigma(t) + t^2 - 2\\
&= 9t^2 + 3t^2 + t^2 - 2\\
&= 13t^2 - 2,\\
g(t) + h\mu(t)g^{\Delta}(t) &= t^3 - 2t + h(2t)(13t^2 - 2)\\
&= t^3 - 2t + (26t^3 - 4t)h,
\end{aligned}$$

$$f'(g(t)+h\mu(t)g^{\Delta}(t)) = \frac{2(t^3-2t)+2(26t^3-4t)h}{1+(t^3-2t+(26t^3-4t)h)^2}$$
$$= 2\frac{t^3-2t+(26t^3-4t)h}{1+(t^3-2t+(26t^3-4t)h)^2}, \quad t \in \mathbb{T}.$$

Then we conclude that $f \circ g$ is delta differentiable and

$$(f \circ g)^{\Delta}(t) = 2\int_0^1 \frac{t^3-2t+(26t^3-4t)h}{1+(t^3-2t+(26t^3-4t)h)^2}dh(13t^2-2)$$
$$= \frac{13t^2-2}{13t^3-2t}\int_0^1 \frac{t^3-2t+(26t^3-4t)h}{1+(t^3-2t+(26t^3-4t)h)^2}d(t^3-2+(26t^3-4t)h)$$
$$= \frac{1}{2t}\int_0^1 \frac{d(t^3-2t+(26t^3-4t)h)^2}{1+(t^3-2t+(26t^3-4t)h)^2}$$
$$= \frac{1}{2t}\log(t^3-2t+(26t^3-4t)h)\Big|_{h=0}^{h=1}$$
$$= \frac{1}{2t}(\log(27t^3-6t)-\log(t^3-2t))$$
$$= \frac{1}{2t}\log\frac{27t^3-6t}{t^3-2t} = \frac{1}{2t}\log\frac{27t^2-6}{t^2-2}, \quad t \in \mathbb{T}.$$

Example 2.103. Let $\mathbb{T} = \{2^{2n} : n \in \mathbb{N}_0\}$, $v(t) = t^2$, and $w(t) = t^2+1$, $t \in \mathbb{T}$. Then $v : \mathbb{T} \to \mathbb{R}$ is strictly increasing, and $\tilde{\mathbb{T}} = v(\mathbb{T}) = \{2^{4n} : n \in \mathbb{N}_0\}$ is a time scale. For $t = 2^{2n} \in \mathbb{T}$, $n \in \mathbb{N}_0$, we have

$$\sigma(t) = \inf\{2^{2l} : 2^{2l} > 2^{2n}, l \in \mathbb{N}_0\}$$
$$= 2^{2n+2}$$
$$= 4t,$$
$$v^{\Delta}(t) = \sigma(t) + t$$
$$= 5t.$$

For $t = 2^{4n} \in \tilde{\mathbb{T}}$, $n \in \mathbb{N}_0$, we have

$$\tilde{\sigma}(t) = \inf\{2^{4l} : 2^{4l} > 2^{4n}, l \in \mathbb{N}_0\}$$
$$= 2^{4n+4}$$
$$= 16t.$$

Also, for $t \in \mathbb{T}$, we have

$$(w \circ v)(t) = w(v(t))$$
$$= (v(t))^2 + 1$$

$$
\begin{aligned}
&= t^4 + 1, \\
(w \circ v)^{\Delta}(t) &= (\sigma(t))^3 + t(\sigma(t))^2 + t^2\sigma(t) + t^3 \\
&= 64t^3 + 16t^3 + 4t^3 + t^3 \\
&= 85t^3, \\
(w^{\tilde{\Delta}} \circ v)(t) &= \tilde{\sigma}(v(t)) + v(t) \\
&= 16v(t) + v(t) \\
&= 17v(t) \\
&= 17t^2, \\
(w^{\tilde{\Delta}} \circ v)(t)v^{\Delta}(t) &= 17t^2(5t) \\
&= 85t^3.
\end{aligned}
$$

Consequently,

$$
(w \circ v)^{\Delta}(t) = (w^{\tilde{\Delta}} \circ v(t))v^{\Delta}(t), \quad t \in \mathbb{T}^{\kappa}.
$$

Example 2.104. Let $\mathbb{T} = \{n+1 : n \in \mathbb{N}_0\}$, $v(t) = t^2$, and $w(t) = t$, $t \in \mathbb{T}$. Then $v : \mathbb{T} \to \mathbb{R}$ is strictly increasing, and $\tilde{\mathbb{T}} = \{(n+1)^2 : n \in \mathbb{N}_0\}$ is a time scale. For $t = n+1 \in \mathbb{T}$, $n \in \mathbb{N}_0$, we have

$$
\begin{aligned}
\sigma(t) &= \inf\{l + 1 : l + 1 > n + 1, l \in \mathbb{N}_0\} \\
&= n + 2 \\
&= t + 1, \\
v^{\Delta}(t) &= \sigma(t) + t \\
&= t + 1 + t \\
&= 2t + 1.
\end{aligned}
$$

For $t = (n+1)^2 \in \tilde{\mathbb{T}}$, $n \in \mathbb{N}_0$, we have

$$
\begin{aligned}
\tilde{\sigma}(t) &= \{(l+1)^2 : (l+1)^2 > (n+1)^2, l \in \mathbb{N}_0\} \\
&= (n+2)^2 \\
&= (n+1)^2 + 2(n+1) + 1 \\
&= t + 2\sqrt{t} + 1.
\end{aligned}
$$

Hence, for $t \in \mathbb{T}$, we get

$$
\begin{aligned}
(w^{\tilde{\Delta}} \circ v)(t) &= 1, \\
(w^{\tilde{\Delta}} \circ v)(t)v^{\Delta}(t) &= 1(2t + 1) \\
&= 2t + 1,
\end{aligned}
$$

$$\begin{aligned} w \circ v(t) &= v(t) \\ &= t^2, \\ (w \circ v)^\Delta(t) &= \sigma(t) + t \\ &= 2t + 1. \end{aligned}$$

Consequently,

$$(w \circ v)^\Delta(t) = (w^{\tilde{\Delta}} \circ v(t))v^\Delta(t), \quad t \in \mathbb{T}^\kappa.$$

Example 2.105. Let $\mathbb{T} = \{2^n : n \in \mathbb{N}_0\}$, $v(t) = t$, and $w(t) = t^2$, $t \in \mathbb{T}$. Then $v : \mathbb{T} \to \mathbb{R}$ is strictly increasing, and $v(\mathbb{T}) = \mathbb{T}$. For $t = 2^n \in \mathbb{T}$, $n \in \mathbb{N}_0$, we have

$$\begin{aligned} \sigma(t) &= \inf\{2^l : 2^l > 2^n, l \in \mathbb{N}_0\} \\ &= 2^{n+1} \\ &= 2t, \\ v^\Delta(t) &= 1, \\ (w \circ v)(t) &= w(v(t)) \\ &= (v(t))^2 \\ &= t^2, \\ (w \circ v)^\Delta(t) &= \sigma(t) + t \\ &= 2t + t \\ &= 3t, \\ (w^\Delta \circ v)(t) &= \sigma(v(t)) + v(t) \\ &= 2v(t) + v(t) \\ &= 3v(t) \\ &= 3t, \\ (w^\Delta \circ v)(t)v^\Delta(t) &= 3t. \end{aligned}$$

Consequently,

$$(w \circ v)^\Delta(t) = (w^{\tilde{\Delta}} \circ v)(t)v^\Delta(t), \quad t \in \mathbb{T}^\kappa.$$

Example 2.106. Let $U = \{\frac{1}{2^n} : n \in \mathbb{N}\}$, let

$$\mathbb{T} = U \cup (1 - U) \cup (1 + U) \cup (2 - U) \cup (2 + U) \cup \{0, 1, 2\},$$

and let

$$f(t) = \cos t,$$

$$g(t) = (2 + t^2)(1 - t^2), \quad t \in \mathbb{T}.$$

We will find $(f \circ g)^{\Delta}(t)$, $t \in \mathbb{T}^{\kappa}$. We have

$$f'(t) = -\sin t, \quad t \in \mathbb{T}.$$

We have the following cases.

1. $t = \frac{1}{2}$. By Example 2.22 we have

$$g^{\Delta}\left(\frac{1}{2}\right) = -\frac{145}{64}.$$

By Example 1.19 we get

$$\mu\left(\frac{1}{2}\right) = \frac{3}{4} - \frac{1}{2} = \frac{1}{4}.$$

Next,

$$g\left(\frac{1}{2}\right) = \left(2 + \frac{1}{4}\right)\left(1 - \frac{1}{4}\right) = \frac{9}{4} \cdot \frac{3}{4} = \frac{27}{16}.$$

Then

$$\begin{aligned}(f \circ g)^{\Delta}(t) &= -g^{\Delta}\left(\frac{1}{2}\right) \int_0^1 \sin\left(g\left(\frac{1}{2}\right) + h\mu\left(\frac{1}{2}\right)g^{\Delta}\left(\frac{1}{2}\right)\right) dh \\ &= \frac{145}{64} \int_0^1 \sin\left(\frac{27}{16} + h\left(\frac{1}{4}\right)\left(-\frac{145}{64}\right)\right) dh \\ &= \frac{145}{64} \int_0^1 \sin\left(\frac{27}{16} - \frac{145}{256}h\right) dh \\ &= -4 \int_0^1 \sin\left(\frac{27}{16} - \frac{145}{256}h\right) d\left(\frac{27}{16} - \frac{145}{256}h\right) \\ &= 4\cos\left(\frac{27}{16} - \frac{145}{256}h\right)\Bigg|_{h=0}^{h=1} \\ &= 4\left(\cos\left(\frac{287}{256}\right) - \cos\left(\frac{27}{16}\right)\right).\end{aligned}$$

2. $t = \frac{3}{2}$. By Example 2.22 we have

$$g^{\Delta}\left(\frac{3}{2}\right) = -\frac{1215}{64}.$$

By Example 1.19 we find

$$\begin{aligned}\mu\left(\frac{3}{2}\right) &= \sigma\left(\frac{3}{2}\right) - \frac{3}{2}\\ &= \frac{7}{4} - \frac{3}{2}\\ &= \frac{1}{4}.\end{aligned}$$

Next, we get

$$\begin{aligned}g\left(\frac{3}{2}\right) &= \left(2 + \frac{9}{4}\right)\left(1 - \frac{9}{4}\right)\\ &= \frac{17}{4}\left(-\frac{5}{4}\right)\\ &= -\frac{85}{16}.\end{aligned}$$

Then

$$\begin{aligned}(f\circ g)^{\Delta}\left(\frac{3}{2}\right) &= -g^{\Delta}\left(\frac{3}{2}\right)\int_0^1 \sin\left(g\left(\frac{3}{2}\right) + h\mu\left(\frac{3}{2}\right)g^{\Delta}\left(\frac{3}{2}\right)\right)dh\\ &= \frac{1215}{64}\int_0^1 \sin\left(-\frac{85}{16} + \frac{1}{4}\left(-\frac{1215}{64}\right)h\right)dh\\ &= \frac{1215}{64}\int_0^1 \sin\left(-\frac{85}{16} - \frac{1215}{256}h\right)dh\\ &= \frac{1215}{64}\int_0^1 \sin\left(-\frac{85}{16} - \frac{1215}{256}h\right)d\left(-\frac{85}{16} - \frac{1215}{256}h\right)\\ &= 4\cos\left(-\frac{85}{16} - \frac{1215}{256}h\right)\Bigg|_{h=0}^{h=1}\\ &= 4\left(-\cos\left(\frac{85}{16}\right) + \cos\left(\frac{2575}{256}\right)\right).\end{aligned}$$

3. $t = 0$. By Example 2.22 we have

$$g^{\Delta}(0) = 0.$$

Hence

$$\begin{aligned}(f \circ g)^{\Delta}(0) &= g^{\Delta}(0) \int_0^1 \sin(g(0) + h\mu(0)g^{\Delta}(0))dh \\ &= 0.\end{aligned}$$

4. $t = 1$. By Example 2.22 we have

$$g^{\Delta}(1) = -6.$$

By Example 1.19 we get

$$\begin{aligned}\mu(1) &= \sigma(1) - 1 \\ &= 1 - 1 \\ &= 0.\end{aligned}$$

Next, we have

$$g(1) = 0.$$

Then

$$\begin{aligned}(f \circ g)^{\Delta}(1) &= -g^{\Delta}(1) \int_0^1 \sin(g(1) + h\mu(1)g^{\Delta}(1))dh \\ &= 6 \int_0^1 \sin 0 dh \\ &= 0.\end{aligned}$$

5. $t = 2$. By Example 2.22 we have

$$g^{\Delta}(2) = -36.$$

By Example 1.19 we get

$$\begin{aligned}\mu(2) &= \sigma(2) - 2 \\ &= 2 - 2 \\ &= 0.\end{aligned}$$

Next, we find

$$\begin{aligned}g(2) &= (2 + 4)(1 - 4) \\ &= -18.\end{aligned}$$

Hence

$$(f \circ g)^{\Delta}(2) = -g^{\Delta}(2) \int_0^2 \sin(g(2) + h\mu(2)g^{\Delta}(2))dh$$
$$= 36 \int_0^1 \sin(-18)dh$$
$$= -36 \sin(18).$$

6. $t \in U\backslash\{\frac{1}{2}\}$. By Example 2.22 we have

$$g^{\Delta}(t) = -15t^3 - 3t.$$

By Example 1.19 we find

$$\begin{aligned}\mu(t) &= \sigma(t) - t\\ &= 2t - t\\ &= t.\end{aligned}$$

Hence

$$\begin{aligned}(f \circ g)^{\Delta}(t) &= -g^{\Delta}(t) \int_0^1 \sin(g(t) + h\mu(t)g^{\Delta}(t))dh\\ &= -\frac{1}{\mu(t)} \int_0^1 \sin(g(t) + h\mu(t)g^{\Delta}(t))d(g(t) + h\mu(t)g^{\Delta}(t))\\ &= \frac{1}{\mu(t)} \cos(g(t) + h\mu(t)g^{\Delta}(t))\Big|_{h=0}^{h=1}\\ &= \frac{1}{\mu(t)}(\cos(g(t) + \mu(t)g^{\Delta}(t)) - \cos(g(t)))\\ &= \frac{1}{t}(\cos((2+t^2)(1-t^2) - t(15t^3 + 3t)) - \cos((2+t^2)(1-t^2))).\end{aligned}$$

7. $t \in (1-U)\backslash\{\frac{1}{2}\}$. By Example 2.22 we have

$$g^{\Delta}(t) = -\frac{15t^3 + 11t^2 + 17t + 5}{8}.$$

By Example 1.19 we find

$$\begin{aligned}\mu(t) &= \sigma(t) - t\\ &= \frac{1+t}{2} - t\\ &= \frac{1-t}{2}.\end{aligned}$$

Hence

$$
\begin{aligned}
(f \circ g)^{\Delta}(t) &= -g^{\Delta}(t) \int_0^1 \sin(g(t) + h\mu(t)g^{\Delta}(t))dh \\
&= -\frac{1}{\mu(t)} \int_0^1 \sin(g(t) + h\mu(t)g^{\Delta}(t))d(g(t) + h\mu(t)g^{\Delta}(t)) \\
&= \frac{1}{\mu(t)} \cos(g(t) + h\mu(t)g^{\Delta}(t))\Big|_{h=0}^{h=1} \\
&= \frac{1}{\mu(t)}(\cos(g(t) + \mu(t)g^{\Delta}(t)) - \cos(g(t))) \\
&= \frac{2}{1-t}\Bigg(\cos\Bigg((2+t^2)(1-t^2) - \frac{(1-t)(15t^3 + 11t^2 + 17t + 5)}{16}\Bigg) \\
&\quad - \cos((2+t^2)(1-t^2))\Bigg).
\end{aligned}
$$

8. $t \in (1 + U)\backslash\{\frac{3}{2}\}$. By Example 2.22 we have

$$g^{\Delta}(t) = -15t^3 + 17t^2 - 10t + 2.$$

By Example 1.19 we find

$$
\begin{aligned}
\mu(t) &= \sigma(t) - t \\
&= 2t - 1 - t \\
&= t - 1.
\end{aligned}
$$

Hence

$$
\begin{aligned}
(f \circ g)^{\Delta}(t) &= -g^{\Delta}(t) \int_0^1 \sin(g(t) + h\mu(t)g^{\Delta}(t))dh \\
&= -\frac{1}{\mu(t)} \int_0^1 \sin(g(t) + h\mu(t)g^{\Delta}(t))d(g(t) + h\mu(t)g^{\Delta}(t)) \\
&= \frac{1}{\mu(t)} \cos(g(t) + h\mu(t)g^{\Delta}(t))\Big|_{h=0}^{h=1} \\
&= \frac{1}{\mu(t)}(\cos(g(t) + \mu(t)g^{\Delta}(t)) - \cos(g(t))) \\
&= \frac{1}{t-1}\Big(\cos\Big((2+t^2)(1-t^2) + (t-1)(-15t^3 + 17t^2 - 10t + 2)\Big) \\
&\quad - \cos((2+t^2)(1-t^2))\Big).
\end{aligned}
$$

9. $t \in (2-U)\backslash\{\frac{3}{2}\}$. By Example 2.22 we have

$$g^{\Delta}(t) = -\frac{15t^3 + 22t^2 + 32t + 8}{8}.$$

By Example 1.19 we find

$$\begin{aligned}\mu(t) &= \sigma(t) - t\\ &= \frac{t+2}{2} - t\\ &= \frac{2-t}{2}.\end{aligned}$$

Hence

$$\begin{aligned}(f \circ g)^{\Delta}(t) &= -g^{\Delta}(t)\int_0^1 \sin(g(t) + h\mu(t)g^{\Delta}(t))dh\\ &= -\frac{1}{\mu(t)}\int_0^1 \sin(g(t) + h\mu(t)g^{\Delta}(t))d(g(t) + h\mu(t)g^{\Delta}(t))\\ &= \frac{1}{\mu(t)}\cos(g(t) + h\mu(t)g^{\Delta}(t))\Big|_{h=0}^{h=1}\\ &= \frac{1}{\mu(t)}(\cos(g(t) + \mu(t)g^{\Delta}(t)) - \cos(g(t)))\\ &= \frac{2}{2-t}\left(\cos\left((2+t^2)(1-t^2) - \frac{(2-t)(15t^3 + 22t^2 + 32t + 8)}{16}\right)\right.\\ &\quad \left.- \cos((2+t^2)(1-t^2))\right).\end{aligned}$$

10. $t \in (2+U)\backslash\{\frac{3}{2}\}$. By Example 2.22 we have

$$g^{\Delta}(t) = -15t^3 + 34t^2 - 31t + 10.$$

By Example 1.19 we find

$$\begin{aligned}\mu(t) &= \sigma(t) - t\\ &= 2t - 2 - t\\ &= t - 2.\end{aligned}$$

Hence

$$(f \circ g)^{\Delta}(t) = -g^{\Delta}(t)\int_0^1 \sin(g(t) + h\mu(t)g^{\Delta}(t))dh$$

$$
\begin{aligned}
&= -\frac{1}{\mu(t)} \int_0^1 \sin(g(t) + h\mu(t)g^\Delta(t)) d(g(t) + h\mu(t)g^\Delta(t)) \\
&= \frac{1}{\mu(t)} \cos(g(t) + h\mu(t)g^\Delta(t))\Big|_{h=0}^{h=1} \\
&= \frac{1}{\mu(t)} (\cos(g(t) + \mu(t)g^\Delta(t)) - \cos(g(t))) \\
&= \frac{1}{t-2} (\cos((2+t^2)(1-t^2) + (t-2)(-15t_3^3 4t^2 - 31t + 10)) \\
&\quad - \cos((2+t^2)(1-t^2))).
\end{aligned}
$$

Example 2.107 (Derivative of the inverse). Let $v : \mathbb{T} \to \mathbb{R}$ be strictly increasing, and let $\tilde{\mathbb{T}} = v(\mathbb{T})$ be a time scale. Then applying the chain rule to $w = v^{-1}$, we get

$$((v^{-1})^{\tilde{\Delta}} \circ v)(t) = \frac{1}{v^\Delta(t)}$$

for all $t \in \mathbb{T}^\kappa$ such that $v^\Delta(t) \neq 0$.

Example 2.108. Let $\mathbb{T} = \mathbb{N}$, and let $f(t) = t^2 + 1, t \in \mathbb{T}$. Then $\sigma(t) = t + 1, t \in \mathbb{T}, f : \mathbb{T} \to \mathbb{R}$ is strictly increasing, and

$$
\begin{aligned}
f^\Delta(t) &= \sigma(t) + t \\
&= 2t + 1, \quad t \in \mathbb{T}.
\end{aligned}
$$

Hence

$$
\begin{aligned}
((f^{-1})^{\tilde{\Delta}} \circ f)(t) &= \frac{1}{f^\Delta(t)} \\
&= \frac{1}{2t+1}, \quad t \in \mathbb{T}.
\end{aligned}
$$

Example 2.109. Let $\mathbb{T} = \{n + 3 : n \in \mathbb{N}_0\}$, and let $f(t) = t^2$, $t \in \mathbb{T}$. Then $f : \mathbb{T} \to \mathbb{R}$ is strictly increasing, $\sigma(t) = t + 1$, and

$$
\begin{aligned}
f^\Delta(t) &= \sigma(t) + t \\
&= 2t + 1, \quad t \in \mathbb{T}.
\end{aligned}
$$

Hence

$$
\begin{aligned}
((f^{-1})^{\tilde{\Delta}} \circ f)(t) &= \frac{1}{f^\Delta(t)} \\
&= \frac{1}{2t+1}, \quad t \in \mathbb{T}.
\end{aligned}
$$

Example 2.110. Let $\mathbb{T} = \{2^{n^2} : n \in \mathbb{N}_0\}$, and let $f(t) = t^3$, $t \in \mathbb{T}$. Then $f : \mathbb{T} \to \mathbb{R}$ is strictly increasing, and for $t = 2^{n^2} \in \mathbb{T}$, $n = (\log_2 t)^{\frac{1}{2}} \in \mathbb{N}_0$, we have

$$\begin{aligned}\sigma(t) &= \inf\{2^{l^2} : 2^{l^2} > 2^{n^2}, l \in \mathbb{N}_0\}\\ &= 2^{(n+1)^2}\\ &= 2^{n^2} 2^{2n+1}\\ &= t 2^{2(\log_2 t)^{\frac{1}{2}}+1}.\end{aligned}$$

Then

$$\begin{aligned}f^{\Delta}(t) &= (\sigma(t))^2 + t\sigma(t) + t^2\\ &= t^2 2^{4(\log_2 t)^{\frac{1}{2}}+2} + t^2 2^{2(\log_2 t)^{\frac{1}{2}}+1} + t^2, \quad t \in \mathbb{T}.\end{aligned}$$

Hence

$$((f^{-1})^{\tilde{\Delta}} \circ f)(t) = \frac{1}{t^2 2^{4(\log_2 t)^{\frac{1}{2}}+2} + t^2 2^{2(\log_2 t)^{\frac{1}{2}}+1} + t^2}, \quad t \in \mathbb{T}.$$

Exercise 2.52. Let $\mathbb{T} = \mathbb{Z}$, $f(t) = t^2 + 2t + 1$, and $g(t) = t^2 - 3t$, $t \in \mathbb{T}$. Find a constant $c \in [1, \sigma(1)]$ such that

$$(f \circ g)^{\Delta}(1) = f'(g(c))g^{\Delta}(1).$$

Exercise 2.53. Let

$$\mathbb{T} = \mathbb{Z}, \quad f(t) = \cos t, \quad \text{and} \quad g(t) = t^2 + t, \quad t \in \mathbb{T}.$$

Find $(f \circ g)^{\Delta}(t)$, $t \in \mathbb{T}$.

Exercise 2.54. Let $\mathbb{T} = \{2^{3^n} : n \in \mathbb{N}_0\}$, $v(t) = t^2$, and $w(t) = t$, $t \in \mathbb{T}$. Prove that

$$(w \circ v)^{\Delta}(t) = (w^{\tilde{\Delta}} \circ v)(t)v^{\Delta}(t), \quad t \in \mathbb{T}^{\kappa}.$$

Exercise 2.55. Let $\mathbb{T} = P_{1,3}$, and let

$$\begin{aligned}f(t) &= \frac{1-t}{1+t},\\ g(t) &= t^2, \quad t \in \mathbb{T}.\end{aligned}$$

Find $(f \circ g)^{\Delta}(t)$, $t \in \mathbb{T}^{\kappa}$.

Exercise 2.56. Let $\mathbb{T} = \{-\frac{1}{n} : n \in \mathbb{N}\} \cup \mathbb{N}_0$, and let

$$f(t) = \frac{1+t}{1+7\sqrt{t}},$$
$$g(t) = t^2, \quad t \in \mathbb{T}.$$

Find $(f \circ g)^{\Delta}(t)$, $t \in \mathbb{T}^{\kappa}$.

2.12 L'Hôpital's rule

Let $\overline{\mathbb{T}} = \mathbb{T} \cup \{\sup \mathbb{T}\} \cup \{\inf \mathbb{T}\}$.

Definition 2.12. If $\infty \in \overline{\mathbb{T}}$, then we call ∞ left-dense. If $-\infty \in \overline{\mathbb{T}}$, then we call $-\infty$ right-dense.

For a left-dense point $t_0 \in \overline{\mathbb{T}}$ and for $\varepsilon > 0$, we set

$$L_\varepsilon(t_0) = \{t \in \mathbb{T} : 0 < t - t_0 < \varepsilon\}.$$

If $t_0 \in \overline{\mathbb{T}}$ is left-dense, then $L_\varepsilon(t_0)$ is nonempty.

If $\infty \in \overline{\mathbb{T}}$, then we have

$$L_\varepsilon(\infty) = \left\{t \in \mathbb{T} : t > \frac{1}{\varepsilon}\right\}.$$

For a right-dense point $t_1 \in \overline{\mathbb{T}}$, we define

$$R_\varepsilon(t_1) = \{t \in \mathbb{T} : 0 < t - t_1 < \varepsilon\}.$$

For every right-dense point $t_1 \in \overline{\mathbb{T}}$, the set $R_\varepsilon(t_1)$ is nonempty.

If $-\infty \in \overline{\mathbb{T}}$, then

$$R_\varepsilon(-\infty) = \left\{t \in \mathbb{T} : t < -\frac{1}{\varepsilon}\right\}.$$

Definition 2.13. Let $h : \mathbb{T} \to \mathbb{R}$.
1. For a left-dense point $t_0 \in \overline{\mathbb{T}}$, we define
$$\liminf_{t \to t_0-} h(t) = \lim_{\varepsilon \to 0+} \inf_{t \in L_\varepsilon(t_0)} h(t), \quad \limsup_{t \to t_0-} h(t) = \lim_{\varepsilon \to 0+} \sup_{t \in L_\varepsilon(t_0)} h(t).$$
2. For a right-dense point $t_1 \in \overline{\mathbb{T}}$, we define
$$\liminf_{t \to t_1-} h(t) = \lim_{\varepsilon \to 0+} \inf_{t \in R_\varepsilon(t_1)} h(t), \quad \limsup_{t \to t_0-} h(t) = \lim_{\varepsilon \to 0+} \sup_{t \in R_\varepsilon(t_1)} h(t).$$

The L'Hôpital's rules in the framework of time scale analysis read as follows.

- Let f and g be differentiable on $\mathbb{T}$, and let

$$\lim_{t\to t_0-} f(t) = \lim_{t\to t_0-} g(t) = 0 \quad \text{for some left-dense point } t_0 \in \overline{\mathbb{T}}.$$

Suppose also that there is $\varepsilon > 0$ such that

$$g(t) > 0, \quad g^{\Delta}(t) < 0 \quad \text{for all } t \in L_{\varepsilon}(t_0).$$

Then

$$\liminf_{t\to t_0-} \frac{f^{\Delta}(t)}{g^{\Delta}(t)} \le \liminf_{t\to t_0-} \frac{f(t)}{g(t)} \le \limsup_{t\to t_0-} \frac{f(t)}{g(t)} \le \liminf_{t\to t_0-} \frac{f^{\Delta}(t)}{g^{\Delta}(t)}.$$

- Let f and g be differentiable on $\mathbb{T}$, and let

$$\lim_{t\to t_0+} f(t) = \lim_{t\to t_0+} g(t) = 0 \quad \text{for some right-dense point } t_0 \in \overline{\mathbb{T}}.$$

Suppose also that there is $\varepsilon > 0$ such that

$$g(t) > 0, \quad g^{\Delta}(t) < 0 \quad \text{for all } t \in R_{\varepsilon}(t_0).$$

Then

$$\liminf_{t\to t_0+} \frac{f^{\Delta}(t)}{g^{\Delta}(t)} \le \liminf_{t\to t_0+} \frac{f(t)}{g(t)} \le \limsup_{t\to t_0+} \frac{f(t)}{g(t)} \le \liminf_{t\to t_0+} \frac{f^{\Delta}(t)}{g^{\Delta}(t)}.$$

- Let f and g be differentiable on $\mathbb{T}$, and let

$$\lim_{t\to t_0-} g(t) = \infty \quad \text{for some left-dense point } t_0 \in \overline{\mathbb{T}}.$$

Suppose also that there is $\varepsilon > 0$ such that

$$g(t) > 0, \quad g^{\Delta}(t) > 0 \quad \text{for all } t \in L_{\varepsilon}(t_0).$$

Then

$$\lim_{t\to t_0-} \frac{f^{\Delta}(t)}{g^{\Delta}(t)} = r \in \overline{\mathbb{R}}$$

implies

$$\lim_{t\to t_0-} \frac{f(t)}{g(t)} = r.$$

– Let f and g be differentiable on $\mathbb{T}$, and let

$$\lim_{t\to t_0+} g(t) = \infty \quad \text{for some right-dense point } t_0 \in \overline{\mathbb{T}}.$$

Suppose also that there is $\varepsilon > 0$ such that

$$g(t) > 0, \quad g^{\Delta}(t) > 0 \quad \text{for all } t \in \mathbb{R}_{\varepsilon}(t_0).$$

Then

$$\lim_{t\to t_0+} \frac{f^{\Delta}(t)}{g^{\Delta}(t)} = r \in \overline{\mathbb{R}}$$

implies

$$\lim_{t\to t_0+} \frac{f(t)}{g(t)} = r.$$

Example 2.111. Let $\mathbb{T} = [-1, 1) \cup \{1, 2, 4, \ldots\}$. We will compute

$$l = \lim_{t\to 1} \frac{t^3 - 2t^2 - t + 2}{t^3 - 7t + 6}.$$

We have that $t = 1$ is a left-dense point. Then

$$\begin{aligned}
l &= \lim_{t\to 1} \frac{t^2(t-2) - (t-2)}{t^3 - t - 6t + 6} \\
&= \lim_{t\to 1} \frac{(t^2-1)(t-2)}{t(t-1)(t+1) - 6(t-1)} \\
&= \lim_{t\to 1} \frac{(t-1)(t+1)(t-2)}{(t-1)(t^2+t-6)} \\
&= \lim_{t\to 1} \frac{(t+1)(t-2)}{t^2+t-6} \\
&= \frac{-2}{-4} \\
&= \frac{1}{2}.
\end{aligned}$$

Example 2.112. Let $\mathbb{T} = [-3, 0) \cup \mathbb{N}_0$. We will find

$$l = \lim_{t\to 0} \frac{t\cos t - \sin t}{t}.$$

Note that $t = 0$ is left-dense. Hence

$$l = \lim_{t\to 0} \frac{(t\cos t - \sin t)'}{t'}$$

$$\begin{aligned}&= \lim_{t\to 0}(\cos t - t\sin t - \cos t)\\&= \lim_{t\to 0}(-t\sin t)\\&= 0.\end{aligned}$$

Example 2.113. Let $\mathbb{T} = [-3, 1) \cup 2^{\mathbb{N}_0}$. We will compute

$$l = \lim_{t\to 0}\frac{\tan t - \sin t}{t - \sin t}.$$

Here $t = 0$ is left-dense. Then

$$\begin{aligned}l &= \lim_{t\to 0}\frac{(\tan t - \sin t)'}{(t - \sin t)'}\\&= \lim_{t\to 0}\frac{\frac{1}{(\cos t)^2} - \cos t}{1 - \cos t}\\&= \lim_{t\to 0}\frac{1 - (\cos t)^3}{(\cos t)^2(1 - \cos t)}\\&= \lim_{t\to 0}\frac{(1 - \cos t)(1 + \cos t + (\cos t)^2)}{(\cos t)^2(1 - \cos t)}\\&= \lim_{t\to 0}\frac{1 + \cos t + (\cos t)^2}{(\cos t)^2}\\&= 3.\end{aligned}$$

Example 2.114. Let $\mathbb{T} = -3^{\mathbb{N}_0} \cup [1, 5]$. We will find

$$l = \lim_{t\to 1}\frac{t^5 - 1}{2t^3 - t - 1}.$$

We have that $t = 1$ is right-dense. Then

$$\begin{aligned}l &= \lim_{t\to 1}\frac{5t^4}{6t^2 - 1}\\&= \frac{5}{5}\\&= 1.\end{aligned}$$

Example 2.115. Let $\mathbb{T} = \{0\} \cup \{\frac{1}{n}\}_{n\in\mathbb{N}} \cup [3, \infty)$. We will find

$$l = \lim_{t\to\infty}\frac{(\log t)^2}{t^2}.$$

We have

$$l = \lim_{t\to\infty}\frac{2\log t}{2t^2}$$

$$\begin{aligned} &= \lim_{t\to\infty} \frac{\log t}{t^2} \\ &= \lim_{t\to\infty} \frac{1}{2t^2} \\ &= 0. \end{aligned}$$

Exercise 2.57. Let $\mathbb{T} = [-3,3) \cup \{3,9,27,\dots\}$, where $[-3,3)$ is a real-valued interval. Compute

1. $\lim_{t\to 0} \frac{\tan t - \sin t}{t - \sin t}$,
2. $\lim_{t\to \frac{\pi}{2}} \frac{\tan t}{\tan(5t)}$,
3. $\lim_{t\to 0} \frac{\log(\sin(10t))}{\log(\sin t)}$,
4. $\lim_{t\to 0}(1 - \cos t)\cot t$,
5. $\lim_{t\to 3}(\frac{1}{t-3} - \frac{5}{t^2-t-6})$,
6. $\lim_{t\to \frac{\pi}{2}}(\frac{t}{\cot t} - \frac{\pi}{2\cos t})$.

2.13 Advanced practical problems

Problem 2.1. Let $\mathbb{T} = (-5\mathbb{N}_0) \cup 4^{\mathbb{N}_0}$, and let

$$f(t) = t^3 - 3, \quad t \in \mathbb{T}.$$

Prove that

$$f^\Delta(t) = \begin{cases} 3t^2 + 15t + 25 & \text{if } t \in (-5\mathbb{N}), \\ 1 & \text{if } t = 0, \\ 21t^2 & \text{if } t \in 4^{\mathbb{N}_0}. \end{cases}$$

Problem 2.2. Let $\mathbb{T} = [-1,3] \cup [7,9] \cup [10,\infty)$, and let

$$f(t) = 1 + 4t + 9t^2, \quad t \in \mathbb{T}.$$

Prove that

$$f^\Delta(t) = \begin{cases} 4 + 18t & \text{if } t \in [-1,3) \cup [7,9) \cup [10,\infty), \\ 94 & \text{if } t = 3, \\ 175 & \text{if } t = 9. \end{cases}$$

Problem 2.3. Let $\mathbb{T} = \{0\} \cup \{1 - \frac{1}{4^n}\}_{n\in\mathbb{N}_0} \cup 2^{\mathbb{N}_0}$, and let

$$f(t) = \frac{1+t^2}{2+t^2}, \quad t \in \mathbb{T}.$$

Prove that

$$f^{\Delta}(t) = \begin{cases} \frac{4(3+5t)}{(2+t^2)(41+6t+t^2)} & \text{if } t \in \{1 - \frac{1}{4^n}\}_{n\in\mathbb{N}}, \\ \frac{6}{41} & \text{if } t = 0, \\ \frac{3t}{(2+t^2)(2+4t^2)} & \text{if } t \in 2^{\mathbb{N}_0}. \end{cases}$$

Problem 2.4. Let $\mathbb{T} = \{2\} \cup \{2 + \frac{1}{n}\}_{n\in\mathbb{N}} \cup [4, 9] \cup \{9 + \frac{1}{n^2}\}_{n\in\mathbb{N}}$, and let

$$f(t) = 1 - 3t - t^2, \quad t \in \mathbb{T}.$$

Prove that

$$f^{\Delta}(t) = \begin{cases} -3 - 2t & \text{if } t \in \{2\} \cup [4, 9] \cup \{10\}, \\ \frac{-13+t+t^2}{3-t} & \text{if } t \in \{2 + \frac{1}{n}\}_{n\in\mathbb{N}, n\geq 2}, \\ -10 & \text{if } t = 3, \\ -12 - t - \frac{t-9}{(\sqrt{t-9}-1)^2} & \text{if } t \in \{9 + \frac{1}{n^2}\}_{n\in\mathbb{N}, n\geq 2}. \end{cases}$$

Problem 2.5. Let $\mathbb{T} = [1, 3] \cup 5^{\mathbb{N}}$, and let

$$f(t) = t^4 - 3t^2, \quad t \in \mathbb{T}.$$

Prove that

$$f^{\Delta}(t) = \begin{cases} 4t^3 - 6t & \text{if } t \in [1, 3), \\ 248 & \text{if } t = 3, \\ 156t^3 - 18t & \text{if } t \in 5^{\mathbb{N}}. \end{cases}$$

Problem 2.6. Let $\mathbb{T} = q^{\mathbb{N}}$, $q > 1$. Prove that

$$\begin{aligned} t &= \frac{1}{1+q} f_2^{\Delta}(t), \\ t^2 &= \frac{1}{1+q+q^2} f_3^{\Delta}(t), \\ t^3 &= \frac{1}{1+q+q^2+q^3} f_4^{\Delta}(t), \\ t^4 &= \frac{1}{1+q+q^2+q^3+q^4} f_5^{\Delta}(t), \\ t^5 &= \frac{1}{1+q+q^2+q^3+q^4+q^5} f_6^{\Delta}(t), \quad t \in \mathbb{T}. \end{aligned}$$

Problem 2.7. Let $\mathbb{T} = \{n^3 : n \in \mathbb{N}_0\}$, and let $f(t) = t^2 + 2t$, $t \in \mathbb{T}$. Find $f^{\Delta}(t)$, $t \in \mathbb{T}^{\kappa}$.

Problem 2.8. Let $\mathbb{T} = 2^{\mathbb{N}_0}$. Prove that

$$t + t^2 + t^3 = \frac{1}{3} f_2^{\Delta}(t) + \frac{1}{7} f_3^{\Delta}(t) + \frac{1}{15} f_4^{\Delta}(t), \quad t \in \mathbb{T}.$$

Problem 2.9. Let $\mathbb{T} = 2^{\mathbb{N}}$. Find $f^{\Delta}(t)$, where

1. $f(t) = t^2 - 3t + 2$,
2. $f(t) = \frac{t^3 - t^2}{t+1}$,
3. $f(t) = \frac{t-1}{t+1}$, $t \in \mathbb{T}$.

Problem 2.10. Let $\mathbb{T} = (-5\mathbb{N}_0) \cup 4^{\mathbb{N}_0}$. Find $f^{\Delta}(t)$, $t \in \mathbb{T}^{\kappa}$, where

$$f(t) = \frac{3-t}{2+9t}, \quad t \in \mathbb{T}.$$

Problem 2.11. Let $\mathbb{T} = \{2\} \cup \{2 + \frac{1}{n}\}_{n\in\mathbb{N}} \cup [4, 9] \cup \{9 + \frac{1}{n^2}\}_{n\in\mathbb{N}}$. Find $f^{\Delta}(t)$, $t \in \mathbb{T}^{\kappa}$, where

$$f(t) = t^3 + t, \quad t \in \mathbb{T}.$$

Problem 2.12. Let $\mathbb{T} = (-5\mathbb{N}_0) \cup 4^{\mathbb{N}_0}$. Find $f^{\Delta^2}(t)$, $t \in \mathbb{T}$, where

$$f(t) = \frac{1+t}{2+t}, \quad t \in \mathbb{T}.$$

Problem 2.13. Let $\mathbb{T} = [-1, 3] \cup [7, 9] \cup [10, \infty)$. Find $f^{\Delta^2}(t)$, $t \in \mathbb{T}$, where

$$f(t) = \frac{1-2t}{1+t}, \quad t \in \mathbb{T}.$$

Problem 2.14. Let $\mathbb{T} = \{0\} \cup \{1 - \frac{1}{4^n}\}_{n\in\mathbb{N}_0} \cup 2^{\mathbb{N}_0}$. Find $f^{\Delta^2}(t)$, $t \in \mathbb{T}$, where

$$f(t) = 2t^2 - 3t + 1, \quad t \in \mathbb{T}.$$

Problem 2.15. Let $\mathbb{T} = [1, 3] \cup 5^{\mathbb{N}}$. Find $f^{\Delta^2}(t)$, $t \in \mathbb{T}$, where

$$f(t) = \frac{1+3t}{3+t}, \quad t \in \mathbb{T}.$$

Problem 2.16. Let $\mathbb{T} = h\mathbb{Z} + k$, $h > 0$, $k \in \mathbb{R}$. Find $f^{\Delta^2}(t)$, $t \in \mathbb{T}$, where

$$f(t) = t^4 + t, \quad t \in \mathbb{T}.$$

Problem 2.17. Let $\mathbb{T} = \{0\} \cup 11^{\mathbb{N}_0}$, $a = 0$, $b = 121$, and

$$f(t) = t^2 + 7t - 5, \quad t \in \mathbb{T}.$$

Prove that there are $\xi_1, \xi_2 \in (a, b)_{\mathbb{T}}$ such that

$$f^{\Delta}(\xi_1)(b-a) \le f(b) - f(a) \le f^{\Delta}(\xi_2)(b-a).$$

Problem 2.18. Let $\mathbb{T} = [1,2] \cup [3,4] \cup [7,8] \cup 9^{\mathbb{N}}$, $a = 1$, $b = 9$, and

$$f(t) = 2t^3 + 4t^2 + 1, \quad t \in \mathbb{T}.$$

Prove that there are $\xi_1, \xi_2 \in (a,b)_{\mathbb{T}}$ such that

$$f^{\Delta}(\xi_1)(b-a) \le f(b) - f(a) \le f^{\Delta}(\xi_2)(b-a).$$

Problem 2.19. Let $\mathbb{T} = (-5\mathbb{N}_0) \cup 4^{\mathbb{N}_0}$, $a = -10$, $b = 16$, and

$$f(t) = t^2 + 4t + 13, \quad t \in \mathbb{T}.$$

Prove that there are $\xi_1, \xi_2 \in (a,b)_{\mathbb{T}}$ such that

$$f^{\Delta}(\xi_1)(b-a) \le f(b) - f(a) \le f^{\Delta}(\xi_2)(b-a).$$

Problem 2.20. Let $\mathbb{T} = [-1,3] \cup [7,9] \cup [10,\infty)$, $a = -1$, $b = 11$, and

$$f(t) = \frac{1-t}{2+7t}, \quad t \in \mathbb{T}.$$

Prove that there are $\xi_1, \xi_2 \in (a,b)_{\mathbb{T}}$ such that

$$f^{\Delta}(\xi_1)(b-a) \le f(b) - f(a) \le f^{\Delta}(\xi_2)(b-a).$$

Problem 2.21. Let $\mathbb{T} = \{0\} \cup \{1 - \frac{1}{4^n}\}_{n \in \mathbb{N}_0} \cup 2^{\mathbb{N}_0}$, $a = 0$, $b = 16$, and

$$f(t) = 4t^3 - 2t^2 + 7t + 8, \quad t \in \mathbb{T}.$$

Prove that there are $\xi_1, \xi_2 \in (a,b)_{\mathbb{T}}$ such that

$$f^{\Delta}(\xi_1)(b-a) \le f(b) - f(a) \le f^{\Delta}(\xi_2)(b-a).$$

Problem 2.22. Let $\mathbb{T}$ and f be as in Problem 2.9. Investigate where f is increasing or decreasing.

Problem 2.23. Let $\mathbb{T}$ and f be as in Problem 2.10. Investigate where f is increasing or decreasing.

Problem 2.24. Let $\mathbb{T}$ and f be as in Problem 2.11. Investigate where f is increasing or decreasing.

Problem 2.25. Let $\mathbb{T}$ and f be as in Problem 2.12. Determine where f is convex or concave.

Problem 2.26. Let $\mathbb{T}$ and f be as in Problem 2.13. Determine where f is convex or concave.

Problem 2.27. Let $\mathbb{T}$ and f be as in Problem 2.14. Determine where f is convex or concave.

Problem 2.28. Let $\mathbb{T}$ and f be as in Problem 2.15. Determine where f is convex or concave.

Problem 2.29. Let $\mathbb{T}$ and f be as in Problem 2.16. Determine where f is convex or concave.

Problem 2.30. Let $\mathbb{T}$ and f be as in Problem 2.7. Check if f is completely delta differentiable at all points of $\mathbb{T}$.

Problem 2.31. Let $\mathbb{T} = 2^{\mathbb{N}_0}$ and $f(t) = 1 + t + t^2$, $t \in \mathbb{T}$. Check if f is completely delta differentiable at all points of $\mathbb{T}$.

Problem 2.32. Let $\mathbb{T}$ and f be as in Problem 2.9. Check if f is completely delta differentiable at all points of $\mathbb{T}$.

Problem 2.33. Let $\mathbb{T}$ and f be as in Problem 2.10. Check if f is completely delta differentiable at all points of $\mathbb{T}$.

Problem 2.34. Let $\mathbb{T}$ and f be as in Problem 2.11. Check if f is completely delta differentiable at all points of $\mathbb{T}$.

Problem 2.35. Let

$$f(t) = \begin{cases} 3t - 5 & \text{if } t \in [-3, 0), \\ -2t - 5 & \text{if } t \in \{0\} \cup \{\frac{1}{n}\}_{n \in \mathbb{N}}. \end{cases}$$

Check if $f_-^{\Delta}(0) = f_+^{\Delta}(0)$.

Problem 2.36. Let

$$f(t) = \begin{cases} -2t + 1 & \text{if } t \in \{0, 1, 2\}, \\ 2t^2 + 9 & \text{if } t \in \{2 + \frac{1}{n}\}_{n \in \mathbb{N}}. \end{cases}$$

Check if $f_-^{\Delta}(2) = f_+^{\Delta}(2)$.

Problem 2.37. Let

$$f(t) = \begin{cases} \frac{t^3 - 2}{t + 1} & \text{if } t \in (-2^{\mathbb{N}_0}) \cup (-1, 1), \\ 5 & \text{if } t \in 3^{\mathbb{N}_0}. \end{cases}$$

Check if $f_-^{\Delta}(1) = f_+^{\Delta}(1)$.

Problem 2.38. Let

$$f(t) = \begin{cases} at^2 - 2t & \text{if } t \in \left\{2 - \frac{1}{n}\right\}_{n \in \mathbb{N}} \cup \{2\}, \\ 4a - 2t & \text{if } t \in (2, 10], \end{cases}$$

where a is a real parameter. Find the values of a such that $f_-^{\Delta}(2) = f_+^{\Delta}(2)$.

Problem 2.39. Let

$$f(t) = \begin{cases} t^2 & \text{if } t \in [-4, 0], \\ at^2 - 2a + 1 & \text{if } t \in \left\{\frac{1}{n}\right\}_{n\in\mathbb{N}} \cup 4^{\mathbb{N}}, \end{cases}$$

where a is a real parameter. Find the values of a such that $f_-^{\Delta}(0) = f_+^{\Delta}(0)$.

Problem 2.40. Let $\mathbb{T} = \{n + 2 : n \in \mathbb{N}_0\}$, $f(t) = t^2 + 2$, and $g(t) = t^2$, $t \in \mathbb{T}$. Find a constant

$$c \in [2, \sigma(2)]$$

such that

$$(f \circ g)^{\Delta}(2) = f'(g(c))g^{\Delta}(2).$$

Problem 2.41. Let $\mathbb{T} = \mathbb{N}$, $f(t) = e^t$, and $g(t) = t^3$, $t \in \mathbb{T}$. Find $(f \circ g)^{\Delta}(t)$, $t \in \mathbb{T}$.

Problem 2.42. Let $\mathbb{T} = \{2^{4n+2} : n \in \mathbb{N}_0\}$, $v(t) = t^3$, and $w(t) = t^2 + t$, $t \in \mathbb{T}$. Prove that

$$(w \circ v)^{\Delta}(t) = (w^{\tilde{\Delta}} \circ v(t))v^{\Delta}(t), \quad t \in \mathbb{T}^{\kappa}.$$

Problem 2.43. Let $\mathbb{T} = \{n + 9 : n \in \mathbb{N}_0\}$, and let $v(t) = t^2 + 7t + 8$, $t \in \mathbb{T}$. Find $((v^{-1})^{\tilde{\Delta}} \circ v)(t)$, $t \in \mathbb{T}$.

Problem 2.44. Let $\mathbb{T} = P_{1,3}$, and let

$$f(t) = \frac{1 - \sqrt[3]{t}}{1 + \sqrt[3]{t}},$$
$$g(t) = t^3, \quad t \in \mathbb{T}.$$

Find $(f \circ g)^{\Delta}(t)$, $t \in \mathbb{T}^{\kappa}$.

Problem 2.45. Let $\mathbb{T} = \{(\frac{1}{2})^{2^n} : n \in \mathbb{N}_0\} \cup \{0, 1\}$, and let

$$f(t) = \frac{1 + 7\sqrt{t}}{4 + 5\sqrt{t}},$$
$$g(t) = t^2, \quad t \in \mathbb{T}.$$

Find $(f \circ g)^{\Delta}(t)$, $t \in \mathbb{T}^{\kappa}$.

Problem 2.46. Let $\mathbb{T} = [-2, 2) \cup \{7, 10, 13, \ldots\}$. Compute
1. $\lim_{t\to 1}\left(\frac{1}{2(1-\sqrt{t})} - \frac{1}{3(1-\sqrt[3]{t})}\right)$,
2. $\lim_{t\to \frac{\pi}{2}}\left(\frac{t}{\cot t} - \frac{\pi}{2\cos t}\right)$,
3. $\lim_{t\to 0} t^t$.

3 Delta integration

In this chapter, we introduce the main concepts for regulated, delta rd-continuous and delta predifferentiable functions. The indefinite delta integral and the Riemann delta integral are defined, and some of their properties are derived. The basic delta monomials are defined and investigated. In the chapter, we present different variants of Taylor's formula. Improper integrals of the first and second kinds are defined, and some of their properties are derived.

3.1 Regulated, rd-continuous, and delta predifferentiable functions

Definition 3.1. A function $f : \mathbb{T} \to \mathbb{R}$ is called regulated if its finite right-sided limits exist at all right-dense points in $\mathbb{T}$ and its finite left-sided limits exist at all left-dense points in $\mathbb{T}$.

Example 3.1. Let $\mathbb{T} = \mathbb{N} \cup \left\{1 - \frac{1}{n}\right\}_{n\in\mathbb{N}}$, and let

$$f(t) = \frac{t^2}{t-1}, \quad g(t) = \frac{t}{t+1}, \quad t \in \mathbb{T}.$$

Note that all points of $\mathbb{T}$ are right-scattered, the points $t \in \mathbb{T}$, $t \neq 1$, are left-scattered, and the point $t = 1$ is left-dense. Also, $\lim_{t\to 1-} f(t)$ is not finite, whereas $\lim_{t\to 1-} g(t)$ exists and is finite. Therefore the function f is not regulated, whereas the function g is.

Example 3.2. Let $\mathbb{T} = \mathbb{R}$, and let

$$f(t) = \begin{cases} 0 & \text{for } t = 0, \\ \frac{1}{t} & \text{for } t \in \mathbb{R}\backslash\{0\}. \end{cases}$$

We have that all points of $\mathbb{T}$ are dense and $\lim_{t\to 0-} f(t)$ and $\lim_{t\to 0+} f(t)$ are not finite. Therefore the function f is not regulated.

Example 3.3. Let $\mathbb{T} = \{1 - \frac{1}{n}\}_{n\in\mathbb{N}} \cup 3^{\mathbb{N}_0}$, and let

$$f(t) = \frac{t^2 - 4}{t^2 - t - 2}, \quad t \in \mathbb{T}.$$

Note that all points of $\mathbb{T}$ are right-scattered and $t = 1$ is left-dense. We have

$$\begin{aligned} \lim_{t\to 1-} f(t) &= \lim_{t\to 1-} \frac{t^2 - 4}{t^2 - t - 2} \\ &= \lim_{t\to 1-} \frac{(t-2)(t+2)}{(t+1)(t-2)} \end{aligned}$$

https://doi.org/10.1515/9783112232088-003

$$= \lim_{t\to 1-} \frac{t+2}{t+1}$$
$$= \frac{3}{2}.$$

Therefore f is regulated.

Example 3.4. Let $\mathbb{T} = P_{1,3}$, and let

$$f(t) = \frac{t-16}{\sqrt{t}-4}, \quad t \in \mathbb{T}.$$

Note that at all points $t \in \bigcup_{k=0}^{\infty}(4k, 4k+1)$, the right- and left-sided limits of f exist. At $t = 0$ the right-sided limit of f exists. At $t \in \bigcup_{k=0,k\neq 4}\{4k\}$ the right-sided limits of f exist, and at $t \in \bigcup_{k=1}^{\infty}\{4k+1\}$, the left-sided limits of f exist. Next, we have

$$\lim_{t\to 16+} f(t) = \lim_{t\to 16+} \frac{t-16}{\sqrt{t}-4}$$
$$= \lim_{t\to 16+} \frac{(\sqrt{t}-4)(\sqrt{t}+4)}{\sqrt{t}-4}$$
$$= \lim_{t\to 16+} (\sqrt{t}+4)$$
$$= 8,$$

i. e., at $t = 16$ the right-sided limit of f exists. Therefore f is regulated.

Example 3.5. Let $\mathbb{T} = \{-\frac{1}{n} : n \in \mathbb{N}\} \cup \mathbb{N}_0$, and let

$$f(t) = \frac{\sqrt[5]{32+t}-2}{t}, \quad t \in \mathbb{T}.$$

Note that $t = 0$ is left-dense and right-scattered, $t = -1$ is right-scattered, and all other points of $\mathbb{T}$ are isolated. For the left-sided limit of f at $t = 0$, we have

$$\lim_{t\to 0-} f(t) = \lim_{t\to 0-} \frac{\sqrt[5]{32+t}-2}{t}$$
$$= \lim_{t\to 0-} \frac{(\sqrt[5]{32+t}-2)(\sqrt[5]{(32+t)^4}+2\sqrt[5]{(32+t)^3}+4\sqrt[5]{(32+t)^2}+8\sqrt[5]{(32+t)}+16)}{t(\sqrt[5]{(32+t)^4}+2\sqrt[5]{(32+t)^3}+4\sqrt[5]{(32+t)^2}+8\sqrt[5]{(32+t)}+16)}$$
$$= \lim_{t\to 0-} \frac{32+t-32}{t(\sqrt[5]{(32+t)^4}+2\sqrt[5]{(32+t)^3}+4\sqrt[5]{(32+t)^2}+8\sqrt[5]{(32+t)}+16)}$$
$$= \lim_{t\to 0-} \frac{1}{\sqrt[5]{(32+t)^4}+2\sqrt[5]{(32+t)^3}+4\sqrt[5]{(32+t)^2}+8\sqrt[5]{(32+t)}+16}$$
$$= \frac{1}{2^4+2^4+2^4+2^4+2^4}$$

$$= \frac{1}{80},$$

i. e., the left-sided limit of f at $t = 0$ exists. Thus f is regulated.

Example 3.6. Let $\mathbb{T} = \{(\frac{1}{2})^{2^n} : n \in \mathbb{N}_0\} \cup \{0, 1\}$, and let

$$f(t) = \frac{2\sqrt{t^2 + t + 1} - 2 - t}{t^2}, \quad t \in \mathbb{T}.$$

Note that $t = 0$ is right-dense, $t = 1$ is left-scattered, and all other points of $\mathbb{T}$ are isolated. For the right-sided limit of f at $t = 0$, we get

$$\begin{aligned}
\lim_{t\to 0+} f(t) &= \lim_{t\to 0+} \frac{2\sqrt{t^2 + t + 1} - 2 - t}{t^2} \\
&= \lim_{t\to 0+} \frac{(2\sqrt{t^2 + t + 1} - 2 - t)(2\sqrt{t^2 + t + 1} + 2 + t)}{t^2(2\sqrt{t^2 + t + 1} + 2 + t)} \\
&= \lim_{t\to 0+} \frac{4t^2 + 4t + 4 - 4 - 4t - t^2}{t^2(2\sqrt{t^2 + t + 1} + 2 + t)} \\
&= \lim_{t\to 0+} \frac{3t^2}{t^2(2\sqrt{t^2 + t + 1} + 2 + t)} \\
&= \lim_{t\to 0+} \frac{3}{2\sqrt{t^2 + t + 1} + 2 + t} \\
&= \frac{3}{2 + 2} \\
&= \frac{3}{4},
\end{aligned}$$

i. e., the right-sided limit of f at $t = 0$ exists. Thus f is regulated.

Example 3.7. Let $U = \{\frac{1}{2^n} : n \in \mathbb{N}\}$, let

$$\mathbb{T} = U \cup (1 - U) \cup (1 + U) \cup (2 - U) \cup (2 + U) \cup \{0, 1, 2\},$$

and let

$$f(t) = \frac{2}{2t - t^2} + \frac{1}{t^2 - 3t + 2}, \quad t \in \mathbb{T}.$$

We have that $t = 0$ is right-dense, $t = 1$ and $t = 2$ are dense, $t = \frac{5}{2}$ is left-scattered, and all other points of $\mathbb{T}$ are isolated. Note that

$$\begin{aligned}
f(t) &= \frac{2}{t(2 - t)} + \frac{1}{(t - 2)(t - 1)} \\
&= \frac{1}{(t - 2)(t - 1)} - \frac{2}{t(t - 2)}
\end{aligned}$$

$$= \frac{t-2(t-1)}{t(t-1)(t-2)}$$
$$= \frac{2-t}{t(t-1)(t-2)}$$
$$= -\frac{1}{t(t-1)}, \quad t \in \mathbb{T}.$$

Since

$$\lim_{t\to 0+} f(t) \quad \text{and} \quad \lim_{t\to 1} f(t)$$

do not exist, we conclude that f is not regulated.

Exercise 3.1. Let $\mathbb{T} = \mathbb{Z}$. Check if the function

$$f(t) = \frac{t^2+4t-5}{t^2-1}, \quad t \in \mathbb{T},$$

is regulated.

Exercise 3.2. Let $\mathbb{T} = (-2\mathbb{N}_0) \cup 2^{\mathbb{N}_0}$. Check if the function

$$f(t) = \frac{t^2+t-2}{2t^2-t-1}, \quad t \in \mathbb{T},$$

is regulated.

Exercise 3.3. Let $\mathbb{T} = P_{3,7} \cup [4,6]$. Check if the function

$$f(t) = \frac{\sqrt{t^2-2t+6} - \sqrt{t^2+2t-6}}{t^2-4t+3}, \quad t \in \mathbb{T},$$

is regulated.

Exercise 3.4. Let $\mathbb{T} = \{0\} \cup 11^{\mathbb{N}_0}$. Check if the function

$$f(t) = \frac{1-\sqrt[3]{t}}{1-\sqrt[5]{t}}, \quad t \in \mathbb{T},$$

is regulated.

Exercise 3.5. Let $\mathbb{T} = [1,2] \cup [3,4] \cup [7,8] \cup 9^{\mathbb{N}}$. Check if the function

$$f(t) = \frac{1+14t}{2t+\sqrt[3]{t^2}}, \quad t \in \mathbb{T},$$

is regulated.

Exercise 3.6. Let $\mathbb{T} = \mathbb{R}$, and let

$$f(t) = \begin{cases} 11 & \text{for } t = 1, \\ \frac{1}{t-1} & \text{for } t \in \mathbb{R}\backslash\{1\}. \end{cases}$$

Determine if f is regulated.

Definition 3.2. A continuous function $f : \mathbb{T} \to \mathbb{R}$ is called delta predifferentiable with region of differentiation region of differentiation D if

1. $D \subset \mathbb{T}^\kappa$,
2. $\mathbb{T}^\kappa \backslash D$ is countable and contains no right-scattered elements of $\mathbb{T}$, and
3. f is delta differentiable at each $t \in D$.

Example 3.8. Let $\mathbb{T} = P_{a,b} = \bigcup_{k=0}^{\infty}[k(a+b), k(a+b)+a]$ for $a > b > 0$, and let $f : \mathbb{T} \to \mathbb{R}$ be defined by

$$f(t) = \begin{cases} 0 & \text{if } t \in \bigcup_{k=0}^{\infty}[k(a+b), k(a+b)+b], \\ t-(a+b)k-b & \text{if } t \in [(a+b)k+b, (a+b)k+a]. \end{cases}$$

Then f is delta predifferentiable with $D = \mathbb{T}\backslash \bigcup_{k=0}^{\infty}\{(a+b)k+b\}$.

Example 3.9. Let $\mathbb{T} = \mathbb{R}$, and let

$$f(t) = \begin{cases} 0 & \text{if } t = 3, \\ \frac{1}{t-3} & \text{if } t \in \mathbb{R}\backslash\{3\}. \end{cases}$$

Since $f : \mathbb{T} \to \mathbb{R}$ is not continuous at $t = 3$, the function f is not delta predifferentiable.

Example 3.10. Let $\mathbb{T} = \mathbb{N}_0 \cup \{1 - \frac{1}{n} : n \in \mathbb{N}\}$, and let

$$f(t) = \begin{cases} 0 & \text{if } t \in \mathbb{N}, \\ t & \text{otherwise}. \end{cases}$$

Since f is not continuous at $t = 1$, it is not delta predifferentiable.

Example 3.11. Let $\mathbb{T} = \{-\frac{1}{n} : n \in \mathbb{N}\} \cup \mathbb{N}_0$, and let

$$f(t) = \begin{cases} t+2 & \text{if } t \in \{-\frac{1}{n} : n \in \mathbb{N}\}, \\ t^2 - 3 & \text{if } t \in \mathbb{N}_0. \end{cases}$$

Then f is delta predifferentiable with region $D = \mathbb{T}$.

Example 3.12. Let $U = \{\frac{1}{2^n} : n \in \mathbb{N}\}$, and let

$$f(t) = \begin{cases} 1 & \text{if } t \in U, \\ 0 & \text{if } t = 0, \\ -t & \text{if } t \in \{1,2\}, \\ t+4 & \text{if } t \in (1-U) \cup (1+U) \cup (2-u) \cup (2+U). \end{cases}$$

Then f is delta predifferentiable with $D = \mathbb{T}^\kappa \backslash \{0,1,2\}$.

Exercise 3.7. Let $\mathbb{T} = P_{3,7} \cup [4,6]$, and let

$$f(t) = \begin{cases} -1 & \text{if } t \in \bigcup_{k=0}^{\infty} [10k, 10k+3], \\ t^2 + 2 & \text{if } t \in \bigcup_{k=0}^{\infty} [10k+3, 10k+7]. \end{cases}$$

Check if f is delta predifferentiable and if it is, then find its region D.

Exercise 3.8. Let $\mathbb{T} = [1,2] \cup [3,4] \cup [7,8] \cup 9^{\mathbb{N}}$, and let

$$f(t) = \begin{cases} t+2 & \text{if } t \in [1,2] \cup (3,4] \cup (7,8] \cup 9^{\mathbb{N}}, \\ t^2 + 10 & \text{if } t \in \{3,7\}. \end{cases}$$

Check if f is delta predifferentiable and if it is, then find its region D.

Exercise 3.9. Let $\mathbb{T} = \mathbb{R}$, and let

$$f(t) = \begin{cases} 0 & \text{if } t = -3, \\ \frac{1}{t+3} & \text{if } t \in \mathbb{R} \backslash \{-3\}. \end{cases}$$

Check if $f : \mathbb{T} \to \mathbb{R}$ is delta predifferentiable and if it is, then find the region.

Definition 3.3. A function $f : \mathbb{T} \to \mathbb{R}$ is called rd-continuous if it is continuous at right-dense points in $\mathbb{T}$ and its finite left-sided limits exist at left-dense points in $\mathbb{T}$. The set of rd-continuous functions $f : \mathbb{T} \to \mathbb{R}$ will be denoted by $\mathscr{C}_{rd}(\mathbb{T})$.

The set of functions $f : \mathbb{T} \to \mathbb{R}$ that are differentiable with rd-continuous derivatives is denoted by $\mathscr{C}_{rd}^1(\mathbb{T})$. Analogously, we define the sets $\mathscr{C}_{rd}^k(\mathbb{T})$ for arbitrary $k \in \mathbb{N}$.

Exercise 3.10. Let $f : \mathbb{T} \to \mathbb{R}$.
1. If f is continuous, then prove that f is rd-continuous.
2. If f is rd-continuous, then prove that f is regulated.
3. Prove that the jump operator σ is rd-continuous.
4. If f is regulated or rd-continuous, then prove that so is f^σ.
5. Assume f is continuous. If $g : \mathbb{T} \to \mathbb{R}$ is regulated or rd-continuous, then prove that $f \circ g$ also is.

Example 3.13. We will show that every regulated function on a compact interval is bounded. For this aim, assume that $f : [a,b]_{\mathbb{T}} \to \mathbb{R}$, $[a,b]_{\mathbb{T}} \subset \mathbb{T}$, is unbounded. Then for each $n \in \mathbb{N}$, there exists $t_n \in [a,b]_{\mathbb{T}}$ such that $|f(t_n)| > n$. Because $\{t_n\}_{n\in\mathbb{N}} \subset [a,b]_{\mathbb{T}}$, there exists a subsequence $\{t_{n_k}\}_{k\in\mathbb{N}} \subset \{t_n\}_{n\in\mathbb{N}}$ such that

$$\lim_{k\to\infty} t_{n_k} = t_0.$$

Since $\mathbb{T}$ is closed, we have that $t_0 \in \mathbb{T}$. Also, t_0 is a right- or left-dense point. Since f is regulated, we get

$$\left|\lim_{k\to\infty} f(t_{n_k})\right| \neq \infty,$$

is a contradiction.

Example 3.14. Let f and g be real-valued functions defined on $\mathbb{T}$, both delta predifferentiable with region of differentiation region of differentiation D. We will prove that

$$|f^{\Delta}(t)| \leq |g^{\Delta}(t)| \quad \text{for all } t \in D$$

implies

$$|f(s) - f(r)| \leq g(s) - g(r) \quad \text{for all } r, s \in \mathbb{T},\ r \leq s. \tag{3.1}$$

Let $r, s \in \mathbb{T}$ with $r \leq s$. Let also

$$[r,s)_{\mathbb{T}} \backslash D = \{t_n : n \in \mathbb{N}\}.$$

Take $\varepsilon > 0$. We consider the statements

$$S(t) : |f(t) - f(r)| \leq g(t) - g(r) + \varepsilon\left(t - r + \sum_{t_n<t} 2^{-n}\right)$$

for $t \in [r,s]_{\mathbb{T}}$.

We will prove, using the induction principle, that $S(t)$ is true for all $t \in [r,s]_{\mathbb{T}}$.

1. $S(r) : 0 \leq \varepsilon \sum_{t_n<r} 2^{-n}$ is true.
2. Let $t \in [r,s]_{\mathbb{T}}$ be right-scattered, and let $S(t)$ hold. Then we have

$$\begin{aligned}
|f(\sigma(t)) - f(r)| &= |f(t) + \mu(t)f^{\Delta}(t) - f(r)| \\
&\leq \mu(t)|f^{\Delta}(t)| + |f(t) - f(r)| \\
&\leq \mu(t)|f^{\Delta}(t)| + g(t) - g(r) + \varepsilon\left(t - r + \sum_{t_n<t} 2^{-n}\right) \\
&\leq \mu(t)g^{\Delta}(t) + g(t) - g(r) + \varepsilon\left(t - r + \sum_{t_n<t} 2^{-n}\right)
\end{aligned}$$

$$\le g(\sigma(t)) - g(r) + \varepsilon\Big(t - r + \sum_{t_n<t} 2^{-n}\Big) \quad (t < \sigma(t))$$

$$< g(\sigma(t)) - g(r) + \varepsilon\Big(\sigma(t) - r + \sum_{t_n<\sigma(t)} 2^{-n}\Big),$$

i. e., $S(\sigma(t))$ holds.

3. Let $t \in [r,s)_{\mathbb{T}}$ be right-dense.
   a. Let $t \in D$. Then f and g are delta differentiable at t. Hence there exists a neighborhood U of t such that

$$|f(t) - f(\tau) - f^{\Delta}(t)(t-\tau)| \le \frac{\varepsilon}{2}|t-\tau|$$

for all $\tau \in U$ and

$$|g(t) - g(\tau) - g^{\Delta}(t)(t-\tau)| \le \frac{\varepsilon}{2}|t-\tau|$$

for all $\tau \in U$. Thus

$$|f(t) - f(\tau)| \le \Big(|f^{\Delta}(t)| + \frac{\varepsilon}{2}\Big)|t-\tau|$$

for all $\tau \in U$, and

$$g(\tau) - g(t) + g^{\Delta}(t)(t-\tau) \ge -\frac{\varepsilon}{2}|t-\tau|$$

for all $\tau \in U$, or

$$g(\tau) - g(t) - g^{\Delta}(t)(\tau-t) \ge -\frac{\varepsilon}{2}|t-\tau|$$

for all $\tau \in U$.
Hence, for all $\tau \in U \cap (t,\infty)_{\mathbb{T}}$,

$$\begin{aligned}
|f(\tau) - f(r)| &= |f(\tau) - f(t) + f(t) - f(r)| \\
&\le |f(\tau) - f(t)| + |f(t) - f(r)| \\
&\le \Big(|f^{\Delta}(t)| + \frac{\varepsilon}{2}\Big)|t-\tau| + g(t) - g(r) + \varepsilon\Big(t - r + \sum_{t_n<t} 2^{-n}\Big) \\
&\le \Big(g^{\Delta}(t) + \frac{\varepsilon}{2}\Big)|t-\tau| + g(t) - g(r) + \varepsilon\Big(t - r + \sum_{t_n<t} 2^{-n}\Big) \\
&= g^{\Delta}(t)(\tau - t) + \frac{\varepsilon}{2}(\tau - t) + g(t) - g(r) + \varepsilon\Big(t - r + \sum_{t_n<t} 2^{-n}\Big) \\
&\le g(\tau) - g(t) + \frac{\varepsilon}{2}|t-\tau| + \frac{\varepsilon}{2}(\tau - t) + g(t) - g(r) + \varepsilon\Big(t - r + \sum_{t_n<t} 2^{-n}\Big)
\end{aligned}$$

$$= g(\tau) - g(r) + \varepsilon(\tau - t) + \varepsilon\left(t - r + \sum_{t_n<t} 2^{-n}\right)$$
$$= g(\tau) - g(r) + \varepsilon\left(\tau - r + \sum_{t_n<t} 2^{-n}\right).$$

Therefore $S(\tau)$ follows for all $\tau \in U \cap (t, \infty)_{\mathbb{T}}$.

b. Let $t \notin D$. Then $t = t_m$ for some $m \in \mathbb{N}$. Since f and g are delta predifferentiable, they both are continuous. Therefore there exists a neighborhood U of t such that

$$|f(\tau) - f(t)| \le \frac{\varepsilon}{2} 2^{-m} \quad \text{for all } \tau \in U$$

and

$$|g(\tau) - g(t)| \le \frac{\varepsilon}{2} 2^{-m} \quad \text{for all } \tau \in U.$$

Therefore

$$g(\tau) - g(t) \ge -\frac{\varepsilon}{2} 2^{-m} \quad \text{for all } \tau \in U.$$

Consequently,

$$\begin{aligned} |f(\tau) - f(r)| &= |f(\tau) - f(t) + f(t) - f(r)| \\ &\le |f(\tau) - f(t)| + |f(t) - f(r)| \\ &\le \frac{\varepsilon}{2} 2^{-m} + g(t) - g(r) + \varepsilon\left(t - r + \sum_{t_n<t} 2^{-n}\right) \\ &\le \frac{\varepsilon}{2} 2^{-m} + g(\tau) + \frac{\varepsilon}{2} 2^{-m} - g(r) + \varepsilon\left(t - r + \sum_{t_n<t} 2^{-n}\right) \\ &= \varepsilon 2^{-m} + g(\tau) - g(r) + \varepsilon\left(t - r + \sum_{t_n<t} 2^{-n}\right) \\ &\le \varepsilon 2^{-m} + g(\tau) - g(r) + \varepsilon\left(\tau - r + \sum_{t_n<\tau} 2^{-n}\right), \end{aligned}$$

so $S(\tau)$ follows for all $\tau \in U \cap (t, \infty)_{\mathbb{T}}$.

4. Let t be left-dense and suppose $S(\tau)$ is true for $\tau \in [r, t)_{\mathbb{T}}$. Then

$$\begin{aligned} \lim_{\tau \to t-} |f(\tau) - f(r)| &\le \lim_{\tau \to t-} \left\{ g(\tau) - g(r) + \varepsilon\left(\tau - r + \sum_{t_n<\tau} 2^{-n}\right) \right\} \\ &\le \lim_{\tau \to t-} \left\{ g(\tau) - g(r) + \varepsilon\left(\tau - r + \sum_{t_n<t} 2^{-n}\right) \right\} \end{aligned}$$

implies $S(t)$ since f and g are continuous at t.

Hence by the induction principle it follows that $S(t)$ is true for all $t \in [r,s]_{\mathbb{T}}$. Consequently, (3.1) holds for all $r \le s$, $r, s \in \mathbb{T}$.

Example 3.15. Suppose $f : \mathbb{T} \to \mathbb{R}$ is delta predifferentiable with region of differentiation D. For a compact interval U with endpoints $r, s \in \mathbb{T}$, we will show that

$$|f(s) - f(r)| \le \Big\{ \sup_{t \in U^{\kappa} \cap D} |f^{\Delta}(t)| \Big\} |s - r|.$$

Without loss of generality, we suppose that $r \le s$. We set

$$g(t) = \Big\{ \sup_{t \in U^{\kappa} \cap D} |f^{\Delta}(t)| \Big\} (t - r), \quad t \in \mathbb{T}.$$

Then

$$g^{\Delta}(t) = \Big\{ \sup_{t \in U^{\kappa} \cap D} |f^{\Delta}(t)| \Big\} \ge |f^{\Delta}(t)|$$

for all $t \in D \cap [r,s]_{\mathbb{T}}^{\kappa}$.

Hence by Example 3.14 it follows that

$$|f(t) - f(r)| \le g(t) - g(r) \quad \text{for all } t \in [r,s]_{\mathbb{T}},$$

whereupon

$$|f(s) - f(r)| \le g(s) - g(r) = g(s) = \Big\{ \sup_{t \in U^{\kappa} \cap D} |f^{\Delta}(t)| \Big\} (s - r).$$

Example 3.16. Let f be delta predifferentiable with region of differentiation D. If $f^{\Delta}(t) = 0$ for all $t \in D$, then we will show that f is a constant function. For this aim, let U be a compact interval with endpoints $r, s \in \mathbb{T}$. From Example 3.15 it follows that for all $r, s \in \mathbb{T}$,

$$|f(s) - f(r)| \le \Big\{ \sup_{t \in U^{\kappa} \cap D} |f^{\Delta}(t)| \Big\} |s - r| = 0,$$

i. e., $f(s) = f(r)$. Therefore f is a constant function.

Example 3.17. Let f and g be delta predifferentiable with region of differentiation D. If $f^{\Delta}(t) = g^{\Delta}(t)$ for all $t \in D$, then we will show that

$$g(t) = f(t) + C \quad \text{for all } t \in \mathbb{T},$$

where C is a constant. For this aim, let $h(t) = f(t) - g(t)$, $t \in \mathbb{T}$. Then

$$h^{\Delta}(t) = f^{\Delta}(t) - g^{\Delta}(t) = 0 \quad \text{for all } t \in D.$$

Hence by Example 3.16 it follows that h is a constant function.

Example 3.18. Suppose $f_n : \mathbb{T} \to \mathbb{R}$ is delta predifferentiable with D for each $n \in \mathbb{N}$. Assume that for each $t \in \mathbb{T}^\kappa$, there exists a compact interval neighborhood $U(t)$ such that the sequence $\{f_n^\Delta\}_{n\in\mathbb{N}}$ converges uniformly on $U(t) \cap D$. If $\{f_n\}_{n\in\mathbb{N}}$ converges at some $t_0 \in U(t)$ for some $t \in \mathbb{T}^\kappa$, we will prove that it converges uniformly on $U(t)$.

Indeed, since $\{f_n^\Delta\}_{n\in\mathbb{N}}$ converges uniformly on $U(t) \cap D$, there exists $N \in \mathbb{N}$ such that

$$\sup_{s\in U(t)\cap D} |(f_m - f_n)^\Delta(s)|$$

is finite for all $m, n \in \mathbb{N}$. Let $m, n \geq N$ and $r \in U(t)$. Then

$$\begin{aligned} |f_n(r) - f_m(r)| &= |f_n(r) - f_m(r) - (f_n(t_0) - f_m(t_0)) + (f_n(t_0) - f_m(t_0))| \\ &\leq |f_n(t_0) - f_m(t_0)| + \Big\{ \sup_{s\in U(t)\cap D} |(f_n - f_m)^\Delta(s)| \Big\} |r - t_0|. \end{aligned}$$

Hence $\{f_n\}_{n\in\mathbb{N}}$ converges uniformly on $U(t)$, i. e., $\{f_n\}_{n\in\mathbb{N}}$ is locally uniformly convergent sequence.

Example 3.19. Suppose $f_n : \mathbb{T} \to \mathbb{R}$ is delta predifferentiable with D for each $n \in \mathbb{N}$. Assume that for each $t \in \mathbb{T}^\kappa$, there exists a compact neighborhood $U(t)$ such that the sequence $\{f_n^\Delta\}_{n\in\mathbb{N}}$ converges uniformly on $U(t) \cap D$. If $\{f_n\}_{n\in\mathbb{N}}$ converges at some $t_0 \in \mathbb{T}$, then we will prove that it converges uniformly on $U(t)$ for all $t \in \mathbb{T}^\kappa$. For this aim, set

$$S(t) : \{f_n(t)\}_{n\in\mathbb{N}} \quad \text{converges.}$$

1. $S(t_0) : \{f_n(t_0)\}$ converges is true.
2. Let t be right-scattered and suppose $S(t)$ holds. Then

   $$f_n(\sigma(t)) = f_n(t) + \mu(t) f_n^\Delta(t)$$

   converges by the assumption, i. e., $S(\sigma(t))$ holds.
3. Let t be right-dense and suppose $S(t)$ holds. Then, by (i), $\{f_n\}_{n\in\mathbb{N}}$ converges on $U(t)$, and so $S(r)$ holds for all $r \in U(t) \cap (t, \infty)_\mathbb{T}$.
4. Let t be left-dense and suppose $S(r)$ holds for all $t_0 \leq r < t$. Since $U(t) \cap [t_0, t)_\mathbb{T} \neq \emptyset$, using again part (i), we have that $\{f_n\}_{n\in\mathbb{N}}$ converges on $U(t)$ and, in particular, $S(t)$ is true.

Consequently, $S(t)$ is true for all $t \in [t_0, \infty)_\mathbb{T}$. Using the dual version of the induction principle for the negative direction, we have that $S(t)$ is also true for all $t \in (-\infty, t_0]_\mathbb{T}$ (note that the first part of this has already been shown, the second part follows by $f_n(\rho(t)) = f_n(t) - \mu(\rho(t)) f_n^\Delta(\rho(t))$, and the third and fourth parts follow again by (i)).

Example 3.20. Suppose $f_n : \mathbb{T} \to \mathbb{R}$ is delta predifferentiable with D for each $n \in \mathbb{N}$. Assume that for each $t \in \mathbb{T}^\kappa$, there exists a compact interval neighborhood $U(t)$ such that the sequence $\{f_n^\Delta\}_{n\in\mathbb{N}}$ converges uniformly on $U(t) \cap D$. We will prove that the limit mapping $f = \lim_{n\to\infty} f_n$ is delta predifferentiable with D and

$$f^\Delta(t) = \lim_{n\to\infty} f_n^\Delta(t) \quad \text{for all } t \in D.$$

Let $t \in D$. Without loss of generality, we can assume that $\sigma(t) \in U(t)$. Take arbitrary $\varepsilon > 0$. By (i) there exists $N \in \mathbb{N}$ such that

$$|(f_n - f_m)(r) - (f_n - f_m)(\sigma(t))| \le \Big\{ \sup_{s\in U(t)\cap D} |(f_n - f_m)^\Delta(s)| \Big\} |r - \sigma(t)|$$

for all $r \in U(t)$ and $m, n \ge N$. Since

$$\{f_n^\Delta\}_{n\in\mathbb{N}}$$

converges uniformly on $U(t) \cap D$, there exists $N_1 \ge N$ such that

$$\sup_{s\in U(t)\cap D} |(f_n - f_m)^\Delta(s)| \le \frac{\varepsilon}{3} \quad \text{for all } m, n \ge N_1.$$

Hence

$$|(f_n - f_m)(r) - (f_n - f_m)(\sigma(t))| \le \frac{\varepsilon}{3} |r - \sigma(t)|$$

for all $r \in U(t)$ and $m, n \ge N_1$. Now letting $m \to \infty$, we have

$$|(f_n - f)(r) - (f_n - f)(\sigma(t))| \le \frac{\varepsilon}{3} |r - \sigma(t)|$$

for all $r \in U(t)$ and $n \ge N_1$. Let

$$g = \lim_{n\to\infty} f_n^\Delta.$$

Then there exists $M \ge N_1$ such that

$$|f_M^\Delta(t) - g(t)| \le \frac{\varepsilon}{3}$$

and since f_M is differentiable at t, there also exists a neighborhood W of t such that

$$|f_M(\sigma(t)) - f_M(r) - f_M^\Delta(t)(\sigma(t) - r)| \le \frac{\varepsilon}{3} |\sigma(t) - r|$$

for all $r \in W$.

From this it follows that for all $r \in U(t) \cap W$,

$$|f(\sigma(t)) - f(r) - g(t)|\sigma(t) - r|| \le |(f_M - f)(\sigma(t)) - (f_M - f)(r)|$$

$$\begin{aligned}
&+ \left|f_M^{\Delta}(t) - g(t)\right|\left|\sigma(t) - r\right| \\
&+ \left|f_M(\sigma(t)) - f_M(r) - f_M^{\Delta}(t)\left|\sigma(t) - r\right|\right| \\
&\le \frac{\varepsilon}{3}\left|\sigma(t) - r\right| + \frac{\varepsilon}{3}\left|\sigma(t) - r\right| + \frac{\varepsilon}{3}\left|\sigma(t) - r\right| = \varepsilon\left|\sigma(t) - r\right|.
\end{aligned}$$

Consequently, f is differentiable at t with $f^{\Delta}(t) = g(t)$.

Example 3.21. Let $t_0 \in \mathbb{T}$ and $x_0 \in \mathbb{R}$, and let $f : \mathbb{T}^{\kappa} \to \mathbb{R}$ be a regulated map. We will prove that there exists exactly one delta predifferentiable function F with D satisfying

$$F^{\Delta}(t) = f(t) \quad \text{for all } t \in D, \quad F(t_0) = x_0.$$

Let $n \in \mathbb{N}$, and let

$$S(t) : \begin{cases} \text{there exists a predifferentiable } (F_{nt}, D_{nt}), \\ F_{nt} : [t_0, t] \to \mathbb{R} \quad \text{with } F_{nt}(t_0) = x_0, \quad \text{and} \\ |F_{nt}^{\Delta}(s) - f(s)| \le \frac{1}{n} \quad \text{for } s \in D_{nt}. \end{cases}$$

1. Let $t = t_0$, $D_{nt_0} = \emptyset$, and $F_{nt_0}(t_0) = x_0$. Then the statement $S(t_0)$ follows.
2. Let t be right-scattered and suppose $S(t)$ is true. Define

$$D_{n\sigma(t)} = D_{nt} \cup \{t\}$$

and $F_{n\sigma(t)}$ on $[t_0, \sigma(t)]$ by

$$F_{n\sigma(t)}(s) = \begin{cases} F_{nt}(s) & \text{if } s \in [t_0, t], \\ F_{nt}(t) + \mu(t)f(t) & \text{if } s = \sigma(t). \end{cases}$$

Then

$$\begin{aligned}
F_{n\sigma(t)}(t_0) &= F_{nt}(t_0) = x_0, \\
\left|F_{n\sigma(t)}^{\Delta}(s) - f(s)\right| &= \left|F_{nt}^{\Delta}(s) - f(s)\right| \le \frac{1}{n} \quad \text{for } s \in D_{nt},
\end{aligned}$$

and

$$\begin{aligned}
\left|F_{n\sigma(t)}^{\Delta}(t) - f(t)\right| &= \left|\frac{F_{n\sigma(t)}(\sigma(t)) - F_{n\sigma(t)}(t)}{\mu(t)} - f(t)\right| \\
&= \left|\frac{F_{nt}(t) + \mu(t)f(t) - F_{n\sigma(t)}(t)}{\mu(t)} - f(t)\right| \\
&= \left|\frac{F_{nt}(t) + \mu(t)f(t) - F_{nt}(t)}{\mu(t)} - f(t)\right| \\
&= \left|\frac{\mu(t)f(t)}{\mu(t)} - f(t)\right|
\end{aligned}$$

$$= 0$$
$$\leq \frac{1}{n}.$$

Therefore the statement $S(\sigma(t))$ is true.

3. Suppose t is right-dense and $S(t)$ is true. Since t is right-dense and $f(t)$ is regulated,

$$f(t^+) = \lim_{s \to t, s > t} f(s) \quad \text{exists.}$$

Hence there is a neighborhood U of t such that

$$|f(s) - f(t^+)| \leq \frac{1}{n} \quad \text{for all } s \in U \cap (t, \infty)_{\mathbb{T}}. \tag{3.2}$$

Let $r \in U \cap (t, \infty)_{\mathbb{T}}$. Define

$$D_{nr} = (D_{nt} \backslash \{t\}) \cup [t, r]_{\mathbb{T}}^{\kappa}$$

and F_{nr} on $[t_0, r]_{\mathbb{T}}$ by

$$F_{nr}(s) = \begin{cases} F_{nt}(s) & \text{if } s \in [t_0, t]_{\mathbb{T}}, \\ F_{nt}(t) + f(t^+)(s - t) & \text{if } s \in (t, r]_{\mathbb{T}}. \end{cases}$$

Then F_{nr} is continuous at t and hence on $[t_0, r]_{\mathbb{T}}$. Also, F_{nr} is differentiable on $(t, r]_{\mathbb{T}}^{\kappa}$ with

$$F_{nr}^{\Delta}(s) = f(t^+) \quad \text{for all } s \in (t, r]_{\mathbb{T}}^{\kappa}.$$

Hence F_{nr} is predifferentiable on $[t_0, t)_{\mathbb{T}}$. Since t is right-dense, we have that F_{nr} is predifferentiable with D_{nr}. From here and from (3.2) we also have

$$|F_{nr}^{\Delta}(s) - f(s)| \leq \frac{1}{n} \quad \text{for all } s \in D_{nr}.$$

Therefore the statement $S(r)$ is true for all $r \in U \cap (t, \infty)_{\mathbb{T}}$.

4. Now suppose that t is left-dense and the statement $S(r)$ is true for $r < t$. Since $f(t)$ is regulated,

$$f(t^-) = \lim_{s \to t, s < t} f(s) \quad \text{exists.} \tag{3.3}$$

Hence there exists a neighborhood U of t such that

$$|f(s) - f(t^-)| \leq \frac{1}{n} \quad \text{for all } s \in U \cap (-\infty, t)_{\mathbb{T}}.$$

Fix some $r \in U \cap (-\infty, t)_{\mathbb{T}}$ and define

$$D_{nt} = \begin{cases} D_{nr} \cup (r,t) & \text{if } r \text{ is right-dense,} \\ D_{nr} \cup [r,t) & \text{if } r \text{ is right-scattered,} \end{cases}$$

and F_{nt} on $[t_0, t]_{\mathbb{T}}$ by

$$F_{nt}(s) = \begin{cases} F_{nr}(s) & \text{if } s \in (t_0, r]_{\mathbb{T}}, \\ F_{nr}(t) + f(t^-)(s-r) & \text{if } s \in (r,t]_{\mathbb{T}}. \end{cases}$$

Note that F_{nt} is continuous at r and hence in $[t_0, t]_{\mathbb{T}}$. Since

$$F_{nt}^{\Delta}(s) = f(t^-) \quad \text{for all } s \in (r,t]_{\mathbb{T}},$$

F_{nt} is delta predifferentiable with D_{nt}, and

$$|F_{nt}^{\Delta}(s) - f(s)| \leq \frac{1}{n} \quad \text{for all } s \in D_{nt}.$$

Hence the statement $S(t)$ holds.

By the induction principle it follows that $S(t)$ is true for all $t \geq t_0$, $t \in \mathbb{T}$. Similarly, we can show that $S(t)$ is true for $t \leq t_0$. Hence F_n is delta predifferentiable with D_n, $F_n(t_0) = x_0$, and

$$|F_n^{\Delta}(t) - f(t)| \leq \frac{1}{n} \quad \text{for all } t \in D_n.$$

Now let

$$F = \lim_{n\to\infty} F_n \quad \text{and} \quad D = \bigcap_{n\in\mathbb{N}} D_n.$$

Then $F(t_0) = x_0$, F is delta predifferentiable on D, and

$$F^{\Delta}(t) = \lim_{n\to\infty} F_n^{\Delta}(t) = f(t) \quad \text{for all } t \in D.$$

3.2 Delta indefinite integral

Definition 3.4. Let $f : \mathbb{T} \to \mathbb{R}$ be a regulated function. Any function F by Example 3.21 is called a delta preantiderivative of f. We define the delta indefinite integral of a regulated function f by

$$\int f(t)\Delta t = F(t) + c,$$

where c is an arbitrary constant and F is a delta preantiderivative of f. We define the Cauchy delta integral by

$$\int_{\tau}^{s} f(t)\Delta t = F(s) - F(\tau) \quad \text{for all } \tau, s \in \mathbb{T}.$$

A function $F : \mathbb{T} \to \mathbb{R}$ is called a delta antiderivative of $f : \mathbb{T} \to \mathbb{R}$ if

$$F^{\Delta}(t) = f(t) \quad \text{for all } t \in \mathbb{T}^{\kappa}.$$

Example 3.22. Let $\mathbb{T} = h\mathbb{Z}$, $h > 0$, and let

$$f(t) = 4t^3 + 3(2h + 11)t^2 + (4h^2 + 3h + 57)t + h^3 + 10h^2 + 45h + 60, \quad t \in \mathbb{T}.$$

Set

$$F(t) = (t + 1)(t + 2)(t + 3)(t + 4), \quad t \in \mathbb{T}.$$

By Example 2.13 it follows that

$$F^{\Lambda}(t) - f(t), \quad t \in \mathbb{T},$$

and then

$$\int (4t^3+3(2h+11)t^2+(4h^2+3h+57)t+h^3+10h^2+45h+60)\Delta t = (t+1)(t+2)(t+3)(t+4)+C, \quad t \in \mathbb{T},$$

where C is a constant.

Example 3.23. Let $\mathbb{T} = 3^{\mathbb{N}_0}$, and let

$$f(t) = 13t^2 + 4t + 1, \quad t \in \mathbb{T}.$$

Set

$$F(t) = t^3 + t^2 + t + 1, \quad t \in \mathbb{T}.$$

Then applying Example 2.14, we find

$$F^{\Delta}(t) = f(t), \quad t \in \mathbb{T},$$

and then

$$\int (13t^3 + 4t + 1)\Delta t = t^3 + t^2 + t + C, \quad t \in \mathbb{T},$$

where C is a constant.

Example 3.24. Let $\mathbb{T} = \mathbb{N}_0^2$, and let

$$f(t) = \frac{-t^3 + 11t^2 + 2t + 18t\sqrt[3]{t^2} + 18t\sqrt[3]{t} + 3\sqrt[3]{t^2} + 3\sqrt[3]{t} - 40}{(t^2 + t + 7)(t^2 + 15t\sqrt[3]{t} + 6t\sqrt[3]{t^2} + 18\sqrt[3]{t^2} + 9\sqrt[3]{t} + 21t + 9)}, \quad t \in \mathbb{T}.$$

Set

$$F(t) = \frac{t^2 - 5t + 6}{t^2 + t + 7}, \quad t \in \mathbb{T}.$$

By Example 2.15 we have

$$F^{\Delta}(t) = f(t), \quad t \in \mathbb{T},$$

and then

$$\int \frac{-t^3 + 11t^2 + 2t + 18t\sqrt[3]{t^2} + 18t\sqrt[3]{t} + 3\sqrt[3]{t^2} + 3\sqrt[3]{t} - 40}{(t^2 + t + 7)(t^2 + 15t\sqrt[3]{t} + 6t\sqrt[3]{t^2} + 18\sqrt[3]{t^2} + 9\sqrt[3]{t} + 21t + 9)} \Delta t = \frac{t^2 - 5t + 6}{t^2 + t + 7} + C, \quad t \in \mathbb{T},$$

where C is a constant.

Example 3.25. Let $\mathbb{T} = \{H_n : n \in \mathbb{N}_0\}$, and let

$$f(t) = -\frac{t\sigma(t) + 12}{(t^2 - 4t - 12)((\sigma(t))^2 - 4\sigma(t) - 12)}, \quad t \in \mathbb{T}.$$

Set

$$F(t) = \frac{t}{t^2 - 4t - 12}, \quad t \in \mathbb{T}.$$

By Example 2.16 we get

$$F^{\Delta}(t) = f(t), \quad t \in \mathbb{T}.$$

Then

$$-\int \frac{t\sigma(t) + 12}{(t^2 - 4t - 12)((\sigma(t))^2 - 4\sigma(t) - 12)} \Delta t = \frac{t}{t^2 - 4t - 12} + C, \quad t \in \mathbb{T},$$

where C is a constant.

Example 3.26. Let $\mathbb{T} = P_{1,3}$, and let

$$f(t) = \begin{cases} -\frac{2}{(1+t)(4+t)} & \text{if } t \text{ is right-scattered,} \\ -\frac{2}{(1+t)^2} & \text{if } t \text{ is right-dense.} \end{cases}$$

Set

$$F(t) = \frac{1-t}{1+t}, \quad t \in \mathbb{T}.$$

By Example 2.17 we get

$$F^{\Delta}(t) = f(t), \quad t \in \mathbb{T}.$$

Therefore

$$\int f(t)\Delta t = \frac{1-t}{1+t} + C, \quad t \in \mathbb{T},$$

where C is a constant.

Example 3.27. Let $\mathbb{T} = C$, where C is the Cantor set, and let

$$f(t) = \begin{cases} \frac{-4t-\frac{2}{3^{m+1}}}{(1+t^2)\left(1+\frac{1}{3^{2m+2}}+\frac{2}{3^{m+1}}t+t^2\right)} & \text{if } t \in C_1, \\ -\frac{4t}{(1+t^2)^2} & t \in \mathbb{T}\backslash C_1. \end{cases}$$

Set

$$F(t) = \frac{1-t^2}{1+t^2}, \quad t \in \mathbb{T}.$$

By Example 2.18 we get

$$F^{\Delta}(t) = f(t), \quad t \in \mathbb{T}.$$

Then

$$\int f(t)\Delta t = \frac{1-t^2}{1+t^2} + C, \quad t \in \mathbb{T},$$

where C is a constant.

Example 3.28. Let $\mathbb{T} = \{\sum_{k=0}^{n-1} a_k,\ n \in \mathbb{N},\ a_k > 0,\ k \in \mathbb{N}_0\}$, and let

$$f(t) = \frac{2}{(3+4t)(3+4\sigma(t))}, \quad t \in \mathbb{T}.$$

Set

$$F(t) = \frac{1+2t}{3+4t}, \quad t \in \mathbb{T}.$$

By Example 2.19 we get

$$F^{\Delta}(t) = f(t), \quad t \in \mathbb{T}.$$

Therefore

$$\int \frac{2}{(3+4t)(3+4\sigma(t))}\Delta t = \frac{1+2t}{3+4t} + C, \quad t \in \mathbb{T},$$

where C is a constant.

Example 3.29. Let $\mathbb{T} = \{-\frac{1}{n} : n \in \mathbb{N}\} \cup \mathbb{N}_0$, and let

$$f(t) = \begin{cases} \frac{6t^2+9t-7}{(1+7t)(1+6t)} & \text{if } t \in \{-\frac{1}{n} : n \in \mathbb{N}\}, \\ \frac{7t^2+9t-6}{(1+7t)(8+7t)} & \text{if } t \in \mathbb{N}_0. \end{cases}$$

Set

$$\frac{1+t^2}{1+7t}, \quad t \in \mathbb{T}.$$

By Example 2.20 we get

$$F^{\Delta}(t) = f(t), \quad t \in \mathbb{T}.$$

Therefore

$$\int f(t)\Delta t = \frac{1+t^2}{1+7t} + C, \quad t \in \mathbb{T},$$

where C is a constant.

Example 3.30. Let $\mathbb{T} = \{(\frac{1}{2})^{2^n} : n \in \mathbb{N}_0\} \cup \{0, 1\}$, and let

$$f(t) = \begin{cases} \frac{23}{(4+5t)(4+5\sqrt{t})} & \text{if } t \in \{(\frac{1}{2})^{2^n} : n \in \mathbb{N}\}, \\ \frac{46}{117} & \text{if } t = \frac{1}{2}, \\ \frac{23}{16} & \text{if } t = 0. \end{cases}$$

Set

$$F(t) = \frac{1+7t}{4+5t}, \quad t \in \mathbb{T}.$$

By Example 2.21 we get

$$F^{\Delta}(t) = f(t), \quad t \in \mathbb{T}.$$

Therefore

$$\int f(t)\Delta t = \frac{1+7t}{4+5t} + C, \quad t \in \mathbb{T},$$

where C is a constant.

Example 3.31. Let $U = \{\frac{1}{2^n} : n \in \mathbb{N}\}$, let

$$\mathbb{T} = U \cup (1-U) \cup (1+U) \cup (2-U) \cup (2+U) \cup \{0,1,2\},$$

and let

$$f(t) = \begin{cases} -\frac{145}{64} & \text{if } t = \frac{1}{2}, \\ -\frac{1215}{64} & \text{if } t = \frac{3}{2}, \\ -15t^3 - 3t & \text{if } t \in U \backslash \{\frac{1}{2}\}, \\ -\frac{15t^3+11t^2+17t+5}{8} & \text{if } t \in (1-U) \backslash \{\frac{1}{2}\}, \\ -15t^3 + 17t^2 - 10t + 2 & \text{if } t \in (1+U) \backslash \{\frac{3}{2}\}, \\ -\frac{15t^3+22t^2+32t+8}{8} & \text{if } t \in (2-U) \backslash \{\frac{3}{2}\}, \\ -15t^3 + 34t^2 - 31t + 10 & \text{if } t \in (2+U) \backslash \{\frac{5}{2}\}, \\ 0 & \text{if } t = 0, \\ -6 & \text{if } t = 1, \\ 36 & \text{if } t = 2. \end{cases}$$

Set

$$F(t) = (2+t^2)(1-t^2), \quad t \in \mathbb{T}.$$

By Example 2.22 we get

$$F^{\Delta}(t) = f(t), \quad t \in \mathbb{T}^{\kappa}.$$

Therefore

$$\int f(t)\Delta t = (2+t^2)(1-t^2) + C, \quad t \in \mathbb{T}^{\kappa},$$

where C is a constant.

Example 3.32. Let $\mathbb{T} = \mathbb{Z}$. Then

$$\sigma(t) = t + 1, \quad t \in \mathbb{T}.$$

Let also

$$f(t) = 3t^2 + 5t + 2, \quad t \in \mathbb{T}.$$

Since

$$(t^3 + t^2)^{\Delta} = (\sigma(t))^2 + t\sigma(t) + t^2 + \sigma(t) + t$$

$$\begin{aligned}&= (t+1)^2 + t(t+1) + t^2 + t + 1 + t\\&= t^2 + 2t + 1 + t^2 + t + t^2 + 2t + 1\\&= 3t^2 + 5t + 2, \quad t \in \mathbb{T},\end{aligned}$$

we get

$$\int (3t^2 + 5t + 2)\Delta t = t^3 + t^2 + C, \quad t \in \mathbb{T},$$

where C is a constant.

Example 3.33. Let $\mathbb{T} = 2^{\mathbb{N}}$, and let $f : \mathbb{T} \to \mathbb{R}$ be defined by

$$f(t) = 2\sin\frac{t}{2}\cos\frac{3t}{2}, t \in \mathbb{T}.$$

In this case, we have that

$$\sigma(t) = 2t, \quad t \in \mathbb{T}.$$

Since

$$\begin{aligned}(\sin t)^{\Delta} &= \frac{\sin\sigma(t) - \sin t}{\sigma(t) - t}\\&= \frac{\sin(2t) - \sin t}{t}\\&= \frac{2}{t}\sin\frac{t}{2}\cos\frac{3t}{2}, \quad t \in \mathbb{T},\end{aligned}$$

we get

$$\int \frac{2}{t}\sin\frac{t}{2}\cos\frac{3t}{2}\Delta t = \sin t + C, \quad t \in \mathbb{T},$$

where C is a constant.

Example 3.34. Let $\mathbb{T} = \mathbb{N}_0^2$, and let $f : \mathbb{T} \to \mathbb{R}$ be defined by

$$f(t) = \frac{1}{1 + 2\sqrt{t}}\log\frac{(\sqrt{t} + 1)^2}{t}, \quad t \in \mathbb{T}.$$

Since

$$\sigma(t) = (\sqrt{t} + 1)^2, \quad t \in \mathbb{T},$$

and

$$(\log t)^{\Delta} = \frac{\log\sigma(t) - \log t}{\sigma(t) - t}$$

$$
\begin{aligned}
&= \frac{\log(\sqrt{t}+1)^2 - \log t}{(\sqrt{t}+1)^2 - t} \\
&= \frac{1}{1+2\sqrt{t}} \log \frac{(\sqrt{t}+1)^2}{t}, \quad t \in \mathbb{T},
\end{aligned}
$$

we get

$$
\int \frac{1}{1+2\sqrt{t}} \log \frac{(1+\sqrt{t})^2}{t} \Delta t = \log t + C, \quad t \in \mathbb{T},
$$

where C is a constant.

Example 3.35. We will show that every rd-continuous function $f : \mathbb{T} \to \mathbb{R}$ has a delta antiderivative. In particular, if $t_0 \in \mathbb{T}$, then F defined by

$$
F(t) = \int_{t_0}^{t} f(\tau)\Delta\tau \quad \text{for } t \in \mathbb{T},
$$

is a delta antiderivative of f. Indeed, since f is rd-continuous, it is regulated. Let F be a function guaranteed to exist by Example 3.21, together with D, satisfying

$$
F^{\Delta}(t) = f(t) \quad \text{for } t \in D.
$$

We have that F is delta predifferentiable with D.

Let $t \in \mathbb{T}^{\kappa} \backslash D$. Then t is right-dense. Since f is rd-continuous, f is continuous at t. Take arbitrary $\varepsilon > 0$. Then there exists a neighborhood U of t such that

$$
|f(s) - f(t)| \le \varepsilon \quad \text{for all } s \in U.
$$

Define

$$
h(\tau) = F(\tau) - f(t)(\tau - t_0) \quad \text{for } \tau \in \mathbb{T}.
$$

Then h is delta predifferentiable with D, and

$$
\begin{aligned}
h^{\Delta}(\tau) &= F^{\Delta}(\tau) - f(t) \\
&= f(\tau) - f(t) \quad \text{for all } \tau \in D.
\end{aligned}
$$

Hence

$$
\begin{aligned}
|h^{\Delta}(s)| &= |f(s) - f(t)| \\
&\le \varepsilon \quad \text{for all } s \in D \cap U.
\end{aligned}
$$

Therefore

$$\sup_{s\in D\cap U} |h^{\Delta}(s)| \leq \varepsilon,$$

whereupon

$$\begin{aligned}|F(t) - F(r) - f(t)(t-r)| &= |h(t) + f(t)(t-t_0) - (h(r) \\ &\quad + f(t)(r-t_0)) - f(t)(t-r)| \\ &= |h(t) - h(r)| \\ &\leq \Big\{\sup_{s\in D\cap U} |h^{\Delta}(s)|\Big\}|t-r| \\ &\leq \varepsilon|t-r|,\end{aligned}$$

which shows that F is delta differentiable at t and $F^{\Delta}(t) = f(t)$.

Exercise 3.11. Let $\mathbb{T} = \mathbb{N}_0^3$. Prove that

$$\int \left(2t + 3\sqrt[3]{t^2} + 3\sqrt[3]{t} + 2\right)\Delta t = t^2 + t + C, \quad t \in \mathbb{T},$$

where C is a constant.

Exercise 3.12. Let $\mathbb{T} = h\mathbb{Z} + k$, $h > 0$, $k \in \mathbb{R}$. Prove that

$$\int \left(4 - 8h - 16t\right)\Delta t = 4t - 8t^2 + C, \quad t \in \mathbb{T},$$

where C is a constant.

Exercise 3.13. Let $\mathbb{T} = (-2\mathbb{N}_0) \cup 3^{\mathbb{N}_0}$, and let

$$f(t) = \begin{cases} 3t^2 + 6t + 3 & \text{if } t \in (-2\mathbb{N}), \\ 0 & \text{if } t = 0, \\ 13t^2 - 1 & \text{if } t \in 3^{\mathbb{N}_0}. \end{cases}$$

Prove that

$$\int f(t)\Delta t = t^3 - t + C, \quad t \in \mathbb{T},$$

where C is a constant.

Exercise 3.14. Let $\mathbb{T} = P_{3,7} \cup [4,6]$, and let

$$f(t) = \begin{cases} \frac{1+20t-2t^2}{(1+2t^2)^2} & \text{if } t \in \left(\bigcup_{k=0}^{\infty}[10k, 10k+3)\right) \cup [4,6); \quad \frac{47}{627} \quad \text{if} \quad t = 3 \\ \frac{71+6t-2t^2}{(1+2t^2)(99+28t+2t^2)} & \text{if } t = 10k+3; \quad \frac{164}{14673} \quad \text{if} \quad t = 6. \end{cases}$$

Prove that

$$\int f(t)\Delta t = \frac{t-5}{1+2t^2}, \quad t \in \mathbb{T},$$

where C is a constant.

Exercise 3.15. Let $\mathbb{T} = \{0\} \cup 11^{\mathbb{N}_0}$, and let

$$f(t) = \begin{cases} \frac{39}{5} & \text{if } t = 0, \\ \frac{39}{(5-4t)(5-44t)} & \text{if } t \in 11^{\mathbb{N}_0}. \end{cases}$$

Prove that

$$\int f(t)\Delta t = \frac{1+7t}{5-4t} + C, \quad t \in \mathbb{T},$$

where C is a constant.

Exercise 3.16. Let $\mathbb{T} = [1,2] \cup [3,4] \cup [7,8] \cup 9^{\mathbb{N}}$, and let

$$f(t) = \begin{cases} -8-4t & \text{if } t \in [1,2) \cup [3,4) \cup [7,8), \\ -8-20t & \text{if } t \in 9^{\mathbb{N}}, \\ -18 & \text{if } t = 2, \\ -30 & \text{if } t = 4, \\ -42 & \text{if } t = 8. \end{cases}$$

Prove that

$$\int f(t)\Delta t = -8t - 2t^2 + C, \quad t \in \mathbb{T},$$

where C is a constant.

3.3 The Darboux delta integral

We choose a couple of points $a, b \in \mathbb{T}$, $a < b$. Consider the segment $[a,b]_{\mathbb{T}}$ and a real-valued function $f : [a,b]_{\mathbb{T}} \to \mathbb{R}$.

Definition 3.5. A partition of $[a,b]_{\mathbb{T}}$ is any finite ordered subset

$$P = \{t_0, t_1, \dots, t_n\} \subset [a,b]_{\mathbb{T}},$$

where

$$a = t_0 < t_1 < \cdots < t_n = b.$$

The intervals $[t_{j-1}, t_j)_{\mathbb{T}}$, $j \in \{1, \dots, n\}$, are called the subintervals of the partition P. The set of all partitions of $[a,b]_{\mathbb{T}}$ is denoted by $\mathscr{P}_{\mathbb{T}} = \mathscr{P}(a,b)_{\mathbb{T}}$.

We set

$$m_j = \inf_{t\in[t_{j-1},t_j)_{\mathbb{T}}} f(t),$$
$$M_j = \sup_{t\in[t_{j-1},t_j)_{\mathbb{T}}} f(t), \quad j \in \{1,\dots,n\},$$

and

$$m = \inf_{t\in[a,b)_{\mathbb{T}}} f(t),$$
$$M = \sup_{t\in[a,b)_{\mathbb{T}}} f(t).$$

Definition 3.6. The upper Darboux Δ-sum $U_{\mathbb{T}}(f,P)$ and the lower Darboux Δ-sum $L_{\mathbb{T}}(f,P)$ of the function f with respect to the partition $P \in \mathscr{P}(a,b)_{\mathbb{T}}$ are defined by

$$U_{\mathbb{T}}(f,P) = \sum_{j=1}^{n} M_j(t_j - t_{j-1}),$$
$$L_{\mathbb{T}}(f,P) = \sum_{j=1}^{n} m_j(t_j - t_{j-1}).$$

By this definition we have

$$L_{\mathbb{T}}(f,P) \le U_{\mathbb{T}}(f,P),$$
$$\begin{aligned} L_{\mathbb{T}}(f,P) &= \sum_{j=1}^{n} m_j(t_j - t_{j-1}) \\ &\ge m\sum_{j=1}^{n}(t_j - t_{j-1}) \\ &= m(b-a), \end{aligned}$$

and

$$\begin{aligned} U_{\mathbb{T}}(f,P) &= \sum_{j=1}^{n} M_j(t_j - t_{j-1}) \\ &\le M\sum_{j=1}^{n}(t_j - t_{j-1}) \\ &= M(b-a). \end{aligned}$$

So we have the inequalities

$$m(b-a) \le L_{\mathbb{T}}(f,P) \le U_{\mathbb{T}}(f,P) \le M(b-a). \tag{3.4}$$

Definition 3.7. The upper Darboux Δ-integral $U_{\mathbb{T}}(f)$ of the function f from a to b is defined as

$$U_{\mathbb{T}}(f) = \inf_{P \in \mathscr{P}(a,b)_{\mathbb{T}}} U_{\mathbb{T}}(f, P),$$

and the lower Darboux Δ-integral $L_{\mathbb{T}}(f)$ of the function f from a to b is defined as

$$L_{\mathbb{T}}(f) = \sup_{P \in \mathscr{P}(a,b)_{\mathbb{T}}} L_{\mathbb{T}}(f, P).$$

By inequalities (3.4) we conclude that $U_{\mathbb{T}}(f)$ and $L_{\mathbb{T}}(f)$ are finite numbers and

$$m(b-a) \le L_{\mathbb{T}}(f) \le U_{\mathbb{T}}(f) \le M(b-a).$$

Definition 3.8. We say that the function f is Δ-integrable or delta integrable (shortly integrable) from a to b (or on $[a, b]_{\mathbb{T}}$) if

$$U_{\mathbb{T}}(f) = L_{\mathbb{T}}(f).$$

In this case, we write

$$\int_a^b f(t)\Delta t = U_{\mathbb{T}}(f) = L_{\mathbb{T}}(f),$$

and we call this integral the Darboux Δ-integral or shortly the Darboux integral.

Definition 3.9. Let $P, Q \in \mathscr{P}(a, b)_{\mathbb{T}}$ be such that $P \subset Q$. Then we say that the partition Q is a refinement of the partition Q.

Example 3.36. Let $P, Q \in \mathscr{P}(a, b)_{\mathbb{T}}$ be such that $P \subset Q$, and let f be a bounded function on $[a, b]_{\mathbb{T}}$. We will show by induction that

$$L_{\mathbb{T}}(f, P) \le L_{\mathbb{T}}(f, Q) \le U_{\mathbb{T}}(f, Q) \le U_{\mathbb{T}}(f, P). \tag{3.5}$$

1. Suppose that Q has one more point, say τ_1, than P. If

   $$P = \{a = t_0 < t_1 < \cdots < t_{n-1} < t_n = b\},$$

   then there is $k \in \{1, \ldots, n\}$ such that

   $$t_{k-1} < \tau_1 < t_k.$$

   Set

   $$m_{k1} = \inf_{t \in [t_{k-1}, \tau_1)_{\mathbb{T}}} f(t),$$

$$\begin{aligned} m_{k2} &= \inf_{t\in[\tau_1,t_k)_{\mathbb{T}}} f(t), \\ M_{k1} &= \sup_{t\in[t_{k-1},\tau_1)_{\mathbb{T}}} f(t), \\ M_{k2} &= \sup_{t\in[\tau_1,t_k)_{\mathbb{T}}} f(t). \end{aligned}$$

Then we have the following inequalities:

$$\begin{aligned} m_{k1} &\ge m_k, \\ m_{k2} &\ge m_k, \\ M_{k1} &\le M_k, \\ M_{k2} &\le M_k. \end{aligned}$$

Hence

$$\begin{aligned} L_{\mathbb{T}}(f,Q) - L_{\mathbb{T}}(f,P) &= m_{k1}(\tau_1 - t_{k-1}) + m_{k2}(t_k - \tau_1) - m_k(t_k - t_{k-1}) \\ &\ge m_k(\tau_1 - t_{k-1}) + m_k(t_k - \tau_1) - m_k(t_k - t_{k-1}) \\ &= 0, \end{aligned}$$

and

$$\begin{aligned} U_{\mathbb{T}}(f,P) - U_{\mathbb{T}}(f,Q) &= M_k(t_k - t_{k-1}) - M_{k2}(t_k - \tau_1) - M_{k1}(\tau_1 - t_{k-1}) \\ &\ge M_k(t_k - t_{k-1}) - M_k(t_k - \tau_1) - M_k(\tau_1 - t_{k-1}) \\ &= 0. \end{aligned}$$

Thus we get inequalities (3.5).

2. Assume that the statement is true when Q has m more points than P for some $m \in \mathbb{N}$.
3. We will prove that the statement is true when Q has $m+1$ more points than P, say $\tau_1, \dots, \tau_m, \tau_{m+1}$. Let

$$Q_1 = Q \backslash \{\tau_{m+1}\}.$$

Then Q has one more point than Q_1. Applying the induction assumption for $P = Q_1$ and Q, we arrive at the inequalities

$$L_{\mathbb{T}}(f,Q_1) \le L_{\mathbb{T}}(f,Q) \le U_{\mathbb{T}}(f,Q) \le U_{\mathbb{T}}(f,Q_1). \tag{3.6}$$

On the other hand, by the induction assumption we have

$$L_{\mathbb{T}}(f,P) \le L_{\mathbb{T}}(f,Q_1) \le U_{\mathbb{T}}(f,Q_1) \le U_{\mathbb{T}}(f,Q).$$

Hence, using (3.6), we obtain inequalities (3.5).

Example 3.37. If f is bounded on $[a,b]_{\mathbb{T}}$ and $P, Q \in \mathscr{P}(a,b)_{\mathbb{T}}$, then we will show that

$$L_{\mathbb{T}}(f,P) \le U_{\mathbb{T}}(f,Q).$$

Since $P, Q \in \mathscr{P}(a,b)_{\mathbb{T}}$, we have that $P \cup Q \subset \mathscr{P}(a,b)_{\mathbb{T}}$ and $P, Q \subset P \cup Q$. Now applying Example 3.36, we get the chain

$$L_{\mathbb{T}}(f,P) \le L_{\mathbb{T}}(f,P\cup Q) \le U_{\mathbb{T}}(f,P\cup Q) \le U_{\mathbb{T}}(f,Q).$$

Example 3.38. If f is bounded on $[a,b]_{\mathbb{T}}$, then we will show that $L_{\mathbb{T}}(f) \le U_{\mathbb{T}}(f)$. Indeed, let $P \in \mathscr{P}(a,b)_{\mathbb{T}}$. By Theorem 3.36 it follows that $L_{\mathbb{T}}(f,P)$ is a lower bound of the set

$$\{U_{\mathbb{T}}(f,Q) : Q \in \mathscr{P}(a,b)_{\mathbb{T}}\}.$$

Hence, applying the definition of $U_{\mathbb{T}}(f)$, we arrive at the inequality

$$L_{\mathbb{T}}(f,P) \le U_{\mathbb{T}}(f). \tag{3.7}$$

Therefore $U_{\mathbb{T}}(f)$ is an upper bound of the set

$$\{L_{\mathbb{T}}(f,Q) : Q \in \mathscr{P}(a,b)_{\mathbb{T}}\}.$$

Now applying the definition of $L_{\mathbb{T}}(f)$, we get

$$L_{\mathbb{T}}(f) \le U_{\mathbb{T}}(f).$$

Therefore for any two partitions $P, Q \in \mathscr{P}(a,b)_{\mathbb{T}}$, we have the inequalities

$$L_{\mathbb{T}}(f,Q) \le L_{\mathbb{T}}(f) \le U_{\mathbb{T}}(f) \le U_{\mathbb{T}}(f,P).$$

In particular, we have the inequalities

$$L_{\mathbb{T}}(f,P) \le L_{\mathbb{T}}(f) \le U_{\mathbb{T}}(f) \le U_{\mathbb{T}}(f,P). \tag{3.8}$$

Example 3.39. If $L_{\mathbb{T}}(f,P) = U_{\mathbb{T}}(f,P)$ for some $P \in \mathscr{P}(a,b)_{\mathbb{T}}$, then we will show that f is integrable on $[a,b]_{\mathbb{T}}$. Indeed, by (3.8) we get $U_{\mathbb{T}}(f) = L_{\mathbb{T}}(f)$. Thus f is integrable on $[a,b]$.

The next example is the Cauchy criterion for integrability.

Example 3.40 (The Cauchy criterion). We will show that a bounded function f on $[a,b]_{\mathbb{T}}$ is integrable if and only if for each $\varepsilon > 0$, there exists a partition $P \in \mathscr{P}(a,b)_{\mathbb{T}}$ such that

$$U_{\mathbb{T}}(f,P) - L_{\mathbb{T}}(f,P) < \varepsilon. \tag{3.9}$$

For this aim, we will consider the following cases.

1. Let f be integrable on $[a,b]_{\mathbb{T}}$, and let $\varepsilon > 0$ be arbitrarily chosen. Using the definition of supremum and infimum, we conclude that there are partitions $P_1, P_2 \in \mathscr{P}(a,b)_{\mathbb{T}}$ such that

$$L_{\mathbb{T}}(f,P_1) > L_{\mathbb{T}}(f) - \frac{\varepsilon}{2}$$

and

$$U_{\mathbb{T}}(f,P_2) < U_{\mathbb{T}}(f) + \frac{\varepsilon}{2}.$$

Set

$$P = P_1 \cup P_2.$$

By Example 3.36 we obtain

$$U_{\mathbb{T}}(f,P) \le U_{\mathbb{T}}(f,P_2) \quad \text{and} \quad L_{\mathbb{T}}(f,P_1) \le L_{\mathbb{T}}(f,P). \tag{3.10}$$

Because f is integrable on $[a,b]_{\mathbb{T}}$, we have

$$U_{\mathbb{T}}(f) = L_{\mathbb{T}}(f).$$

By the last equation and the chain (3.10) we get

$$\begin{aligned} U_{\mathbb{T}}(f,P) - L_{\mathbb{T}}(f,P) &\le U_{\mathbb{T}}(f,P_1) - L_{\mathbb{T}}(f,P_1) \\ &< U_{\mathbb{T}}(f) + \frac{\varepsilon}{2} - L_{\mathbb{T}}(f) + \frac{\varepsilon}{2} \\ &= \varepsilon, \end{aligned}$$

i. e., (3.9) holds.

2. Suppose that for each $\varepsilon > 0$, there exists a partition $P \in \mathscr{P}(a,b)_{\mathbb{T}}$ such that (3.9) holds. Then

$$\begin{aligned} U_{\mathbb{T}}(f) &\le U_{\mathbb{T}}(f,P) \\ &= U_{\mathbb{T}}(f,P) - L_{\mathbb{T}}(f,P) + L_{\mathbb{T}}(f,P) \\ &< \varepsilon + L_{\mathbb{T}}(f,P) \\ &\le \varepsilon + L_{\mathbb{T}}(f), \end{aligned}$$

i. e.,

$$0 \le U_{\mathbb{T}}(f) - L_{\mathbb{T}}(f) \le \varepsilon.$$

Because $\varepsilon > 0$ was arbitrarily chosen, by the last inequalities we obtain

$$U_{\mathbb{T}}(f) = L_{\mathbb{T}}(f).$$

Therefore f is integrable on $[a,b]_{\mathbb{T}}$. This completes the proof.

Example 3.41. We will show that for each $\delta > 0$, there exists a partition $P \in \mathscr{P}(a,b)_{\mathbb{T}}$ such that

$$a = t_0 < t_1 < \cdots < t_{n-1} < t_n = b$$

and for each $j \in \{1,\ldots,n\}$, either

$$t_j - t_{j-1} \le \delta,$$

or

$$t_j - t_{j-1} > \delta \quad \text{and} \quad \rho(t_j) = t_{j-1}.$$

Take arbitrary $\delta > 0$. We will define the desired partition as follows: take $t_0 = a$ and

$$t_j = \begin{cases} \sup((t_{j-1}, t_{j-1}+\delta] \cap [a,b]_{\mathbb{T}}) & \text{if } (t_{j-1}, t_{j-1}+\delta] \cap [a,b]_{\mathbb{T}} \neq \emptyset, \\ \sigma(t_{j-1}) & \text{if } (t_{j-1}, t_{j-1}+\delta] \cap [a,b]_{\mathbb{T}} = \emptyset. \end{cases}$$

Then

$$a = t_0 < t_1 < \cdots,$$

and

$$\begin{aligned} t_{j+1} - t_{j-1} &= \begin{cases} \sup((t_j, t_j+\delta] \cap [a,b]_{\mathbb{T}}) - t_{j-1} & \text{if } (t_j, t_j+\delta] \cap [a,b]_{\mathbb{T}} \neq \emptyset, \\ \sigma(t_j) - t_{j-1} & \text{if } (t_j, t_j+\delta] \cap [a,b]_{\mathbb{T}} = \emptyset \end{cases} \\ &\ge \begin{cases} t_j - t_{j-1} & \text{if } (t_j, t_j+\delta] \cap [a,b]_{\mathbb{T}} \neq \emptyset, \\ t_+\delta - t_{j-1} & \text{if } (t_j, t_j+\delta] \cap [a,b]_{\mathbb{T}} = \emptyset \end{cases} \\ &\ge \begin{cases} t_{j-1} + \delta - t_{j-1} & \text{if } (t_j, t_j+\delta] \cap [a,b]_{\mathbb{T}} \neq \emptyset, \\ t_{j-1} + \delta - t_{j-1} & \text{if } (t_j, t_j+\delta] \cap [a,b]_{\mathbb{T}} = \emptyset \end{cases} \\ &= \delta. \end{aligned}$$

Therefore there is $n \in \mathbb{N}$ such that $t_n = b$. Note that the inequality $t_{j-1} < t_j$ with $\sigma(t_{j-1}) = t_j$ for some $j \in \{1,\ldots,n\}$ is equivalent to the inequality $t_{j-1} < t_j$ with $\rho(t_j) = t_{j-1}$. This completes the proof.

Definition 3.10. The set of all partitions $P \in \mathscr{P}(a,b)_{\mathbb{T}}$ that possess the property indicated by Example 3.41 will be denoted by $\mathscr{P}_\delta(a,b)_{\mathbb{T}}$.

Now using Definition 3.10, we will give an analogue of the classical Cauchy criterion.

Example 3.42 (The Cauchy criterion). We will show that a bounded function f on $[a,b]_{\mathbb{T}}$ is integrable from a to b if and only if for each $\varepsilon > 0$, there exists $\delta > 0$ such that

$$P \in \mathscr{P}_\delta(a,b)_{\mathbb{T}} \quad \text{implies} \quad U_{\mathbb{T}}(f,P) - L_{\mathbb{T}}(f,P) < \varepsilon. \tag{3.11}$$

For this aim, we consider the following cases.

1. Let f be integrable from a to b. Without loss of generality, suppose that $f \not\equiv 0$ on $[a,b]_{\mathbb{T}}$. Take arbitrary $\varepsilon > 0$. Since f is integrable over $[a,b]_{\mathbb{T}}$, there is a partition $P_0 \in \mathscr{P}(a,b)_{\mathbb{T}}$ given by

$$a = \tau_0 < \tau_1 < \cdots < \tau_l = b$$

such that

$$U_{\mathbb{T}}(f,P_0) - L_{\mathbb{T}}(f,P_0) < \frac{\varepsilon}{2}. \tag{3.12}$$

Let

$$\delta = \frac{\varepsilon}{8lB}$$

and

$$B = \sup_{t\in[a,b]_{\mathbb{T}}} |f(t)|.$$

Consider any partition $P \in \mathscr{P}_\delta(a,b)_{\mathbb{T}}$ given by

$$a = t_0 < t_1 < \cdots < t_n = b$$

and set

$$Q = P \cup P_0.$$

Let Q have one more element, say τ, than P. Then $\tau \in (t_{k-1},t_k)_{\mathbb{T}}$ for some $k \in \{1,\dots,n\}$. If $t_k - t_{k-1} > \delta$, then using the condition $P \in \mathscr{P}_\delta(a,b)_{\mathbb{T}}$, it follows that $\rho(t_k) = t_{k-1}$ and $(t_{k-1},t_k)_{\mathbb{T}} = \emptyset$. Thus $t_k - t_{k-1} \le \delta$. Let

$$m_{k1} = \inf_{t\in[t_{k-1},\tau)_{\mathbb{T}}} f(t),$$
$$m_{k2} = \sup_{t\in[\tau,t_k)_{\mathbb{T}}} f(t).$$

Then

$$\begin{aligned}L_{\mathbb{T}}(f,Q)-L_{\mathbb{T}}(f,P)&=m_{k1}(\tau-t_{k-1})+m_{k2}(t_k-\tau)-m_k(t_k-t_{k-1})\\&\le N(\tau-t_{k-1})+B(t_k-\tau)-m_k(t_k-t_{k-1})\\&=B(t_k-t_{k-1})-m_k(t_k-t_{k-1})\\&=(B-m_k)(t_k-t_{k-1})\\&\le 2B(t_k-t_{k-1})\\&\le 2B\delta.\end{aligned}$$

An induction argument shows that if Q has $l-1$ elements that are not in P, then

$$\begin{aligned}L_{\mathbb{T}}(f,Q)-L_{\mathbb{T}}(f,P)&\le 2(l-1)B\delta\\&=2(l-1)B\frac{\varepsilon}{8lB}\\&<\frac{\varepsilon}{4}.\end{aligned}$$

By Example 3.36 we get

$$L_{\mathbb{T}}(f,P_0)\le L_{\mathbb{T}}(f,Q).$$

Therefore

$$\begin{aligned}L_{\mathbb{T}}(f,P_0)-L_{\mathbb{T}}(f,P)&\le L_{\mathbb{T}}(f,Q)-L_{\mathbb{T}}(f,P)\\&<\frac{\varepsilon}{4}.\end{aligned}$$

As above,

$$U_{\mathbb{T}}(f,P)-U_{\mathbb{T}}(f,P_0)<\frac{\varepsilon}{4}.$$

Now applying (3.12), we get

$$\begin{aligned}U_{\mathbb{T}}(f,P)-L_{\mathbb{T}}(f,P)&=U_{\mathbb{T}}(f,P)-U_{\mathbb{T}}(f,P_0)+L_{\mathbb{T}}(f,P_0)-L_{\mathbb{T}}(f,P)\\&\quad+U_{\mathbb{T}}(f,P_0)-L_{\mathbb{T}}(f,P_0)\\&<\frac{\varepsilon}{4}+\frac{\varepsilon}{4}+\frac{\varepsilon}{2}\\&=\varepsilon.\end{aligned}$$

So (3.11) holds.

2. Suppose that for each $\varepsilon>0$, there is $\delta>0$ such that (3.11) holds. Then for each $\varepsilon>0$, there is a partition $P\in\mathscr{P}(a,b)_{\mathbb{T}}$ such that

$$U_{\mathbb{T}}(f,P)-L_{\mathbb{T}}(f,P)<\varepsilon.$$

From this by the Cauchy criterion, Example 3.37, we conclude that f is integrable from a to b.

3.4 The Riemann delta integral

In this section, we define the Riemann delta integral.

Definition 3.11. Let f be a bounded function on $[a,b]_{\mathbb{T}}$, and let $P \in \mathscr{P}_\delta(a,b)_{\mathbb{T}}$ be given by

$$a = t_0 < t_1 < \cdots < t_n = b.$$

In each interval $[t_{j-1}, t_j)_{\mathbb{T}}, j \in \{1,\ldots,n\}$, choose an arbitrary point ξ_j and form the sum

$$S_T(f,P) = \sum_{j=1}^{n} f(\xi_j)(t_j - t_{j-1}).$$

We say that $S_{\mathbb{T}}(f,P)$ is a Riemann Δ-sum or a Riemann delta sum (shortly Riemann sum) of the function f corresponding to the partition $P \in \mathscr{P}(a,b)_{\mathbb{T}}$. We say that f is Riemann delta integrable or Riemann Δ-integrable (shortly Riemann integrable) from a to b (or on $[a,b]_{\mathbb{T}}$) if there exists a number I with the following property: for each $\varepsilon > 0$, there exists $\delta > 0$ such that

$$|S_{\mathbb{T}}(f,P) - I| < \varepsilon$$

for any Riemann sum $S_{\mathbb{T}}(f,P)$ corresponding to any partition $P \in \mathscr{P}_\delta(a,b)_{\mathbb{T}}$ and independent of the choice $\xi_j \in [t_{j-1}, t_j)_{\mathbb{T}}, j \in \{1,\ldots,n\}$. This number I is unique and is called the Riemann delta integral or Riemann Δ-integral (shortly the Riemann integral) of the function f from a to b (or on $[a,b]_{\mathbb{T}}$).

Example 3.43. We will show that a bounded function f on $[a,b]_{\mathbb{T}}$ is Riemann integrable if and only if it is Darboux integrable, in which case the values of the integrals are equal. For this aim, we will consider the following cases.

1. Suppose that f is Darboux integrable from a to b in the sense of Definitions 3.5–3.7. Take arbitrary $\varepsilon > 0$. By Example 3.42 it follows that there is $\delta > 0$ such that (3.11) holds. Let also $P \in \mathscr{P}_\delta(a,b)_{\mathbb{T}}$, and let $S_{\mathbb{T}}(f,P)$ be the Riemann sum of f corresponding to the partition P. Then clearly we have

   $$L_{\mathbb{T}}(f,P) \le S_{\mathbb{T}}(f,P) \le U_{\mathbb{T}}(f,P).$$

   By inequality (3.11) we get

   $$\begin{aligned} U_{\mathbb{T}}(f,P) &< L_{\mathbb{T}}(f,P) + \varepsilon \\ &\le L_{\mathbb{T}}(f) + \varepsilon \\ &= \int_a^b f(t)\Delta t + \varepsilon \end{aligned}$$

and

$$\begin{aligned} L_{\mathbb{T}}(f,P) &> U_{\mathbb{T}}(f,P) - \varepsilon \\ &\geq U_T(f) - \varepsilon \\ &= \int_a^b f(t)\Delta t - \varepsilon. \end{aligned}$$

Hence

$$\begin{aligned} S_{\mathbb{T}}(f,P) - \int_a^b f(t)\Delta t &\leq U_{\mathbb{T}}(f,P) - \int_a^b f(t)\Delta t \\ &< \varepsilon \end{aligned}$$

and

$$\begin{aligned} S_{\mathbb{T}}(f,P) - \int_a^b f(t)\Delta t &\geq L_{\mathbb{T}}(f,P) - \int_a^b f(t)\Delta t \\ &> \varepsilon, \end{aligned}$$

i. e.,

$$\left| S_{\mathbb{T}}(f,P) - \int_a^b f(t)\Delta t \right| < \varepsilon.$$

Thus f is Riemann integrable on $[a,b]_{\mathbb{T}}$, and

$$I = \int_a^b f(t)\Delta t.$$

2. Now suppose that f is Riemann integrable on $[a,b]_{\mathbb{T}}$ in the sense of Definition 3.11. Take arbitrary $\varepsilon > 0$. Let $\delta > 0$ and I be as in Definition 3.11. Let also arbitrary $P \in \mathscr{P}_\delta(a,b)_{\mathbb{T}}$ be given by

$$a = t_0 < t_1 < \cdots < t_n = b,$$

and let $\xi_j \in [t_{j-1}, t_j)_{\mathbb{T}}, j \in \{1, \ldots, n\}$, be chosen such that

$$M_j - \varepsilon < f(\xi_j) < m_j + \varepsilon, \quad j \in \{1, \ldots, n\},$$

where

$$m_j = \inf_{t \in [t_{j-1}, t_j)_{\mathbb{T}}} f(t),$$

$$M_j = \sup_{t \in [t_{j-1}, t_j)_{\mathbb{T}}} f(t).$$

Then

$$\begin{aligned}
|S_{\mathbb{T}}(f,P) - I| &< \varepsilon, \\
S_{\mathbb{T}}(f,P) &= \sum_{j=1}^{n} f(\xi_j)(t_j - t_{j-1}) \\
&< \sum_{j=1}^{n} (m_j + \varepsilon)(t_j - t_{j-1}) \\
&= \sum_{j=1}^{n} m_j(t_j - t_{j-1}) + \varepsilon \sum_{j=1}^{n} (t_j - t_{j-1}) \\
&= L_{\mathbb{T}}(f,P) + \varepsilon(b-a),
\end{aligned}$$

and

$$\begin{aligned}
S_{\mathbb{T}}(f,P) &= \sum_{j=1}^{n} f(\xi_j)(t_j - t_{j-1}) \\
&> \sum_{j=1}^{n} (M_j + \varepsilon)(t_j - t_{j-1}) \\
&= \sum_{j=1}^{n} M_j(t_j - t_{j-1}) + \varepsilon \sum_{j=1}^{n} (t_j - t_{j-1}) \\
&= U_{\mathbb{T}}(f,P) + \varepsilon(b-a).
\end{aligned}$$

Hence

$$\begin{aligned}
L_{\mathbb{T}}(f) &\geq L_{\mathbb{T}}(f,P) \\
&> S_{\mathbb{T}}(f,P) - \varepsilon(b-a) \\
&> I - \varepsilon - \varepsilon(b-a),
\end{aligned}$$

and

$$\begin{aligned}
U_{\mathbb{T}}(f) &\geq U_{\mathbb{T}}(f,P) \\
&< S_{\mathbb{T}}(f,P) + \varepsilon(b-a) \\
&< I + \varepsilon + \varepsilon(b-a).
\end{aligned}$$

Because $\varepsilon > 0$ was arbitrarily chosen, we get

$$I \leq L_{\mathbb{T}}(f) \leq U_{\mathbb{T}}(f) \leq I,$$

whereupon f is Darboux integrable on $[a,b]_{\mathbb{T}}$, and

$$I = \int_a^b f(t)\Delta t.$$

This completes the proof.

Remark 3.1. In the definition of $\int_a^b f(t)\Delta t$, we have assumed that $b > a$. We remove this restriction by the following definitions:

$$\int_a^a f(t)\Delta t = 0$$

and

$$\int_a^b f(t)\Delta t = -\int_b^a f(t)\Delta t \quad \text{if } a > b.$$

Example 3.44. We will show that

$$\int_a^b c\Delta t = c(b-a) \tag{3.13}$$

for any constant $c \in \mathbb{R}$. Let

$$f(t) = c, \quad t \in [a,b]_T.$$

Take a partition $P \in \mathscr{P}(a,b)_{\mathbb{T}}$ given by

$$a = t_0 < t_1 < \cdots < t_n = b.$$

Then

$$\begin{aligned} M_j &= \sup_{t\in[t_{j-1},t_j)_{\mathbb{T}}} f(t) \\ &= c, \\ m_j &= \sup_{t\in[t_{j-1},t_j)_{\mathbb{T}}} f(t) \\ &= c, \quad j \in \{1,\ldots,n\}. \end{aligned}$$

Therefore

$$\begin{aligned} U_{\mathbb{T}}(f,P) &= \sum_{j=1}^{n} M_j(t_j - t_{j-1}) \\ &= c\sum_{j=1}^{n}(t_j - t_{j-1}) \\ &= c(b-a), \end{aligned}$$

and

$$\begin{aligned} L_{\mathbb{T}}(f,P) &= \sum_{j=1}^{n} m_j(t_j - t_{j-1}) \\ &= c\sum_{j=1}^{n}(t_j - t_{j-1}) \\ &= c(b-a). \end{aligned}$$

Consequently,

$$U_{\mathbb{T}}(f,P) = L_{\mathbb{T}}(f,P) = c(b-a).$$

Because $P \in \mathscr{P}(a,b)_{\mathbb{T}}$ was arbitrarily chosen, we get

$$U_{\mathbb{T}}(f) = L_{\mathbb{T}}(f).$$

Consequently, f is integrable on $[a,b]_{\mathbb{T}}$, and (3.13) holds.

Example 3.45. Let $f : \mathbb{T} \to \mathbb{R}$, and let $t \in \mathbb{T}$. We will prove that the function f is integrable from t to $\sigma(t)$ and

$$\int_{t}^{\sigma(t)} f(s)\Delta s = \mu(t)f(t). \tag{3.14}$$

If $t = \sigma(t)$, then equality (3.14) is obvious. If $\sigma(t) > t$, then $\mathscr{P}(t,\sigma(t))_{\mathbb{T}}$ only contains one single element given by

$$t = s_0 < s_1 = \sigma(t).$$

Since $[t,\sigma(t))_{\mathbb{T}} = \{t\}$, we have that

$$\begin{aligned} L_{\mathbb{T}}(f,P) &= U_{\mathbb{T}}(f,P) \\ &= f(t)(\sigma(t) - t) \\ &= \mu(t)f(t). \end{aligned}$$

Hence, applying Example 3.39, we conclude that the function f is integrable from t to $\sigma(t)$ and (3.14) holds.

Example 3.46. Let $\mathbb{T} = \mathbb{R}$, let $f : \mathbb{T} \to \mathbb{R}$ be integrable, and let $a, b \in \mathbb{T}$, $a < b$. Then

$$\int_a^b f(t)\Delta t = \int_a^b f(t)dt.$$

Exercise 3.17. Let $\mathbb{T} = \mathbb{Z}$, $f : \mathbb{T} \to \mathbb{R}$, and $a, b \in \mathbb{T}$, $a < b$. Prove that f is integrable on $[a, b]_{\mathbb{T}}$ and

$$\int_a^b f(t)\Delta t = \begin{cases} \sum_{k=a}^{b-1} f(t) & \text{if } a < b, \\ 0 & \text{if } a = b, \\ -\sum_{k=b}^{a-1} f(t) & \text{if } a > b. \end{cases}$$

Exercise 3.18. Let $\mathbb{T} = h\mathbb{Z}$ for some $h > 0$, let $f : \mathbb{T} \to \mathbb{R}$, and let $a, b \in \mathbb{T}$, $a < b$. Prove that f is integrable from a to b and

$$\int_a^b f(t)\Delta t = \begin{cases} \sum_{k=\frac{a}{h}}^{\frac{b}{h}-1} f(kh)h & \text{if } a < b, \\ 0 & \text{if } a = b, \\ -\sum_{k=\frac{b}{h}}^{\frac{a}{h}-1} f(kh)h & \text{if } a > b. \end{cases}$$

3.5 Another definition of the Riemann delta integral

In this section, we give another definition of the Riemann integral and prove that both definitions of the Riemann integral are equivalent.

Definition 3.12. For a function $f : [a, b]_{\mathbb{T}} \to \mathbb{R}$, we define the extension $\tilde{f} : [a, b] \to \mathbb{R}$ by

$$\tilde{f}(t) = f(\sup[a, t]_{\mathbb{T}}) \quad \text{for all } t \in [a, b].$$

By this definition it follows that

$$\tilde{f}(t) = f(t), \quad t \in [a, b]_{\mathbb{T}}.$$

If $t \in [a, b] \backslash [a, b]_{\mathbb{T}}$ and s is the nearest left-hand point to the point t such that $s \in \mathbb{T}$, then

$$\tilde{f}(t) = f(s).$$

Example 3.47. We will show that the function $\tilde{f}$ is well defined by Definition 3.12. For this aim, for $t \in [a, b]_{\mathbb{T}}$, define

$$S = [a, b]_{\mathbb{T}} \cap [a, b].$$

We have that S is a compact set. Fix $t \in [a,b]_{\mathbb{T}}$. Consider the function

$$\phi(s) = t - s, \quad s \in S.$$

Note that $\phi \in \mathscr{C}(S)$. Let $s_1, s_2 \in S$ be such that $s_1 \leq s_2$. Then

$$\begin{aligned}\phi(s_1) &= t - s_1\\ &\geq t - s_2\\ &= \phi(s_2).\end{aligned}$$

Thus ϕ is a nonincreasing function on S. Since S is a compact set, we conclude that there is a unique point $s^* \in S$ such that

$$\phi(s^*) = \inf_{s\in S} \phi(s).$$

Because ϕ is a nonincreasing function on S, we have that s^* is the nearest left-hand point to the point t such that $s^* \in \mathbb{T}$. Therefore

$$\tilde{f}(t) = f(s^*).$$

Example 3.48. We will show that the function $f : [a,b]_{\mathbb{T}} \to \mathbb{R}$ is Riemann integrable on $[a,b]_{\mathbb{T}}$ if and only if the function $\tilde{f} : [a,b] \to \mathbb{R}$ is Riemann integrable on $[a,b]$, and then their integrals are equal:

$$\int_a^b f(t)\Delta t = \int_a^b \tilde{f}(t)dt. \tag{3.15}$$

Let $P \in \mathscr{P}(a,b)_{\mathbb{T}}$ be a partition of the interval $[a,b]_{\mathbb{T}}$ given by

$$a = t_0 < t_1 < \cdots < t_n = b.$$

Let also

$$\begin{aligned}m_j &= \inf_{t\in[t_{j-1},t_j)_{\mathbb{T}}} f(t),\\ M_j &= \sup_{t\in[t_{j-1},t_j)_{\mathbb{T}}} f(t), \quad j \in \{1,\ldots,n\},\end{aligned}$$

and

$$\begin{aligned}\widetilde{m}_j &= \inf_{t\in[t_{j-1},t_j)} \tilde{f}(t),\\ \widetilde{M}_j &= \sup_{t\in[t_{j-1},t_j)} \tilde{f}(t), \quad j \in \{1,\ldots,n\}.\end{aligned}$$

Then

$$L_{\mathbb{T}}(f,P) = \sum_{j=1}^{n} m_j(t_j - t_{j-1}),$$
$$U_{\mathbb{T}}(f,P) = \sum_{j=1}^{n} M_j(t_j - t_{j-1}).$$

Note that the partition P of the interval $[a,b]_{\mathbb{T}}$ is also a partition of the interval $[a,b]$. Thus

$$L_{\mathbb{T}}(\tilde{f},P) = \sum_{j=1}^{n} \tilde{m}_j(t_j - t_{j-1}),$$
$$U_{\mathbb{T}}(\tilde{f},P) = \sum_{j=1}^{n} \tilde{M}_j(t_j - t_{j-1}).$$

Since the set of the values of the function $\tilde{f}$ on $[t_{j-1},t_j)$ coincides with the set of the values of the function f on $[t_{j-1},t_j)_{\mathbb{T}}$ for all $j \in \{1,\dots,n\}$, we have that

$$\begin{aligned} m_j &= \tilde{m}_j, \\ M_j &= \tilde{M}_j, \quad j \in \{1,\dots,n\}. \end{aligned}$$

Therefore

$$\begin{aligned} U_{\mathbb{T}}(f,P) &= U(\tilde{f},P), \\ L_{\mathbb{T}}(f,P) &= L(\tilde{f},P). \end{aligned}$$

Since

$$\mathscr{P}(a,b)_{\mathbb{T}} \subset \mathscr{P}(a,b),$$

we get

$$\begin{aligned} L_{\mathbb{T}}(f) &= \sup_{P \in \mathscr{P}(a,b)_{\mathbb{T}}} L(f,P) \\ &\le \sup_{\tilde{P} \in \mathscr{P}(a,b)} L(\tilde{f},\tilde{P}) \\ &= L(\tilde{f}) \\ &\le U(\tilde{f}) \\ &= \inf_{\tilde{P} \in \mathscr{P}(a,b)_{\mathbb{T}}} U(\tilde{f},\tilde{P}) \\ &\le \inf_{P \in \mathscr{P}(a,b)} U(f,P) \end{aligned}$$

$$= U_{\mathbb{T}}(f, P),$$

i. e.,

$$L(f) \leq L(\tilde{f}) \leq U(\tilde{f}) \leq U(f). \tag{3.16}$$

By relations (3.16) it follows that if f is Δ-Riemann integrable on $[a,b]_{\mathbb{T}}$, then $\tilde{f}$ is Riemann integrable on $[a,b]$, and (3.15) holds. Now suppose that $\tilde{f}$ is Riemann integrable on $[a,b]$. We will prove that f is delta Riemann integrable on $[a,b]_{\mathbb{T}}$ and (3.15) holds. For this aim, take arbitrary $\varepsilon > 0$. Since $\tilde{f}$ is Riemann integrable on $[a,b]$, by the classical Cauchy criterion it follows that there exists a partition $\tilde{P} \in \mathscr{P}(a,b)$ such that

$$U(\tilde{f}, \tilde{P}) - L(\tilde{f}, \tilde{P}) < \varepsilon.$$

Let the partition $\tilde{P}$ be given by

$$a = t_0 < t_1 < \cdots < t_n = b.$$

We will construct a subpartition of $\tilde{P}$. If $t_1 \in [a,b]_{\mathbb{T}}$, then we pass to the next point of $\tilde{P}$. If $t_1 \notin [a,b]_{\mathbb{T}}$, then there exist points $s_1, s_2 \in [a,b]_{\mathbb{T}}$, possibly $s_1 = a$ and $s_2 = b$, such that

$$a \leq s_1 < t_1 < s_2 \leq b,$$

and then

$$\tilde{f}(t) = f(s_1), \quad t \in [s_1, s_2).$$

Without loss of generality, we choose $s_2 < t_2$. We add the points s_1 and s_2 to the partition $\tilde{P}$. Then we have

$$\begin{aligned} \inf_{t \in [s_1, t_1)} \tilde{f}(t) &= \inf_{t \in [t_1, s_2)} \tilde{f}(t) \\ &= f(s_1). \end{aligned}$$

Therefore

$$m_1 s_1 + f(s_1)(t_1 - s_1) + f(s_1)(s_2 - t_1) = m_1 s_1 + f(s_1)(s_2 - s_1).$$

Thus the part of the lower Darboux sum for $\tilde{f}$ with respect to the points $a \leq s_1 < t_1 < s_2$ of the partition of $[a,b]$ coincides with the part of the lower Darboux sum for f with respect to the points $a < s_1 < s_2$. From s_2 to t_2 we repeat this procedure, and so on, until we go over all points of the partition $\tilde{P}$. As a result, we get a subpartition $\tilde{Q}$ of the partition $\tilde{P}$ constructed by adding the points $\{s_j\}$ to the points $\{t_j\}$. By Q we will denote the partition that contains the points $\{s_j\}$. By the construction of Q and $\tilde{Q}$ we get

$$L(\tilde{f},\tilde{Q}) = L_{\mathbb{T}}(f,Q).$$

As above,

$$U(\tilde{f},\tilde{Q}) = U_{\mathbb{T}}(f,Q).$$

Since $\tilde{Q}$ is a subpartition of $\tilde{P}$, we have the following chain:

$$\begin{aligned} L(\tilde{f},\tilde{P}) &\leq L(\tilde{f},\tilde{Q}) \\ &\leq U(\tilde{f},\tilde{Q}) \\ &\leq U(\tilde{f},\tilde{P}). \end{aligned}$$

Hence

$$\begin{aligned} U_{\mathbb{T}}(f,Q) - L_{\mathbb{T}}(f,Q) &= U(\tilde{f},\tilde{Q}) - L(\tilde{f},\tilde{Q}) \\ &\leq U(\tilde{f},\tilde{P}) - L(\tilde{f},\tilde{P}) \\ &< \varepsilon. \end{aligned}$$

Now applying Example 3.40, we conclude that f is delta Riemann integrable on $[a,b]_{\mathbb{T}}$.

3.6 Properties of the Riemann integral

In this section, we establish some basic properties of the Riemann integral and show that monotone, continuous, piecewise continuous, and regulated functions are Riemann integrable. We start with the following example.

Example 3.49. We will show that any monotone function on $[a,b]_{\mathbb{T}}$ is integrable. Indeed, without loss of generality, suppose that $f : [a,b]_{\mathbb{T}} \to \mathbb{R}$ is nondecreasing. Then

$$f(a) \leq f(t) \leq f(b), \quad t \in [a,b]_{\mathbb{T}}.$$

Thus the function f is bounded on $[a,b]_{\mathbb{T}}$. Take arbitrary $\varepsilon > 0$ and

$$\delta = \frac{\varepsilon}{f(b) - f(a) + 1}.$$

Let also $P \in \mathscr{P}_{\delta}(a,b)_{\mathbb{T}}$ be given by

$$a = t_0 < t_1 < \cdots < t_n = b.$$

Then

$$M_j = \sup_{t \in [t_{j-1},t_j)_{\mathbb{T}}} f(t)$$

$$\leq f(\rho(t_j)), \quad j \in \{1, \dots, n\},$$

and

$$\begin{aligned} m_j &= \inf_{t\in[t_{j-1},t_j)_{\mathbb{T}}} f(t) \\ &= f(t_{j-1}), \quad j \in \{1, \dots, n\}. \end{aligned}$$

Hence

$$\begin{aligned} U_{\mathbb{T}}(f,P) - L_{\mathbb{T}}(f,P) &= \sum_{j=1}^{n} M_j(t_j - t_{j-1}) - \sum_{j=1}^{n} m_j(t_j - t_{j-1}) \\ &= \sum_{j=1}^{n} (M_j - m_j)(t_j - t_{j-1}) \\ &\leq \sum_{j=1}^{n} (f(\rho(t_j)) - f(t_{j-1}))(t_j - t_{j-1}) \\ &= \sum_{t_j - t_{j-1} \leq \delta} (f(\rho(t_j)) - f(t_{j-1}))(t_j - t_{j-1}) \\ &\quad + \sum_{t_j - t_{j-1} > \delta} (f(\rho(t_j)) - f(t_{j-1}))(t_j - t_{j-1}) \\ &= \sum_{t_j - t_{j-1} \leq \delta} (f(\rho(t_j)) - f(t_{j-1}))(t_j - t_{j-1}), \end{aligned}$$

where we have used that if $t_j - t_{j-1} > \delta$, then $\rho(t_j) = t_{j-1}$. Therefore

$$\begin{aligned} U_{\mathbb{T}}(f,P) - L_{\mathbb{T}}(f,P) &= \sum_{t_j - t_{j-1} \leq \delta} (f(\rho(t_j)) - f(t_{j-1}))(t_j - t_{j-1}) \\ &\leq \delta \sum_{t_j - t_{j-1} \leq \delta} (f(\rho(t_j)) - f(t_{j-1})) \\ &= \delta(f(b) - f(a)) \\ &= \frac{\varepsilon}{f(b) - f(a) + 1}(f(b) - f(a)) \\ &< \varepsilon. \end{aligned}$$

Now applying Example 3.40, we conclude that f is integrable on $[a, b]_{\mathbb{T}}$.

Example 3.50. We will show that every continuous function f on $[a, b]_{\mathbb{T}}$ is Riemann integrable. Indeed, take arbitrary $\varepsilon > 0$. Since f is a continuous function on $[a, b]_{\mathbb{T}}$, it is uniformly continuous on $[a, b]_{\mathbb{T}}$. Therefore there exists $\delta > 0$ such that

$$t, \tau \in [a, b]_{\mathbb{T}} \quad \text{and} \quad |t - \tau| \leq \delta \quad \text{imply} \quad |f(t) - f(\tau)| < \frac{\varepsilon}{b - a + 1}. \tag{3.17}$$

Let $P \in \mathscr{P}_{\delta}(a, b)_{\mathbb{T}}$ be given by

$$a = t_0 < t_1 < \cdots < t_n = b,$$

and let

$$\begin{aligned} m_j &= \int\limits_{t\in[t_{j-1},t_j)_{\mathbb{T}}} f(t), \\ \widetilde{m}_j &= \inf_{t\in[t_{j-1},\rho(t_j)]_{\mathbb{T}}} f(t), \\ M_j &= \sup_{t\in[t_{j-1},t_j)_{\mathbb{T}}} f(t), \\ \widetilde{M}_j &= \sup_{t\in[t_{j-1},\rho(t_j)]_{\mathbb{T}}} f(t), \quad j \in \{1,\ldots,n\}. \end{aligned}$$

Since

$$[t_{j-1}, t_j)_{\mathbb{T}} \subset [t_{j-1}, \rho(t_j)]_{\mathbb{T}}, \quad j \in \{1,\ldots,n\},$$

we have

$$\widetilde{m}_j \le m_j \le M_j \le \widetilde{M}_j, \quad j \in \{1,\ldots,n\}.$$

Consequently,

$$\begin{aligned} U_{\mathbb{T}}(f,P) - L_{\mathbb{T}}(f,P) &= \sum_{j=1}^{n} M_j(t_j - t_{j-1}) - \sum_{j=1}^{n} m_j(t_j - t_{j-1}) \\ &= \sum_{j=1}^{n}(M_j - m_j)(t_j - t_{j-1}) \\ &\le \sum_{j=1}^{n}(\widetilde{M}_j - \widetilde{m}_j)(t_j - t_{j-1}) \\ &\le \sum_{t_j-t_{j-1}\le\delta} (\widetilde{M}_j - \widetilde{m}_j)(t_j - t_{j-1}) \\ &\quad + \sum_{t_j-t_{j-1}>\delta} (\widetilde{M}_j - \widetilde{m}_j)(t_j - t_{j-1}) \\ &= \sum_{t_j-t_{j-1}\le\delta} (\widetilde{M}_j - \widetilde{m}_j)(t_j - t_{j-1}) \\ &\le \frac{\varepsilon}{b-a+1}\sum_{j=1}^{n}(t_j - t_{j-1}) \\ &= \frac{\varepsilon}{b-a+1}(b-a) \\ &< \varepsilon, \end{aligned}$$

where we have used that if $t_j - t_{j-1} \ge \delta$, then $\rho(t_j) = t_{j-1}$ and $\widetilde{m}_j = \widetilde{M}_j, j \in \{1,\ldots,n\}$. Now applying Theorem 3.40, we conclude that f is integrable on $[a,b]_{\mathbb{T}}$.

Example 3.51. We will show that every bounded function f on $[a,b]_{\mathbb{T}}$ with only finitely many discontinuity points is Riemann integrable. Indeed, let

$$m = \inf_{t\in[a,b)_{\mathbb{T}}} f(t),$$
$$M = \sup_{t\in[a,b)_{\mathbb{T}}} f(t).$$

If $m = M$, then f is a constant function and thus is integrable on $[a,b]_{\mathbb{T}}$. Let $M > m$. Take arbitrary $\varepsilon > 0$. Note that f is continuous at all isolated points of $[a,b]_{\mathbb{T}}$ and also at the endpoints a and b if a is right-scattered and b is left-scattered. Let the discontinuity points of f be

$$a \le \tau_1 < \tau_2 < \cdots < \tau_p \le b.$$

For $j \in \{1,\dots,p\}$, define the sets V_j as follows:

1. if τ_j is both left-dense and right-dense, then

$$V_j = (\alpha_j,\beta_j)_{\mathbb{T}}, \quad \alpha_j < \tau_j < \beta_j, \quad \alpha_j,\beta_j \in [a,b]_{\mathbb{T}},$$

   and

$$\beta_j - \alpha_j < \frac{\varepsilon}{2(p+1)(M-m)}.$$

2. if τ_j is left-dense and right-scattered, or $\tau_j = b$, then

$$V_j = (\alpha_j,\tau_j]_{\mathbb{T}}, \quad \tau_j > \alpha_j \in [a,b]_{\mathbb{T}},$$

   and

$$\tau_j - \alpha_j < \frac{\varepsilon}{2(p+1)(M-m)}.$$

3. if τ_j is left-scattered and right-dense, or $\tau_j = a$, then

$$V_j = [\tau_j,\beta_j)_{\mathbb{T}}, \quad \tau_j < \beta_j \in [a,b]_{\mathbb{T}},$$

   and

$$\beta_j - \tau_j < \frac{\varepsilon}{2(p+1)(M-m)}.$$

   Set

$$N = [a,b]_{\mathbb{T}} \setminus \bigcup_{j=1}^{p} V_j.$$

Note that N consists of finitely many closed intervals $J_j, j \in \{1, \dots, p+1\}$. Thus N is a compact subset of $[a,b]_{\mathbb{T}}$. Since f is continuous on N, it is uniformly continuous on N. Therefore there is $\delta_0 > 0$ such that

$$t, \tau \in N \quad \text{and} \quad |t-\tau| \le \delta_0 \quad \text{imply} \quad |f(t) - f(\tau)| < \frac{\varepsilon}{2(b-a+1)}.$$

For each interval $J_j, j \in \{1, \dots, p+1\}$, we pick $P_j \in \mathscr{P}_{\delta_0}(J_j)_{\mathbb{T}}, j \in \{1, \dots, p+1\}$, and define

$$P = \left(\bigcup_{j=1}^{p+1} P_j \right) \cup \{\tau_j : \tau_j \text{ is left-scattered or right-scattered}\}.$$

Then $P \in \mathscr{P}(a,b)_{\mathbb{T}}$. Denote

$$A_1 = \{j \in \{1, \dots, n\} : t_{j-1} = \inf V_j,\ t_j = \sup V_j\},$$
$$A_2 = \{1, \dots, n\} \backslash A_1.$$

We have

$$\begin{aligned}
U_{\mathbb{T}}(f,P) - L_{\mathbb{T}}(f,P) &= \sum_{j=1}^{n} (M_j - m_j)(t_j - t_{j-1}) \\
&= \sum_{A_1} (M_j - m_j)(t_j - t_{j-1}) + \sum_{A_2} (M_j - m_j)(t_j - t_{j-1}) \\
&\le (M-m) \sum_{A_1} (t_j - t_{j-1}) + \frac{\varepsilon}{2(b-a+1)} \sum_{A_2} (t_j - t_{j-1}) \\
&\le (M-m) \sum_{A_1} \frac{\varepsilon}{2(p+1)(M-m)} + \frac{\varepsilon}{2(b-a+1)}(b-a) \\
&< \frac{\varepsilon}{2} + \frac{\varepsilon}{2} \\
&= \varepsilon.
\end{aligned}$$

Now applying Example 3.42, we conclude that f is Riemann integrable on $[a,b]_{\mathbb{T}}$.

Definition 3.13. A function $f : [a,b]_{\mathbb{T}} \to \mathbb{R}$ is said to be regulated if its finite right-sided limits exist at all right-dense points of $[a,b)_{\mathbb{T}}$ and its finite left-sided limits exist at all left-dense points of $(a,b]_{\mathbb{T}}$.

Example 3.52. We will show that any regulated function $f : [a,b]_{\mathbb{T}} \to \mathbb{R}$ is bounded. Indeed, assume that the function f is unbounded. Then, for each $n \in \mathbb{N}$, there is $t_n \in [a,b]_{\mathbb{T}}$ such that

$$|f(t_n)| > n.$$

Since $\{t_n\}_{n\in\mathbb{N}} \subset [a,b]_{\mathbb{T}}$ and $[a,b]_{\mathbb{T}}$ is a compact set, it follows that there is a subsequence $\{t_{n_k}\}_{k\in\mathbb{N}}$ of $\{t_n\}_{n\in\mathbb{N}}$ that converges to $t_0 \in [a,b]_{\mathbb{T}}$. Hence t_0 is not isolated, and there exists either a subsequence that tends to t_0 from below or a subsequence that tends to t_0 from above, and in either case the limit of $f(t)$ as $t \to t_0$ is finite. This is a contradiction. Therefore the function f is bounded. This completes the proof.

Example 3.53. We will show that every regulated function $f : [a,b]_{\mathbb{T}} \to \mathbb{R}$ is integrable on $[a,b]_{\mathbb{T}}$. Indeed, since f is regulated, by Example 3.52 it follows that it is bounded. Denote

$$m = \inf_{t\in[a,b)_{\mathbb{T}}} f(t),$$
$$M = \sup_{t\in[a,b)_{\mathbb{T}}} f(t).$$

If $m = M$, then f is a constant function and thus is integrable on $[a,b]_{\mathbb{T}}$. Suppose that $M > m$. Take arbitrary $\varepsilon > 0$. If $p \in (a,b]_{\mathbb{T}}$ is a left-dense point and $q \in [a,b)_{\mathbb{T}}$ is a right-dense point, then the finite limits

$$f(p^-) = \lim_{t\to p, t<p} f(t)$$

and

$$f(q^+) = \lim_{t\to q, t>q} f(t)$$

exist. Thus, for each $\tau \in [a,b]_{\mathbb{T}}$, there is $\delta(\tau) > 0$ such that

$$\begin{aligned} &|r-s| < \delta(\tau) \quad \text{and} \quad r,s<\tau \quad \text{or} \quad r,s>\tau \\ &\text{imply} \quad |f(r)-f(s)| < \frac{\varepsilon}{2(b-a+1)}. \end{aligned} \tag{3.18}$$

In (3.18), we assume only $r,s > a$ for $\tau = a$ and only $r,s < b$ for $\tau = b$. If τ is an isolated or one-sided scattered endpoint, then we choose $U_{\delta(\tau)} = \{\tau\}$, where

$$U_{\delta(\tau)} = \{t \in \mathbb{T} : |t-\tau| < \delta(\tau)\}.$$

Thus the collection

$$\{U_{\delta(\tau)} : \tau \in [a,b]_{\mathbb{T}}\}$$

forms an open covering of the compact subset $[a,b]_{\mathbb{T}}$ of $\mathbb{T}$. By the Heine–Borel theorem it follows that there is a finite open covering

$$\{U_{\delta(\tau_1)}, \dots, U_{\delta(\tau_p)}\}.$$

Now we use the points $\tau_1, \dots, \tau_p$ to define the intervals $V_1, \dots, V_p \subset [a,b]_{\mathbb{T}}$ as in the proof of Example 3.51 and letting $V_j = \emptyset$ if τ_j is an isolated or one-sided scattered endpoint. The set

$$N = [a,b]_{\mathbb{T}} \setminus \bigcup_{j=1}^{p} V_j$$

consists of finitely many closed intervals $J_1, \dots, J_{p+1}$. Without loss of generality, assume that

$$\tau_1 < \tau_2 < \cdots < \tau_p.$$

Choose $\delta_0 > 0$ such that

$$\delta_0 < \min_{j\in\{1,\dots,p-1\}} (\inf U_{\delta(\tau_{j-1})} - \sup U_{\delta(\tau_j)}).$$

Therefore

$$t, s \in \mathbb{T} \quad \text{and} \quad |t - s| < \delta_0 \quad \text{imply} \quad |f(t) - f(s)| < \frac{\varepsilon}{2(b - a + 1)}.$$

We construct a partition P as in the proof of Theorem 3.51 so that

$$U_{\mathbb{T}}(f, P) - L_{\mathbb{T}}(f, P) < \varepsilon.$$

Now applying Example 3.40, we conclude that f is integrable on $[a,b]_{\mathbb{T}}$. This completes the proof.

Definition 3.14. We say that a function $f : [a,b]_{\mathbb{T}} \to \mathbb{R}$ satisfies the Lipschitz condition with constant $B > 0$ if

$$|f(x) - f(y)| \le B|x - y|, \quad x, y \in [a,b]_{\mathbb{T}}.$$

The number B is called a Lipschitz constant.

Example 3.54. Let $f : [a,b]_{\mathbb{T}} \to \mathbb{R}$ be a Riemann-integrable function. Denote

$$M = \sup_{t\in[a,b)_{\mathbb{T}}} f(t),$$
$$m = \inf_{t\in[a,b)_{\mathbb{T}}} f(t).$$

Let $\phi : [m, M] \to \mathbb{R}$ satisfy the Lipschitz condition with constant B. We will show that the composite function $h = \phi \circ f$ is integrable on $[a,b]_{\mathbb{T}}$. For this aim, take arbitrary $\varepsilon > 0$. Since $f : [a,b]_{\mathbb{T}} \to \mathbb{R}$ is integrable on $[a,b]_{\mathbb{T}}$, by Example 3.40 it follows that there is a partition $P \in \mathscr{P}(a,b)_{\mathbb{T}}$ given by

$$a = t_0 < t_1 < \cdots < t_n = b$$

such that

$$U_{\mathbb{T}}(f, P) - L_{\mathbb{T}}(f, P) < \frac{\varepsilon}{B}.$$

Define

$$\begin{aligned} M_j &= \sup_{t \in [t_{j-1}, t_j)_{\mathbb{T}}} f(t), \\ m_j &= \inf_{t \in [t_{j-1}, t_j)_{\mathbb{T}}} f(t), \quad j \in \{1, \ldots, n\}, \end{aligned}$$

and

$$\begin{aligned} M_j^* &= \sup_{t \in [t_{j-1}, t_j)_{\mathbb{T}}} h(t), \\ m_j^* &= \inf_{t \in [t_{j-1}, t_j)_{\mathbb{T}}} h(t), \quad j \in \{1, \ldots, n\}. \end{aligned}$$

Since ϕ satisfies the Lipschitz condition on $[a, b]_{\mathbb{T}}$ with constant B, we get

$$\begin{aligned} h(s) - h(r) &\le |h(s) - h(r)| \\ &= |\phi(f(s)) - \phi(f(r))| \\ &\le B|f(s) - f(r)| \\ &\le B(M_j - m_j) \end{aligned}$$

for all $s, \tau \in [t_{j-1}, t_j)_{\mathbb{T}}$ and $j \in \{1, \ldots, n\}$. Hence

$$M_j^* - m_j^* \le B(M_j - m_j), \quad j \in \{1, \ldots, n\}.$$

Consequently,

$$\begin{aligned} U_{\mathbb{T}}(h, P) - L_{\mathbb{T}}(h, P) &= \sum_{j=1}^{n} (M_j^* - m_j^*)(f_j - t_{j-1}) \\ &\le B \sum_{j=1}^{n} (M_j - m_j)(t_j - t_{j-1}) \\ &= B\big(U_{\mathbb{T}}(f, P) - L_{\mathbb{T}}(f, P)\big) \\ &< B\frac{\varepsilon}{B} \\ &= \varepsilon. \end{aligned}$$

Now applying Example 3.40, we conclude that h is integrable on $[a, b]_{\mathbb{T}}$.

Example 3.55. Let $f : [a,b]_{\mathbb{T}} \to \mathbb{R}$ be integrable, and let

$$M = \sup_{tr\in[a,b]_{\mathbb{T}}} f(t),$$
$$m = \inf_{t\in[a,b]_{\mathbb{T}}} f(t).$$

Let also $\phi : [m,M] \to \mathbb{R}$ be a continuous function. We will show that the composite function $h = \phi \circ f : [a,b]_{\mathbb{T}} \to \mathbb{R}$ is integrable on $[a,b]_{\mathbb{T}}$. Indeed, since $\phi : [m,M] \to \mathbb{R}$ is a continuous function, there exists a constant $B \geq 0$ such that

$$\max_{x\in[m,M]} |\phi(x)| = B.$$

Take arbitrary $\varepsilon > 0$ and

$$\varepsilon_1 = \frac{\varepsilon}{b - a + 2B}.$$

Because $\phi : [m,M] \to \mathbb{R}$ is a continuous function and the interval $[m,M]$ is a compact set, it is uniformly continuous on $[m,M]$. Then there exists $\delta \in (0,\varepsilon_1)$ such that

$$x,y \in [m,M] \quad \text{and} \quad |x-y| < \delta \quad \text{imply} \quad |\phi(x) - \phi(y)| < \varepsilon_1.$$

Since $f : [a,b]_{\mathbb{T}} \to \mathbb{R}$ is integrable on $[a,b]_{\mathbb{T}}$, by Example 3.40 it follows that there is a partition $P \in \mathscr{P}(a,b)_{\mathbb{T}}$ given by

$$a = t_0 < t_1 < \cdots < t_n = b$$

such that

$$U_{\mathbb{T}}(f,P) - L_{\mathbb{T}}(f,P) < \delta^2.$$

Denote

$$M_j = \sup_{t\in[t_{j-1},t_j)_{\mathbb{T}}} f(t),$$
$$m_j = \inf_{t\in[t_{j-1},t_j)_{\mathbb{T}}} f(t), \quad j \in \{1,\ldots,n\},$$

and let M_j^* and m_j^*, $j \in \{1,\ldots,n\}$, be the corresponding numbers for the function h. Let also

$$J_1 = \{j \in \{1,\ldots,n\} : M_j - m_j < \delta\}$$

and

$$J_2 = \{j \in \{1,\ldots,n\} : M_j - m_j \geq \delta\}.$$

Then, for $j \in J_1$, we have

$$\begin{aligned}|h(s) - h(\tau)| &= |\phi(f(s)) - \phi(f(\tau))| \\ &< \varepsilon, \quad s, \tau \in [t_{j-1}, t_j),\end{aligned}$$

and hence

$$M_j^* - m_j^* < \varepsilon_1.$$

Note that

$$M_j^* - m_j^* \le 2B, \quad j \in \{1, \dots, n\}.$$

Then

$$\begin{aligned}U_{\mathbb{T}}(h, P) - L_{\mathbb{T}}(h, P) &= \sum_{j=1}^{n} (M_j^* - m_j^*)(t_j - t_{j-1}) \\ &= \sum_{j \in J_1} (M_j^* - m_j^*)(t_j - t_{j-1}) + \sum_{j \in J_2} (M_j^* - m_j^*)(t_j - t_{j-1}) \\ &< \varepsilon_1 \sum_{j=1}^{n} (t_j - t_{j-1}) + 2B \sum_{j \in J_2} (t_j - t_{j-1}) \\ &= \varepsilon_1 (b - a) + \frac{2B}{\delta} \delta \sum_{j \in J_1} (t_j - t_{j-1}) \\ &\le \varepsilon_1 (b - a) + \frac{2B}{\delta} \sum_{j \in J_1} (M_j^* - m_j^*)(t_j - t_{j-1}) \\ &\le \varepsilon_1 (b - a) + \frac{2B}{\delta} \sum_{j=1}^{n} (M_j^* - m_j^*)(t_j - t_{j-1}) \\ &= \varepsilon_1 (b - a) + \frac{2B}{\delta} (U_{\mathbb{T}}(f, P) - l_{\mathbb{T}}(f, P)) \\ &< \varepsilon_1 (b - a) + \frac{2B}{\delta} \delta^2 \\ &= \varepsilon_1 (b - a) + 2B\delta \\ &< \varepsilon_1 (b - a + 2B) \\ &= \frac{\varepsilon}{b - a + 2B} (b - a + 2B) \\ &= \varepsilon.\end{aligned}$$

Now applying Example 3.40, we conclude that h is integrable on $[a, b]_{\mathbb{T}}$.

Exercise 3.19. Let $f : [a, b]_{\mathbb{T}} \to \mathbb{R}$ be integrable on $[a, b]_{\mathbb{T}}$. Prove that for any $\alpha > 0$, the function $|f|^{\alpha} : [a, b]_{\mathbb{T}} \to \mathbb{R}$ is integrable on $[a, b]_{\mathbb{T}}$.

Exercise 3.20. Let $f : [a, b]_{\mathbb{T}} \to \mathbb{R}$ be a bounded function that is integrable on $[a, b]_{\mathbb{T}}$. Prove that the function f is integrable on any subinterval $[c, d]_{\mathbb{T}}$ of the interval $[a, b]_{\mathbb{T}}$.

Exercise 3.21. Let f be integrable on $[a, b]_{\mathbb{T}}$, and let $c \in \mathbb{R}$. Prove that cf is integrable on $[a, b]_{\mathbb{T}}$ and

$$\int_a^b (cf)(t)\Delta t = c\int_a^b f(t)\Delta t.$$

Exercise 3.22. Let $f, g : [a, b]_{\mathbb{T}} \to \mathbb{R}$ be integrable on $[a, b]_{\mathbb{T}}$. Prove that $f + g : [a, b]_{\mathbb{T}} \to \mathbb{R}$ is Riemann integrable on $[a, b]_{\mathbb{T}}$. Moreover,

$$\int_a^b (f+g)(t)\Delta t = \int_a^b f(t)\Delta t + \int_a^b g(t)\Delta t.$$

Exercise 3.23. Let $f, g : [a, b]_{\mathbb{T}} \to \mathbb{R}$ be integrable functions on $[a, b]_{\mathbb{T}}$. Prove that the function $fg : [a, b]_{\mathbb{T}} \to \mathbb{R}$ is integrable on $[a, b]_{\mathbb{T}}$.

Exercise 3.24. Let $f : [a, b]_{\mathbb{T}} \to \mathbb{R}$, and let $a < c < b$, $c \in \mathbb{T}$. If f is bounded and integrable on $[a, c]_{\mathbb{T}}$ and $[c, b]_{\mathbb{T}}$, then prove that f is integrable on $[a, b]_{\mathbb{T}}$ and

$$\int_a^b f(t)\Delta t = \int_a^c f(t)\Delta t + \int_c^b f(t)\Delta t. \tag{3.19}$$

Exercise 3.25. Let $f, g : [a, b]_{\mathbb{T}} \to \mathbb{R}$ be integrable functions on $[a, b]_{\mathbb{T}}$ such that $f(t) \le g(t)$, $t \in [a, b]_{\mathbb{T}}$. Prove that

$$\int_a^b f(t)\Delta t \le \int_a^b g(t)\Delta t.$$

Exercise 3.26. Let $f :, [a, b]_{\mathbb{T}} \to \mathbb{R}$ be integrable on $[a, b]_{\mathbb{T}}$. Prove that so is $|f|$ and

$$\left|\int_a^b f(t)\Delta t\right| \le \int_a^b |f(t)|\Delta t.$$

Exercise 3.27. Let $f, g : [a, b]_{\mathbb{T}} \to \mathbb{R}$ be integrable on $[a, b]_{\mathbb{T}}$. Prove that

$$\begin{aligned}\left|\int_a^b f(t)g(t)\Delta t\right| &\le \int_a^b |f(t)||g(t)|\Delta t\\ &\le \left(\sup_{t\in[a,b)} |f(t)|\right)\left(\int_a^b |g(t)|\Delta t\right).\end{aligned}$$

Example 3.56. Let $\{f_k\}_{k\in\mathbb{N}}$ be a sequence of integrable functions on $[a,b]_{\mathbb{T}}$ such that $f_k \to f$ as $k \to \infty$, uniformly on $[a,b]_{\mathbb{T}}$ for a function f defined on $[a,b]_{\mathbb{T}}$. We will show that f is integrable on $[a,b]_{\mathbb{T}}$ and

$$\int_a^b f(t)\Delta t = \lim_{k\to\infty} \int_a^b f_k(t)\Delta t. \tag{3.20}$$

Take arbitrary $\varepsilon > 0$. Consider a partition $P \in \mathscr{P}(a,b)_{\mathbb{T}}$ given by

$$a = t_0 < t_1 < \cdots < t_n = b.$$

Let also

$$\begin{aligned} M_j &= \sup_{t\in[t_{j-1},t_j)_{\mathbb{T}}} f(t), \\ m_j &= \inf_{t\in[t_{j-1},t_j)_{\mathbb{T}}} f(t), \\ M_j^k &= \sup_{t\in[t_{j-1},t_j)_{\mathbb{T}}} f(t), \\ m_j^k &= \inf_{t\in[t_{j-1},t_j)_{\mathbb{T}}} f(t), \quad j \in \{1,\dots,n\},\ k \in \mathbb{N}. \end{aligned}$$

Note that for all $k \in \mathbb{N}, j \in \{1,\dots,n\}$, and $t^1, t^2 \in [t_{j-1}, t_j)$, we have

$$\begin{aligned} f(t^1) - f(t^2) &= (f(t^1) - f_k(t^1)) + (f_k(t^1) - f_k(t^2)) \\ &\quad + (f_k(t^2) - f(t^2)), \end{aligned}$$

whereupon

$$\begin{aligned} |f(t^1) - f(t^2)| &= |(f(t^1) - f_k(t^1)) + (f_k(t^1) - f_k(t^2)) \\ &\quad + (f_k(t^2) - f(t^2))| \\ &\le |f(t^1) - f_k(t^1)| + |f_k(t^1) - f_k(t^2)| \\ &\quad + |f_k(t^2) - f(t^2)|. \end{aligned} \tag{3.21}$$

Since $f_k \to f$ as $k \to \infty$ uniformly on $[a,b]_{\mathbb{T}}$, there exists $l \in \mathbb{N}$ such that

$$|f_k(t) - f(t)| < \frac{\varepsilon}{4(b-a)} \tag{3.22}$$

for all $t \in [a,b]_{\mathbb{T}}$ and $k > l$. Now applying (3.21), we get

$$|f(t^1) - f(t^2)| < |f_k(t^1) - f_k(t^2)| + \frac{\varepsilon}{2(b-a)}, \quad t^1, t^2 \in [t_{j-1}, t_j)_{\mathbb{T}},$$

$j \in \{1,\dots,n\}$, $k \in \mathbb{N}$, $k > l$. By the last inequality we get

$$M_j - m_j \le M_j^k - m_j^k + \frac{\varepsilon}{2(b-a)}, \quad j \in \{1, \dots, n\},\ k \in \mathbb{N},\ k > l. \tag{3.23}$$

Hence

$$\begin{aligned} U(f,P) - L(f,P) &= \sum_{j=1}^{n} (M_j - m_j)(t_j - t_{j-1}) \\ &\le \sum_{j=1}^{n} \left(M_j^k - m_j^k + \frac{\varepsilon}{2(b-a)} \right)(t_j - t_{j-1}) \\ &= \sum_{j=1}^{n} (M_j^k - m_j^k)(t_j - t_{j-1}) + \sum_{j=1}^{n} \frac{\varepsilon}{2(b-a)}(t_j - t_{j-1}) \\ &= U(f_k, P) - L(f_k, P) + \frac{\varepsilon}{2}, \quad k \in \mathbb{N},\ k > l, \end{aligned}$$

i. e.,

$$U(f,P) - L(f,P) \le U(f_k,P) - L(f_k,P) + \frac{\varepsilon}{2}, \quad k \in \mathbb{N},\ k > l. \tag{3.24}$$

Fix $k_1 \in \mathbb{N}$ such that $k_1 > l$. Then applying (3.24), we obtain

$$U(f,P) - L(f,P) \le U(f_{k_1},P) - L(f_{k_1},P) + \frac{\varepsilon}{2}. \tag{3.25}$$

Since f_{k_1} is integrable on $[a,b]_{\mathbb{T}}$, by Example 3.40 it follows that there is a partition $P_1 \in \mathscr{P}(a,b)_{\mathbb{T}}$ such that

$$U(f_{k_1},P_1) - L(f_{k_1},P_1) < \frac{\varepsilon}{2}.$$

From this and from (3.25) we get

$$\begin{aligned} U(f,P) - L(f,P) &< \frac{\varepsilon}{2} + \frac{\varepsilon}{2} \\ &= \varepsilon. \end{aligned}$$

From here, applying Example 3.40, we conclude that the function f is integrable on $[a,b]_{\mathbb{T}}$. Now applying inequality (3.22), we arrive at

$$\begin{aligned} \left| \int_a^b f_k(t)\Delta t - \int_a^b f(t)\Delta t \right| &= \left| \int_a^b (f_k(t) - f(t))\Delta t \right| \\ &\le \int_a^b |f_k(t) - f(t)|\Delta t \\ &< \frac{\varepsilon}{4(b-a)} \int_a^b \Delta t \end{aligned}$$

$$
\begin{aligned}
&= \frac{\varepsilon}{4(b-a)}(b-a) \\
&= \frac{\varepsilon}{4} \\
&< \varepsilon
\end{aligned}
$$

for all $k > l$. Since $\varepsilon > 0$ was arbitrarily chosen, we obtain (3.20). This completes the proof.

Exercise 3.28. Suppose that $f = \sum_{k=1}^{\infty} f_k$ is a series of integrable functions on $[a,b]_{\mathbb{T}}$ that converges uniformly to f on $[a,b]_{\mathbb{T}}$. Then f is integrable on $[a,b]_{\mathbb{T}}$, and

$$\int_a^b f(t)\Delta t = \sum_{k=1}^{\infty} \int_a^b f_k(t)\Delta t.$$

Now we will represent some versions of the fundamental theorem of calculus.

Example 3.57 (Fundamental theorem of calculus I). Let $g : [a,b]_{\mathbb{T}} \to \mathbb{R}$ be a continuous function that is differentiable on $[a,b)_{\mathbb{T}}$. We will show that if g^{Δ} is integrable on $[a,b)_{\mathbb{T}}$, then

$$\int_a^b g^{\Delta}(t)\Delta t = g(b) - g(a).$$

Take arbitrary $\varepsilon > 0$. Since g^{Δ} is integrable on $[a,b)_{\mathbb{T}}$, there is a partition $P \in \mathscr{P}(a,b)_{\mathbb{T}}$ given by

$$a = t_0 < t_1 < \cdots < t_n = b$$

such that

$$U_{\mathbb{T}}(g^{\Delta}, P) - L_{\mathbb{T}}(g^{\Delta}, P) < \varepsilon. \tag{3.26}$$

By the mean value theorem it follows that for any $j \in \{1, \ldots, n\}$, there are $\tau_j, \xi_j \in [t_{j-1}, t_j)_{\mathbb{T}}$ such that

$$g^{\Delta}(\tau_j)(t_j - t_{j-1}) \le g(t_j) - g(t_{j-1}) \le g^{\Delta}(\xi_j)(t_j - t_{j-1}),$$

whereupon

$$\sum_{j=1}^{n} g^{\Delta}(\tau_j)(t_j - t_{j-1}) \le \sum_{j=1}^{n}(g(t_j) - g(t_{j-1})) \le \sum_{j=1}^{n} g^{\Delta}(\xi_j)(t_j - t_{j-1}),$$

or

$$\sum_{j=1}^{n} g^{\Delta}(\tau_j)(t_j - t_{j-1}) \le g(b) - g(a) \le \sum_{j=1}^{n} g^{\Delta}(\xi_j)(t_j - t_{j-1}),$$

and we get the estimate

$$L_{\mathbb{T}}(g^{\Delta}, P) \le g(b) - g(a) \le U_{\mathbb{T}}(g^{\Delta}, P).$$

On the other hand, we have

$$L_{\mathbb{T}}(g^{\Delta}, P) \le \int_a^b g^{\Delta}(t)\Delta t \le U_{\mathbb{T}}(g^{\Delta}, P).$$

Hence

$$\int_a^b g^{\Delta}(t)\Delta t - (g(b) - g(a)) \le U_{\mathbb{T}}(g^{\Delta}, P) - L_{\mathbb{T}}(g^{\Delta}, P),$$

and

$$\int_a^b g^{\Delta}(t)\Delta t - (g(b) - g(a)) \ge L_{\mathbb{T}}(g^{\Delta}, P) - U_{\mathbb{T}}(g^{\Delta}, P).$$

Therefore, applying (3.26), we arrive at the chain

$$\begin{aligned}\left|\int_a^b g^{\Delta}(t)\Delta t - (g(b) - g(a))\right| &\le U_{\mathbb{T}}(g^{\Delta}, P) - L_{\mathbb{T}}(g^{\Delta}, P)\\ &< \varepsilon.\end{aligned}$$

Example 3.58 (Integration by parts). Let u and v be continuous functions on $[a, b]_{\mathbb{T}}$ and differentiable on $[a, b)_{\mathbb{T}}$. We will show that if u^{Δ} and v^{Δ} are integrable on $[a, b)_{\mathbb{T}}$, then

$$\int_a^b u^{\Delta}(t)v(t)\Delta t + \int_a^b u^{\sigma}(t)v^{\Delta}(t)\Delta t = u(b)v(b) - u(a)v(a).$$

Let $g = uv$. Then

$$g^{\Delta} = u^{\Delta}v + u^{\sigma}v^{\Delta},$$

and g^{Δ} is integrable on $[a, b)_{\mathbb{T}}$. Now applying Example 3.57, we arrive at the following chain:

$$\int_a^b u^{\Delta}(t)v(t)\Delta t + \int_a^b u^{\sigma}(t)v^{\Delta}(t)\Delta t = \int_a^b (u^{\Delta}(t)v(t) + u^{\sigma}(t)v^{\Delta}(t))\Delta t$$

$$
\begin{aligned}
&= \int_a^b g^{\Delta}(t)\Delta t \\
&= g(b) - g(a) \\
&= u(b)v(b) - u(a)v(a).
\end{aligned}
$$

Example 3.59 (Fundamental theorem of calculus II). Let f be integrable on $[a,b]_{\mathbb{T}}$, and let

$$
F(t) = \int_a^t f(\tau)\Delta\tau, \quad t \in [a,b]_{\mathbb{T}}.
$$

We will show that F is uniformly continuous on $[a,b]_{\mathbb{T}}$. Moreover, if $t_0 \in [a,b)_{\mathbb{T}}$ and f is continuous at a right-dense point t_0, then F is differentiable at t_0 and

$$
F^{\Delta}(t_0) = f(t_0).
$$

Let $B > 0$ be such that

$$
|f(t)| \le B, \quad t \in [a,b)_{\mathbb{T}}.
$$

Let also $s, t \in [a,b]_{\mathbb{T}}$ be such that

$$
|t - s| < \frac{\varepsilon}{B}.
$$

Without loss of generality, suppose that $t < s$. Then

$$
\begin{aligned}
|F(s) - F(t)| &= \left| \int_a^s f(\tau)\Delta\tau - \int_a^t f(\tau)\Delta\tau \right| \\
&= \left| \int_t^s f(\tau)\Delta\tau \right| \\
&\le \int_t^s |f(\tau)|\Delta\tau \\
&\le B(s - t) \\
&< B\frac{\varepsilon}{B} \\
&= \varepsilon.
\end{aligned}
$$

Thus F is uniformly continuous ion $[a,b]_{\mathbb{T}}$. Let now $t_0 \in [a,b)_{\mathbb{T}}$. We will consider the following two cases.

1. t_0 is right-scattered. Then

$$
F^{\Delta}(t_0) = \frac{F(\sigma(t_0)) - F(t_0)}{\sigma(t_0) - t_0}
$$

$$
\begin{aligned}
&= \frac{\int_a^{\sigma(t_0)} f(\tau)\Delta\tau - \int_a^{t_0} f(\tau)\Delta\tau}{\sigma(t_0) - t_0} \\
&= \frac{\int_{t_0}^{\sigma(t_0)} f(\tau)\Delta\tau}{\mu(t_0)} \\
&= \frac{\mu(t_0) f(t_0)}{\mu(t_0)} \\
&= f(t_0).
\end{aligned}
$$

2. t_0 is right-dense. Take arbitrary $\varepsilon > 0$. Since f is continuous at t_0, there is $\delta > 0$ such that

$$
s \in [a,b]_{\mathbb{T}} \quad \text{and} \quad |s - t_0| < \delta \quad \text{imply} \quad |f(s) - f(t_0)| < \varepsilon.
$$

Then for all $t \in [a,b]_{\mathbb{T}}$ such that $0 < |t - t_0| < \delta$, we have

$$
\begin{aligned}
\left|\frac{F(t) - F(t_0)}{t - t_0} - f(t_0)\right| &= \left|\frac{1}{t - t_0}\left(\int_a^t f(\tau)\Delta\tau - \int_a^{t_0} f(\tau)\Delta\tau\right) - f(t_0)\right| \\
&= \left|\frac{1}{t - t_0}\int_{t_0}^t f(\tau)\Delta\tau - f(t_0)\right| \\
&= \left|\frac{1}{t - t_0}\int_{t_0}^t (f(\tau) - f(t_0))\Delta\tau\right| \\
&\le \frac{1}{|t - t_0|}\left|\int_{t_0}^t |f(\tau) - f(t_0)|\Delta\tau\right| \\
&< \frac{1}{|t - t_0|}\varepsilon|t - t_0| \\
&= \varepsilon,
\end{aligned}
$$

whereupon

$$
F^{\Delta}(t_0) = f(t_0).
$$

Example 3.60. Let f and f^{Δ} be continuous on $[a,b]_{\mathbb{T}}$. We will show that

$$
\left(\int_a^t f(t,s)\Delta s\right)^{\Delta} = f(\sigma(t),t) + \int_a^t f^{\Delta}(t,s)\Delta s, \quad t \in [a,b]_{\mathbb{T}}.
$$

Define the function

$$g(t) = \int_a^t f(t,s)\Delta s, \quad t \in [a,b]_{\mathbb{T}}.$$

Take arbitrary $t \in [a,b]_{\mathbb{T}}$. We will consider the following two cases.

1. t is right-scattered. Then

$$\begin{aligned}
g^{\Delta}(t) &= \frac{g(\sigma(t)) - g(t)}{\sigma(t) - t} \\
&= \frac{\int_a^{\sigma(t)} f(\sigma(t),s)\Delta s - \int_a^t f(t,s)\Delta s}{\sigma(t) - t} \\
&= \int_a^{\sigma(t)} \frac{f(\sigma(t),s) - f(t,s)}{\sigma(t) - t}\Delta s - \int_a^t \frac{f(t,s)}{\sigma(t) - t}\Delta s \\
&= \int_a^t \frac{f(\sigma(t),s)}{\sigma(t) - t}\Delta s + \int_t^{\sigma(t)} \frac{f(\sigma(t),s)}{\sigma(t) - t}\Delta s - \int_a^t \frac{f(t,s)}{\sigma(t) - t}\Delta s \\
&= \int_a^t \frac{f(\sigma(t),s) - f(t,s)}{\sigma(t) - t}\Delta s + \mu(t)\frac{f(\sigma(t),t)}{\sigma(t) - t} \\
&= f(\sigma(t),t) + \int_a^t f^{\Delta}(t,s)\Delta s.
\end{aligned}$$

2. t is right-dense. Take arbitrary $\varepsilon > 0$. Since f and f^{Δ} are continuous on $[a,b]_{\mathbb{T}}$, they are uniformly continuous on $[a,b]_{\mathbb{T}}$. Then there is a neighborhood U of the point t such that

$$|f^{\Delta}(\tau_1,s) - f^{\Delta}(\xi_1,s)| < \frac{\varepsilon}{2},$$
$$|f(r,s_1) - f(t,t)| < \frac{\varepsilon}{2}$$

for all $\tau_1, \xi_1, r, s_1 \in U$ and for any fixed $s \in [a,b]_{\mathbb{T}}$. On the other hand, for any fixed $s \in [a,b]_{\mathbb{T}}$ and for any $r \in U\backslash\{t\}$, by the mean value theorem it follows that there are $\tau, \xi \in U\backslash\{t\}$ such that

$$f^{\Delta}(\tau,s) \le \frac{f(t,s) - f(r,s)}{t - r} \le f^{\Delta}(\xi,s).$$

Hence there is a neighborhood $U_1 \subset U\backslash\{t\}$ such that for all $r \in U_1$, we have

$$\left|\frac{f(t,s) - f(r,s)}{t - r} - f^{\Delta}(t,s)\right| < \frac{\varepsilon}{2(b-a)}.$$

Let now $r \in U_1$. Without loss of generality, suppose that $t > r$. Then we have

$$
\begin{aligned}
&\left|\frac{g(t)-g(r)}{t-r} - \int_a^t f^{\Delta}(t,s) - f(t,t)\right| \\
&= \left|\frac{1}{t-r}\int_a^t f(t,s)\Delta s - \frac{1}{t-r}\int_a^r f(r,s)\Delta s - \int_a^t f^{\Delta}(t,s)\Delta s - f(t,t)\right| \\
&= \left|\frac{1}{t-r}\int_a^t f(t,s)\Delta s - \frac{1}{t-r}\int_a^t f(r,s)\Delta s + \frac{1}{t-r}\int_r^t f(r,s)\Delta s - f(t,t) - \int_a^t f^{\Delta}(t,s)\Delta s\right| \\
&= \left|\frac{1}{t-r}\int_a^t (f(t,s)-f(r,s))\Delta s - \int_a^t f^{\Delta}(t,s)\Delta s + \frac{1}{t-r}\int_r^t ((r,s)-(t,t))\Delta s\right| \\
&\le \left|\frac{1}{t-r}\int_a^t (f(t,s)-f(r,s))\Delta s - \int_a^t f^{\Delta}(t,s)\Delta s\right| + \frac{1}{t-r}\left|\int_r^t ((r,s)-(t,t))\Delta s\right| \\
&\le \left|\int_a^t \left(\frac{f(t,s)-f(r,s)}{t-r} - f^{\Delta}(t,s)\right)\Delta s\right| + \int_r^t \frac{|f(r,s)-f(t,t)|}{t-r}\Delta s \\
&\le \int_a^t \left|\frac{f(t,s)-f(r,s)}{t-r} - f^{\Delta}(t,s)\right|\Delta s + \frac{\varepsilon}{2}\frac{1}{t-r}(t-r) \\
&< \frac{\varepsilon}{2(b-a)}\int_a^b \Delta s + \frac{\varepsilon}{2} \\
&= \frac{\varepsilon}{2} + \frac{\varepsilon}{2} \\
&= \varepsilon.
\end{aligned}
$$

Example 3.61. Let f be integrable on $[a,b]_{\mathbb{T}}$. We will show that if f has a preantiderivative F with differentiation region D, then

$$\int_a^b f(t)\Delta t = F(b) - F(a).$$

Take arbitrary $\varepsilon > 0$. Since f is integrable on $[a,b]_{\mathbb{T}}$, there is a partition $P \in \mathscr{P}(a,b)_{\mathbb{T}}$ given by

$$a = t_0 < t_1 < \cdots < t_n = b$$

such that

$$U_{\mathbb{T}}(f,P) - L_{\mathbb{T}}(f,P) < \varepsilon. \tag{3.27}$$

Applying the mean value theorem to $F : [t_{j-1}, t_j]_{\mathbb{T}} \to \mathbb{R}$, $j \in \{1, \dots, n\}$, we obtain the existence of $\xi_j, \tau_j \in [t_{j-1}, t_j]_{\mathbb{T}}$, $j \in \{1, \dots, n\}$, such that

$$(t_j - t_{j-1})F^{\Delta}(\tau_j) \le F(t_j) - F(t_{j-1}) \le (t_j - t_{j-1})F^{\Delta}(\xi_j), \quad j \in \{1, \dots, n\},$$

and since

$$F^{\Delta}(t) = f(t), \quad t \in D,$$

we have

$$(t_j - t_{j-1})f(\tau_j) \le F(t_j) - F(t_{j-1}) \le (t_j - t_{j-1})f(\xi_j), \quad j \in \{1, \dots, n\}.$$

Summing the last inequality yields

$$\sum_{j=1}^{n}(t_j - t_{j-1})f(\tau_j) \le \sum_{j=1}^{n}(F(t_j) - F(t_{j-1})) \le \sum_{j=1}^{n}(t_j - t_{j-1})f(\xi_j),$$

or

$$U_{\mathbb{T}}(f, P) \le F(b) - F(a) \le L_{\mathbb{T}}(f, P).$$

Hence applying (3.27), we get

$$\begin{aligned}\int_a^b f(t)\Delta t - (F(b) - F(a)) &\le U_{\mathbb{T}}(f, P) - L_{\mathbb{T}}(f, P)\\ &< \varepsilon\end{aligned}$$

and

$$\begin{aligned}\int_a^b f(t)\Delta t - (F(b) - F(a)) &\ge L_{\mathbb{T}}(f, P) - U_{\mathbb{T}}(f, P)\\ &> -\varepsilon.\end{aligned}$$

Therefore

$$\left|\int_a^b f(t)\Delta t - (F(b) - F(a))\right| < \varepsilon.$$

Definition 3.15. A function $f : \mathbb{T} \to \mathbb{R}$ is said to be locally Δ-integrable or locally delta integrable (shortly locally integrable) if it is integrable on any finite interval.

Example 3.62 (Change of variables). Let $g : \mathbb{T} \to \mathbb{R}$ be a strictly increasing function such that $g(\mathbb{T}) = \tilde{T}$ is a time scale with delta differentiation operator $\tilde{\Delta}$. Let also $f : \mathbb{T} \to \mathbb{R}$ be locally integrable on T and suppose that g is differentiable with locally integrable derivative. We will show that if g^{Δ} possesses an antiderivative and if $a, b \in \mathbb{T}$, then

$$\int_a^b f(t)g^{\Delta}(t)\Delta t = \int_{g(a)}^{g(b)} (f(g^{-1}))(s)\tilde{\Delta}s.$$

By Exercise 3.23 we have that the product of two integrable functions on $[a, b]_{\mathbb{T}}$ is integrable on $[a, b]_{\mathbb{T}}$. Therefore fg^{Δ} possesses an antiderivative F, i. e.,

$$F^{\Delta} = fg^{\Delta}.$$

Then applying Example 3.57, we get

$$\begin{aligned}
\int_a^b f(t)g^{\Delta}(t)\Delta t &= \int_a^b F^{\Delta}(t)\Delta t \\
&- \Gamma(b) - \Gamma(a) \\
&= (F(g^{-1}))(g(b)) - (F(g^{-1}))(g(a)) \\
&= \int_{g(a)}^{g(b)} (F(g^{-1}))^{\tilde{\Delta}}(s)\tilde{\Delta}s \\
&= \int_{g(a)}^{g(b)} (F^{\Delta}(g^{-1}))(s)(g^{-1})^{\tilde{\Delta}}(s)\tilde{\Delta}s \\
&= \int_{g(a)}^{g(b)} ((fg^{\Delta})(g^{-1}))(s)(g^{-1})^{\tilde{\Delta}}(s)\tilde{\Delta}s \\
&= \int_{g(a)}^{g(b)} (f(g^{-1}))(s)(g^{\Delta}(g^{-1}))(s)(g^{-1})^{\tilde{\Delta}}(s)\tilde{\Delta}s \\
&= \int_{g(a)}^{g(b)} (f(g^{-1}))(s)\tilde{\Delta}s.
\end{aligned}$$

Example 3.63. Let $\mathbb{T} = \mathbb{Z}$. Then $\sigma(t) = t + 1$, $\mu(t) = 1$, and

$$\int_t^{t+1} (\tau^3 + \tau^2 + \tau + 1)\Delta\tau = t^3 + t^2 + t + 1, \quad t \in \mathbb{T}.$$

Example 3.64. Let $\mathbb{T} = \mathbb{N}_0^4$. Then $\sigma(t) = (\sqrt[4]{t}+1)^4$, $\mu(t) = (\sqrt[4]{t}+1)^4 - t$, and

$$\int_t^{(\sqrt[4]{t}+1)^4} \tau^2 \Delta\tau = ((\sqrt[4]{t}+1)^4 - t)t^2, \quad t \in \mathbb{T}.$$

Example 3.65. Let $\mathbb{T} = 3^{\mathbb{N}_0}$. Then $\sigma(t) = 3t$, $\mu(t) = 2t$, and

$$\int_t^{3t} \sin \tau \Delta\tau = 2t \sin t, \quad t \in \mathbb{T}.$$

Example 3.66. Let $\mathbb{T} = \mathbb{Z}$. We will compute

$$I = \int_{-2}^{3} (t^2 + t + 1)\Delta t, \quad t \in \mathbb{T}.$$

1. First way. We have

$$\begin{aligned} t^2 &= \frac{1}{3}(t^3)^\Delta - \frac{1}{2}(t^2)^\Delta + \frac{1}{6} \\ &= \left(\frac{1}{3}t^3 - \frac{1}{2}t^2 + \frac{t}{6}\right)^\Delta, \\ t &= \frac{1}{2}(t^2)^\Delta - \frac{1}{2} \\ &= \left(\frac{1}{2}t^2 - \frac{t}{2}\right)^\Delta, \\ 1 &= t^\Delta, \quad t \in \mathbb{T}. \end{aligned}$$

Hence

$$\begin{aligned} t^2 + t + 1 &= \left(\frac{1}{3}t^3 - \frac{1}{2}t^2 + \frac{t}{6}\right)^\Delta + \left(\frac{1}{2}t^2 - \frac{1}{2}t\right)^\Delta + t^\Delta \\ &= \left(\frac{1}{3}t^3 - \frac{1}{2}t^2 + \frac{t}{6} + \frac{1}{2}t^2 - \frac{t}{2} + t\right)^\Delta \\ &= \left(\frac{1}{3}t^3 + \frac{2}{3}t\right)^\Delta, \quad t \in \mathbb{T}. \end{aligned}$$

Therefore

$$I = \int_{-2}^{3} \left(\frac{1}{3}t^3 + \frac{2}{3}t\right)^\Delta \Delta t$$

$$
\begin{aligned}
&= \left(\frac{1}{3}t^3 + \frac{2}{3}t\right)\Big|_{t=-2}^{t=3} \\
&= \left(\frac{27}{3} + 2\right) - \left(-\frac{8}{3} - \frac{4}{3}\right) \\
&= 11 + 4 \\
&= 15.
\end{aligned}
$$

2. Second way. Since all points of $\mathbb{T}$ are isolated and $\mu(t) = 1$, $t \in \mathbb{T}$, we have that

$$
\begin{aligned}
I &= \sum_{t=-2}^{2} (t^2 + t + 2) \\
&= (4 - 2 + 1) + (1 - 1 + 1) + 1 + (1 + 1 + 1) + (4 + 2 + 1) \\
&= 3 + 1 + 1 + 3 + 7 \\
&= 15.
\end{aligned}
$$

Example 3.67. Let $\mathbb{T} = 2^{\mathbb{N}_0}$. We will compute

$$
I = \int_1^4 \left(\frac{\sin\frac{t}{2}\sin\frac{3t}{2}}{t} + t^2\right) dt.
$$

1. First way. Note that $\sigma(t) = 2t$, $\mu(t) = t$, and

$$
\begin{aligned}
(\cos t)^{\Delta} &= -2\frac{\sin\frac{t}{2}\sin\frac{3t}{2}}{t}, \\
t^2 &= \frac{1}{7}(t^3)^{\Delta}, \quad t \in \mathbb{T}.
\end{aligned}
$$

Hence

$$
\begin{aligned}
\frac{\sin\frac{t}{2}\sin\frac{3t}{2}}{t} + t^2 &= -\frac{1}{2}(\cos t)^{\Delta} + \frac{1}{7}(t^3)^{\Delta} \\
&= \left(-\frac{1}{2}\cos t + \frac{1}{7}t^3\right)^{\Delta}, \quad t \in \mathbb{T}.
\end{aligned}
$$

Therefore

$$
\begin{aligned}
I &= \int_1^4 \left(-\frac{1}{2}\cos t + \frac{1}{7}t^3\right)^{\Delta} \Delta t \\
&= \left(-\frac{1}{2}\cos t + \frac{1}{7}t^3\right)\Big|_{t=1}^{t=4} \\
&= -\frac{1}{2}\cos 4 + \frac{64}{7} + \frac{1}{2}\cos 1 - \frac{1}{7}
\end{aligned}
$$

$$
\begin{aligned}
&= -\frac{1}{2}(\cos 4 - \cos 1) + 9\\
&= \sin\frac{3}{2}\sin\frac{5}{2} + 9.
\end{aligned}
$$

2. Second way. Since all points of $\mathbb{T}$ are isolated, we obtain

$$
\begin{aligned}
I &= \sum_{t=1,2} \mu(t)\left(\frac{\sin\frac{t}{2}\sin\frac{3t}{2}}{t} + t^2\right)\\
&= \sin\frac{1}{2}\sin\frac{3}{2} + 1 + 2\left(\frac{\sin 1 \sin 3}{2} + 4\right)\\
&= \sin\frac{1}{2}\sin\frac{3}{2} + \sin 1 \sin 3 + 9\\
&= \frac{1}{2}(\cos 1 - \cos 2 + \cos 2 - \cos 4) + 9\\
&= \frac{1}{2}(\cos 1 - \cos 4) + 9\\
&= -\sin\frac{-3}{2}\sin\frac{5}{2} + 9\\
&= \sin\frac{3}{2}\sin\frac{5}{2} + 9.
\end{aligned}
$$

Example 3.68. Let $\mathbb{T} = [-1, 0] \cup 3^{\mathbb{N}_0}$, where $[-1, 0]$ is a real-valued interval. Let also

$$
f(t) = \begin{cases} \frac{1}{(t+2)^3} - \frac{1}{8} & \text{for } t \in [-1, 0),\\ 0 & \text{for } t = 0,\\ t^2 - t & \text{for } t \in 3^{\mathbb{N}_0}. \end{cases}
$$

We will compute

$$
I = \int_{-1}^{3} f(t)\Delta t.
$$

We have

$$
\begin{aligned}
I &= \int_{-1}^{0} f(t)\Delta t + \int_{1}^{3} f(t)\Delta t\\
&= \int_{-1}^{0}\left(\frac{1}{(t+2)^3} - \frac{1}{8}\right)dt + \int_{1}^{3}(t^2 - t)\Delta t\\
&= -\frac{1}{2(t+2)^2}\Big|_{t=-1}^{t=0} - \frac{1}{8} + 2t(t^2 - t)\big|_{t=1}
\end{aligned}
$$

$$= -\frac{1}{4} + \frac{1}{2}$$
$$= \frac{1}{4}.$$

Example 3.69. Let $\mathbb{T} = P_{1,3}$. We will compute

$$I = \int_0^8 \frac{1-t}{1+t} \Delta t.$$

We have

$$\begin{aligned}
I &= \int_0^1 \frac{1-t}{1+t} dt + \int_4^5 \frac{1-t}{1+t} dt \\
&= \int_0^1 \frac{2-1-t}{1+t} dt + \int_4^5 \frac{2-1-t}{1+t} dt \\
&= 2\int_0^1 \frac{dt}{1+t} - \int_0^1 dt + 2\int_4^5 \frac{dt}{1+t} - \int_4^5 dt \\
&= 2\log(1+t)\Big|_{t=0}^{t=1} - 2\log(1+t)\Big|_{t=4}^{t=5} - 2 \\
&= 2\log 2 - 2(\log 6 - \log 5) - 2 \\
&= 2\log 2 - 2\log\frac{6}{5} - 2 \\
&= 2\log\frac{5}{3} - 2.
\end{aligned}$$

Example 3.70. Let $\mathbb{T} = \{-\frac{1}{n} : n \in \mathbb{N}\} \cup \mathbb{N}_0$. We will compute

$$I = \int_{-1}^2 \frac{1+t^2}{7+t} \Delta t.$$

We have

$$\begin{aligned}
I &= \int_{-1}^0 \frac{1+t^2}{7+t} \Delta t + \int_0^2 \frac{1+t^2}{7+t} \Delta t \\
&= \sum_{n=1}^{\infty} \frac{1+\frac{1}{n^2}}{7-\frac{1}{n}} + \frac{1+0^2}{1+7\cdot 0} + \frac{1+1^2}{7+1\cdot 1} \\
&= \sum_{n=1}^{\infty} \frac{n^2+1}{n(7n-1)} + 1 + \frac{2}{8} \\
&= \sum_{n=1}^{\infty} \frac{n^2+1}{n(7n-1)} + \frac{5}{4}.
\end{aligned}$$

Example 3.71. Let $U = \{\frac{1}{2^n} : n \in \mathbb{N}\}$, and let

$$\mathbb{T} = U \cup (1 - U) \cup (1 + U) \cup (2 - U) \cup (2 + U) \cup \{0, 1, 2\}.$$

We will find

$$I = \int_0^{\frac{3}{2}} (2 + t^2)(1 - t^2)\Delta t.$$

We have

$$\begin{aligned} I &= \int_0^{\frac{1}{2}} (2 + t^2)(1 - t^2)\Delta t + \int_{\frac{1}{2}}^{1} (2 + t^2)(1 - t^2)\Delta t + \int_1^{\frac{3}{2}} (2 + t^2)(1 - t^2)\Delta t \\ &= \sum_{n=2}^{\infty} \left(1 + \frac{1}{2^{2n}}\right)\left(1 - \frac{1}{2^{2n}}\right) + \sum_{n=2}^{\infty} \left(2 + \left(1 - \frac{1}{2^n}\right)^2\right)\left(1 - \left(1 - \frac{1}{2^n}\right)^2\right) \\ &\quad + \sum_{n=2}^{\infty} \left(2 + \left(1 + \frac{1}{2^n}\right)^2\right)\left(1 - \left(1 + \frac{1}{2^n}\right)^2\right) \\ &= \sum_{n=2}^{\infty} \left(2 + \frac{1}{2^{2n}}\right)\left(1 + \frac{1}{2^{2n}}\right) + \sum_{n=2}^{\infty} \left(3 - \frac{1}{2^{n-1}} + \frac{1}{2^{2n}}\right)\left(\frac{1}{2^{n-1}} - \frac{1}{2^{2n}}\right) \\ &\quad + \sum_{n=2}^{\infty} \left(3 + \frac{1}{2^{n-1}} + \frac{1}{2^{2n}}\right)\left(-\frac{1}{2^{n-1}} + \frac{1}{2^{2n}}\right). \end{aligned}$$

Exercise 3.29. Let $\mathbb{T} = 2\mathbb{Z}$. Prove that

$$\int_{-2}^{2} (t^2 + t)\Delta t = -4.$$

Exercise 3.30. Let $\mathbb{T} = 3^{\mathbb{N}_0}$. Prove that

$$\int_t^{3t} \left(\tau^2 - 3\tau + 4 \sin \tau\right)\Delta\tau = 2t\left(t^2 - 3t + 4 \sin t\right), \quad t \in \mathbb{T}.$$

Exercise 3.31. Let $\mathbb{T} = (-2\mathbb{N}_0) \cup 3^{\mathbb{N}_0}$. Prove that

$$\int_{-4}^{9} (1 - t + t^2)\Delta t = 101.$$

Exercise 3.32. Let $\mathbb{T} = \{0\} \cup 11^{\mathbb{N}_0}$. Prove that

$$\int_0^{121} \frac{1-t}{2+3t}\Delta t = -\frac{433}{14}.$$

Exercise 3.33. Let $\mathbb{T} = [1,2] \cup [3,4] \cup [7,8] \cup 9^{\mathbb{N}}$. Prove that

$$\int_4^{81} (t^2 + t)\Delta t = \frac{39623}{6}.$$

Example 3.72 (First mean value theorem). Let f and g be bounded integrable functions on $[a,b]_{\mathbb{T}}$. Let also g be nonnegative or nonpositive on $[a,b]_{\mathbb{T}}$, and let

$$m = \inf_{t\in[a,b)_{\mathbb{T}}} f(t),$$
$$M = \sup_{t\in[a,b)_{\mathbb{T}}} f(t).$$

We will prove that there exists a constant $A \in [m, M]$ such that

$$\int_a^b f(t)g(t)\Delta t = A\int_a^b f(t)\Delta t. \tag{3.28}$$

For this aim, without loss of generality, assume that g is nonnegative on $[a,b]_{\mathbb{T}}$. Note that

$$m \le f(t) \le M, \quad t \in [a,b]_{\mathbb{T}},$$

and

$$mg(t) \le f(t)g(t) \le Mg(t), \quad t \in [a,b]_{\mathbb{T}}. \tag{3.29}$$

By Exercise 3.21 we have that mf and Mf are integrable on $[a,b]_{\mathbb{T}}$. By Exercise 3.23 we obtain that fg is integrable on $[a,b]_{\mathbb{T}}$. Then applying Exercise 3.25 and using inequalities (3.29), we arrive at the inequalities

$$m\int_a^b g(t)\Delta t \le \int_a^b f(t)g(t)\Delta t \le M\int_a^b g(t)\Delta t. \tag{3.30}$$

If $g \equiv 0$ on $[a,b]_{\mathbb{T}}$, then

$$\int_a^b g(t)\Delta t = 0,$$

and

$$\int_a^b f(t)g(t)\Delta t = 0,$$

and inequality (3.28) holds for any $A \in [m, M]$. Let now $g \not\equiv 0$ on $[a, b]_{\mathbb{T}}$. Then

$$\int_a^b g(t)\Delta t > 0,$$

and applying (3.30), we get

$$m \leq \frac{\int_a^b f(t)g(t)\Delta t}{\int_a^b g(t)\Delta t} \leq M.$$

Set

$$A = \frac{\int_a^b f(t)g(t)\Delta t}{\int_a^b g(t)\Delta t}.$$

Then $A \in [m, M]$, and thus inequality (3.28) holds. We leave to the reader the case where g is nonpositive as an exercise.

Exercise 3.34 (First mean value theorem). Let f be bounded integrable function on $[a, b]_{\mathbb{T}}$, and let

$$m = \inf_{t\in[a,b)_{\mathbb{T}}} f(t),$$
$$M = \sup_{t\in[a,b)_{\mathbb{T}}} f(t).$$

Prove that there exists a constant $A \in [m, M]$ such that

$$\int_a^b f(t)\Delta t = A(b - a). \tag{3.31}$$

Example 3.73 (The Abel lemma). Let the nonnegative numbers p_j, $j \in \{1, \dots, n\}$, be such that

$$p_1 \geq p_2 \geq \cdots \geq p_n \geq 0$$

and the numbers

$$S_k = \sum_{j=1}^{k} q_j, \quad k \in \{1, \dots, n\},$$

satisfy the inequalities

$$m \leq S_k \leq M, \quad k \in \{1, \dots, n\},$$

where $q_j, j \in \{1, \dots, n\}$, m, and M are real numbers. We will prove that

$$mp_1 \leq \sum_{j=1}^{n} p_j q_j \leq Mp_1.$$

We have

$$\begin{aligned} mp_1 &\leq p_1 q_1 \\ &\leq \sum_{j=1}^{n} p_j q_j \\ &\leq p_1 \sum_{j=1}^{n} q_j \\ &= p_1 S_n \\ &\leq Mp_1. \end{aligned}$$

Example 3.74 (Second mean value theorem I). Let f be bounded integrable function on $[a,b]_{\mathbb{T}}$,

$$F(t) = \int_a^t f(s)\Delta s, \quad t \in [a,b]_{\mathbb{T}},$$

and

$$\begin{aligned} m_F &= \inf_{t \in [a,b]_{\mathbb{T}}} f(t), \\ M_F &= \sup_{t \in [a,b]_{\mathbb{T}}} f(t). \end{aligned}$$

We will prove that if a function $g : [a,b]_{\mathbb{T}} \to \mathbb{R}$ is nonincreasing and nonnegative on $[a,b]_{\mathbb{T}}$, then there is a constant $A \in [m_F, M_F]$ such that

$$\int_a^b f(t)g(t)\Delta t = Ag(a). \tag{3.32}$$

Take arbitrary $\varepsilon > 0$. Since g is nonincreasing on $[a,b]_{\mathbb{T}}$, by Example 3.49 we get that the function g is integrable on $[a,b]_{\mathbb{T}}$. Because f and g are integrable on $[a,b]_{\mathbb{T}}$, by Exercise 3.23 it follows that the function fg is integrable on $[a,b]_{\mathbb{T}}$. Since f and fg are integrable on $[a,b]_{\mathbb{T}}$, by Example 3.40 it follows that there is a partition $P \in \mathscr{P}(a,b)_{\mathbb{T}}$ given by

$$a = t_0 < t_1 < \cdots < t_n = b$$

such that

$$\sum_{j=1}^{n} (M_j - m_j)(t_j - t_{j-1}) < \varepsilon \tag{3.33}$$

and

$$\left| \sum_{j=1}^{n} f(t_{j-1}) g(t_{j-1})(t_j - t_{j-1}) - \int_a^b f(t) g(t) \Delta t \right| < \varepsilon, \tag{3.34}$$

where

$$m_j = \inf_{t \in [t_{j-1}, t_j]_{\mathbb{T}}} f(t),$$
$$M_j = \sup_{t \in [t_{j-1}, t_j]_{\mathbb{T}}} f(t), \quad j \in \{1, \ldots, n\}.$$

Since $g(t_{j-1}) \geq 0, j \in \{1, \ldots, n\}$, and

$$m_j \leq f(t_{j-1}) \leq M_j, \quad j \in \{1, \ldots, n\},$$

we obtain

$$\begin{aligned} \sum_{j=1}^{n} m_j g(t_{j-1})(t_j - t_{j-1}) &\leq \sum_{j=1}^{n} (fg)(t_{j-1})(t_j - t_{j-1}) \\ &\leq \sum_{j=1}^{n} M_j g(t_{j-1})(t_j - t_{j-1}). \end{aligned} \tag{3.35}$$

By Exercise 3.34 it follows that there are constants $A_j \in [m_j, M_j], j \in \{1, \ldots, n\}$, such that

$$\int_{t_{j-1}}^{t_j} f(t) \Delta t = A_j (t_j - t_{j-1}), \quad j \in \{1, \ldots, n\}.$$

Consider the numbers

$$S_k = \sum_{j=1}^{k} A_j (t_j - t_{j-1}), \quad k \in \{1, \ldots, n\}.$$

We have

$$S_k = \sum_{j=1}^{k} \int_{t_{j-1}}^{t_j} f(t) \Delta t$$

$$
\begin{aligned}
&= \int_a^{t_1} f(t)\Delta t + \int_{t_1}^{t_2} f(t)\Delta t + \cdots + \int_{t_{k-1}}^{t_k} f(t)\Delta t \\
&= \int_a^{t_k} f(t)\Delta t, \quad k \in \{1, \ldots, n\},
\end{aligned}
$$

and

$$
m_F \le S_k \le M_F, \quad k \in \{1, \ldots, n\}.
$$

Set

$$
\begin{aligned}
p_j &= g(t_{j-1}), \\
q_j &= A_j(t_j - t_{j-1}), \quad j \in \{1, \ldots, n\}.
\end{aligned}
$$

Since g is nonincreasing on $[a, b]_{\mathbb{T}}$, we have

$$
p_1 \ge p_2 \ge \cdots \ge p_n \ge 0.
$$

We apply the Abel lemma for p_j, S_j, q_j, $j \in \{1, \ldots, n\}$, and we find

$$
m_F g(a) \le \sum_{j=1}^{n} g(t_{j-1}) A_j (t_j - t_{j-1}) \le M_F g(a). \tag{3.36}
$$

On the other hand,

$$
\begin{aligned}
\sum_{j=1}^{n} m_j g(t_{j-1})(t_j - t_{j-1}) &\le \sum_{j=1}^{n} g(t_{j-1}) A_j (t_j - t_{j-1}) \\
&\le \sum_{j=1}^{n} M_j g(t_{j-1})(t_j - t_{j-1}).
\end{aligned}
$$

By the last inequality, (3.35), and (3.36) we obtain

$$
\begin{aligned}
\left| \sum_{j=1}^{n} g(t_{j-1})(f(t_{j-1}) - A_j)(t_j - t_{j-1}) \right| &\le \sum_{j=1}^{n} g(t_{j-1}) |f(t_{j-1}) - A_j| (t_j - t_{j-1})| \\
&\le \sum_{j=1}^{n} g(t_{j-1})(M_j - m_j)(t_j - t_{j-1}) \\
&\le g(a) \sum_{j=1}^{n} (M_j - m_j)(t_j - t_{j-1}) \\
&\le g(a)\varepsilon.
\end{aligned}
$$

Hence, applying (3.34), we get

$$\begin{aligned}
&\left|\int_a^b f(t)g(t)\Delta t - \sum_{j=1}^n g(t_{j-1})A_j(t_j - t_{j-1})\right| \\
&\quad = \left|\int_a^b f(t)g(t)\Delta t - \sum_{j=1}^n f(t_{j-1})g(t_{j-1})(t_j - t_{j-1}) + \sum_{j=1}^n g(t_{j-1})(f(t_{j-1}) - A_j)(t_j - t_{j-1})\right| \\
&\quad \le \left|\int_a^b f(t)g(t)\Delta t - \sum_{j=1}^n f(t_{j-1})g(t_{j-1})(t_j - t_{j-1})\right| + \left|\sum_{j=1}^n g(t_{j-1})(f(t_{j-1}) - A_j)(t_j - t_{j-1})\right| \\
&\quad \le \varepsilon + g(a)\varepsilon,
\end{aligned}$$

whereupon

$$-\varepsilon - g(a)\varepsilon \le \int_a^b f(t)g(t)\Delta t - \sum_{j=1}^n g(t_{j-1})A_j(t_j - t_{j-1}) \le \varepsilon + g(a)\varepsilon,$$

or

$$\begin{aligned}
-\varepsilon - g(a)\varepsilon + \sum_{j=1}^n g(t_{j-1})A_j(t_j - t_{j-1}) &\le \int_a^b f(t)g(t)\Delta t \\
&\le \varepsilon + g(a)\varepsilon + \sum_{j=1}^n g(t_{j-1})A_j(t_j - t_{j-1}).
\end{aligned}$$

Now applying inequality (3.36), we get

$$-\varepsilon - g(a)\varepsilon + m_F g(a) \le \int_a^b f(t)g(t)\Delta t \le \varepsilon + g(a)\varepsilon + M_F g(a).$$

Because $\varepsilon > 0$ was arbitrarily chosen, by the last inequality we get

$$m_F g(a) \le \int_a^b f(t)g(t)\Delta t \le M_F g(a).$$

If $g(a) = 0$, then since g is nonincreasing and nonnegative on $[a, b]_{\mathbb{T}}$, we get

$$g(t) = 0, \quad t \in [a, b]_{\mathbb{T}},$$

and then

$$\int_a^b f(t)g(t)\Delta t = 0,$$

so that (3.32) holds. Now let $g(a) \neq 0$. Set

$$A = \frac{\int_a^b f(t)g(t)\Delta t}{g(a)}.$$

Then $A \in [m_F, M_F]$, and (3.32) holds.

Example 3.75 (Second mean value theorem II). Let f be bounded integrable function on $[a,b]_{\mathbb{T}}$, let

$$F(t) = \int_a^t f(s)\Delta s, \quad t \in [a,b]_{\mathbb{T}},$$

and let

$$m_F = \inf_{t\in[a,b]_{\mathbb{T}}} f(t),$$
$$M_F = \sup_{t\in[a,b]_{\mathbb{T}}} f(t).$$

We will prove that if $g : [a,b]_{\mathbb{T}} \to \mathbb{R}$ is a monotone function, then there is a constant $A \in [m_F, M_F]$ such that

$$\int_a^b f(t)g(t)\Delta t = A(g(a) - g(b)) + g(b)\int_a^b f(t)\Delta t. \tag{3.37}$$

Firstly, suppose that g is a nonincreasing function on $[a,b]_{\mathbb{T}}$. Define the function

$$h(t) = g(t) - g(b), \quad t \in [a,b]_{\mathbb{T}}.$$

Then h is nonincreasing on $[a,b]_{\mathbb{T}}$. Now applying Example 3.74, we obtain that there is a constant $A \in [m_F, M_F]$ such that

$$\int_a^b f(t)h(t)\Delta t = Ah(a),$$

or

$$\begin{aligned} A(g(a)-g(b)) &= \int_a^b f(t)(g(t)-g(b))\Delta t \\ &= \int_a^b f(t)g(t)\Delta t - g(b)\int_a^b f(t)\Delta t, \end{aligned}$$

whereupon we get (3.37). If g is a nondecreasing function, then setting $g_1 = -g$, we get the desired result.

Example 3.76 (The Abel lemma). Let nonnegative numbers $p_j, j \in \{1, \ldots, n\}$, be such that

$$0 \le p_1 \le p_2 \le \cdots \le p_n.$$

Suppose that the numbers

$$S_k = \sum_{j=1}^{k} q_j, \quad k \in \{1, \ldots, n\},$$

satisfy the inequalities

$$m \le S_k \le M, \quad k \in \{1, \ldots, n\},$$

where $q_j, j \in \{1, \ldots, n\}$, m, and M are real numbers. We will prove that

$$mp_n \le \sum_{j=1}^{n} p_j q_j \le Mp_n.$$

We have

$$\begin{aligned} mp_n &\le p_n q_n \\ &\le \sum_{j=1}^{n} p_j q_j \\ &\le p_n \sum_{j=1}^{n} q_j \\ &= p_n S_n \\ &\le Mp_n. \end{aligned}$$

Example 3.77 (Second mean value theorem III). Let f be bounded integrable function on $[a, b]_{\mathbb{T}}$, let

$$\Phi(t) = \int_t^b f(s)\Delta s, \quad t \in [a, b]_{\mathbb{T}},$$

and let

$$m_\Phi = \inf_{t\in[a,b]_{\mathbb{T}}} f(t),$$
$$M_\Phi = \sup_{t\in[a,b]_{\mathbb{T}}} f(t).$$

We will prove that if $g : [a,b]_{\mathbb{T}} \to \mathbb{R}$ is a nondecreasing and nonnegative function, then there is a constant $A \in [m_F, M_F]$ such that

$$\int_a^b f(t)g(t)\Delta t = Ag(b). \tag{3.38}$$

Take arbitrary $\varepsilon > 0$. Because g is nonincreasing on $[a,b]_{\mathbb{T}}$, applying Example 3.49, we conclude that g is integrable on $[a,b]_{\mathbb{T}}$. Since f and g are integrable on $[a,b]_{\mathbb{T}}$, applying Exercise 3.23, we get that the function fg is integrable on $[a,b]_{\mathbb{T}}$. Since f and fg are integrable on $[a,b]_{\mathbb{T}}$, applying Example 3.40, we get that there is a partition $P \in \mathscr{P}(a,b)_{\mathbb{T}}$ given by

$$a = t_0 < t_1 < \cdots < t_n = b$$

such that

$$\sum_{j=1}^{n}(M_j - m_j)(t_j - t_{j-1}) < \varepsilon \tag{3.39}$$

and

$$\left|\sum_{j=1}^{n} f(t_{j-1})g(t_{j-1})(t_j - t_{j-1}) - \int_a^b f(t)g(t)\Delta t\right| < \varepsilon, \tag{3.40}$$

where

$$m_j = \inf_{t\in[t_{j-1},t_j]_{\mathbb{T}}} f(t),$$
$$M_j = \sup_{t\in[t_{j-1},t_j]_{\mathbb{T}}} f(t), \quad j \in \{1,\dots,n\}.$$

Since $g(t_{j-1}) \geq 0, j \in \{1,\dots,n\}$, and

$$m_j \leq f(t_{j-1}) \leq M_j, \quad j \in \{1,\dots,n\},$$

we arrive at

$$\begin{aligned}\sum_{j=1}^{n} m_j g(t_{j-1})(t_j - t_{j-1}) &\le \sum_{j=1}^{n} (fg)(t_{j-1})(t_j - t_{j-1}) \\ &\le \sum_{j=1}^{n} M_j g(t_{j-1})(t_j - t_{j-1}).\end{aligned} \tag{3.41}$$

Applying Exercise 3.34, we get that there exist constants $A_j \in [m_j, M_j]$, $j \in \{1, \dots, n\}$, such that

$$\int_{t_{n-j+1}}^{t_{n-j}} f(t)\Delta t = A_j(t_{n-j+1} - t_{n-j}), \quad j \in \{1, \dots, n\}.$$

Define

$$S_{n-k} = \sum_{j=1}^{k} A_j(t_{n-j+1} - t_{n-j}), \quad k \in \{1, \dots, n\}.$$

We have

$$\begin{aligned}S_{n-k} &= \sum_{j=1}^{k} \int_{t_{n-j+1}}^{t_{n-j}} f(t)\Delta t \\ &= \int_{t_{n-1}}^{b} f(t)\Delta t + \int_{t_{n-2}}^{t_{n-1}} f(t)\Delta t + \dots + \int_{t_{n-k}}^{t_{n-k+1}} f(t)\Delta t \\ &= \int_{t_{n-k}}^{b} f(t)\Delta t, \quad k \in \{1, \dots, n\},\end{aligned}$$

and

$$m_\Phi \le S_{n-k} \le M_\Phi, \quad k \in \{1, \dots, n\}.$$

Denote

$$\begin{aligned}p_{n-j+1} &= g(t_{n-j+1}), \\ q_{n-j+1} &= A_j(t_{n-j+1} - t_{n-j}), \quad j \in \{1, \dots, n\}.\end{aligned}$$

Since g is nondecreasing on $[a, b]_{\mathbb{T}}$, we get

$$0 \le p_1 \le p_2 \le \dots \le p_n.$$

Applying the Abel lemma, Example 3.76, for $p_j, S_j, q_j, j \in \{1, \dots, n\}$, we get

$$m_F g(b) \le \sum_{j=1}^{n} g(t_{n-j})A_j(t_{n-j+1} - t_{n-j}) \le M_F g(b). \tag{3.42}$$

On the other hand,

$$\begin{aligned}\sum_{j=1}^{n} m_{n-j+1} g(t_{n-j})(t_{n-j+1} - t_{n-j}) &\le \sum_{j=1}^{n} g(t_{n-j})A_j(t_{n-j+1} - t_{n-j})\\ &\le \sum_{j=1}^{n} M_{n-j+1} g(t_{n-j})(t_{n-j+1} - t_{n-j}).\end{aligned}$$

By the last inequality, (3.41), and (3.42) we obtain

$$\begin{aligned}\left|\sum_{j=1}^{n} g(t_{n-j})(f(t_{n-j}) - A_j)(t_{n-j+1} - t_{n-j})\right| &\le \sum_{j=1}^{n} g(t_{n-j})|f(t_{n-j}) - A_j|(t_{n-j+1} - t_{n-j})|\\ &\le \sum_{j=1}^{n} g(t_{n-j})(M_{n-j+1} - m_{n-j+1})(t_{n-j+1} - t_{n-j})\\ &\le g(b)\sum_{j=1}^{n}(M_{n-j+1} - m_{n-j+1})(t_{n-j+1} - t_{n-j})\\ &\le g(b)\varepsilon.\end{aligned}$$

Hence, applying (3.40), we get

$$\begin{aligned}&\left|\int_a^b f(t)g(t)\Delta t - \sum_{j=1}^{n} g(t_{n-j})A_j(t_{n-j+1} - t_{n-j})\right|\\ &\quad= \left|\int_a^b f(t)g(t)\Delta t - \sum_{j=1}^{n} f(t_{n-j})g(t_{n-j})(t_{n-j+1} - t_{n-j}) + \sum_{j=1}^{n} g(t_{n-j})(f(t_{n-j}) - A_j)(t_{n-j+1} - t_{n-j})\right|\\ &\quad\le \left|\int_a^b f(t)g(t)\Delta t - \sum_{j=1}^{n} f(t_{n-j})g(t_{n-j})(t_{n-j+1} - t_{n-j})\right| + \left|\sum_{j=1}^{n} g(t_{n-j})(f(t_{n-j}) - A_j)(t_{n-j+1} - t_{n-j})\right|\\ &\quad\le \varepsilon + g(b)\varepsilon,\end{aligned}$$

whereupon

$$-\varepsilon - g(b)\varepsilon \le \int_a^b f(t)g(t)\Delta t - \sum_{j=1}^{n} g(t_{n-j})A_j(t_{n-j+1} - t_{n-j}) \le \varepsilon + g(b)\varepsilon,$$

or

$$-\varepsilon - g(b)\varepsilon + \sum_{j=1}^{n} g(t_{n-j})A_j(t_{n-j+1} - t_{n-j}) \le \int_a^b f(t)g(t)\Delta t$$

$$\le \varepsilon + g(b)\varepsilon + \sum_{j=1}^{n} g(t_{n-j})A_j(t_{n-j+1} - t_{n-j}).$$

Now applying inequality (3.42), we get

$$-\varepsilon - g(b)\varepsilon + m_F g(b) \le \int_a^b f(t)g(t)\Delta t \le \varepsilon + g(b)\varepsilon + M_F g(b).$$

Because $\varepsilon > 0$ was arbitrarily chosen, by the last inequality we get

$$m_F g(b) \le \int_a^b f(t)g(t)\Delta t \le M_F g(b).$$

If $g(b) = 0$, then since g is nondecreasing and nonnegative on $[a, b]_{\mathbb{T}}$, we get

$$g(t) = 0, \quad t \in [a, b]_{\mathbb{T}},$$

and therefore

$$\int_a^b f(t)g(t)\Delta t = 0,$$

and (3.38) holds. Let $g(b) \ne 0$. Set

$$A = \frac{\int_a^b f(t)g(t)\Delta t}{g(b)}.$$

Then $A \in [m_F, M_F]$, and thus (3.38) holds. This completes the proof.

Example 3.78 (Second mean value theorem IV). Let f be bounded integrable function on $[a, b]_{\mathbb{T}}$, let

$$\Phi(t) = \int_t^b f(s)\Delta s, \quad t \in [a, b]_{\mathbb{T}},$$

and let

$$m_\Phi = \inf_{t \in [a,b]_{\mathbb{T}}} f(t),$$
$$M_\Phi = \sup_{t \in [a,b]_{\mathbb{T}}} f(t).$$

We will prove that if $g : [a, b]_{\mathbb{T}} \to \mathbb{R}$ is a monotone function on $[a, b]_{\mathbb{T}}$, then there is a constant $A \in [m_\Phi, M_\Phi]$ such that

$$\int_a^b f(t)g(t)\Delta t = A(g(b) - g(a)) + g(a)\int_a^b f(t)\Delta t. \tag{3.43}$$

Firstly, suppose that g is a nondecreasing function on $[a,b]_{\mathbb{T}}$. Define the function

$$h(t) = g(t) - g(a), \quad t \in [a,b]_{\mathbb{T}}.$$

Then h is nondecreasing on $[a,b]_{\mathbb{T}}$. Now applying Example 3.77, we obtain that there is a constant $A \in [m_\Phi, M_\Phi]$ such that

$$\int_a^b f(t)h(t)\Delta t = Ah(b)$$

or

$$\begin{aligned} A(g(b) - g(a)) &= \int_a^b f(t)(g(t) - g(a))\Delta t \\ &= \int_a^b f(t)g(t)\Delta t - g(a)\int_a^b f(t)\Delta t, \end{aligned}$$

whereupon we get (3.43). If g is any nonincreasing function, then setting $g_1 = -g$, we get the desired result.

3.7 Improper delta Integrals of the first kind

Assume that $\mathbb{T}$ is unbounded above.

Definition 3.16. Suppose that a real-valued function f is defined on $[a,\infty)$ and is integrable from a to any point $A \in \mathbb{T}, A \geq a$. If the integral

$$F(A) = \int_a^A f(t)\Delta t$$

has a finite limit as $A \to \infty$, then we call this limit the improper integral of the first kind from a to ∞ and write

$$\int_a^\infty f(t)\Delta t = \lim_{A\to\infty}\left\{\int_a^A f(t)\Delta t\right\}. \tag{3.44}$$

In such a case, we say that the improper integral

$$\int_a^\infty f(t)\Delta t \tag{3.45}$$

exists or that it converges. If limit (3.44) does not exist, then we say that the improper integral (3.45) does not exist or that it diverges.

Example 3.79. Let $\mathbb{T} = \mathbb{Z}$. We consider

$$I = \int_1^\infty \frac{3t^2 + 3t + 1}{t^3(t+1)^3}\Delta t.$$

We have $\sigma(t) = t + 1$ and

$$\begin{aligned}
\left(\frac{1}{t^3}\right)^\Delta &= -\frac{(t^3)^\Delta}{t^3(\sigma(t))^3} \\
&= -\frac{(\sigma(t))^2 + t\sigma(t) + t^2}{t^3(t+1)^3} \\
&= -\frac{(t+1)^2 + t(t+1) + t^2}{t^3(t+1)^3} \\
&= -\frac{t^2 + 2t + 1 + t^2 + t + t^2}{t^3(t+1)^3} \\
&= -\frac{3t^2 + 3t + 1}{t^3(t+1)^3}, \quad t \in \mathbb{T}.
\end{aligned}$$

Therefore

$$\begin{aligned}
\int_a^\infty \frac{3t^2 + 3t + 1}{t^3(t+1)^3}\Delta t &= \lim_{A\to\infty} \int_1^A \frac{3t^2 + 3t + 1}{t^3(t+1)^3}\Delta t \\
&= -\lim_{A\to\infty} \int_1^A \left(\frac{1}{t^3}\right)^\Delta \Delta t \\
&= -\lim_{A\to\infty} \frac{1}{t^3}\bigg|_{t=1}^{t=A} \\
&= -\lim_{A\to\infty} \left(\frac{1}{A^3} - 1\right) \\
&= 1.
\end{aligned}$$

Example 3.80. Let $\mathbb{T} = 2^{\mathbb{N}_0}$. We consider

$$I = \int_1^\infty \frac{1}{t^3}\Delta t.$$

Here $\sigma(t) = 2t$, and

$$\begin{aligned}\left(\frac{1}{t^2}\right)^{\Delta} &= -\frac{(t^2)^{\Delta}}{t^2(\sigma(t))^2}\\ &= -\frac{\sigma(t)+t}{4t^4}\\ &= -\frac{2t+t}{4t^4}\\ &= -\frac{3}{4t^3},\end{aligned}$$

whereupon

$$\frac{1}{t^3} = -\frac{4}{3}\left(\frac{1}{t^2}\right)^{\Delta}, \quad t \in \mathbb{T}.$$

Therefore

$$\begin{aligned}I &= \lim_{A\to\infty}\int_1^A \frac{1}{t^2}\Delta t\\ &= -\frac{4}{3}\lim_{A\to\infty}\int_1^A\left(\frac{1}{t^2}\right)^{\Delta}\Delta t\\ &= -\frac{4}{3}\lim_{A\to\infty}\frac{1}{t^2}\Big|_{t=1}^{t=A}\\ &= -\frac{4}{3}\lim_{A\to\infty}\left(\frac{1}{A^2}-1\right)\\ &= \frac{4}{3}.\end{aligned}$$

Example 3.81. Let $\mathbb{T} = 3^{\mathbb{N}_0}$. We consider

$$I = \int_1^{\infty}\frac{1}{\sqrt{t}}\Delta t.$$

Here $\sigma(t) = 3t$, and

$$\begin{aligned}(\sqrt{t})^{\Delta} &= \frac{\sqrt{\sigma(t)}-\sqrt{t}}{\sigma(t)-t}\\ &= \frac{\sqrt{\sigma(t)}-\sqrt{t}}{(\sqrt{\sigma(t)}-\sqrt{t})(\sqrt{\sigma(t)}+\sqrt{t})}\\ &= \frac{1}{\sqrt{\sigma(t)}+\sqrt{t}}\end{aligned}$$

$$= \frac{1}{(1+\sqrt{3})\sqrt{t}},$$

whereupon

$$\frac{1}{\sqrt{t}} = (1+\sqrt{3})(\sqrt{t})^{\Delta}, \quad t \in \mathbb{T}.$$

Therefore

$$\begin{aligned} I &= \lim_{A\to\infty} \int_1^A \frac{1}{\sqrt{t}}\Delta t \\ &= (1+\sqrt{3}) \lim_{A\to\infty} \int_1^A (\sqrt{t})^{\Delta}\Delta t \\ &= (1+\sqrt{3}) \lim_{A\to\infty} \sqrt{t}\Big|_{t=1}^{t=A} \\ &= (1+\sqrt{3}) \lim_{A\to\infty} (\sqrt{A}-1) \\ &= +\infty. \end{aligned}$$

Consequently, this improper integral diverges to $+\infty$.

Example 3.82. Let $\mathbb{T} = \{-\frac{1}{n} : n \in \mathbb{N}\} \cup \mathbb{N}_0$, and let

$$f(t) = \begin{cases} \frac{6t^2+9t-7}{(1+7t)(1+6t)} & \text{if } t \in \{-\frac{1}{n} : n \in \mathbb{N}\}, \\ \frac{7t^2+9t-6}{(1+7t)(8+7t)} & \text{if } t \in \mathbb{N}_0. \end{cases}$$

We will investigate for convergence the integral

$$\int_{-1}^{\infty} f(t)\Delta t.$$

Let

$$g(t) = \frac{1+t^2}{1+7t}, \quad t \in \mathbb{T}.$$

By Example 2.20 we have

$$g^{\Delta}(t) = f(t), \quad t \in \mathbb{T}.$$

Then

$$\int_{-1}^{\infty} f(t)\Delta t = \lim_{A\to\infty} \int_{-1}^{A} f(t)\Delta t$$

$$
\begin{aligned}
&= \lim_{A\to\infty} \frac{1+t^2}{1+7t}\Bigg|_{t=-1}^{t=A} \\
&= \lim_{A\to\infty} \left(\frac{1+A^2}{1+7A} + \frac{1}{3}\right) \\
&= \infty.
\end{aligned}
$$

Thus the considered integral diverges.

Example 3.83. Let $\mathbb{T} = \{(\frac{1}{2})^{2^n} : n \in \mathbb{N}_0\} \cup \{0, 1\}$. Consider the integral

$$I = -\int_0^1 \frac{\sqrt{t}+t}{t^2} \Delta t.$$

Let

$$
\begin{aligned}
f(t) &= \frac{1}{t^2}, \\
g(t) &= t^2, \quad t \in \mathbb{T}.
\end{aligned}
$$

We have

$$\sigma(t) = \sqrt{t} \quad \text{if } t \in \left\{\left(\frac{1}{2}\right)^{2^n} : n \in \mathbb{N}\right\}$$

and

$$\sigma\left(\frac{1}{2}\right) = 1.$$

Then

$$
\begin{aligned}
g^{\Delta}(t) &= \sigma(t) + t \\
&= \sqrt{t} + t, \quad t \in \left\{\left(\frac{1}{2}\right)^{2^n} : n \in \mathbb{N}\right\},
\end{aligned}
$$

and

$$
\begin{aligned}
f^{\Delta}(t) &= -\frac{g^{\Delta}(t)}{g(t)g(\sigma(t))} \\
&= -\frac{\sqrt{t}+t}{t^3}, \quad t \in \left\{\left(\frac{1}{2}\right)^{2^n} : n \in \mathbb{N}\right\}.
\end{aligned}
$$

Hence

$$I = -\int_0^1 \frac{\sqrt{t}+t}{t^3} \Delta t$$

$$
\begin{aligned}
&= -\int_0^{\frac{1}{2}} \frac{\sqrt{t}+1}{t^3}\Delta t - \int_{\frac{1}{2}}^{1} \frac{\sqrt{t}+t}{t^3}\Delta t \\
&= -\sum_{n=1}^{\infty} \frac{\sqrt{(\frac{1}{2})^{2^n}}+1}{((\frac{1}{2})^{2^n})^3} - \frac{1}{2}\frac{\sqrt{\frac{1}{2}}+\frac{1}{2}}{\frac{1}{8}} \\
&= -\sum_{n=1}^{\infty} \frac{(\frac{1}{2})^{2^{n-1}}}{(\frac{1}{2})^{3\cdot 2^n}} - \sum_{n=1}^{\infty} \frac{1}{\left(\frac{1}{2}\right)^{3\cdot 2^n}} - 4\left(\frac{\sqrt{2}}{2}+\frac{1}{2}\right) \\
&= -\sum_{n=1}^{\infty} \frac{1}{(\frac{1}{2})^{5\cdot 2^{n-1}}} - \sum_{n=1}^{\infty} 2^{3\cdot 2^n} - 2(1+\sqrt{2}) \\
&= -\sum_{n=1}^{\infty} 2^{5\cdot 2^{n-1}} - \sum_{n=1}^{\infty} 2^{3\cdot 2^n} - 2(1+\sqrt{2}).
\end{aligned}
$$

Let

$$
a_n = 2^{5\cdot 2^{n-1}}, \quad n \in \mathbb{N}.
$$

Then

$$
\begin{aligned}
\frac{a_{n+1}}{a_n} &= \frac{2^{5\cdot 2^n}}{2^{5\cdot 2^{n-1}}} \\
&= 2^{5\cdot 2^{n-1}} \\
&> 1, \quad n \in \mathbb{N}.
\end{aligned}
$$

Thus the series

$$
\sum_{n=1}^{\infty} 2^{5\cdot 2^{n-1}}
$$

diverges, and the considered integral also diverges.

Exercise 3.35. Let $\mathbb{T} = 2\mathbb{Z}$. Investigate for convergence and divergence the following integrals:
1. $\int_1^{\infty} \frac{2t+1}{(t^2+1)(t^2+4t+5)}\Delta t$,
2. $\int_2^{\infty} \frac{1}{4t^2+12t+5}\Delta t$,
3. $\int_0^{\infty} \frac{1}{t^2+10t+24}\Delta t$.

The Cauchy criterion for the existence of an improper delta integral of the first kind reads as follows.

Example 3.84 (The Cauchy criterion). For the existence of integral (3.44), it is necessary and sufficient that for each $\varepsilon > 0$, there exists $A_0 > a$ such that

$$\left| \int_{A_1}^{A_2} f(t)\Delta t \right| < \varepsilon \tag{3.46}$$

for all $A_1, A_2 \in \mathbb{T}$ such that $A_1 > A_0$ and $A_2 > A_0$. To prove this, observe that the convergence of integral (3.44) is equivalent to the existence of the limit $\lim_{A\to\infty} F(A)$. Using the Cauchy criterion for the existence of the limit of a function, it follows that the existence of integral (3.44) is equivalent to condition (3.46).

Example 3.85. Let $\mathbb{T} = 3\mathbb{Z}$. We will prove that the integral

$$\int_1^\infty \frac{1}{t^2 + 11t + 28} \Delta t$$

converges.

Note that $\sigma(t) = t + 3$ and

$$\begin{aligned} \left(\frac{1}{t+4} \right)^\Delta &= -\frac{(t+4)^\Delta}{(t+4)(\sigma(t)+4)} \\ &= -\frac{1}{(t+4)(t+7)} \\ &= -\frac{1}{t^2 + 11t + 28}, \quad t \in \mathbb{T}. \end{aligned}$$

Take arbitrary $\varepsilon > 0$ and

$$A > \max\left\{ \frac{2 - 4\varepsilon}{\varepsilon}, 1 \right\}.$$

Then for all $A_1, A_2 > A$, we have

$$\begin{aligned} \left| \int_{A_1}^{A_2} \frac{1}{t^2 + 11t + 28} \Delta t \right| &= \left| -\int_{A_1}^{A_2} \left(\frac{1}{t+4} \right)^\Delta \Delta t \right| \\ &= \left| \frac{1}{t+4} \Big|_{t=A_1}^{t=A_2} \right| \\ &= \left| \frac{1}{A_2 + 4} - \frac{1}{A_1 + 4} \right| \\ &\le \frac{1}{A_2 + 4} + \frac{1}{A_1 + 4} \\ &< \frac{2}{A + 4} \\ &< \varepsilon. \end{aligned}$$

Hence by the Cauchy criterion we conclude that the considered integral converges.

Example 3.86. Let $\mathbb{T} = 2^{\mathbb{N}_0}$. We will investigate for convergence the integral

$$\int_1^\infty \frac{3t+7}{2(t+2)(t+3)(t+4)(2t+3)} \Delta t.$$

Here $\sigma(t) = 2t$, and

$$\begin{aligned}\left(\frac{1}{(t+3)(t+4)}\right)^\Delta &= -\frac{((t+3)(t+4))^\Delta}{(t+3)(t+4)(\sigma(t)+3)(\sigma(t)+4)}\\ &= -\frac{(t+3)^\Delta(t+4)+(\sigma(t)+3)(t+4)^\Delta}{2(t+2)(t+3)(t+4)(2t+3)}\\ &= -\frac{t+4+2t+3}{2(t+2)(t+3)(t+4)(2t+3)}\\ &= -\frac{3t+7}{2(t+2)(t+3)(t+4)(2t+3)}, \qquad t \in \mathbb{T}.\end{aligned}$$

Take arbitrary $\varepsilon > 0$ and $A > 0$ such that

$$A^2 + 7A + 12 - \frac{2}{\varepsilon} > 0.$$

Then for all $A_1, A_2 > A$, we have

$$\begin{aligned}\left|\int_{A_1}^{A_2} \frac{3t+7}{2(t+2)(t+3)(t+4)(2t+3)} \Delta t\right| &= \left|-\int_{A_1}^{A_2}\left(\frac{1}{(t+3)(t+4)}\right)^\Delta \Delta t\right|\\ &= \left|-\frac{1}{(t+3)(t+4)}\Big|_{t=A_1}^{t=A_2}\right|\\ &= \left|-\frac{1}{(A_2+3)(A_2+4)} + \frac{1}{(A_1+3)(A_1+4)}\right|\\ &\le \frac{1}{(A_2+3)(A_2+4)} + \frac{1}{(A_1+3)(A_1+4)}\\ &< \frac{2}{(A+3)(A+4)}\\ &< \varepsilon.\end{aligned}$$

Hence by the Cauchy criterion it follows that the considered integral converges.

Example 3.87. Let $\mathbb{T} = 2^{\mathbb{N}_0}$. We will investigate for convergence the integral

$$\int_1^\infty \frac{\sin t}{2t^2} \Delta t.$$

Here $\sigma(t) = 2t$, and

$$
\begin{aligned}
\left(\frac{\sin t}{t}\right)^{\Delta} &= \frac{t(\sin t)^{\Delta} - \sin(\sigma(t))t^{\Delta}}{t\sigma(t)} \\
&= \frac{t\frac{\sin(2t)-\sin t}{2t-t} - \sin(2t)}{2t^2} \\
&= -\frac{\sin t}{2t^2}, \qquad t \in \mathbb{T}.
\end{aligned}
$$

Take arbitrary $\varepsilon > 0$ and $A > \frac{2}{\varepsilon}$. Then for all $A_1, A_2 > A$, we have

$$
\begin{aligned}
\left|\int_{A_1}^{A_1} \frac{\sin t}{2t^2}\Delta t\right| &= \left|-\int_{A_1}^{A_2}\left(\frac{\sin t}{t}\right)^{\Delta}\Delta t\right| \\
&= \left|-\frac{\sin t}{t}\Big|_{t=A_1}^{t=A_2}\right| \\
&= \left|-\frac{\sin A_2}{A_2} + \frac{\sin A_1}{A_1}\right| \\
&\le \frac{|\sin A_2|}{A_2} + \frac{|\sin A_1|}{A_1} \\
&\le \frac{1}{A_2} + \frac{1}{A_1} \\
&< \frac{2}{A} \\
&< \varepsilon.
\end{aligned}
$$

Hence by the Cauchy criterion we conclude that the considered integral converges.

Exercise 3.36. Let $\mathbb{T} = 3\mathbb{Z}$. Using the Cauchy criterion, prove that the integral

$$\int_{1}^{\infty} \frac{1}{(2t+1)(2t+7)}\Delta t$$

converges.

Definition 3.17. An integral of type (3.45) is said to absolutely converge if the integral

$$\int_{a}^{\infty} |f(t)|\Delta t \tag{3.47}$$

of the modulus of the function f converges.

If an integral of type (3.45) converges but does not absolutely converge, then we say that it conditionally converges.

Example 3.88. We will show that if integral (3.45) absolutely converges, then it converges. For this aim, let (3.45) absolutely converge. Then integral (3.47) converges. Take also arbitrary $\varepsilon > 0$. Hence by the Cauchy criterion it follows that there exists $A > a$ such that for all $A_1, A_2 > A$, we have

$$\left|\int_{A_1}^{A_2} |f(t)|\Delta t\right| < \varepsilon.$$

From here, for all $A_1, A_2 > A$, we have

$$\left|\int_{A_1}^{A_2} f(t)\Delta t\right| \le \left|\int_{A_1}^{A_2} |f(t)|\Delta t\right| < \varepsilon,$$

which completes the proof.

Example 3.89. We will show that integral (3.45) with $f(t) \ge 0$ for all $t \ge a$ converges if and only if there exists a constant $M > 0$ such that

$$\int_a^A f(t)\Delta t \le M \quad \text{for } A \ge a.$$

For this aim, we consider the following cases.

1. Let $F(A) = \int_a^A f(t)\Delta t \le M$ for $A \ge a$. Then

$$\int_a^\infty f(t)\Delta t = \lim_{A\to\infty} F(A) \le M.$$

   Therefore integral (3.45) converges.
2. Suppose integral (3.45) converges. Assume that the function $F(A)$, $A \ge a$, is unbounded. Then

$$\int_a^\infty f(t)\Delta t = \lim_{A\to\infty} F(A) = \infty,$$

   a contradiction.

Example 3.90. Let the inequalities $0 \le f(t) \le g(t)$ be satisfied for all $t \in [a,\infty)_{\mathbb{T}}$. We will show that the convergence of the improper integral

$$\int_a^\infty g(t)\Delta t \tag{3.48}$$

implies the convergence of the improper integral (3.45), whereas the divergence of the improper integral (3.45) implies the divergence of the improper integral (3.48). Indeed, since $0 \le f(t) \le g(t)$ for all $t \in [a, \infty)_{\mathbb{T}}$, we get

$$0 \le \int_a^A f(t)\Delta t \le \int_a^A g(t)\Delta t \quad \text{for all } A \in [a, \infty)_{\mathbb{T}},$$

which is the desired result.

Example 3.91. Let $\mathbb{T} = \mathbb{Z}$. Consider the integral

$$I = \int_1^\infty \log \frac{t+1}{t}(t^6 + 7t^3 + 100)\Delta t.$$

Here $\sigma(t) = t + 1$,

$$\log \frac{t+1}{t}(t^6 + 7t^3 + 100) \ge \log \frac{t+1}{t} \quad \text{for all } t \in [1, \infty)_{\mathbb{T}},$$

and

$$I \ge \int_1^\infty \log \frac{t+1}{t}\Delta t.$$

Note that

$$\begin{aligned}(\log t)^\Delta &= \frac{\log \sigma(t) - \log t}{\sigma(t) - t}\\ &= \log(t+1) - \log t\\ &= \log \frac{t+1}{t}, \quad t \in \mathbb{T}.\end{aligned}$$

Therefore

$$\begin{aligned}\int_1^\infty \log \frac{t+1}{t}\Delta t &= \lim_{A\to\infty} \int_1^A (\log t)^\Delta \Delta t\\ &= \lim_{A\to\infty} \log t\Big|_{t=1}^{t=A}\\ &= \infty.\end{aligned}$$

Hence the improper integral I diverges.

Example 3.92. Let $\mathbb{T} = 2^{\mathbb{N}_0}$. Consider the integral

$$I = \int_1^\infty \frac{1}{t^3(t^2+5)(t^2+7t+1)}\Delta t.$$

Here $\sigma(t) = 2t$, and

$$\frac{1}{t^3(t^2+5)(t^2+7t+1)} \le \frac{1}{t^3} \quad \text{for all } t \in [1,\infty)_{\mathbb{T}}.$$

Also,

$$\begin{aligned}\left(\frac{1}{t^2}\right)^\Delta &= -\frac{(t^2)^\Delta}{t^2(\sigma(t))^2}\\ &= -\frac{\sigma(t)+t}{4t^4}\\ &= -\frac{3}{4t^3}, \quad t \in \mathbb{T},\end{aligned}$$

whereupon

$$\frac{1}{t^3} = -\frac{4}{3}\left(\frac{1}{t^2}\right)^\Delta, \quad t \in \mathbb{T}.$$

Hence

$$\begin{aligned}\int_1^\infty \frac{1}{t^3}\Delta t &= \lim_{A\to\infty}\int_1^A \frac{1}{t^3}\Delta t\\ &= -\frac{4}{3}\lim_{A\to\infty}\int_1^A \left(\frac{1}{t^2}\right)^\Delta \Delta t\\ &= -\frac{4}{3}\lim_{A\to\infty}\frac{1}{t^2}\Big|_{t=1}^{A}\\ &= \frac{4}{3}.\end{aligned}$$

Therefore the integral I converges.

Example 3.93. Let $\mathbb{T} = \{-\frac{1}{n} : n \in \mathbb{N}\} \cup \mathbb{N}_0$, and let

$$g(t) = \begin{cases} \frac{6t^2+9t+7}{(1+7t)(1+6t)} & \text{if } t \in \{-\frac{1}{n} : n \in \mathbb{N}\},\\ \frac{7t^2+9t+6}{(1+7t)(8+7t)} & \text{if } t \in \mathbb{N}_0.\end{cases}$$

We have that

$$g(t) \geq \begin{cases} \frac{6t^2+9t-7}{(1+7t)(1+6t)} & \text{if } t \in \{-\frac{1}{n} : n \in \mathbb{N}\}, \\ \frac{7t^2+9t-6}{(1+7t)(8+7t)} & \text{if } t \in \mathbb{N}_0 \end{cases}$$
$$= f(t), \quad t \in \mathbb{T},$$

where we have used the notation in Example 3.82. By Example 3.82 we have that

$$\int_{-1}^{\infty} f(t)\Delta t$$

diverges. Therefore

$$\int_{-1}^{\infty} g(t)\Delta t$$

diverges.

Example 3.94. Let $\mathbb{T} = \{(\frac{1}{2})^{2^n} : n \in \mathbb{N}_0\} \cup \{0, 1\}$. Consider the integral

$$-2\int_0^1 \frac{\Delta t}{t^{\frac{5}{2}}}.$$

We have

$$-2\int_0^1 \frac{\Delta t}{t^{\frac{5}{2}}} = -2\int_0^1 \frac{\sqrt{t}}{t^3}\Delta t$$
$$\leq -\int_0^1 \frac{\sqrt{t}+t}{t^3}\Delta t.$$

By Example 3.83 we have that

$$-\int_0^1 \frac{\sqrt{t}+t}{t^3}\Delta t = -\infty.$$

Therefore

$$-2\int_0^1 \frac{\Delta t}{t^{\frac{5}{2}}} = -\infty.$$

Exercise 3.37. Investigate for convergence and divergence the following integrals:
1. $\int_1^t \log \frac{t+1}{t}(t^7 + 11t^6 + 12t^5 + 13t + 4)\Delta t$, $\mathbb{T} = \mathbb{Z}$,
2. $\int_1^\infty \log \frac{t+3}{t}(t^4 + 11t^2 + 100)\Delta t$, $\mathbb{T} = 3\mathbb{Z}$,
3. $\int_1^\infty \frac{1}{t^2}(t^4 + 5t^3 + 5t^2 + 5t + 5)\Delta t$, $\mathbb{T} = 2^{\mathbb{N}_0}$,
4. $\int_1^\infty \frac{1}{t^2(t^3+11t^2+12t+13)(t+1)}\Delta t$, $\mathbb{T} = 3^{\mathbb{N}_0}$,
5. $\int_1^\infty \frac{1}{(t+1)^2(t+3)^3(t^4+10t+10)}\Delta t$, $\mathbb{T} = \mathbb{Z}$,
6. $\int_1^\infty t^2(t^6 + 5t + 1)\Delta t$, $\mathbb{T} = 3^{\mathbb{N}_0}$.

Example 3.95. Let $|f(t)| \le g(t)$ for all $t \in \mathbb{T}$ such that $t \ge a$. We will show that the convergence of the integral $\int_a^\infty g(t)\Delta t$ implies the convergence of the integral $\int_a^\infty f(t)\Delta t$.

Indeed, since $\int_a^\infty g(t)\Delta t$ converges, using Example 3.90, we have that the integral $\int_a^\infty |f(t)|\Delta t$ converges. Therefore the integral $\int_a^\infty f(t)\Delta t$ absolutely converges. From here and from (3.46) it follows that the integral $\int_a^\infty f(t)\Delta t$ converges.

Example 3.96 (Comparison criterion). Let $\int_a^\infty f(t)\Delta t$ and $\int_a^\infty g(t)\Delta t$ be improper integrals of the first kind with positive integrands, and suppose that the finite limit

$$\lim_{t\to\infty} \frac{f(t)}{g(t)} = L \tag{3.49}$$

exists and is not zero. We will show that the integrals simultaneously converge or diverge.

For this aim, take arbitrary $\varepsilon \in (0, L)$. From (3.49) it follows that there exists $A_0 > a$ such that

$$L - \varepsilon \le \frac{f(t)}{g(t)} \le L + \varepsilon \quad \text{for all } t \ge A_0,$$

from which

$$(L - \varepsilon)g(t) \le f(t) \le (L + \varepsilon)g(t) \quad \text{for all } t \ge A_0.$$

Hence

$$(L - \varepsilon)\int_{A_0}^\infty g(t)\Delta t \le \int_{A_0}^\infty f(t)\Delta t \le (L + \varepsilon)\int_{A_0}^\infty g(t)\Delta t. \tag{3.50}$$

1. Suppose $\int_a^\infty g(t)\Delta t$ converges. Then $\int_{A_0}^\infty g(t)\Delta t$ converges. Hence

$$(L + \varepsilon)\int_{A_0}^\infty g(t)\Delta t$$

converges. From here and from Example 3.90, using (3.50), we obtain that $\int_{A_0}^{\infty} f(t)\Delta t$ converges. Therefore $\int_a^{\infty} f(t)\Delta t$ converges.

2. Suppose $\int_a^{\infty} f(t)\Delta t$ converges. Then $\int_{A_0}^{\infty} f(t)\Delta t$ converges. From here and from Example 3.90, using (3.50), we obtain that $(L-\varepsilon)\int_{A_0}^{\infty} g(t)\Delta t$ converges. Therefore $\int_{A_0}^{\infty} g(t)\Delta t$ converges, and $\int_a^{\infty} g(t)\Delta t$ converges.
3. Suppose $\int_a^{\infty} f(t)\Delta t$ diverges. Hence $\int_{A_0}^{\infty} f(t)\Delta t$ diverges. Hence by (3.50) it follows that $\int_{A_0}^{\infty} g(t)\Delta t$ diverges. Therefore $\int_a^{\infty} g(t)\Delta t$ diverges.
4. Suppose $\int_a^{\infty} g(t)\Delta t$ diverges. Hence $\int_{A_0}^{\infty} g(t)\Delta t$ diverges. Hence by (3.50) it follows that $\int_{A_0}^{\infty} f(t)\Delta t$ diverges. Therefore $\int_a^{\infty} f(t)\Delta t$ diverges.

Example 3.97. Let $\mathbb{T} = \mathbb{Z}$. Consider the integral

$$I = \int_1^{\infty} \frac{t^4}{(t^2+11t+30)(t^4+t^3+t^2+1)}\Delta t.$$

Let

$$f(t) = \frac{t^4}{(t^2+11t+30)(t^4+t^3+t^2+1)}, \quad g(t) = \frac{1}{t^2+11t+30}, \quad t \in \mathbb{T}$$

$$J = \int_1^{\infty} \frac{1}{t^2+11t+30}\Delta t.$$

We have

$$\begin{aligned}\lim_{t\to\infty}\frac{f(t)}{g(t)} &= \lim_{t\to\infty}\frac{\frac{t^4}{(t^2+11t+30)(t^4+t^3+t^2+1)}}{\frac{1}{t^2+11t+30}}\\ &= \lim_{t\to\infty}\frac{t^4}{t^4+t^3+t^2+1}\\ &= 1.\end{aligned}$$

Hence by Example 3.96 it follows that the integrals I and J simultaneously converge or diverge.

Note that

$$\begin{aligned}\left(\frac{1}{t+5}\right)^{\Delta} &= -\frac{(t+5)^{\Delta}}{(t+5)(\sigma(t)+5)}\\ &= -\frac{1}{(t+5)(t+6)}\\ &= -\frac{1}{t^2+11t+30}, \quad t \in \mathbb{T}.\end{aligned}$$

Therefore

$$J = -\lim_{A\to\infty}\int_1^A \left(\frac{1}{t+5}\right)^\Delta \Delta t$$

$$= -\lim_{A\to\infty}\frac{1}{t+5}\bigg|_{t=1}^{t=A}$$

$$= \frac{1}{6}.$$

Consequently, the integral I converges.

Example 3.98. Let $\mathbb{T} = 2^{\mathbb{N}_0}$. Consider the integral

$$I = \int_1^\infty (t^2 + 2t + 3)\Delta t.$$

Let

$$f(t) = t^2 + 2t + 3, \quad g(t) = t^2, \quad t \in \mathbb{T}, \quad J = \int_1^\infty g(t)\Delta t.$$

We have that

$$\lim_{t\to\infty}\frac{f(t)}{g(t)} = \lim_{t\to\infty}\frac{t^2 + 2t + 3}{t^2}$$
$$= 1.$$

Hence by Example 3.96 it follows that the integrals I and J simultaneously converge or diverge. Since

$$J = \lim_{A\to\infty}\int_1^A t^2 \Delta t$$

$$= \frac{1}{7}\lim_{A\to\infty}\int_1^A (t^3)^\Delta \Delta t$$

$$= \frac{1}{7}\lim_{A\to\infty} t^3\bigg|_{t=1}^{t=A}$$
$$= \infty,$$

we conclude that I diverges.

Example 3.99. Let $\mathbb{T} = 2^{\mathbb{N}_0}$. Consider the integral

$$I = \int_1^\infty \frac{e^{-2t} - e^{-t}}{t^2}(t+1)\Delta t.$$

We set

$$f(t) = \frac{e^{-2t} - e^{-t}}{t^2}(t+1), \quad g(t) = \frac{e^{-2t} - e^{-t}}{t}, \quad t \in \mathbb{T},$$

$$J = \int_1^\infty \frac{e^{-2t} - e^{-t}}{t} \Delta t.$$

Then

$$\begin{aligned} \lim_{t\to\infty} \frac{f(t)}{g(t)} &= \lim_{t\to\infty} \frac{\frac{e^{-2t}-e^{-t}}{t^2}(t+1)}{\frac{e^{-2t}-e^{-t}}{t}} \\ &= \lim_{t\to\infty} \frac{t+1}{t} \\ &= 1. \end{aligned}$$

Using Example 3.96, we conclude that the integrals I and J simultaneously converge or diverge.

Note that

$$\begin{aligned} (e^{-t})^\Delta &= \frac{e^{-\sigma(t)} - e^{-t}}{\sigma(t) - t} \\ &= \frac{e^{-2t} - e^{-t}}{t}, \quad t \in \mathbb{T}. \end{aligned}$$

Hence

$$\begin{aligned} J &= \lim_{A\to\infty} \int_1^A (e^{-t})^\Delta \Delta t \\ &= \lim_{A\to\infty} e^{-t}\Big|_{t=1}^{t=A} \\ &= -\frac{1}{e}. \end{aligned}$$

Consequently, the integral I converges.

Exercise 3.38. Let $\mathbb{T} = 2^{\mathbb{N}_0}$. Investigate for convergence and divergence of the integral

$$\int_1^\infty \frac{1}{t+1} \operatorname{arctanh} \frac{t}{1-2t^2} \Delta t.$$

Example 3.100. Let f be integrable from a to any point $A \in \mathbb{T}$, $A > a$, and suppose the integral

$$F(A)=\int_a^A f(t)\Delta t$$

is bounded for all $A\geq a$. Let also g be monotonic on $[a,\infty)$ and such that $\lim_{t\to\infty} g(t)=0$. We will prove that the improper integral of the first kind of the form

$$\int_a^\infty f(t)g(t)\Delta t \tag{3.51}$$

converges.

For this aim, let $A_1, A_2\in\mathbb{T}$, $A_2>A_1\geq a$. It follows that there is Λ between $\inf_{A\in[A_1,A_2]} F(A)$ and $\sup_{A\in[A_1,A_2]} F(A)$ such that

$$\int_{A_1}^{A_2} f(t)g(t)\Delta t=(g(A_1)-g(A_2))\Lambda+g(A_2)\int_{A_1}^{A_2} f(t)\Delta t. \tag{3.52}$$

Let $M>0$ be a constant such that

$$|F(A)|\leq M \quad \text{on } [a,\infty).$$

From (3.52) we get

$$\int_{A_1}^{A_2} f(t)g(t)\Delta t=(g(A_1)-g(A_2))\Lambda+g(A_2)(F(A_2)-F(A_1))$$

and

$$\begin{aligned}\left|\int_{A_1}^{A_2} f(t)g(t)\Delta t\right| &= |(g(A_1)-g(A_2))\Lambda+g(A_2)(F(A_2)-F(A_1))|\\ &\leq |g(A_1)||\Lambda|+|g(A_2)||\Lambda|+|g(A_2)|(|F(A_2)|+|F(A_1)|)\\ &\leq M|g(A_1)|+3M|g(A_2)|\\ &= M(|g(A_1)|+3|g(A_2)|).\end{aligned} \tag{3.53}$$

Take arbitrary $\varepsilon>0$. Since $g(t)\to 0$ as $t\to\infty$, there is $A_3>a$ such that

$$|g(A)|<\frac{\varepsilon}{4M} \quad \text{for all } A>A_3.$$

Hence by (3.53), for $A_1, A_2>A_3$, we get

$$\begin{aligned}\left|\int_{A_1}^{A_2} f(t)g(t)\Delta t\right| &< M\left(\frac{\varepsilon}{4M}+\frac{3\varepsilon}{4M}\right)\\ &= \varepsilon.\end{aligned}$$

Thus by the Cauchy criterion it follows that integral (3.51) converges.

Example 3.101. Let $\mathbb{T} = \mathbb{Z}$. Consider the integral

$$I = \int_1^\infty \frac{t\sin t}{(t^2+3t+2)(t^2+1)}\Delta t.$$

Let

$$f(t) = \frac{\sin t}{t^2+3t+2}, \quad g(t) = \frac{t}{t^2+1}, \quad t \in \mathbb{T}.$$

We have

$$\begin{aligned} g^\Delta(t) &= \frac{t^\Delta(t^2+1) - t(t^2+1)^\Delta}{(t^2+1)((\sigma(t))^2+1)}\\ &= \frac{t^2+1-t(\sigma(t)+t)}{(t^2+1)((t+1)^2+1)}\\ &= \frac{t^2+1-t(2t+1)}{(t^2+1)(t^2+2t+2)}\\ &= \frac{t^2+1-2t^2-t}{(t^2+1)(t^2+2t+2)}\\ &= \frac{1-t-t^2}{(t^2+1)(t^2+2t+2)}, \quad t \in \mathbb{T}.\end{aligned}$$

Therefore the function g is monotonic on $[1,\infty)$. Also,

$$\begin{aligned}\lim_{t\to\infty} g(t) &= \lim_{t\to\infty}\frac{t}{t^2+1}\\ &= 0,\\ \left|\int_1^\infty f(t)\Delta t\right| &\le \int_1^\infty |f(t)|\Delta t\\ &= \int_1^\infty \frac{|\sin t|}{(t+1)(t+2)}\Delta t\\ &\le \int_1^\infty \frac{1}{(t+1)(t+2)}\Delta t\end{aligned}$$

$$= -\int_1^\infty \left(\frac{1}{t+1}\right)^\Delta \Delta t$$
$$= -\frac{1}{t+1}\Big|_{t=1}^{t=\infty}$$
$$= \frac{1}{2}.$$

Hence by Example 3.100 it follows that the integral I converges.

Example 3.102. Let $\mathbb{T} = 2^{\mathbb{N}_0}$. Consider the integral

$$I = \int_1^\infty \frac{\sin^2 t + \cos t + 3}{t^2(t^2+1)} \Delta t.$$

Let

$$f(t) = \frac{\sin^2 t + \cos t + 3}{t^2}, \quad g(t) = \frac{1}{t^2+1}, \quad t \in \mathbb{T}.$$

Then

$$\begin{aligned} \left(\frac{1}{t^2+1}\right)^\Delta &= -\frac{(t^2+1)^\Delta}{(t^2+1)((\sigma(t))^2+1)} \\ &= -\frac{\sigma(t)+t}{(t^2+1)(4t^2+1)} \\ &= -\frac{3t}{(t^2+1)(4t^2+1)}, \quad t \in \mathbb{T}. \end{aligned}$$

Therefore the function g is monotonic on $[1, \infty)$. Also,

$$\begin{aligned} \lim_{t\to\infty} g(t) &= \lim_{t\to\infty} \frac{1}{t^2+1} \\ &= 0, \\ \left|\int_1^\infty f(t)\Delta t\right| &\le \int_1^\infty |f(t)|\Delta t \\ &= \int_1^\infty \frac{|\sin^2 t + \cos t + 3|}{t^2} \Delta t \\ &\le \int_1^\infty \frac{\sin^2 t + |\cos t| + 3}{t^2} \Delta t \\ &\le 5\int_1^\infty \frac{1}{t^2} \Delta t \end{aligned}$$

$$
\begin{aligned}
&= -10 \int_1^\infty \left(\frac{1}{t}\right)^\Delta \Delta t \\
&= -10 \frac{1}{t}\Big|_{t=1}^{t=\infty} \\
&= 10.
\end{aligned}
$$

Hence by Example 3.100 it follows that the integral I converges.

Example 3.103. Let $\mathbb{T} = 3^{\mathbb{N}_0}$. Consider the integral

$$
I = \int_1^\infty \frac{1}{t(t^{10} + t^{11} + t^{12} + 1)} \Delta t.
$$

We set

$$
f(t) = \frac{1}{t^{10} + t^{11} + t^{12} + 1}, \quad g(t) = \frac{1}{t}, \quad t \in \mathbb{T}.
$$

Then

$$
\begin{aligned}
g^\Delta(t) &= -\frac{t^\Delta}{t\sigma(t)} \\
&= -\frac{1}{3t^2}, \quad t \in \mathbb{T}.
\end{aligned}
$$

Therefore g is a monotonic function on $[1, \infty)$. Also,

$$
\begin{aligned}
\lim_{t\to\infty} g(t) &= \lim_{t\to\infty} \frac{1}{t} \\
&= 0, \\
\left| \int_1^\infty f(t)\Delta t \right| &\le \int_1^\infty |f(t)| \Delta t \\
&\le \int_1^\infty \frac{1}{t^{10} + t^{11} + t^{12} + 1} \Delta t \\
&\le \int_1^\infty \frac{1}{t^2} \Delta t \\
&= -3 \int_1^\infty \left(\frac{1}{t}\right)^\Delta \Delta t \\
&= -3 \frac{1}{t}\Big|_{t=1}^{t=\infty} \\
&= 3.
\end{aligned}
$$

Hence by Example 3.100 it follows that the integral I converges.

Exercise 3.39. Let $\mathbb{T} = \mathbb{Z}$. Using Theorem 3.100, prove that the integral

$$\int_1^\infty \frac{\sin t - 2\cos t + 10}{(t^2+1)(t^2-3t+5)} \Delta t$$

converges.

Exercise 3.40. Let $\mathbb{T}$ be a time scale of the form

$$\mathbb{T} = \{t_k : k \in \mathbb{N}_0\} \quad \text{with } 0 < t_0 < t_1 < \cdots \text{ and } \lim_{k\to\infty} t_k = \infty. \tag{3.54}$$

Suppose that $f : [t_0, \infty) \to \mathbb{R}$ is nonincreasing with $\int_{t_0}^\infty f(t)\Delta t < \infty$. If $g : \mathbb{T} \to \mathbb{R}_+$ satisfies

$$g(t_k) \le Kf(t_{k+1}) \quad \text{for all } k \in \mathbb{N}_0,$$

where $K > 0$ is a constant, then prove that $\int_{t_0}^\infty g(t)\Delta t < \infty$.

Exercise 3.41. Let $\mathbb{T} = 2^{\mathbb{N}_0}$. Prove that the integral

$$\int_1^\infty \frac{1}{t^p} \Delta t$$

diverges for $p \in [0, 1]$ and converges for $p > 1$.

3.8 Improper integrals of the second kind

Let $\mathbb{T}$ be a time scale, let $a < b$ be fixed points in $\mathbb{T}$, and let b be left-dense. Suppose that the function f is defined in the interval $[a, b)_\mathbb{T}$. Let also f be integrable on any interval $[a, c]_\mathbb{T}$ with $c < b$ and unbounded on $[a, b)_\mathbb{T}$. Then the ordinary Riemann integral of f on $[a, b]_\mathbb{T}$ cannot exist since a Riemann-integrable function from a to b must be bounded on $[a, b)_\mathbb{T}$.

Definition 3.18. The formal expression

$$\int_a^b f(t)\Delta t \tag{3.55}$$

is called an improper integral of the second kind. We say that integral (3.55) is improper at $t = b$. Sometimes, we say that f has singularity at $t = b$. If the left-sided limit

$$\lim_{c\to b-} \int_a^c f(t)\Delta t \tag{3.56}$$

exists and is finite, then we say that the improper integral (3.55) exists or that it converges. In such a case, we call this limit the value of the improper integral (3.55) and write

$$\int_a^b f(t)\Delta t = \lim_{c\to b-}\int_a^c f(t)\Delta t.$$

If limit (3.56) does not exist, we say that integral (3.55) does not exist or that it diverges.

Example 3.104. Let $\mathbb{T} = [0,1] \cup 2^{\mathbb{N}}$, where $[0,1]$ is a real-valued interval. Let also

$$f(t) = \begin{cases} \sqrt{1-t^2} & \text{for } t \in [0,1], \\ t^4 & \text{for } t \in 2^{\mathbb{N}}. \end{cases}$$

Consider the integral

$$I = \int_0^6 \frac{1}{f(t)}\Delta t.$$

We have

$$\begin{aligned} I &= \int_0^1 \frac{1}{f(t)}\Delta t + \int_2^6 \frac{1}{f(t)}\Delta t \\ &= \int_0^1 \frac{1}{\sqrt{1-t^2}}dt + \int_2^6 \frac{1}{t^4}\Delta t \\ &= \lim_{c\to 1-}\int_0^c \frac{1}{\sqrt{1-t^2}}dt + \frac{1}{t^4}\mu(t)\Big|_{t=2} + \frac{1}{t^4}\mu(t)\Big|_{t=4} \\ &= \lim_{c\to 1-} \arcsin t\Big|_{t=0}^{t=c} + \frac{2}{16} + \frac{4}{256} \\ &= \frac{\pi}{2} + \frac{9}{64}. \end{aligned}$$

Therefore the considered integral converges.

Example 3.105. Let $\mathbb{T} = \{-4,-2\} \cup [0,1]$, where $[0,1]$ is a real-valued interval. Consider the integral

$$I = \int_{-4}^1 \frac{\Delta t}{\sqrt{1-t}}.$$

We have

$$\begin{aligned}
I &= \int_{-4}^{-2} \frac{\Delta t}{\sqrt{1-t}} + \int_{0}^{1} \frac{dt}{\sqrt{1-t}} \\
&= \frac{1}{\sqrt{1-t}}\mu(t)\Big|_{t=-4} + \lim_{c\to 1-} \int_{0}^{c} \frac{dt}{\sqrt{1-t}} \\
&= \frac{2}{\sqrt{5}} - 2\lim_{c\to 1-} \sqrt{1-t}\Big|_{t=0}^{t=c} \\
&= \frac{2}{\sqrt{5}} + 2.
\end{aligned}$$

Therefore the considered integral converges.

Example 3.106. Let $\mathbb{T} = \{-1, 0\} \cup [1, 2]$, where $[1, 2]$ is a real-valued interval. Consider the integral

$$I = \int_{-1}^{2} \frac{t^3}{\sqrt{4-t^2}} \Delta t.$$

We have

$$\begin{aligned}
I &= \int_{-1}^{0} \frac{t^3}{\sqrt{4-t^2}} \Delta t + \int_{1}^{2} \frac{t^3}{\sqrt{4-t^2}} dt \\
&= \frac{t^3\mu(t)}{\sqrt{4-t^2}}\Big|_{t=-1} - \lim_{c\to 2-} \int_{1}^{c} t^2 d\sqrt{4-t^2} \\
&= -\frac{1}{\sqrt{3}} - \lim_{c\to 2-} t^2\sqrt{4-t^3}\Big|_{t=1}^{t=c} + 2\lim_{c\to 2-} \int_{1}^{c} t\sqrt{4-t^2}dt \\
&= -\frac{1}{\sqrt{3}} + \sqrt{3} - \lim_{c\to 2-} \int_{1}^{c} \sqrt{4-t^2}d(4-t^2) \\
&= \frac{2\sqrt{3}}{3} - \lim_{c\to 2-} \frac{(4-t^2)^{\frac{3}{2}}}{\frac{3}{2}}\Big|_{t=1}^{t=c} \\
&= \frac{2\sqrt{3}}{3} + 2\sqrt{3} \\
&= \frac{8\sqrt{3}}{3}.
\end{aligned}$$

Therefore the considered integral converges.

Exercise 3.42. Investigate for convergence and divergence the following integrals:
1. $\int_{-3}^{1} \frac{\Delta t}{(1-t)(2t-1)}$, $\mathbb{T} = \{-3, -2, -1, 0\} \cup [\frac{1}{2}, 1]$, where $[\frac{1}{2}, 1]$ is a real-valued interval,
2. $\int_{-3}^{10} \frac{2t}{(t^2-1)^2} \Delta t$, $\mathbb{T} = [-3, 3] \cup \{4, 7, 10\}$, where $[-3, 3]$ is a real-valued interval,
3. $\int_{-3}^{2} \frac{\Delta t}{t\sqrt{3t^2-2t-1}}$, $\mathbb{T} = \{-3, -2, -1, 0\} \cup [1, 2]$, where $[1, 2]$ is a real-valued interval,
4. $\int_{-3}^{3} \frac{\Delta t}{(t-7)\sqrt{t^2-3}}$, $\mathbb{T} = \{-3, -2, -1\} \cup [\sqrt{3}, 3]$, where $[\sqrt{3}, 3]$ is a real-valued interval,
5. $$\int_{-7}^{1} \frac{\Delta t}{\sqrt{t(1-t)}}, \quad \mathbb{T} = \{-7, -4, -1\} \cup [0, 1],$$
   where $[0, 1]$ is a real-valued interval.
6. $$\int_{-\frac{1}{2}}^{1} \frac{\Delta t}{(10-t)\sqrt{1-t^2}},$$
   $$\mathbb{T} = \left\{-\frac{1}{2}, -\frac{1}{4}, 0\right\} \cup \left[\frac{1}{2}, 1\right],$$
   where $[\frac{1}{2}, 1]$ is the real-valued interval.

All results for the improper integrals of the first kind have exact analogues for the improper integrals of the second kind.
1. For the existence of integral (3.55), it is necessary and sufficient that for each $\varepsilon > 0$, there is $b_0 < b$ such that
   $$\left| \int_{c_1}^{c_2} f(t) \Delta t \right| < \varepsilon$$
   for all $c_1, c_2 \in \mathbb{T}$ satisfying the inequalities $b_0 < c_1 < b$ and $b_0 < c_2 < b$.
2. Suppose that $f(t) \geq 0$. Then for all $c \in [a, b]_{\mathbb{T}}$,
   $$F(c) = \int_{a}^{c} f(t) \Delta t$$
   does not decrease as c increases, and integral (3.55) converges if and only if F is bounded, in which case the value of the integral is $\lim_{c \to b-} F(c)$.
3. Let the finite nonzero limit
   $$\lim_{t \to b-} \frac{f(t)}{g(t)} = L$$
   exist. Then the integrals $\int_a^b f(t) \Delta t$ and $\int_a^b g(t) \Delta t$ simultaneously converge or diverge.

Similar definitions are made, and entirely similar results are obtained for integrals of the second kind improper at the lower integration limit.

Example 3.107. Let $\mathbb{T}$ be an arbitrary time scale, let $a, b \in \mathbb{T}$ with $a < b$, and suppose that b is left-dense. Let also $p \geq 1$. We will prove that the integral

$$\int_a^b \frac{\Delta t}{(b-t)^p} \tag{3.57}$$

diverges.

1. Let $p = 1$. Let us choose points $t_n \in \mathbb{T}$ for $n \in \mathbb{N}_0$ such that

$$a = t_0 < t_1 < \cdots < b \quad \text{and} \quad \lim_{n\to\infty} t_n = b. \tag{3.58}$$

We set

$$\tau_n = \frac{1}{b - t_n} \quad \text{for any } n \in \mathbb{N}_0. \tag{3.59}$$

Then $\lim_{n\to\infty} \tau_n = \infty$, $t_n = b - \frac{1}{\tau_n}$, and

$$\begin{aligned} t_{n+1} - t_n &= \frac{1}{\tau_n} - \frac{1}{\tau_{n+1}} \\ &= \frac{\tau_{n+1} - \tau_n}{\tau_n \tau_{n+1}} \quad \text{for all } n \in \mathbb{N}_0. \end{aligned}$$

Hence

$$\begin{aligned} \int_a^b \frac{\Delta t}{b-a} &= \sum_{n=0}^{\infty} \int_{t_n}^{t_{n+1}} \frac{\Delta t}{b-t} \\ &\geq \sum_{n=0}^{\infty} \frac{1}{b-t_n} \int_{t_n}^{t_{n+1}} \Delta t \\ &= \sum_{n=0}^{\infty} \frac{t_{n+1} - t_n}{b - t_n} \\ &= \sum_{n=0}^{\infty} \tau_n \frac{\tau_{n+1} - \tau_n}{\tau_n \tau_{n+1}} \\ &= \sum_{n=0}^{\infty} \frac{\tau_{n+1} - \tau_n}{\tau_{n+1}} \\ &= \infty. \end{aligned}$$

2. Let $p > 1$. There is $d \in [a, b)_{\mathbb{T}}$ such that

$$0 < b - t < 1 \quad \text{for } t \in [d, b)_{\mathbb{T}}.$$

Then

$$(b - t)^p < b - t \quad \text{for } t \in [d, b)_{\mathbb{T}}.$$

Hence

$$\begin{aligned}\int_a^b \frac{\Delta t}{(b-t)^p} &= \int_a^d \frac{\Delta t}{(b-t)^p} + \int_d^b \frac{\Delta t}{(b-t)^p}\\ &> \int_a^d \frac{\Delta t}{(b-t)^p} + \int_d^b \frac{\Delta t}{b-t}\\ &= \infty.\end{aligned}$$

Example 3.108. Let $\mathbb{T}$ be a time scale satisfying (3.58). Let also $p < 1$ and suppose that for some $\alpha \in [1, \frac{1}{p})$,

$$\frac{1}{b - t_{k+1}} = O\left(\frac{1}{(b - t_k)^\alpha}\right) \quad \text{as} \quad k \to \infty.$$

We will prove that the improper integral (3.57) converges.

Let τ_n be defined by (3.59). Then

$$\tau_{k+1} = O(\tau_k^\alpha) \quad \text{as} \quad k \to \infty.$$

Hence $\tau_{k+1} \le K\tau_k^\alpha$ for all $k \in \mathbb{N}_0$, where $K > 0$ is a constant. Then

$$\begin{aligned}\int_a^b \frac{\Delta t}{(b-t)^p} &= \sum_{k=0}^{\infty} \int_{t_k}^{t_{k+1}} \frac{\Delta t}{(b-t)^p}\\ &\le \sum_{k=0}^{\infty} \frac{1}{(b - t_{k+1})^p} \int_{t_k}^{t_{k+1}} \Delta t\\ &= \sum_{k=0}^{\infty} \frac{t_{k+1} - t_k}{(b - t_{k+1})^p}\\ &= \sum_{k=0}^{\infty} \frac{\tau_{k+1} - \tau_k}{\tau_k \tau_{k+1}^{1-p}}\\ &\le K^{\frac{1}{\alpha}} \sum_{k=0}^{\infty} \frac{\tau_{k+1} - \tau_k}{\tau_{k+1}^{\frac{1}{\alpha}+1-p}}\end{aligned}$$

$$= K^{\frac{1}{\alpha}} \sum_{k=0}^{\infty} \int_{\tau_k}^{\tau_{k+1}} \frac{\Delta t}{t^{\frac{1}{\alpha}+1-p}}$$

$$= K^{\frac{1}{\alpha}} \int_{\tau_0}^{\infty} \frac{\Delta t}{t^{\frac{1}{\alpha}+1-p}}$$

$$< \infty.$$

Example 3.109. Let $\mathbb{T}$ be a time scale satisfying (3.58). Consider the integral

$$I = \int_a^b \frac{\Delta t}{(t^4 + t^2 + 1)(b - t)^{\frac{1}{2}}}.$$

We have

$$I \le \int_a^b \frac{\Delta t}{(b - t)^{\frac{1}{2}}} < \infty.$$

Example 3.110. Let $\mathbb{T}$ be a time scale satisfying (3.58). Let also $a = 0$, $b = 2$, and $\frac{1}{2}, 1 \in \mathbb{T}$. Consider the integral

$$I = \int_0^2 \frac{t^{\alpha-1}}{|1 - t|} \Delta t.$$

We have

$$I = \int_0^{\frac{1}{2}} \frac{t^{\alpha-1}}{1 - t} \Delta t + \int_{\frac{1}{2}}^{1} \frac{t^{\alpha-1}}{1 - t} \Delta t + \int_1^2 \frac{t^{\alpha-1}}{t - 1} \Delta t.$$

Since $\int_{\frac{1}{2}}^{1} \frac{t^{\alpha-1}}{1-t} \Delta t$ diverges for all $\alpha \in \mathbb{R}$, we conclude that the integral I diverges.

Example 3.111. Let $\mathbb{T}$ be a time scale satisfying (3.58). Let also, $a = 0$, $b = 1$, and $\frac{1}{2} \in \mathbb{T}$. Consider the integral

$$I = \int_0^1 t^{\alpha-1}(1 - t)^{\beta-1} \Delta t.$$

We have

$$I = \int_0^{\frac{1}{2}} t^{\alpha-1}(1 - t)^{\beta-1} \Delta t + \int_{\frac{1}{2}}^{1} t^{\alpha-1}(1 - t)^{\beta-1} \Delta t$$

$$= I_1 + I_2.$$

Note that I_1 converges for $\alpha > 0$ and diverges for $\alpha \le 0$. Also, I_2 converges for $\beta > 0$ and diverges for $\beta \le 0$.

Therefore I converges for $\alpha > 0$ and $\beta > 0$ and diverges for $\alpha \le 0$ or $\beta \le 0$.

Exercise 3.43. Let $\mathbb{T}$ be a time scale satisfying (3.58). Investigate for convergence and divergence the integral

$$\int_a^b (t-a)^\alpha (b-t)^\beta \Delta t.$$

3.9 Delta monomials

Let $s, t \in \mathbb{T}$. Define the monomials

$$\begin{aligned} g_0(t,s) &= 1, \\ h_0(t,s) &= 1, \\ g_{k+1}(t,s) &= \int_s^t g_k(\sigma(\tau),s)\Delta\tau, \\ h_{k+1}(t,s) &= \int_s^t h_k(\tau,s)\Delta\tau, \quad k = 0,1,2,\dots. \end{aligned}$$

We have

$$\begin{aligned} g_1(t,s) &= \int_s^t g_0(\sigma(\tau),s)\Delta\tau \\ &= \int_s^t \Delta\tau \\ &= t-s, \\ g_2(t,s) &= \int_s^t g_1(\sigma(\tau),s)\Delta\tau \\ &= \int_s^t (\sigma(\tau)-s)\Delta\tau, \\ h_1(t,s) &= \int_s^t h_0(\tau,s)\Delta\tau \end{aligned}$$

$$\begin{aligned} &= \int_s^t \Delta\tau \\ &= t - s, \\ h_2(t,s) &= \int_s^t h_1(\tau,s)\Delta\tau \\ &= \int_s^t (\tau - s)\Delta\tau, \end{aligned}$$

and so on. Also,

$$\begin{aligned} g_k^{\Delta}(t,s) &= g_{k-1}(\sigma(t),s), \\ h_k^{\Delta}(t,s) &= h_{k-1}(t,s), \quad k \in \mathbb{N}. \end{aligned}$$

Example 3.112. We will show that for all $k \in \mathbb{N}_0$,

$$0 \le h_k(t,s) \le \frac{(t-s)^k}{k!}, \quad t \ge s. \tag{3.60}$$

Let

$$g(t) = (t-s)^{k+1}, \quad t,s \in \mathbb{T},\ k \in \mathbb{N}.$$

Then

$$\begin{aligned} g^{\Delta}(t) &= \lim_{y\to t} \frac{g(\sigma(t)) - g(y)}{\sigma(t) - y} \\ &= \lim_{y\to t} \frac{(\sigma(t)-s)^{k+1} - (y-s)^{k+1}}{\sigma(t)-y} \\ &= \lim_{y\to t} \frac{(\sigma(t)-y)\sum_{\nu=0}^{k}(\sigma(t)-s)^{\nu}(y-s)^{k-\nu}}{\sigma(t)-y} \\ &= \lim_{y\to t} \sum_{\nu=0}^{k} (\sigma(t)-s)^{\nu}(y-s)^{k-\nu} \\ &= \sum_{\nu=0}^{k} (\sigma(t)-s)^{\nu}(t-s)^{k-\nu}, \quad t,s \in \mathbb{T},\ k \in \mathbb{N}. \end{aligned}$$

Note that inequalities (3.60) are true for $k = 0$. Assume that inequalities (3.60) are true for some $k \in \mathbb{N}$. We will prove inequalities (3.60) for $k+1$. We have

$$\begin{aligned} 0 &\le h_{k+1}(t,s) \\ &= \int_s^t h_k(\tau.s)\Delta\tau \end{aligned}$$

$$
\begin{aligned}
&\leq \frac{1}{k!}\int_s^t (\tau - s)^k \Delta\tau \\
&= \frac{1}{(k+1)!}\int_s^t \sum_{\nu=0}^{k} (\tau - s)^k \Delta\tau \\
&= \frac{1}{(k+1)!}\int_s^t \sum_{\nu=0}^{k} (\tau - s)^\nu (\tau - s)^{k-\nu} \Delta\tau \\
&\leq \frac{1}{(k+1)!}\int_s^t \sum_{\nu=0}^{k} (\sigma(\tau) - s)^\nu (\tau - s)^{k-\nu} \Delta\tau \\
&= \frac{1}{(k+1)!}\int_s^t g^\Delta(\tau)\Delta\tau \\
&= \frac{1}{(k+1)!} g(\tau)\Big|_{\tau=s}^{\tau=t} \\
&= \frac{1}{(k+1)!}(\tau - s)^{k+1}\Big|_{\tau=s}^{\tau=t} \\
&= \frac{(t-s)^{k+1}}{(k+1)!}, \quad t, s \in \mathbb{T},\ t \geq s.
\end{aligned}
$$

By the principle of mathematical induction it follows that (3.60) is true for all $k \in \mathbb{N}$.

Example 3.113. Let $\mathbb{T} = h\mathbb{Z}$, $h > 0$. We will find

$$h_1(t,s), \quad h_2(t,s), \quad \text{and} \quad h_3(t,s), \quad t, s \in \mathbb{T}.$$

We have

$$h_1(t,s) = t - s, \quad s, t \in \mathbb{T}.$$

Fix $s \in \mathbb{T}$. Let

$$g(t) = \frac{t^2 - ht}{2} - st, \quad t \in \mathbb{T}.$$

Then

$$
\begin{aligned}
g^\Delta(t) &= \frac{\sigma(t) + t - h}{2} - s \\
&= \frac{t + h + t - h}{2} - s \\
&= t - s, \quad t \in \mathbb{T}.
\end{aligned}
$$

Hence

$$
\begin{aligned}
h_2(t,s) &= \int_s^t h_1(\tau,s)\Delta\tau \\
&= \int_s^t (\tau - s)\Delta\tau \\
&= \int_s^t g^{\Delta}(\tau)\Delta\tau \\
&= g(\tau)\Big|_{\tau=s}^{\tau=t} \\
&= \left(\frac{\tau^2 - h\tau}{2} - s\tau\right)\Big|_{\tau=s}^{\tau=t} \\
&= \frac{t^2 - s^2 - h(t-s)}{2} - s(t-s) \\
&= \frac{(t-s)(t+s) - h(t-s)}{2} - s(t-s) \\
&= \frac{(t-s)(t+s-h-2s)}{2} \\
&= \frac{(t-s)(t-s-h)}{2} \\
&= \frac{1}{2}(t^2 - (2s+h)t + s(s+h)), \quad t, s \in \mathbb{T}.
\end{aligned}
$$

Let

$$
f(t) = \frac{1}{2}\left(\frac{1}{3}t^3 - (s+h)t^2 + \left(\frac{h^2}{6} + \frac{(2s+h)h}{2} + s(s+h)\right)t\right), \quad t \in \mathbb{T}.
$$

Then

$$
\begin{aligned}
f^{\Delta}(t) &= \frac{1}{2}\left(\frac{1}{3}((\sigma(t))^2 + t\sigma(t) + t^2) - (s+h)(\sigma(t) + t) + \frac{h^2}{6} + \frac{(2s+h)h}{2} + s(s+h)\right) \\
&= \frac{1}{2}\left(\frac{1}{3}((t+h)^2 + t(t+h) + t^2) - (s+h)(t+h+t) + \frac{h^2}{6} + \frac{(2s+h)h}{2} + s(s+h)\right) \\
&= \frac{1}{2}\left(\frac{1}{3}(t^2 + 2ht + h^2 + t^2 + ht + t^2) - (s+h)(2t+h) + \frac{h^2}{6} + \frac{(2s+h)h}{2} + s(s+h)\right) \\
&= \frac{1}{2}\left(t^2 + ht + \frac{h^2}{3} - 2st - sh - 2ht - h^2 + \frac{h^2}{6} + sh + \frac{h^2}{2} + s^2 + sh\right) \\
&= \frac{1}{2}(t^2 - ht - 2st + s^2 + sh) \\
&= h_2(t,s), \quad t \in \mathbb{T}.
\end{aligned}
$$

Hence

$$\begin{aligned}
h_3(t,s) &= \int_s^t h_2(\tau,s)\Delta\tau \\
&= \int_s^t f^{\Delta}(\tau)\Delta\tau \\
&= f(\tau)\Big|_{\tau=s}^{\tau=t} \\
&= \frac{1}{2}\left(\frac{1}{3}\tau^3 - (s+h)\tau^2 + \left(\frac{h^2}{6} + \frac{(2s+h)h}{2} + s(s+h)\right)\tau\right)\Bigg|_{\tau=s}^{\tau=t} \\
&= \frac{1}{2}\left(\frac{1}{3}t^3 - (s+h)t^2 + \left(\frac{h^2}{6} + \frac{(2s+h)h}{2} + s(s+h)\right)t\right. \\
&\qquad \left. - \frac{1}{3}s^3 - (s+h)s^2 - \left(\frac{h^2}{6} + \frac{(2s+h)h}{2} + s(s+h)\right)s\right) \\
&= \frac{1}{2}\left(\frac{1}{3}(t-s)(t^2+st+s^2) - (s+h)(t-s)(t+s)\right. \\
&\qquad \left. + \left(\frac{h^2}{6} + \frac{(2s+h)h}{2} + s(s+h)\right)(t-s)\right) \\
&= \frac{t-s}{2}\left(\frac{t^2+st+s^2}{3} - (s+h)(t+s) + \frac{h^2}{6} + \frac{(2s+h)h}{2} + s(s+h)\right) \\
&= \frac{t-s}{2}\left(\frac{t^2+st+s^2}{3} - t(s+h) - s(s+h) + \frac{h^2}{6} + \frac{(2s+h)h}{2} + s(s+h)\right) \\
&= \frac{t-s}{2}\left(\frac{t^2+st+s^2}{3} - t(s+h) + \frac{h(3s+2h)}{3}\right) \\
&= \frac{t-s}{6}(t^2+st+s^2 - 3t(s+h) + h(3s+2h)) \\
&= \frac{t-s}{6}(t^2 - t(2s+3h) + h(3s+2h) + s^2) \\
&= \frac{(t-s)(t-s-h)(t-s-2h)}{6}, \quad t,s \in \mathbb{T}.
\end{aligned}$$

Exercise 3.44. Let $\mathbb{T} = h\mathbb{Z}$, $h > 0$. Using Example 3.113, prove that

$$h_k(t,s) = \frac{(t-s)(t-s-h)(t-s-2h)\cdots(t-s-(k-1)h)}{k!}, \quad t,s \in \mathbb{T},\ k \in \mathbb{N}.$$

Example 3.114. Let $\mathbb{T} = q^{\mathbb{N}_0}$. We will find

$$h_1(t,s), \quad h_2(t,s), \quad \text{and} \quad h_3(t,s), \quad t,s \in \mathbb{T}.$$

We have

$$h_1(t,s) = t - s, \quad t,s \in \mathbb{T}.$$

Fix $s \in \mathbb{T}$. Let

$$f(t) = \frac{t^2}{q+1} - st, \quad t \in \mathbb{T}.$$

Then

$$\begin{aligned} f^{\Delta}(t) &= \frac{\sigma(t) + t}{q+1} - s \\ &= \frac{qt + t}{q+1} - s \\ &= t - s \\ &= h_1(t, s), \quad t \in \mathbb{T}. \end{aligned}$$

Hence

$$\begin{aligned} h_2(t, s) &= \int_s^t h_1(\tau, s)\Delta\tau \\ &= \int_s^t f^{\Delta}(\tau)\Delta\tau \\ &= f(\tau)\Big|_{\tau=s}^{\tau=t} \\ &= \left(\frac{\tau^2}{q+1} - s\tau\right)\Big|_{\tau=s}^{\tau=t} \\ &= \frac{t^2}{q+1} - st - \frac{s^2}{q+1} + s^2 \\ &= \frac{(t-s)(t+s)}{q+1} - s(t-s) \\ &= \frac{(t-s)(t+s-(q+1)s)}{q+1} \\ &= \frac{(t-s)(t-qs)}{q+1}, \quad s, t \in \mathbb{T}. \end{aligned}$$

Let

$$g(t) = \frac{t^3}{(1+q)(1+q+q^2)} - \frac{t^2}{1+q}s + \frac{qs^2}{q+1}t, \quad t \in \mathbb{T}.$$

Then we have

$$\begin{aligned} g^{\Delta}(t) &= \frac{(\sigma(t))^2 + t\sigma(t) + t^2}{(1+q)(1+q+q^2)} - \frac{\sigma(t) + t}{1+q} + \frac{qs^2}{q+1} \\ &= \frac{q^2t^2 + qt^2 + t^2}{(1+q)(1+q+q^2)} - \frac{qt+t}{q+1}s + \frac{qs^2}{q+1} \end{aligned}$$

$$\begin{aligned}
&= \frac{t^2}{q+1} - ts + \frac{qs^2}{q+1} \\
&= h_2(t,s), \quad t \in \mathbb{T}.
\end{aligned}$$

Hence

$$\begin{aligned}
h_3(t,s) &= \int_s^t h_2(\tau,s)\Delta\tau \\
&= \int_s^t g^{\Delta}(\tau)\Delta\tau \\
&= g(\tau)\Big|_{\tau=s}^{\tau=t} \\
&= \left(\frac{\tau^3}{(1+q)(1+q+q^2)} - \frac{\tau^2 s}{q+1} + \frac{qs^2}{q+1}\tau\right)\Bigg|_{\tau=s}^{\tau=t} \\
&= \frac{t^3}{(1+q)(1+q+q^2)} - \frac{t^2}{1+q}s + \frac{qs^2}{q+1}t \\
&\quad - \frac{s^3}{(1+q)(1+q+q^2)} + \frac{s^3}{1+q} - \frac{qs^3}{1+q} \\
&= \frac{(t-s)(t^2+st+s^2)}{(q+1)(1+q+q^2)} - \frac{s(t-s)(t+s)}{1+q} + \frac{qs^2(t-s)}{1+q} \\
&= \frac{t-s}{(1+q)(1+q+q^2)}(t^2+st+s^2-(1+q+q^2)s(t+s)+q(1+q+q^2)s^2) \\
&= \frac{t-s}{(1+q)(1+q+q^2)}(t^2+st+s^2-(1+q+q^2)st-(1+q+q^2)s^2+q(1+q+q^2)s^2) \\
&= \frac{t-s}{(1+q)(1+q+q^2)}(t^2-q(1+q)st+(1-1-q-q^2+q+q^2+q^3)s^2) \\
&= \frac{t-s}{(1+q)(1+q+q^2)}(t^2-q(q+q)st+q^3s^2) \\
&= \frac{(t-s)(t-qs)(t-q^2s)}{(1+q)(1+q+q^2)}, \quad t,s \in \mathbb{T}.
\end{aligned}$$

Exercise 3.45. Let $\mathbb{T} = q^{\mathbb{N}_0}$. Using Example 3.114, prove that

$$h_k(t,s) = \frac{(t-s)(t-qs)\cdots(t-q^{k-1}s)}{(1+q)(1+q+q^2)\cdots(1+q+\cdots+q^{k-1})}, \quad t,s \in \mathbb{T}.$$

Example 3.115. Let $\mathbb{T} = (-3\mathbb{N}_0) \cup 2^{\mathbb{N}_0}$. We will find

$$h_1(t,s), \quad h_2(t,s), \quad \text{and} \quad h_3(t,s), \quad t,s \in \mathbb{T}, \quad s \le t.$$

We have the following cases.

1. $s < t \leq 0$. Then

$$h_1(t,s) = t - s,$$
$$h_2(t,s) = \frac{(t-s)(t-s-3)}{2},$$
$$h_3(t,s) = \frac{(t-s)(t-s-3)(t-s-6)}{6}.$$

2. $s \leq 0 < 1 < t$. Then

$$\begin{aligned}
h_1(t,s) &= t - s,\\
h_2(t,s) &= \int_s^t h_1(\tau,s)\Delta\tau\\
&= \int_s^t (\tau - s)\Delta\tau\\
&= \int_s^0 (\tau - s)\Delta\tau + \int_0^1 (\tau - s)\Delta\tau + \int_1^t (\tau - s)\Delta\tau\\
&= \left(\frac{1}{2}\tau^2 - \frac{3}{2}\tau\right)\Big|_{\tau=s}^{\tau=0} - s(-s) + (-s) + \frac{\tau^2}{3}\Big|_{\tau=1}^{\tau=t} - s(t-1)\\
&= -\frac{1}{2}s^2 + \frac{3}{2}s + s^2 - s + \frac{t^2}{3} - \frac{1}{3} - s(t-1)\\
&= \frac{s^2}{2} + \frac{s}{2} + \frac{t^2}{3} - \frac{1}{3} - st + s\\
&= \frac{t^2}{3} - st + \frac{s^2}{2} + \frac{3s}{2} - \frac{1}{3},
\end{aligned}$$

and

$$\begin{aligned}
h_3(t,s) &= \int_s^t h_2(\tau,s)\Delta\tau\\
&= \int_s^t \left(\frac{\tau^2}{3} - s\tau + \frac{s^2}{2} + \frac{3s}{2} - \frac{1}{3}\right)\Delta\tau\\
&= \frac{1}{3}\int_s^t \tau^2\Delta\tau - s\int_s^t \tau\Delta\tau + \left(\frac{s^2}{2} + \frac{3s}{2} - \frac{1}{3}\right)(t-s)\\
&= \frac{1}{3}\int_s^0 \tau^2\Delta\tau + \frac{1}{3}\int_0^1 \tau^2\Delta\tau + \frac{1}{3}\int_1^t \tau^2\Delta\tau
\end{aligned}$$

$$
\begin{aligned}
&-s\int_s^0 \tau\Delta\tau - s\int_0^1 \tau\Delta\tau - s\int_1^t \tau\Delta\tau + \left(\frac{s^2}{2}+\frac{3s}{2}-\frac{1}{3}\right)(t-s)\\
&= \frac{1}{3}\left(\frac{1}{3}\tau^3 - \frac{3}{2}\tau^2 + \frac{3}{2}\tau\right)\Big|_{\tau=s}^{\tau=0} + \frac{\tau^3}{21}\Big|_{\tau=1}^{\tau=t}\\
&\quad - s\left(\frac{1}{2}\tau^2 - \frac{3}{2}\tau\right)\Big|_{\tau=s}^{\tau=0} - s\frac{\tau^2}{3}\Big|_{\tau=1}^{\tau=t} + \left(\frac{s^2}{2}+\frac{3s}{2}-\frac{1}{3}\right)(t-s)\\
&= -\frac{1}{9}s^3 + \frac{1}{2}s^2 - \frac{1}{2}s + \frac{t^3}{21} - \frac{1}{21} + \frac{s^3}{2} - \frac{3}{2}s^2 - s\frac{t^2}{3} + \frac{s}{3}\\
&\quad + \left(\frac{s^2}{2}+\frac{3s}{2}-\frac{1}{3}\right)t - \frac{s^3}{2} - \frac{3s^2}{2} + \frac{s}{3}\\
&= \frac{t^3}{21} - \frac{s}{3}t^2 + \left(\frac{s^2}{2}+\frac{3s}{2}-\frac{1}{3}\right)t - \frac{1}{9}s^3 - \frac{5}{2}s^2 + \frac{1}{6}s - \frac{1}{21}.
\end{aligned}
$$

3. $s \leq 0 < 1 = t$. Then

$$
\begin{aligned}
h_1(1,s) &= 1-s,\\
h_2(1,s) &= \int_s^1 (\tau - s)\Delta\tau\\
&= \int_s^1 \tau\Delta\tau - s(1-s)\\
&= \int_s^0 \tau\Delta\tau + \int_0^1 \tau\Delta\tau - s(1-s)\\
&= \left(\frac{1}{2}\tau^2 - \frac{3}{2}\tau\right)\Big|_{\tau=s}^{\tau=0} - s + s^2\\
&= -\frac{s^2}{2} + \frac{3}{2}s - s + s^2\\
&= \frac{1}{2}(s^2+s),
\end{aligned}
$$

and

$$
\begin{aligned}
h_2(1,s) &= h_3^{\Delta}(1,s)\\
&= \frac{h_3(\sigma(1),s) - h_3(1,s)}{\sigma(1)-1}\\
&= h_3(2,s) - h_3(1,s).
\end{aligned}
$$

By the previous case we get

$$
h_3(2,s) = \frac{8}{21} - \frac{4}{3}s + s^2 + 3s - \frac{2}{3} - \frac{s^3}{9} - \frac{5}{2}s^2 + \frac{s}{6} - \frac{1}{21}
$$

$$= -\frac{s^3}{9} - \frac{3}{2}s^2 + \frac{11}{6}s - \frac{5}{7}.$$

Hence

$$\frac{1}{2}s^2 + \frac{1}{2}s = -\frac{s^3}{9} - \frac{3}{2}s^2 + \frac{11}{6}s - \frac{5}{7} - h_3(1,s),$$

and thus

$$\begin{aligned} h_3(1,s) &= -\frac{1}{9}s^3 - \frac{3}{2}s^2 + \frac{11}{6}s - \frac{5}{7} - \frac{1}{2}s^2 - \frac{1}{2}s \\ &= -\frac{1}{9}s^3 - 2s^2 + \frac{4}{3}s - \frac{5}{7}. \end{aligned}$$

4. $1 \le s < t$. Then

$$\begin{aligned} h_1(t,s) &= t - s, \\ h_2(t,s) &= \frac{(t-s)(t-2s)}{3}, \\ h_3(t,s) &= \frac{(t-s)(t-2s)(t-4s)}{21}. \end{aligned}$$

Example 3.116. Let

$$\mathbb{T} = (-\mathbb{N}_0) \cup \left\{\frac{1}{n}\right\}_{n\in\mathbb{N}} \cup [1,2] \cup 2^{\mathbb{N}}.$$

We will find

$$h_1(t,s), \quad h_2(t,s), \quad \text{and} \quad h_3(t,s), \quad t,s \in \mathbb{T}, \quad s \le t.$$

For this aim, we will consider the following cases.

1. $s \in (-\mathbb{N}_0)$ and $t \in 2^{\mathbb{N}}$, $t \ge 4$. Then

$$\begin{aligned} h_1(t,s) &= t - s, \\ h_2(t,s) &= \int_s^t h_1(\tau,s)\Delta\tau \\ &= \int_s^t (\tau - s)\Delta\tau \\ &= \int_s^t \tau\Delta\tau - s(t-s) \\ &= \int_s^0 \tau\Delta\tau + \int_0^1 \tau\Delta\tau + \int_1^2 \tau\Delta\tau + \int_2^t \tau\Delta\tau - st + s^2 \end{aligned}$$

$$
\begin{aligned}
&= \frac{\tau^2-\tau}{2}\Big|_{\tau=s}^{\tau=0} + \sum_{n=2}^{\infty} \frac{1}{n^2(n-1)} + \frac{\tau^2}{2}\Big|_{\tau=1}^{\tau=2} + \frac{\tau^2}{3}\Big|_{\tau=2}^{\tau=t} - st + s^2 \\
&= \frac{s-s^2}{2} + \sum_{n=2}^{\infty} \frac{1}{n^2(n-1)} + 2 - \frac{1}{2} + \frac{t^2}{3} - \frac{4}{3} - st + s^2 \\
&= \frac{t^2}{2} - st + \frac{s+s^2}{2} + \frac{1}{6} + \sum_{n=2}^{\infty} \frac{1}{n^2(n-1)},
\end{aligned}
$$

and

$$
\begin{aligned}
h_3(t,s) &= \int_s^t h_2(\tau,s)\Delta\tau \\
&= \int_s^t \left(\frac{\tau^2}{3} - s\tau + \frac{s+s^2}{2} + \frac{1}{6} + \sum_{n=2}^{\infty} \frac{1}{n^2(n-1)} \right)\Delta\tau \\
&= \frac{1}{3}\int_s^t \tau^2\Delta\tau - s\int_s^t \tau\Delta\tau + \left(\frac{s+s^2}{2} + \frac{1}{6} + \sum_{n=2}^{\infty} \frac{1}{n^2(n-1)} \right)(t-s) \\
&= \frac{1}{3}\int_s^0 \tau^2\Delta\tau + \frac{1}{3}\int_0^1 \tau^2\Delta\tau + \frac{1}{3}\int_1^2 \tau^2\Delta\tau + \frac{1}{3}\int_2^t \tau^2\Delta\tau \\
&\quad - s\int_s^0 \tau\Delta\tau - s\int_0^1 \tau\Delta\tau - s\int_1^2 \tau\Delta\tau - s\int_2^t \tau\Delta\tau \\
&\quad + \left(\frac{s+s^2}{2} + \frac{1}{6} + \sum_{n=2}^{\infty} \frac{1}{n^2(n-1)} \right)(t-s) \\
&= \frac{1}{3}\left(\frac{1}{3}\tau^3 - \frac{1}{2}\tau^2 + \frac{1}{6}\tau \right)\Big|_{\tau=s}^{\tau=0} + \frac{1}{3}\sum_{n=2}^{\infty} \frac{1}{n^3(n-1)} + \frac{\tau^3}{9}\Big|_{\tau=1}^{\tau=2} \\
&\quad + \frac{\tau^3}{21}\Big|_{\tau=2}^{\tau=t} - s\left(\frac{\tau^2}{2} - \frac{\tau}{2} \right)\Big|_{\tau=s}^{\tau=0} - s\sum_{n=2}^{\infty} \frac{1}{n^2(n-1)} - s\frac{\tau^2}{2}\Big|_{\tau=1}^{\tau=2} \\
&\quad - s\frac{\tau^2}{3}\Big|_{\tau=2}^{\tau=t} + \left(\frac{s+s^2}{2} + \frac{1}{6} + \sum_{n=2}^{\infty} \frac{1}{n^2(n-1)} \right)(t-s) \\
&= -\frac{1}{3}\left(\frac{1}{3}s^3 - \frac{1}{2}s^2 + \frac{1}{6}s \right) + \frac{1}{3}\sum_{n=2}^{\infty} \frac{1}{n^3(n-1)} + \frac{8}{9} - \frac{1}{9} \\
&\quad + \frac{t^3}{21} - \frac{8}{21} + s\left(\frac{s^2}{2} - \frac{s}{2} \right) - s\sum_{n=2}^{\infty} \frac{1}{n^2(n-1)} - 2s + \frac{1}{2}s \\
&\quad - s\frac{t^2}{3} + \frac{4}{3}s + \left(\frac{s+s^2}{2} + \frac{1}{6} + \sum_{n=2}^{\infty} \frac{1}{n^2(n-1)} \right)(t-s) \\
&= \frac{t^3}{21} - \frac{s}{3}t^2 + \left(\frac{s+s^2}{2} + \frac{1}{6} + \sum_{n=2}^{\infty} \frac{1}{n^2(n-1)} \right)t
\end{aligned}
$$

$$
\begin{aligned}
&-\frac{1}{9}s^3+\frac{1}{6}s^2-\frac{1}{18}s+\frac{1}{3}\sum_{n=2}^{\infty}\frac{1}{n^3(n-1)}+\frac{7}{9}-\frac{8}{21}+\frac{s^3}{2}-\frac{s^2}{2}\\
&-s\sum_{n=2}^{\infty}\frac{1}{n^2(n-1)}-\frac{3}{2}s+\frac{4}{3}s-\frac{s^2}{2}-\frac{s^3}{2}-\frac{s}{6}-s\sum_{n=2}^{\infty}\frac{1}{n^2(n-1)}\\
&=\frac{t^3}{21}-\frac{s}{3}t^2+\left(\frac{s+s^2}{2}+\frac{1}{6}+\sum_{n=2}^{\infty}\frac{1}{n^2(n-1)}\right)t+\frac{1}{3}\sum_{n=2}^{\infty}\frac{1}{n^3(n-1)}\\
&-2s\sum_{n=2}^{\infty}\frac{1}{n^2(n-1)}-\frac{1}{9}s^3-\frac{5}{6}s^2-\frac{7}{18}.
\end{aligned}
$$

2. $s\in(-\mathbb{N}_0)$ and $t\in[1,2]$. Then

$$
\begin{aligned}
h_1(t,s)&=t-s,\\
h_2(t,s)&=\int_s^t h_1(\tau,s)\Delta\tau\\
&=\int_s^t(\tau-s)\Delta\tau\\
&=\int_s^t \tau\Delta\tau-st+s^2\\
&=\int_s^0 \tau\Delta\tau+\int_0^1 \tau\Delta\tau+\int_1^t \tau\Delta\tau-st+s^2\\
&=\frac{\tau^2-\tau}{2}\Big|_{\tau=s}^{\tau=0}+\sum_{n=2}^{\infty}\frac{1}{n^2(n-1)}+\frac{\tau^2}{2}\Big|_{\tau=1}^{\tau=t}-st+s^2\\
&=-\frac{s^2-s}{2}+\sum_{n=2}^{\infty}\frac{1}{n^2(n-1)}+\frac{t^2}{2}-\frac{1}{2}-st+s^2\\
&=\frac{t^2}{2}-st+\sum_{n=2}^{\infty}\frac{1}{n^2(n-1)}-\frac{1}{2}+\frac{s+s^2}{2},
\end{aligned}
$$

and

$$
\begin{aligned}
h_3(t,s)&=\int_s^t h_2(\tau,s)\Delta\tau\\
&=\int_s^t\left(\frac{\tau^2}{2}-s\tau+\sum_{n=2}^{\infty}\frac{1}{n^2(n-1)}-\frac{1}{2}+\frac{s^2+s}{2}\right)\Delta\tau\\
&=\frac{1}{2}\int_s^t \tau^2\Delta\tau-s\int_s^t \tau\Delta\tau+\left(\sum_{n=2}^{\infty}\frac{1}{n^2(n-1)}-\frac{1}{2}+\frac{s^2+s}{2}\right)(t-s)
\end{aligned}
$$

$$
\begin{aligned}
&= \frac{1}{2}\int_s^0 \tau^2 \Delta\tau + \frac{1}{2}\int_0^1 \tau^2\Delta\tau + \frac{1}{2}\int_1^t \tau^2\Delta\tau \\
&\quad - s\int_s^0 \tau\Delta\tau - s\int_0^1 \tau\Delta\tau - s\int_1^t \tau\Delta\tau \\
&\quad + \left(\sum_{n=2}^{\infty}\frac{1}{n^2(n-1)} - \frac{1}{2} + \frac{s^2+s}{2}\right)(t-s) \\
&= \frac{1}{2}\left(\frac{1}{3}\tau^3 - \frac{1}{2}\tau^2 + \frac{1}{6}\tau\right)\Bigg|_{\tau=s}^{\tau=0} + \frac{1}{2}\sum_{n=2}^{\infty}\frac{1}{n^3(n-1)} + \frac{\tau^3}{6}\Bigg|_{\tau=1}^{\tau=t} \\
&\quad - s\frac{\tau^2-\tau}{2}\Bigg|_{\tau=s}^{\tau=0} - s\sum_{n=2}^{\infty}\frac{1}{n^2(n-1)} - s\frac{\tau^2}{2}\Bigg|_{\tau=1}^{\tau=t} \\
&\quad + \left(\sum_{n=2}^{\infty}\frac{1}{n^2(n-1)} - \frac{1}{2} + \frac{s^2+s}{2}\right)(t-s) \\
&= -\frac{1}{6}s^3 + \frac{1}{4}s^2 - \frac{1}{12}s + \frac{1}{2}\sum_{n=2}^{\infty}\frac{1}{n^3(n-1)} + \frac{t^3}{6} - \frac{1}{6} \\
&\quad + \frac{s^3}{2} - \frac{s^2}{2} - s\sum_{n=2}^{\infty}\frac{1}{n^2(n-1)} - s\frac{t^2}{2} + \frac{s}{2} \\
&\quad + \left(\sum_{n=2}^{\infty}\frac{1}{n^2(n-1)} - \frac{1}{2} + \frac{s^2+s}{2}\right)t - s\sum_{n=2}^{\infty}\frac{1}{n^2(n-1)} + \frac{s}{2} - \frac{s^2}{2} - \frac{s^3}{2} \\
&= \frac{t^3}{6} - \frac{s}{2}t^2 + \left(\sum_{n=2}^{\infty}\frac{1}{n^2(n-1)} - \frac{1}{2} + \frac{s^2+s}{2}\right)t - \frac{1}{6}s^3 - \frac{3}{4}s^2 \\
&\quad - \frac{1}{12}s - 2s\sum_{n=2}^{\infty}\frac{1}{n^2(n-1)} + \frac{1}{2}\sum_{n=2}^{\infty}\frac{1}{n^3(n-1)}.
\end{aligned}
$$

3. $s \in (-\mathbb{N}_0)$ and $t = \frac{1}{n}$ for some $n \in \mathbb{N}$, $n \geq 2$. Then

$$
\begin{aligned}
h_1(t,s) &= t - s, \\
h_2(t,s) &= \int_s^t (\tau - s)\Delta\tau \\
&= \int_s^t \tau\Delta\tau - s(t-s) \\
&= \int_s^0 \tau\Delta\tau + \int_0^{\frac{1}{n}} \tau\Delta\tau - st + s^2 \\
&= \frac{1}{2}(\tau^2 - \tau)\Bigg|_{\tau=s}^{\tau=0} + \sum_{k=n}^{\infty}\frac{1}{k^2(k-1)} + s^2 - st
\end{aligned}
$$

$$
\begin{aligned}
&= -\frac{s^2-s}{2} + \sum_{k=n}^{\infty} \frac{1}{k^2(k-1)} + s^2 - st \\
&= \sum_{k=\frac{1}{t}}^{\infty} \frac{1}{k^2(k-1)} - st + \frac{s+s^2}{2},
\end{aligned}
$$

and

$$
\begin{aligned}
h_3(t,s) &= \int_s^t h_2(\tau,s)\Delta\tau \\
&= \int_s^t \left(\sum_{k=\frac{1}{\tau}}^{\infty} \frac{1}{k^2(k-1)} - s\tau + \frac{s+s^2}{2} \right) \Delta\tau \\
&= \int_s^t \sum_{k=\frac{1}{\tau}}^{\infty} \frac{1}{k^2(k-1)} \Delta\tau - s\int_s^t \tau\Delta\tau + \frac{s+s^2}{2}(t-s) \\
&= \int_s^0 \sum_{k=\frac{1}{\tau}}^{\infty} \frac{1}{k^2(k-1)} \Delta\tau + \int_0^t \sum_{k=\frac{1}{\tau}}^{\infty} \frac{1}{k^2(k-1)} \Delta\tau \\
&\quad - s\int_s^0 \tau\Delta\tau - s\int_0^t \tau\Delta\tau + \frac{s+s^2}{2}t - \frac{s^2+s^3}{2} \\
&= \sum_{\tau=s}^{-1} \sum_{k=\frac{1}{\tau}}^{\infty} \frac{1}{k^2(k-1)} + \sum_{\tau=\frac{1}{n}}^{\infty} \frac{\tau^2}{1-\tau} \sum_{k=\frac{1}{\tau}}^{\infty} \frac{1}{k^2(k-1)} - s\frac{\tau^2-\tau}{2}\Big|_{\tau=s}^{\tau=0} \\
&\quad - s\sum_{k=\frac{1}{t}} \frac{1}{k^2(k-1)} + \frac{s+s^2}{2}t - \frac{s^2+s^3}{2} \\
&= \sum_{\tau=s}^{-1} \sum_{k=\frac{1}{\tau}}^{\infty} \frac{1}{k^2(k-1)} + \sum_{\tau=\frac{1}{n}}^{\infty} \frac{\tau^2}{1-\tau} \sum_{k=\frac{1}{\tau}}^{\infty} \frac{1}{k^2(k-1)} + \frac{s^3-s^2}{2} \\
&\quad - s\sum_{k=\frac{1}{t}} \frac{1}{k^2(k-1)} + \frac{s+s^2}{2}t - \frac{s^2+s^3}{2} \\
&= \sum_{\tau=s}^{-1} \sum_{k=\frac{1}{\tau}}^{\infty} \frac{1}{k^2(k-1)} + \sum_{\tau=\frac{1}{n}}^{\infty} \frac{\tau^2}{1-\tau} \sum_{k=\frac{1}{\tau}}^{\infty} \frac{1}{k^2(k-1)} + \frac{s+s^2}{2}t \\
&\quad - s\sum_{k=\frac{1}{t}}^{\infty} \frac{1}{k^2(k-1)} - s^2.
\end{aligned}
$$

4. $s,t \in (-\mathbb{N}_0)$, $s \le t$. Then

$$
h_1(t,s) = t - s,
$$

$$h_2(t,s) = \frac{(t-s)(t-s-1)}{2},$$
$$h_3(t,s) = \frac{(t-s)(t-s-1)(t-s-2)}{6}.$$

5. $s = 0$ and $t \in 2^{\mathbb{N}}$, $t \geq 4$. Then

$$\begin{aligned}
h_1(t,0) &= t,\\
h_2(t,0) &= \int_0^t h_1(\tau,0)\Delta\tau\\
&= \int_0^t \tau\Delta\tau\\
&= \int_0^1 \tau\Delta\tau + \int_1^2 \tau\Delta\tau + \int_2^t \tau\Delta\tau\\
&= \sum_{k=2}^{\infty} \frac{1}{k^2(k-1)} + \frac{\tau^2}{2}\Big|_{\tau=1}^{\tau=2} + \frac{\tau^2}{3}\Big|_{\tau=2}^{\tau=t}\\
&= \sum_{k=2}^{\infty} \frac{1}{k^2(k-1)} + 2 - \frac{1}{2} + \frac{t^2}{3} - \frac{4}{3}\\
&= \frac{t^2}{3} + \sum_{k=2}^{\infty} \frac{1}{k^2(k-1)} + \frac{1}{6},
\end{aligned}$$

and

$$\begin{aligned}
h_3(t,0) &= \int_0^t h_2(\tau,0)\Delta\tau\\
&= \int_0^t \left(\frac{1}{3}\tau^2 + \sum_{k=2}^{\infty} \frac{1}{k^2(k-1)} + \frac{1}{6}\right)\Delta\tau\\
&= \frac{1}{3}\int_0^t \tau^2\Delta\tau + \left(\sum_{k=2}^{\infty} \frac{1}{k^2(k-1)} + \frac{1}{6}\right)t\\
&= \frac{1}{3}\int_0^1 \tau^2\Delta\tau + \frac{1}{3}\int_1^2 \tau^2\Delta\tau + \frac{1}{3}\int_2^t \tau^2\Delta\tau + \left(\sum_{k=2}^{\infty} \frac{1}{k^2(k-1)} + \frac{1}{6}\right)t\\
&= \frac{1}{3}\sum_{k=2}^{\infty} \frac{1}{k^3(k-1)} + \frac{\tau^3}{9}\Big|_{\tau=1}^{\tau=2} + \frac{1}{21}\tau^3\Big|_{\tau=2}^{\tau=t}\\
&\quad + \left(\sum_{k=2}^{\infty} \frac{1}{k^2(k-1)} + \frac{1}{6}\right)t
\end{aligned}$$

$$= \frac{t^3}{21} + \left(\sum_{k=2}^{\infty} \frac{1}{k^2(k-1)} + \frac{1}{6} \right) t + \frac{1}{3} \sum_{k=2}^{\infty} \frac{1}{k^3(k-1)} + \frac{25}{63}.$$

6. $s = 0$ and $t \in [1, 2]$. Then

$$\begin{aligned} h_1(t,0) &= t, \\ h_2(t,0) &= \int_0^t h_1(\tau,0)\Delta\tau \\ &= \int_0^t \tau\Delta\tau \\ &= \int_0^1 \tau\Delta\tau + \int_1^t \tau\Delta\tau \\ &= \sum_{k=2}^{\infty} \frac{1}{k^2(k-1)} + \frac{\tau^2}{3}\Big|_{\tau=1}^{\tau=t} \\ &= \frac{t^2}{3} + \sum_{k=2}^{\infty} \frac{1}{k^2(k-1)} - \frac{1}{3}, \end{aligned}$$

and

$$\begin{aligned} h_3(t,0) &= \int_0^t h_2(\tau,0)\Delta\tau \\ &= \int_0^t \left(\frac{\tau^2}{3} + \left(\sum_{k=2}^{\infty} \frac{1}{k^2(k-1)} - \frac{1}{3} \right) \right) \Delta\tau \\ &= \frac{1}{3} \int_0^t \tau^2 \Delta\tau + \left(\sum_{k=2}^{\infty} \frac{1}{k^2(k-1)} - \frac{1}{3} \right) t \\ &= \frac{1}{3} \int_0^1 \tau^2 \Delta\tau + \frac{1}{3} \int_1^t \tau^2 \Delta\tau + \left(\sum_{k=2}^{\infty} \frac{1}{k^2(k-1)} - \frac{1}{3} \right) t \\ &= \frac{1}{3} \sum_{k=2}^{\infty} \frac{1}{k^3(k-1)} + \frac{\tau^3}{9}\Big|_{\tau=1}^{\tau=t} + \left(\sum_{k=2}^{\infty} \frac{1}{k^2(k-1)} - \frac{1}{3} \right) t \\ &= \frac{t^3}{9} + \left(\sum_{k=2}^{\infty} \frac{1}{k^2(k-1)} - \frac{1}{3} \right) t + \frac{1}{3} \sum_{k=2}^{\infty} \frac{1}{k^3(k-1)} - \frac{1}{9}. \end{aligned}$$

7. $s = 0$ and $t = \frac{1}{n}$ for some $n \in \mathbb{N}$. Then

$$h_1(t,0) = t,$$

$$
\begin{aligned}
h_2(t,0) &= \int_0^t h_1(\tau,0)\Delta\tau \\
&= \int_0^t \tau\Delta\tau \\
&= \sum_{k=n}^{\infty} \frac{1}{k^2(k-1)} \\
&= \sum_{k=\frac{1}{t}}^{\infty} \frac{1}{k^2(k-1)},
\end{aligned}
$$

and

$$
\begin{aligned}
h_3(t,0) &= \int_0^t h_2(\tau,0)\Delta\tau \\
&= \int_0^t \sum_{k=\frac{1}{\tau}}^{\infty} \frac{1}{k^2(k-1)}\Delta\tau \\
&= \sum_{\tau=\frac{1}{n}}^{\infty} \frac{\tau^2}{1-\tau} \sum_{k=\frac{1}{\tau}}^{\infty} \frac{1}{k^2(k-1)}.
\end{aligned}
$$

8. $s = \frac{1}{n}$ for some $n \in \mathbb{N}$, $n \geq 2$, and $t \in 2^{\mathbb{N}}$, $t \geq 4$. Then

$$
\begin{aligned}
h_1(t,s) &= t - s, \\
h_2(t,s) &= \int_s^t h_1(\tau,s)\Delta\tau \\
&= \int_s^t (\tau - s)\Delta\tau \\
&= \int_s^t \tau\Delta\tau - s(t-s) \\
&= \int_s^1 \tau\Delta\tau + \int_1^2 \tau\Delta\tau + \int_2^t \tau\Delta\tau - st + s^2 \\
&= \sum_{k=2}^{\infty} \frac{1}{k^2(k-1)} + \frac{\tau^2}{2}\Big|_{\tau=1}^{\tau=2} + \frac{\tau^2}{3}\Big|_{\tau=2}^{\tau=t} - st + s^2 \\
&= \sum_{k=2}^{\infty} \frac{1}{k^2(k-1)} + 2 - \frac{1}{2} + \frac{t^2}{3} - \frac{4}{3} - st + s^2
\end{aligned}
$$

$$= \frac{1}{3}t^2 - st + s^2 + \sum_{k=2}^{\infty} \frac{1}{k^2(k-1)} + \frac{1}{6},$$

and

$$\begin{aligned}
h_3(t,s) &= \int_s^t h_2(\tau,s)\Delta\tau \\
&= \int_s^t \left(\frac{1}{3}\tau^2 - s\tau + s^2 + \sum_{k=2}^{\infty} \frac{1}{k^2(k-1)} + \frac{1}{6}\right)\Delta\tau \\
&= \frac{1}{3}\int_s^t \tau^2\Delta\tau - s\int_s^t \tau\Delta\tau + \left(s^2 + \sum_{k=2}^{\infty} \frac{1}{k^2(k-1)} + \frac{1}{6}\right)(t-s) \\
&= \frac{1}{3}\int_s^1 \tau^2\Delta\tau + \frac{1}{3}\int_1^2 \tau^2\Delta\tau + \frac{1}{3}\int_2^t \tau^2\Delta\tau \\
&\quad - s\int_s^1 \tau\Delta\tau - s\int_1^2 \tau\Delta\tau - s\int_2^t \tau\Delta\tau \\
&\quad + \left(s^2 + \sum_{k=2}^{\infty} \frac{1}{k^2(k-1)} + \frac{1}{6}\right)(t-s) \\
&= \frac{1}{3}\sum_{k=2}^{\infty} \frac{1}{k^3(k-1)} + \frac{\tau^3}{9}\bigg|_{\tau=1}^{\tau=2} + \frac{1}{21}\tau^3\bigg|_{\tau=2}^{\tau=t} \\
&\quad - s\sum_{k=2}^{\infty} \frac{1}{k^2(k-1)} - s\frac{\tau^2}{2}\bigg|_{\tau=1}^{\tau=2} - \frac{s}{3}\tau^2\bigg|_{\tau=2}^{\tau=t} \\
&= \frac{1}{3}\sum_{k=2}^{\infty} \frac{1}{k^3(k-1)} + \frac{8}{9} - \frac{1}{9} + \frac{t^3}{21} - \frac{8}{21} - s\sum_{k=2}^{\infty} \frac{1}{k^2(k-1)} \\
&\quad - 2s + \frac{s}{2} - \frac{1}{3}st^2 + \frac{4}{3}s \\
&= \frac{t^3}{21} - \frac{1}{3}st^2 + \frac{1}{3}\sum_{k=2}^{\infty} \frac{1}{k^3(k-1)} - s\sum_{k=2}^{\infty} \frac{1}{k^2(k-1)} - \frac{1}{6}s + \frac{25}{63}.
\end{aligned}$$

9. $s = \frac{1}{n}$ for some $n \in \mathbb{N}$, $n \geq 2$, and $t \in [1,2]$. Then

$$\begin{aligned}
h_1(t,s) &= t - s, \\
h_2(t,s) &= \int_s^t h_1(\tau,s)\Delta\tau \\
&= \int_s^t (\tau - s)\Delta\tau
\end{aligned}$$

$$
\begin{aligned}
&= \int_s^t \tau \Delta\tau - s(t-s) \\
&= \int_s^1 \tau \Delta\tau + \int_1^t \tau \Delta\tau - st + s^2 \\
&= \sum_{k=n}^{\infty} \frac{1}{k^2(k-1)} + \frac{\tau^2}{2}\Big|_{\tau=1}^{\tau=t} - st + s^2 \\
&= \frac{t^2}{2} - st + \sum_{k=n}^{\infty} \frac{1}{k^2(k-1)} - \frac{1}{2} + s^2,
\end{aligned}
$$

and

$$
\begin{aligned}
h_3(t,s) &= \int_s^t h_2(\tau,s)\Delta\tau \\
&= \int_s^t \left(\frac{\tau^2}{2} - s\tau + \sum_{k=n}^{\infty} \frac{1}{k^2(k-1)} - \frac{1}{2} + s^2 \right) \Delta s \\
&= \frac{1}{2}\int_s^t \tau^2 \Delta\tau - s\int_s^t \tau\Delta\tau + \left(\sum_{k=n}^{\infty} \frac{1}{k^2(k-1)} - \frac{1}{2} + s^2 \right)(t-s) \\
&= \frac{1}{2}\int_s^1 \tau^2\Delta\tau + \frac{1}{2}\int_1^t \tau^2\Delta\tau - s\int_s^1 \tau\Delta\tau - s\int_1^t \tau\Delta\tau \\
&\quad + \left(\sum_{k=n}^{\infty} \frac{1}{k^2(k-1)} - \frac{1}{2} + s^2 \right)(t-s) \\
&= \frac{1}{2}\sum_{k=n}^{\infty} \frac{1}{k^3(k-1)} + \frac{\tau^3}{6}\Big|_{\tau=1}^{\tau=t} - s\sum_{k=n}^{\infty} \frac{1}{k^2(k-1)} - s\frac{\tau^2}{2}\Big|_{\tau=1}^{\tau=t} \\
&\quad + \left(\sum_{k=n}^{\infty} \frac{1}{k^2(k-1)} - \frac{1}{2} + s^2 \right)t - s\sum_{k=n}^{\infty} \frac{1}{k^2(k-1)} + \frac{s}{2} - s^3 \\
&= \frac{t^3}{6} + \left(\sum_{k=n}^{\infty} \frac{1}{k^2(k-1)} - \frac{1}{2} + s^2 \right)t - 2s\sum_{k=n}^{\infty} \frac{1}{k^2(k-1)} + \frac{1}{2}\sum_{k=n}^{\infty} \frac{1}{k^3(k-1)} \\
&\quad - s\frac{t^2}{2} - s^3 - \frac{1}{6} + s.
\end{aligned}
$$

10. $s = \frac{1}{n}$ and $t = \frac{1}{m}$ for some $m, n \in \mathbb{N}$, $m \le n$, $n \ge 2$. Then

$$
\begin{aligned}
h_1(t,s) &= t - s, \\
h_2(t,s) &= \int_s^t h_1(\tau,s)\Delta\tau
\end{aligned}
$$

$$
\begin{aligned}
&= \int_s^t (\tau - s)\Delta\tau \\
&= \int_s^t \tau\Delta\tau - s(t-s) \\
&= \sum_{k=n}^{\frac{1}{t}} \frac{1}{k^2(k-1)} - st + s^2,
\end{aligned}
$$

and

$$
\begin{aligned}
h_3(t,s) &= \int_s^t h_2(\tau,s)\Delta\tau \\
&= \int_s^t \left(\sum_{k=n}^{\frac{1}{\tau}} \frac{1}{k^2(k-1)} - s\tau + s^2 \right)\Delta s \\
&= \int_s^t \sum_{k=n}^{\frac{1}{\tau}} \frac{1}{k^2(k-1)}\Delta\tau - s\int_s^t \tau\Delta\tau + s^2(t-s) \\
&= \sum_{\tau=\frac{1}{n}}^{\frac{1}{t}} \frac{\tau^2}{1-\tau} \sum_{k=n}^{\frac{1}{\tau}} \frac{1}{k^2(k-1)} - s\sum_{k=n}^{\frac{1}{t}} \frac{1}{k^2(k-1)} + s^2 t - s^3.
\end{aligned}
$$

11. $s \in [1,2]$ and $t \in 2^{\mathbb{N}}$, $t \geq 4$. Then

$$
\begin{aligned}
h_1(t,s) &= t - s, \\
h_2(t,s) &= \int_s^t h_1(\tau,s)\Delta\tau \\
&= \int_s^t (\tau - s)\Delta\tau \\
&= \int_s^t \tau\Delta\tau - s(t-s) \\
&= \int_s^2 \tau\Delta\tau + \int_2^t \tau\Delta\tau - st + s^2 \\
&= \frac{\tau^2}{2}\bigg|_{\tau=s}^{\tau=2} + \frac{\tau^2}{3}\bigg|_{\tau=2}^{\tau=t} - st + s^2 \\
&= 2 - \frac{s^2}{2} + \frac{t^2}{3} - \frac{4}{3} - st + s^2 \\
&= \frac{t^2}{3} - st + \frac{s^2}{2} + \frac{2}{3},
\end{aligned}
$$

and

$$
\begin{aligned}
h_3(t,s) &= \int_s^t h_2(\tau,s)\Delta\tau \\
&= \int_s^t \left(\frac{\tau^2}{3} - s\tau + \frac{s^2}{2} + \frac{2}{3}\right)\Delta\tau \\
&= \frac{1}{3}\int_s^t \tau^2\Delta\tau - s\int_s^t \tau\Delta\tau + \left(\frac{s^2}{2} + \frac{2}{3}\right)(t-s) \\
&= \frac{1}{3}\int_s^2 \tau^2\Delta\tau + \frac{1}{3}\int_2^t \tau^2\Delta\tau - s\int_s^2 \tau\Delta\tau - s\int_2^t \tau\Delta\tau \\
&\quad + \left(\frac{s^2}{2} + \frac{2}{3}\right)t - \frac{s^3}{2} - \frac{2}{3}s \\
&= \frac{1}{9}\tau^3\Big|_{\tau=s}^{\tau=2} + \frac{\tau^3}{21}\Big|_{\tau=2}^{\tau=t} - s\frac{\tau^2}{2}\Big|_{\tau=s}^{\tau=2} - s\frac{\tau^2}{3}\Big|_{\tau=2}^{\tau=t} \\
&\quad + \left(\frac{s^2}{2} + \frac{2}{3}\right)t - \frac{s^3}{2} - \frac{2}{3}s \\
&= \frac{8}{9} - \frac{s^3}{9} + \frac{t^3}{21} - \frac{8}{21} - 2s + \frac{s^3}{2} - s\frac{t^2}{3} + \frac{4}{3}s \\
&\quad + \left(\frac{s^2}{2} + \frac{2}{3}\right)t - \frac{s^3}{2} - \frac{2}{3}s \\
&= \frac{t^3}{21} - s\frac{t^2}{3} + \left(\frac{s^2}{2} + \frac{2}{3}\right)t - \frac{1}{9}s^3 - \frac{4}{3}s + \frac{32}{63}.
\end{aligned}
$$

12. $s, t \in [1,2]$. Then

$$
\begin{aligned}
h_1(t,s) &= t - s, \\
h_2(t,s) &= \frac{(t-s)^2}{2}, \\
h_3(t,s) &= \frac{(t-s)^3}{6}.
\end{aligned}
$$

13. $s, t \in 2^{\mathbb{N}}$. Then

$$
\begin{aligned}
h_1(t,s) &= t - s, \\
h_2(t,s) &= \frac{(t-s)(t-2s)}{3}, \\
h_3(t,s) &= \frac{(t-s)(t-2s)(t-4s)}{21}.
\end{aligned}
$$

Exercise 3.46. Let $\mathbb{T} = (-\mathbb{N}_0) \cup 3^{\mathbb{N}_0}$. Find

$$h_1(t,s), \quad h_2(t,s), \quad \text{and} \quad h_3(t,s), \quad s,t \in \mathbb{T}, \quad s \le t.$$

Exercise 3.47. Let $\mathbb{T} = (-2\mathbb{N}_0) \cup \{1 + \frac{1}{n}\}_{n\in\mathbb{N}} \cup 2^{\mathbb{N}_0}$. Find

$$h_1(t,s), \quad h_2(t,s), \quad \text{and} \quad h_3(t,s), \quad s,t \in \mathbb{T}, \quad s \le t.$$

Exercise 3.48. Prove that

$$h_{k+m+1}(t,t_0) = \int_{t_0}^{t} h_k\big(t,\sigma(s)\big)h_m(s,t_0)\Delta s, \quad t,t_0 \in \mathbb{T},\ k,m \in \mathbb{N}_0.$$

3.10 The Taylor formula

Example 3.117. Let $n \in \mathbb{N}$, let f be n times differentiable, and let p_k, $0 \le k \le n-1$, be differentiable at some $t \in \mathbb{T}$ with

$$p^{\Delta}_{k+1}(t) = p^{\sigma}_{k}(t) \quad \text{for all } 0 \le k \le n-2,\ n \ge 2.$$

We will prove that

$$\left(\sum_{k=0}^{n-1}(-1)^k f^{\Delta^k}(t)p_k(t)\right)^{\Delta} = (-1)^{n-1}f^{\Delta^n}(t)p^{\sigma}_{n-1}(t) + f(t)p^{\Delta}_0(t).$$

We have

$$\begin{aligned}
\left(\sum_{k=0}^{n-1}(-1)^k f^{\Delta^k}(t)p_k(t)\right)^{\Delta} &= \sum_{k=0}^{n-1}(-1)^k\big(f^{\Delta^k}(t)p_k(t)\big)^{\Delta} \\
&= \sum_{k=0}^{n-1}(-1)^k\big(f^{\Delta^{k+1}}(t)p^{\sigma}_k(t) + f^{\Delta^k}(t)p^{\Delta}_k(t)\big) \\
&= \sum_{k=0}^{n-1}(-1)^k f^{\Delta^{k+1}}(t)p^{\sigma}_k(t) + \sum_{k=0}^{n-1}(-1)^k f^{\Delta^k}(t)p^{\Delta}_k(t) \\
&= \sum_{k=0}^{n-2}(-1)^k f^{\Delta^{k+1}}(t)p^{\sigma}_k(t) + (-1)^{n-1}f^{\Delta^n}(t)p^{\sigma}_{n-1}(t) \\
&\quad + \sum_{k=0}^{n-1}(-1)^k f^{\Delta^k}(t)p^{\Delta}_k(t) + f^{\Delta^0}(t)p^{\Delta}_0(t) \\
&= \sum_{k=0}^{n-2}(-1)^k f^{\Delta^{k+1}}(t)p^{\Delta}_{k+1}(t) + (-1)^{n-1}f^{\Delta^n}(t)p^{\sigma}_{n-1}(t)
\end{aligned}$$

$$
\begin{aligned}
&+\sum_{k=0}^{n-2}(-1)^{k+1}f^{\Delta^{k+1}}(t)p_{k+1}^{\Delta}(t)+f(t)p_0^{\Delta}(t)\\
&=(-1)^{n-1}f^{\Delta^n}(t)p_{n-1}^{\sigma}(t)+f(t)p_0^{\Delta}(t).
\end{aligned}
$$

Example 3.118. The functions $g_n(t,s)$ satisfy for all $t \in \mathbb{T}$ the relationship

$$g_n(\rho^k(t),t)=0 \quad \text{for all } n \in \mathbb{N} \text{ and } 0 \le k \le n-1.$$

Take arbitrary $n \in \mathbb{N}$. Then

$$
\begin{aligned}
g_n(\rho^0(t),t) &= g_n(t,t)\\
&= \int_t^t g_{n-1}(\sigma(\tau),t)\Delta\tau\\
&= 0, \quad t \in \mathbb{T}.
\end{aligned}
$$

Assume that

$$g_{n-1}(\rho^k(t),t)=0 \quad \text{and} \quad g_n(\rho^k(t),t)=0, \quad t \in \mathbb{T},$$

for some $0 \le k < n-1$.

We will prove that

$$g_n(\rho^{k+1}(t),t)=0, \quad t \in \mathbb{T}.$$

1. Let $\rho^k(t)$ be left-dense. Then

$$\rho^{k+1}(t)=\rho(\rho^k(t))=\rho^k(t).$$

Consequently, using the induction assumption, we have

$$g_n(\rho^{k+1}(t),t)=g_n(\rho^k(t),t)=0.$$

2. Let $\rho^k(t)$ be left-scattered. Then

$$\rho(\rho^k(t))<\rho^k(t),$$

and there is no $s \in \mathbb{T}$ such that $\rho^{k+1}(t)<s<\rho^k(t)$. Hence

$$\sigma(\rho^{k+1}(t))=\rho^k(t).$$

Therefore

$$g_n(\sigma(\rho^{k+1}(t)),t)=g_n(\rho^{k+1}(t),t)+\mu(\rho^{k+1}(t))g_n^{\Delta}(\rho^{k+1}(t),t),$$

or

$$g_n(\rho^k(t), t) = g_n(\rho^{k+1}(t), t) + \mu(\rho^{k+1}(t))g_n^{\Delta}(\rho^{k+1}(t), t),$$

whereupon

$$\begin{aligned} g_n(\rho^{k+1}(t), t) &= g_n(\rho^k(t), t) - \mu(\rho^{k+1}(t))g_n^{\Delta}(\rho^{k+1}(t), t) \\ &= g_n(\rho^k(t), t) - \mu(\rho^{k+1}(t))g_{n-1}(\sigma(\rho^{k+1}(t)), t) \\ &= g_n(\rho^k(t), t) - \mu(\rho^{k+1}(t))g_{n-1}(\rho^k(t), t) \\ &= 0. \end{aligned}$$

Example 3.119. Let $n \in \mathbb{N}$, and suppose that f is $(n-1)$ times differentiable at $\rho^{n-1}(t)$. We will prove that

$$f(t) = \sum_{k=0}^{n-1} (-1)^k f^{\Delta^k}(\rho^{n-1}(t))g_k(\rho^{n-1}(t), t).$$

For this aim, we consider the following cases.

1. $n = 1$. Then

$$\begin{aligned} \sum_{k=0}^{0} (-1)^k f^{\Delta^k}(\rho^0(t))g_k(\rho^0(t), t) &= (-1)^0 f^{\Delta^0}(t)g_0(t, t) \\ &= f(t). \end{aligned}$$

2. Assume that

$$f(t) = \sum_{k=0}^{m-1} (-1)^k f^{\Delta^k}(\rho^{m-1}(t))g_k(\rho^{m-1}(t), t)$$

for some $m \in \mathbb{N}$. We will prove that

$$f(t) = \sum_{k=0}^{m} (-1)^k f^{\Delta^k}(\rho^m(t))g_k(\rho^m(t), t).$$

a. Let $\rho^{m-1}(t)$ be left-dense. Then

$$\rho^m(t) = \rho(\rho^{m-1}(t)) = \rho^{m-1}(t).$$

Hence by the induction assumption we obtain

$$\sum_{k=0}^{m} (-1)^k f^{\Delta^k}(\rho^m(t))g_k(\rho^m(t), t)$$

$$= \sum_{k=0}^{m-1} (-1)^k f^{\Delta^k}(\rho^m(t)) g_k(\rho^m(t), t) + (-1)^m f^{\Delta^m}(\rho^m(t)) g_m(\rho^m(t), t)$$

$$= \sum_{k=0}^{m-1} (-1)^k f^{\Delta^k}(\rho^{m-1}(t)) g_k(\rho^{m-1}(t), t)$$

$$+ (-1)^m f^{\Delta^m}(\rho^{m-1}(t)) g_m(\rho^{m-1}(t), t)$$

now we apply Example 3.118 ($g_m(\rho^{m-1}(t), t) = 0$)

$$= \sum_{k=0}^{m-1} (-1)^k f^{\Delta^k}(\rho^{m-1}(t)) g_k(\rho^{m-1}(t), t)$$

and now we apply the induction assumption

$$= f(t).$$

b. Let $\rho^{m-1}(t)$ be left-scattered. Then

$$\rho^m(t) = \rho(\rho^{m-1}(t)) < \rho^{m-1}(t),$$

and there is no $s \in \mathbb{T}$ such that

$$\rho^m(t) < s < \rho^{m-1}(t).$$

Also,

$$\sigma(\rho^m(t)) = \rho^{m-1}(t).$$

Hence

$$g_k(\sigma(\rho^m(t)), t) = g_k(\rho^{m-1}(t), t).$$

Therefore

$$\begin{aligned} g_k(\rho^{m-1}(t), t) &= g_k(\sigma(\rho^m(t)), t) \\ &= g_k(\rho^m(t), t) + \mu(\rho^m(t)) g_k^{\Delta}(\rho^m(t), t) \\ &= g_k(\rho^m(t), t) + \mu(\rho^m(t)) g_{k-1}(\sigma(\rho^m(t)), t) \\ &= g_k(\rho^m(t), t) + \mu(\rho^m(t)) g_{k-1}(\rho^{m-1}(t), t), \end{aligned}$$

whereupon

$$g_k(\rho^m(t), t) = g_k(\rho^{m-1}(t), t) - \mu(\rho^m(t)) g_{k-1}(\rho^{m-1}(t), t).$$

Consequently,

$$\sum_{k=0}^{m} (-1)^k f^{\Delta^k}(\rho^m(t)) g_k(\rho^m(t), t)$$

$$\begin{aligned}
&= f(\rho^m(t)) + \sum_{k=1}^{m}(-1)^k f^{\Delta^k}(\rho^m(t))g_k(\rho^m(t),t)\\
&= f(\rho^m(t)) + \sum_{k=1}^{m}(-1)^k f^{\Delta^k}(\rho^m(t))g_k(\rho^{m-1}(t),t)\\
&\quad + \sum_{k=1}^{m}(-1)^{k-1} f^{\Delta^k}(\rho^m(t))\mu(\rho^m(t))g_{k-1}(\rho^{m-1}(t),t)\\
&= f(\rho^m(t)) + \sum_{k=1}^{m-1}(-1)^k f^{\Delta^k}(\rho^m(t))g_k(\rho^{m-1}(t),t)\\
&\quad + (-1)^m f^{\Delta^m}(\rho^m(t))g_m(\rho^{m-1}(t),t)\\
&\quad + \sum_{k=0}^{m-1}(-1)^k f^{\Delta^{k-1}}(\rho^m(t))\mu(\rho^m(t))g_k(\rho^{m-1}(t),t)\\
&= \sum_{k=0}^{m-1}(-1)^k f^{\Delta^k}(\rho^m(t))g_k(\rho^{m-1}(t),t)\\
&\quad + \sum_{k=0}^{m-1}(-1)^k \mu(\rho^m(t)) f^{\Delta^{k+1}}(\rho^m(t))g_k(\rho^{m-1}(t),t)\\
&= \sum_{k=0}^{m-1}(-1)^k (f^{\Delta^k}(\rho^m(t)) + \mu(\rho^m(t))(f^{\Delta^k})^{\Delta}(\rho^m(t)))g_k(\rho^{m-1}(t),t)\\
&= \sum_{k=0}^{m-1}(-1)^k f^{\Delta^k}(\sigma(\rho^m(t)))g_k(\rho^{m-1}(t),t)\\
&= \sum_{k=0}^{m-1}(-1)^k f^{\Delta^k}(\rho^{m-1}(t))g_k(\rho^{m-1}(t),t)\\
&= f(t).
\end{aligned}$$

Example 3.120 (Taylor's formula). Let $n \in \mathbb{N}$, let f be an n times differentiable function on $\mathbb{T}^{\kappa^n}$, and let $\alpha \in \mathbb{T}^{\kappa^{n-1}}$, $t \in \mathbb{T}$. We will prove that

$$\begin{aligned}
f(t) &= \sum_{k=0}^{n-1}(-1)^k g_k(\alpha,t) f^{\Delta^k}(\alpha)\\
&\quad + \int_{\alpha}^{\rho^{n-1}(t)} (-1)^{n-1} g_{n-1}(\sigma(\tau),t) f^{\Delta^n}(\tau)\Delta\tau.
\end{aligned}$$

Note that applying Example 3.117 for $p_k = g_k$, we have

$$\begin{aligned}
\left(\sum_{k=0}^{n-1}(-1)^k g_k(\tau,t) f^{\Delta^k}(\tau)\right)_{\tau}^{\Delta} &= (-1)^{n-1} f^{\Delta^n}(\tau) g_{n-1}(\sigma(\tau),t) + f(\tau) g_0^{\Delta}(\tau,t)\\
&= (-1)^{n-1} f^{\Delta^n}(\tau) g_{n-1}(\sigma(\tau),t) \quad \text{for all } \tau \in \mathbb{T}^{\kappa^n}.
\end{aligned}$$

Integrating the last relation from a to $\rho^{n-1}(t)$, we get

$$\int_a^{\rho^{n-1}(t)} \left(\sum_{k=0}^{n-1}(-1)^k g_k(\tau,t) f^{\Delta^k}(\tau)\right)_\tau^{\Delta} \Delta\tau = \int_a^{\rho^{n-1}(t)} (-1)^{n-1} f^{\Delta^n}(\tau) g_{n-1}(\sigma(\tau),t)\Delta\tau$$

or

$$\sum_{k=0}^{n-1}(-1)^k g_k(\rho^{n-1}(t),t) f^{\Delta^k}(\rho^{n-1}(t)) - \sum_{k=0}^{n-1}(-1)^k g_k(a,t) f^{\Delta^k}(a) = \int_a^{\rho^{n-1}(t)} (-1)^{n-1} f^{\Delta^n}(\tau) g_{n-1}(\sigma(\tau),t)\Delta\tau.$$

Hence, applying Example 3.119, we have

$$f(t) - \sum_{k=0}^{n-1}(-1)^k g_k(a,t) f^{\Delta^k}(a) = \int_a^{\rho^{n-1}(t)} (-1)^{n-1} f^{\Delta^n}(\tau) g_{n-1}(\sigma(\tau),t)\Delta\tau.$$

Example 3.121. We will show that the functions g_n and h_n satisfy the relationship

$$h_n(t,s) = (-1)^n g_n(s,t)$$

for all $t \in \mathbb{T}$ and $s \in \mathbb{T}^{\kappa^n}$.

Indeed, let $t \in \mathbb{T}$ and $s \in \mathbb{T}^{\kappa^n}$. We apply Example 3.120 for $a = s$ and $f(\tau) = h_n(\tau, s)$. Observe that

$$f^{\Delta^k}(\tau) = h_{n-k}(\tau,s), \quad 0 \le k \le n.$$

Hence

$$f^{\Delta^k}(s) = h_{n-k}(s,s) = 0, \quad 0 \le k \le n-1,$$
$$f^{\Delta^n}(s) = h_0(s,s) = 1, \quad f^{\Delta^{n+1}}(\tau) = 0.$$

From this, using Taylor's formula, we get

$$f(t) = h_n(t,s) = \sum_{k=0}^{n}(-1)^k g_k(a,t) f^{\Delta^k}(a) + \int_a^{\rho^n(t)} (-1)^n g_n(\sigma(\tau),t) f^{\Delta^{n+1}}(\tau)\Delta\tau$$

$$= \sum_{k=0}^{n} (-1)^k g_k(s,t) f^{\Delta^k}(s) + \int_s^{\rho^n(t)} (-1)^n g_n(\sigma(\tau),t) f^{\Delta^{n+1}}(\tau)\Delta\tau$$
$$= \sum_{k=0}^{n-1} (-1)^k g_k(s,t) f^{\Delta^k}(s) + (-1)^n g_n(s,t) f^{\Delta^n}(s)$$
$$= (-1)^n g_n(s,t) f^{\Delta^n}(s)$$
$$= (-1)^n g_n(s,t),$$

i. e.,

$$h_n(t,s) = (-1)^n g_n(s,t).$$

From Examples 3.120 and 3.121 the following follows.

Exercise 3.49 (Taylor's formula). Let $n \in \mathbb{N}$, let f be an n times differentiable function on $\mathbb{T}^{\kappa^n}$, and let $a \in \mathbb{T}^{\kappa^{n-1}}$, $t \in \mathbb{T}$. Prove that

$$f(t) = \sum_{k=0}^{n-1} h_k(t,a) f^{\Delta^k}(a) + \int_a^{\rho^{n-1}(t)} h_{n-1}(t,\sigma(\tau)) f^{\Delta^n}(\tau)\Delta\tau.$$

Exercise 3.50. Let $a \in [a,b]$. Then for $|\lambda| < \infty$,

$$\sum_{k=0}^{\infty} \lambda^k h_k(t,a)$$

absolutely and uniformly converges on the interval $a \le t \le b$.

Now we give another variant of Taylor's formula.

Example 3.122 (Taylor's formula). Let $n \in \mathbb{N}$, let f be an $n+1$ times differentiable function on $\mathbb{T}^{\kappa^{n+1}}$, and let $a \in \mathbb{T}^{\kappa^{n+1}}$, $t \in \mathbb{T}$, and $t > a$. We will prove that

$$f(t) = \sum_{k=0}^{n} h_k(t,a) f^{\Delta^k}(a) + \int_a^t h_n(t,\sigma(\tau)) f^{\Delta^{n+1}}(\tau)\Delta\tau. \tag{3.61}$$

Let

$$g(t) = f^{\Delta^{n+1}}(t).$$

Then f solves the problem

$$x^{\Delta^{n+1}} = g(t), \quad x^{\Delta^k}(a) = f^{\Delta^k}(a), \quad k \in \{0,\ldots,n\}.$$

We have that

$$y(t,s) = h_n(t,\sigma(s))$$

is the Cauchy function for $y^{\Delta^{n+1}} = 0$. Hence it follows that

$$\begin{aligned} f(t) &= u(t) + \int_a^t y(t,\sigma(\tau))g(\tau)\Delta\tau \\ &= u(t) + \int_a^t h_n(t,\sigma(s))g(s)\Delta s, \end{aligned} \tag{3.62}$$

where u solves the initial value problem

$$u^{\Delta^{n+1}} = 0, \quad u^{\Delta^m}(a) = f^{\Delta^m}(a), \quad m \in \{0,\dots,n\}.$$

We set

$$w(t) = \sum_{k=0}^{n} h_k(t,a)f^{\Delta^k}(a). \tag{3.63}$$

We have

$$w^{\Delta^m}(t) = \sum_{k=0}^{n} h_{k-m}(t,a)f^{\Delta^k}(a), \quad m \in \{0,\dots,n\},$$

and hence

$$\begin{aligned} w^{\Delta^m}(a) &= \sum_{k=0}^{n} h_{k-m}(a,a)f^{\Delta^k}(a) \\ &= f^{\Delta^m}(a), \quad m \in \{0,\dots,n\}, \end{aligned}$$

i. e., w solves (3.62). Consequently, $w = u$. Hence by (3.63) we obtain (3.61).

Let

$$R_n(t,a) = \int_a^{\rho^{n-1}(t)} h_{n-1}(t,\sigma(\tau))f^{\Delta^n}(\tau)\Delta\tau.$$

Example 3.123. Let $t \in \mathbb{T}$, $t \geq a$, and

$$M_n(t) = \sup\{|f^{\Delta^n}(\tau)| : \tau \in [a,t]_{\mathbb{T}}\}.$$

We will prove that

$$|R_n(t,\alpha)| \le M_n(t)\frac{(t-\alpha)^n}{(n-1)!}.$$

Let $\tau \in [\alpha, t)_{\mathbb{T}}$. Then $\alpha \le \sigma(\tau) \le t$, and applying (3.60), we get

$$\begin{aligned} 0 &\le h_{n-1}(t,\sigma(\tau)) \\ &\le \frac{(t-\sigma(\tau))^{n-1}}{(n-1)!} \\ &\le \frac{(t-\tau)^{n-1}}{(n-1)!} \\ &\le \frac{(t-\alpha)^{n-1}}{(n-1)!}. \end{aligned}$$

Hence

$$\begin{aligned} |R_n(t,\alpha)| &= \left| \int_\alpha^{\rho^{n-1}(t)} h_{n-1}(t,\sigma(\tau)) f^{\Delta^n}(\tau)\Delta\tau \right| \\ &\le \int_\alpha^t h_{n-1}(t,\sigma(\tau))|f^{\Delta^n}(\tau)|\Delta\tau \\ &\le M_n(t)\int_\alpha^t \frac{(t-\alpha)^{n-1}}{(n-1)!}\Delta\tau \\ &= M_n(t)\frac{(t-\alpha)^n}{(n-1)!}. \end{aligned}$$

If a function $f : \mathbb{T} \to \mathbb{R}$ is infinitely Δ-differentiable at a point $\alpha \in \mathbb{T}^\infty = \bigcap_{n=1}^\infty \mathbb{T}^{\kappa^n}$, then we can formally write

$$\sum_{k=0}^{\infty} h_k(t,\alpha) f^{\Delta^k}(\alpha) = f(\alpha) + h_1(t,\alpha)f^{\Delta}(\alpha) + h_2(t,\alpha)f^{\Delta^2}(\alpha) + \cdots. \tag{3.64}$$

Definition 3.19. Series (3.64) is called Taylor's series for the function f at the point α.

For given values of α and t, Taylor's series can converge or diverge. Taylor's series (3.64) converges if and only if the remainder of Taylor's formula

$$f(t) = \sum_{k=0}^{n-1} h_k(t,\alpha) f^{\Delta^k}(\alpha) + R_n(t,\alpha)$$

tends to zero as $n \to \infty$, that is, $\lim_{n\to\infty} R_n(t,\alpha) = 0$. It may turn out that for some t, series (3.64) converges but its sum is not equal to $f(t)$.

Example 3.124. We will prove that for all $z \in \mathbb{C}$ and $\alpha, R \in \mathbb{T}$, $R > \alpha$, the initial value problem

$$y^{\Delta} = zy, \quad y(a) = 1, \quad t \in [a, R]_{\mathbb{T}}, \tag{3.65}$$

has a unique solution y, which is represented in the form

$$y(t) = \sum_{k=0}^{\infty} z^k h_k(t, a), \quad t \in [a, R]_{\mathbb{T}},$$

and satisfies the inequality

$$|y(t)| \le e^{|z|(t-a)}, \quad t \in [a, R]_{\mathbb{T}}.$$

Problem (3.65) is equivalent to finding a continuous solution of the integral equation

$$y(t) = 1 + \int_a^t y(\tau)\Delta\tau, \quad t \in [a, R]_{\mathbb{T}}. \tag{3.66}$$

We will solve equation (3.66) using the method of successive approximations. Let

$$y_0(t) = 1, \quad y_k(t) = z \int_a^t y_{k-1}(\tau)\Delta\tau, \quad t \in [a, R]_{\mathbb{T}}, \ k \in \mathbb{N}.$$

Note that

$$\begin{aligned} y_0(t) &= h_0(t, a), \\ y_1(t) &= z \int_a^t h_0(\tau, a)\Delta\tau \\ &= z \int_a^t \Delta\tau \\ &= z(t - a) \\ &= z h_1(t, a), \\ y_2(t) &= z \int_a^t y_1(\tau)\Delta\tau \\ &= z^2 \int_a^t h_1(\tau, a)\Delta\tau \\ &= z^2 h_2(t, a), \quad t \in [a, R]_{\mathbb{T}}. \end{aligned}$$

Assume that

$$y_k(t) = z^k h_k(t, a), \quad t \in [a, R]_{\mathbb{T}}, \tag{3.67}$$

for some $k \in \mathbb{N}$. We will prove that

$$y_{k+1}(t) = z^{k+1} h_{k+1}(t, a), \quad t \in [a, R]_{\mathbb{T}}.$$

Indeed, we have

$$\begin{aligned} y_{k+1}(t) &= z \int_a^t y_k(\tau)\Delta\tau \\ &= z^{k+1} \int_a^t h_k(\tau, a)\Delta\tau \\ &= z^{k+1} h_{k+1}(t, a), \quad t \in [a, R]_{\mathbb{T}}. \end{aligned}$$

Therefore (3.67) holds for all $k \in \mathbb{N}$. Now using Example 3.112, we get

$$\begin{aligned} |y_k(t)| &= |z^k h_k(t, a)| \\ &= |z|^k |h_k(t, a)| \\ &\le |z|^k \frac{(t-a)^k}{k!}, \quad k \in \mathbb{N}_0, \end{aligned}$$

and

$$\begin{aligned} \sum_{k=0}^{\infty} |y_k(t)| &\le \sum_{k=0}^{\infty} |z|^k \frac{(t-a)^k}{k!} \\ &= e^{|z|(t-a)}, \quad t \in [a, R]_{\mathbb{T}}. \end{aligned}$$

Therefore the series $\sum_{k=0}^{\infty}$ converges uniformly with respect to $t \in [a, R]_{\mathbb{T}}$. Consequently, its sum is a continuous solution of problem (3.66). Suppose that (3.66) has two solutions y and x. Let $u = y - x$. Then

$$u(t) = z \int_a^t u(\tau)\Delta\tau, \quad t \in [a, R]_{\mathbb{T}}.$$

We set

$$M = \sup\{|u(t)| : t \in [a, R]_{\mathbb{T}}\}.$$

Hence

$$|u(t)| = \left| z \int_a^t u(\tau)\Delta\tau \right|$$

$$\begin{aligned}
&\le |z| \int_a^t |u(\tau)| \Delta\tau \\
&\le M|z|(t-a) \\
&= Mh_1(t,a), \quad t \in [a,R]_{\mathbb{T}},
\end{aligned}$$

and thus

$$\begin{aligned}
|u(t)| &\le |z|^2 M \int_a^t h_1(\tau,a)\Delta\tau \\
&= M|z|^2 h_2(t,a), \quad t \in [a,R]_{\mathbb{T}}.
\end{aligned}$$

Repeating this procedure, we obtain

$$|u(t)| \le M|z|^k h_k(t,a),$$

and applying Example 3.112, we get

$$|u(t)| \le M|z|\frac{(t-a)^k}{k!}, \quad t \in [a,R]_{\mathbb{T}},$$

for all $k \in \mathbb{N}$. Passing to the limit as $k \to \infty$, we get $u(t) = 0$ for all $t \in [a,R]_{\mathbb{T}}$.

3.11 Survey on nabla integrals

Suppose that $\mathbb{T}$ is a time scale with forward jump operator, backward jump operator, delta differentiation operator, and nabla differentiation operator σ, ρ, Δ, and ∇. Take $a, b \in \mathbb{T}$, $a < b$. Consider the segment $[a,b]_{\mathbb{T}}$, the real-valued function $f : [a,b]_{\mathbb{T}} \to \mathbb{R}$, and the partition

$$P = \{t_0, t_1, \dots, t_n\} \subset [a,b]_{\mathbb{T}},$$

where

$$a = t_0 < t_1 < \cdots < t_n = b.$$

Definition 3.20. The intervals $(t_{j-1}, t_j]_{\mathbb{T}}$, $j \in \{1,\dots,n\}$, are called the subintervals of the partition P. The set of all partitions of $[a,b]_{\mathbb{T}}$ is denoted by $\mathscr{P}_{\mathbb{T}} = \mathscr{P}(a,b)_{\mathbb{T}}$.

We set

$$\begin{aligned}
m_j &= \inf_{t\in(t_{j-1},t_j]_{\mathbb{T}}} f(t), \\
M_j &= \sup_{t\in(t_{j-1},t_j]_{\mathbb{T}}} f(t), \quad j \in \{1,\dots,n\},
\end{aligned}$$

and

$$m = \inf_{t \in (a,b]_{\mathbb{T}}} f(t),$$
$$M = \sup_{t \in (a,b]_{\mathbb{T}}} f(t).$$

Definition 3.21. The upper Darboux ∇-sum $U_{\mathbb{T}}(f,P)$ and the lower Darboux ∇-sum $L_{\mathbb{T}}(f,P)$ of the function f with respect to the partition $P \in \mathscr{P}(a,b)_{\mathbb{T}}$ are defined as

$$U_{\mathbb{T}}(f,P) = \sum_{j=1}^{n} M_j(t_j - t_{j-1}),$$
$$L_{\mathbb{T}}(f,P) = \sum_{j=1}^{n} m_j(t_j - t_{j-1}).$$

We have the following inequalities:

$$L_{\mathbb{T}}(f,P) \leq U_{\mathbb{T}}(f,P),$$
$$\begin{aligned} L_{\mathbb{T}}(f,P) &= \sum_{j=1}^{n} m_j(t_j - t_{j-1}) \\ &\geq m \sum_{j=1}^{n} (t_j - t_{j-1}) \\ &= m(b-a), \end{aligned}$$

and

$$\begin{aligned} U_{\mathbb{T}}(f,P) &= \sum_{j=1}^{n} M_j(t_j - t_{j-1}) \\ &\leq M \sum_{j=1}^{n} (t_j - t_{j-1}) \\ &= M(b-a). \end{aligned}$$

Thus

$$m(b-a) \leq L_{\mathbb{T}}(f,P) \leq U_{\mathbb{T}}(f,P) \leq M(b-a).$$

Definition 3.22. The upper Darboux ∇-integral $U_{\mathbb{T}}(f)$ of a function f from a to b is defined as

$$U_{\mathbb{T}}(f) = \inf_{P \in \mathscr{P}(a,b)_{\mathbb{T}}} U_{\mathbb{T}}(f,P).$$

The lower Darboux ∇-integral $L_{\mathbb{T}}(f)$ of a function f from a to b is defined as

$$L_{\mathbb{T}}(f) = \sup_{P \in \mathscr{P}(a,b)_{\mathbb{T}}} L_{\mathbb{T}}(f,P).$$

Note that $U_{\mathbb{T}}(f)$ and $L_{\mathbb{T}}(f)$ are finite numbers and

$$m(b-a) \leq L_{\mathbb{T}}(f) \leq U_{\mathbb{T}}(f) \leq M(b-a).$$

Definition 3.23. We say that a function f is ∇-integrable or nabla integrable (shortly integrable) from a to b (or on $[a,b]_{\mathbb{T}}$) if

$$U_{\mathbb{T}}(f) = L_{\mathbb{T}}(f).$$

In this case, we write

$$\begin{aligned}\int_a^b f(t)\nabla t &= U_{\mathbb{T}}(f)\\ &= L_{\mathbb{T}}(f).\end{aligned}$$

Definition 3.24. In each interval $(t_{j-1}, t_j]_{\mathbb{T}}, j \in \{1, \ldots, n\}$, choose an arbitrary point ξ and form the sum

$$S_{\mathbb{T}}(f, P) = \sum_{j=1}^{n} f(\xi_j)(t_j - t_{j-1}).$$

We say that $S_{\mathbb{T}}(f,P)$ is a Riemann ∇-sum or Riemann nabla sum (shortly Riemann sum) of the function f corresponding to the partition $P \in \mathscr{P}(a,b)_{\mathbb{T}}$. We say that f is Riemann nabla integrable or Riemann ∇-integrable (shortly Riemann integrable) from a to b (or on $[a,b]_{\mathbb{T}}$) if there exists a number I with the following property: for each $\varepsilon > 0$, there exists $\delta > 0$ such that

$$\left|S_{\mathbb{T}}(f,P) - I\right| < \varepsilon$$

for every Riemann sum $S_{\mathbb{T}}(f,P)$ corresponding to any partition $P \in \mathscr{P}_{\delta}(a,b)_{\mathbb{T}}$ and independently of the choice of $\xi_j \in (t_{j-1}, t_j]_{\mathbb{T}}, j \in \{1, \ldots, n\}$. The number I is unique and is called the Riemann nabla integral or Riemann ∇-integral (shortly Riemann integral) of the function f from a to b (or on $[a,b]_{\mathbb{T}}$).

Note that the basic properties of the Riemann nabla integrals are proved in a similar way as the properties of the Riemann delta integrals. Therefore we leave them to the reader as exercises.

Exercise 3.51. Prove that any monotone function on $[a,b]_{\mathbb{T}}$ is Riemann nabla integrable.

Exercise 3.52. Prove that every continuous function f on $[a,b]_{\mathbb{T}}$ is Riemann nabla integrable.

Exercise 3.53. Prove that every bounded function f on $[a,b]_{\mathbb{T}}$ with only finitely many discontinuity points is Riemann nabla integrable.

Definition 3.25. A function $f : [a,b]_{\mathbb{T}} \to \mathbb{R}$ is said to be nabla regulated if its finite left-sided limits exist at all left-dense points of $[a,b)_{\mathbb{T}}$ and its finite right-sided limits exist at all right-dense points of $[a,b)_{\mathbb{T}}$.

Exercise 3.54. Prove that any nabla regulated function $f : [a,b]_{\mathbb{T}} \to \mathbb{R}$ is bounded.

Exercise 3.55. Prove that every nabla regulated function $f : [a,b]_{\mathbb{T}} \to \mathbb{R}$ is Riemann nabla integrable on $[a,b]_{\mathbb{T}}$.

Exercise 3.56. Let $f : [a,b]_{\mathbb{T}} \to \mathbb{R}$ be a Riemann nabla integrable, let

$$M = \sup_{t\in(a,b]_{\mathbb{T}}} f(t),$$
$$m = \inf_{t\in(a,b]_{\mathbb{T}}} f(t),$$

and let $h : [m,M] \to \mathbb{R}$ satisfy the Lipschitz condition with constant B. Prove that the composite function $h = \phi \circ f$ is Riemann nabla integrable on $[a,b]_{\mathbb{T}}$.

Exercise 3.57. Let $f : [a,b]_{\mathbb{T}} \to \mathbb{R}$ be Riemann nabla integrable, let

$$M = \sup_{tr\in[a,b]_{\mathbb{T}}} f(t),$$
$$m = \inf_{t\in[a,b]_{\mathbb{T}}} f(t),$$

and let $\phi : [m,M] \to \mathbb{R}$ be a continuous function. Prove that the composite function $h = \phi \circ f : [a,b]_{\mathbb{T}} \to \mathbb{R}$ is Riemann nabla integrable on $[a,b]_{\mathbb{T}}$.

Exercise 3.58. Let $f : [a,b]_{\mathbb{T}} \to \mathbb{R}$ be Riemann nabla integrable on $[a,b]_{\mathbb{T}}$. Prove that for all $\alpha > 0$, the function $|f|^{\alpha} : [a,b]_{\mathbb{T}} \to \mathbb{R}$ is Riemann nabla integrable on $[a,b]_{\mathbb{T}}$.

Exercise 3.59. Let $f : [a,b]_{\mathbb{T}} \to \mathbb{R}$ be a bounded Riemann nabla integrable function on $[a,b]_{\mathbb{T}}$. Prove that f is Riemann nabla integrable on every subinterval $[c,d]_{\mathbb{T}}$ of $[a,b]_{\mathbb{T}}$.

Exercise 3.60. Let f be Riemann nabla integrable on $[a,b]_{\mathbb{T}}$, and let $c \in \mathbb{R}$. Prove that cf is Riemann nabla integrable on $[a,b]_{\mathbb{T}}$ and

$$\int_a^b (cf)(t)\nabla t = c\int_a^b f(t)\nabla t.$$

Exercise 3.61. Let $f, g : [a,b]_{\mathbb{T}} \to \mathbb{R}$ be Riemann nabla integrable on $[a,b]_{\mathbb{T}}$. Prove that $f+g : [a,b]_{\mathbb{T}} \to \mathbb{R}$ is Riemann nabla integrable on $[a,b]_{\mathbb{T}}$. Moreover,

$$\int_a^b (f+g)(t)\nabla t = \int_a^b f(t)\nabla t + \int_a^b g(t)\nabla t.$$

Exercise 3.62. Let $f, g : [a,b]_{\mathbb{T}} \to \mathbb{R}$ be Riemann nabla integrable functions on $[a,b]_{\mathbb{T}}$. Prove that the function $fg : [a,b]_{\mathbb{T}} \to \mathbb{R}$ is Riemann nabla integrable on $[a,b]_{\mathbb{T}}$.

Exercise 3.63. Let $f : [a,b]_{\mathbb{T}} \to \mathbb{R}$, and let $a < c < b$, $c \in \mathbb{T}$. If f is bounded and Riemann nabla integrable on $[a,c]_{\mathbb{T}}$ and $[c,b]_{\mathbb{T}}$, then prove that f is Riemann nabla integrable on $[a,b]_{\mathbb{T}}$ and

$$\int_a^b f(t)\nabla t = \int_a^c f(t)\nabla t + \int_c^b f(t)\nabla t.$$

Exercise 3.64. Let $f, g : [a,b]_{\mathbb{T}} \to \mathbb{R}$ be Riemann nabla integrable functions on $[a,b]_{\mathbb{T}}$ such that $f(t) \le g(t)$, $t \in [a,b]_{\mathbb{T}}$. Prove that

$$\int_a^b f(t)\nabla t \le \int_a^b g(t)\nabla t.$$

Exercise 3.65. Let $f : [a,b]_{\mathbb{T}} \to \mathbb{R}$ be Riemann nabla integrable on $[a,b]_{\mathbb{T}}$. Prove that so is $|f|$ and

$$\left|\int_a^b f(t)\nabla t\right| \le \int_a^b |f(t)|\nabla t.$$

Exercise 3.66. Let $f, g : [a,b]_{\mathbb{T}} \to \mathbb{R}$ be Riemann nabla integrable on $[a,b]_{\mathbb{T}}$. Prove that

$$\begin{aligned}\left|\int_a^b f(t)g(t)\nabla t\right| &\le \int_a^b |f(t)||g(t)|\nabla t\\ &\le \Big(\sup_{t\in[a,b)} |f(t)|\Big)\Bigg(\int_a^b |g(t)|\nabla t\Bigg).\end{aligned}$$

Exercise 3.67. Let $\{f_k\}_{k\in\mathbb{N}}$ be a sequence of Riemann nabla integrable functions on $[a,b]_{\mathbb{T}}$ such that $f_k \to f$ as $k \to \infty$ uniformly on $[a,b]_{\mathbb{T}}$ for a function f on $[a,b]_{\mathbb{T}}$. Prove that f is Riemann nabla integrable on $[a,b]_{\mathbb{T}}$ and

$$\int_a^b f(t)\nabla t = \lim_{k\to\infty} \int_a^b f_k(t)\nabla t.$$

Exercise 3.68. Let $f = \sum_{k=1}^{\infty} f_k$ be a series of Riemann nabla integrable functions on $[a,b]_{\mathbb{T}}$ that converges uniformly to f on $[a,b]_{\mathbb{T}}$. Then f is Riemann nabla integrable on $[a,b]_{\mathbb{T}}$, and

$$\int_a^b f(t)\nabla t = \sum_{k=1}^{\infty} \int_a^b f_k(t)\nabla t.$$

Exercise 3.69 (Fundamental theorem of calculus I). Let $g : [a,b]_{\mathbb{T}} \to \mathbb{R}$ be a continuous function that is nabla differentiable on $(a,b]_{\mathbb{T}}$ such that g^{Δ} is Riemann nabla integrable on $(a,b]_{\mathbb{T}}$. Prove that

$$\int_a^b g^{\nabla}(t)\nabla t = g(b) - g(a).$$

Exercise 3.70 (Integration by parts). Let u and v be continuous functions on $[a,b]_{\mathbb{T}}$ and nabla differentiable on $[a,b)_{\mathbb{T}}$. Prove that if u^{∇} and v^{∇} are nabla integrable on $(a,b]_{\mathbb{T}}$, then

$$\int_a^b u^{\nabla}(t)v(t)\nabla t + \int_a^b u^{\rho}(t)v^{\nabla}(t)\nabla t = u(b)v(b) - u(a)v(a).$$

Exercise 3.71 (Fundamental theorem of calculus II). Let f be nabla integrable on $[a,b]_{\mathbb{T}}$, and let

$$F(t) = \int_a^t f(\tau)\nabla\tau, \quad t \in [a,b]_{\mathbb{T}}.$$

Prove that F is uniformly continuous on $[a,b]_{\mathbb{T}}$. Moreover, if f is continuous at a left-dense point $t_0 \in (a,b]_{\mathbb{T}}$, then F is nabla differentiable at t_0, and

$$F^{\nabla}(t_0) = f(t_0).$$

Exercise 3.72. Let f and f^{∇} be continuous on $[a,b]_{\mathbb{T}}$. Prove that

$$\left(\int_a^t f(t,s)\nabla s\right)^{\nabla} = f\big(\rho(t),t\big) + \int_a^t f^{\nabla}(t,s)\nabla s, \quad t \in [a,b]_{\mathbb{T}}.$$

Definition 3.26. A function $f : \mathbb{T} \to \mathbb{R}$ is said to be locally ∇-integrable or locally nabla integrable (shortly locally integrable) if it is nabla integrable on any finite interval.

Exercise 3.73 (Change of variables). Let $g : \mathbb{T} \to \mathbb{R}$ be a strictly increasing function such that $g(\mathbb{T}) = \tilde{T}$ is a time scale with nabla differentiation operator $\tilde{\nabla}$. Let also $f : \mathbb{T} \to \mathbb{R}$ be locally nabla integrable on T, and let g be nabla differentiable with locally nabla integrable derivative. Suppose that g^{∇} possesses an antiderivative. Prove that for all $a, b \in \mathbb{T}$,

$$\int_a^b f(t)g^\nabla(t)\nabla t = \int_{g(a)}^{g(b)} \big(f(g^{-1})\big)(s)\tilde{\nabla} s.$$

Exercise 3.74 (First mean value theorem). Let f and g be bounded and Riemann nabla integrable functions on $[a,b]_\mathbb{T}$. Let g be a nonnegative or nonpositive function on $[a,b]_\mathbb{T}$, and let

$$m = \inf_{t\in[a,b)_\mathbb{T}} f(t),$$
$$M = \sup_{t\in[a,b)_\mathbb{T}} f(t).$$

Prove that there exists a constant $A \in [m, M]$ such that

$$\int_a^b f(t)g(t)\nabla t = A\int_a^b f(t)\nabla t.$$

Exercise 3.75 (First mean value theorem). Let f be bounded and Riemann nabla integrable function on $[a,b]_\mathbb{T}$, and let

$$m - \inf_{t\in(a,b]_\mathbb{T}} f(t),$$
$$M = \sup_{t\in(a,b]_\mathbb{T}} f(t).$$

Prove that there exists a constant $A \in [m, M]$ such that

$$\int_a^b f(t)\Delta t = A(b-a).$$

Exercise 3.76 (Second mean value theorem I). Let f be bounded and Riemann nabla integrable function on $[a,b]_\mathbb{T}$, let

$$F(t) = \int_a^t f(s)\nabla s, \quad t \in [a,b]_\mathbb{T},$$

and let

$$m_F = \inf_{t\in(a,b]_\mathbb{T}} f(t),$$
$$M_F = \sup_{t\in(a,b]_\mathbb{T}} f(t).$$

Suppose that $g : [a,b]_\mathbb{T} \to \mathbb{R}$ is nonincreasing and nonnegative a function on $[a,b]_\mathbb{T}$. Prove that there is a constant $A \in [m_F, M_F]$ such that

$$\int_a^b f(t)g(t)\nabla t = Ag(a).$$

Exercise 3.77 (Second mean value theorem II). Let f be bounded and Riemann nabla integrable function on $[a,b]_{\mathbb{T}}$, let

$$F(t) = \int_a^t f(s)\nabla s, \quad t \in [a,b]_{\mathbb{T}},$$

and let

$$m_F = \inf_{t\in(a,b]_{\mathbb{T}}} f(t),$$
$$M_F = \sup_{t\in(a,b]_{\mathbb{T}}} f(t).$$

Let $g : [a,b]_{\mathbb{T}} \to \mathbb{R}$ be a monotone function on $[a,b]_{\mathbb{T}}$. Prove that there is a constant $A \in [m_F, M_F]$ such that

$$\int_a^b f(t)g(t)\nabla t = A\big(g(a) - g(b)\big) + g(b)\int_a^b f(t)\nabla t. \tag{3.68}$$

Exercise 3.78 (Second mean value theorem III). Let f be bounded and Riemann nabla integrable function on $[a,b]_{\mathbb{T}}$, let

$$\Phi(t) = \int_t^b f(s)\nabla s, \quad t \in [a,b]_{\mathbb{T}},$$

and let

$$m_\Phi = \inf_{t\in(a,b]_{\mathbb{T}}} f(t),$$
$$M_\Phi = \sup_{t\in(a,b]_{\mathbb{T}}} f(t).$$

Let $g : [a,b]_{\mathbb{T}} \to \mathbb{R}$ be a nondecreasing nonnegative function on $[a,b]_{\mathbb{T}}$. Prove that there is a constant $A \in [m_F, M_F]$ such that

$$\int_a^b f(t)g(t)\nabla t = Ag(b). \tag{3.69}$$

Exercise 3.79 (Second mean value theorem IV). Let f be bounded and Riemann nabla integrable function on $[a,b]_{\mathbb{T}}$, let

$$\Phi(t) = \int_t^b f(s)\nabla s, \quad t \in [a,b]_{\mathbb{T}},$$

and let

$$m_\Phi = \inf_{t\in(a,b]_{\mathbb{T}}} f(t),$$
$$M_\Phi = \sup_{t\in(a,b]_{\mathbb{T}}} f(t).$$

Let $g : [a,b]_{\mathbb{T}} \to \mathbb{R}$ be a monotone function on $[a,b]_{\mathbb{T}}$. Prove that there is a constant $A \in [m_\Phi, M_\Phi]$ such that

$$\int_a^b f(t)g(t)\nabla t = A\big(g(b) - g(a)\big) + g(a)\int_a^b f(t)\nabla t. \tag{3.70}$$

Example 3.125. Let $\mathbb{T} = \mathbb{Z}$. We will compute the integral

$$\int_{-3}^{1} (3t^2 - 3t + 2)\nabla t.$$

Here

$$\rho(t) = t - 1, \quad t \in \mathbb{T}.$$

Let

$$g(t) = t^3 + t - 1, \quad t \in \mathbb{T}.$$

We have

$$\begin{aligned} g^\nabla(t) &= (\rho(t))^2 + t\rho(t) + t^2 + 1 \\ &= (t-1)^2 + t(t-1) + t^2 + 1 \\ &= t^2 - 2t + 1 + t^2 - t + t^2 + 1 \\ &= 3t^2 - 3t + 2, \quad t \in \mathbb{T}. \end{aligned}$$

Therefore

$$\begin{aligned} \int_{-3}^{1} (3t^2 - 3t + 2)\nabla t &= \int_{-3}^{1} g^\nabla(t)\nabla t \\ &= g(t)\Big|_{t=-3}^{t=1} \\ &= g(1) - g(-3) \\ &= (1^3 + 1 - 1) - ((-3)^3 - 3 - 1) \\ &= 1 - (-27 - 4) \\ &= 1 + 31 \\ &= 32. \end{aligned}$$

Example 3.126. Let $\mathbb{T} = 2^{\mathbb{N}_0}$, and let

$$f(t) = \frac{7}{4}t^2 + \frac{3}{2}t + 1, \quad t \in \mathbb{T}.$$

We will compute

$$\int f(t)\nabla t.$$

Let

$$g(t) = t^3 + t^2 + t + 1, \quad t \in \mathbb{T}.$$

By Example 2.41 we get

$$g^\nabla(t) = f(t), \quad t \in \mathbb{T}_\kappa.$$

Then

$$\begin{aligned}\int f(t)\nabla t &= \int g^\nabla(t)\nabla t\\ &= g(t) + C\\ &= t^3 + t^2 + t + C, \quad t \in \mathbb{T}_\kappa,\end{aligned}$$

where C is a constant.

Example 3.127. Let $\mathbb{T} = P_{1,3}$ and

$$f(t) = \begin{cases} -\frac{2}{(t+1)(t-2)} & \text{if } t \text{ is left-dense,}\\ -\frac{2}{(t+1)^2} & \text{if } t \text{ is left-scattered.}\end{cases}$$

We will compute

$$\int f(t)\nabla t, \quad t \in \mathbb{T}_\kappa.$$

Let

$$g(t) = \frac{1-t}{1+t}, \quad t \in \mathbb{T}.$$

By Example 2.42 we have that

$$g^\nabla(t) = f(t), \quad t \in \mathbb{T}_\kappa.$$

Therefore

$$\begin{aligned}\int f(t)\nabla t &= \int g^\nabla(t)\nabla t\\ &= g(t) + C\\ &= \frac{1-t}{1+t} + C, \quad t \in \mathbb{T}_\kappa,\end{aligned}$$

where C is a constant.

Example 3.128. Let $\mathbb{T} = \{-\frac{1}{n} : n \in \mathbb{N}\} \cup \mathbb{N}_0$, and let

$$f(t) = \begin{cases} \frac{8t^2-5t-7}{(8t+1)(7t+1)} & \text{if } t \in \{-\frac{1}{n} : n \in \mathbb{N},\ n \geq 2\}, \\ -7 & \text{if } t = 0, \\ \frac{7t^2-5t-8}{(7t+1)(7t-6)} & \text{if } t \in \mathbb{N}. \end{cases}$$

We will compute

$$\int f(t)\nabla t, \quad t \in \mathbb{T}_\kappa.$$

Let

$$g(t) = \frac{1+t^2}{1+7t}, \quad t \in \mathbb{T}.$$

By Example 2.43 we have that

$$g^\nabla(t) = f(t), \quad t \in \mathbb{T}_\kappa.$$

Therefore

$$\begin{aligned} \int f(t)\nabla t &= \int g^\nabla(t)\nabla t \\ &= g(t) + C \\ &= \frac{1+t^2}{1+7t} + C, \quad t \in \mathbb{T}_\kappa, \end{aligned}$$

where C is a constant.

Exercise 3.80. Let $U = \{\frac{1}{2^n} : n \in \mathbb{N}\}$, let $\mathbb{T} = \{0\} \cup U \cup (1-U) \cup \{1\}$, and let

$$f(t) = \begin{cases} -\frac{15}{8}t^3 - \frac{3}{2}t & \text{if } t \in U, \\ -15t^3 + 17t^2 - 10t + 2 & \text{if } t \in (1-U)\backslash\{\frac{1}{2}\}, \\ -\frac{63}{4} & \text{if } t = \frac{1}{2}, \\ -6 & \text{if } t = 1. \end{cases}$$

Prove that

$$\int f(t)\nabla t = -t^4 - t^2 + C, \quad t \in \mathbb{T}_\kappa,$$

where C is a constant.

Example 3.129. Let f and f^∇ be continuous on $[a,b]_{\mathbb{T}}$. Then

$$\left(\int_a^t f(t,s)\Delta s\right)^\nabla = f(\rho(t),\rho(t)) + \int_a^t f_t^\nabla(t,s)\Delta s, \quad t\in[a,b]_{\mathbb{T}}.$$

Indeed, define the function

$$g(t) = \int_a^t f(t,s)\Delta s, \quad t\in\mathbb{T}.$$

Take arbitrary $t\in[a,b]_{\mathbb{T}}$. We will consider the following cases.

1. t is left-scattered. Then

$$\begin{aligned} g^\nabla(t) &= \frac{g(\rho(t))-g(t)}{\rho(t)-t} \\ &= \frac{\int_a^{\rho(t)} f(\rho(t),s)\Delta s - \int_a^t f(t,s)\Delta s}{\rho(t)-t} \\ &= \int_a^{\rho(t)} \frac{f(\rho(t),s)}{\rho(t)-t}\Delta s - \int_a^t \frac{f(t,s)}{\rho(t)-t}\Delta s \\ &= \int_a^t \frac{f(\rho(t),s)}{\rho(t)-t}\Delta s + \int_t^{\rho(t)} \frac{f(\rho(t),s)}{\rho(t)-t}\Delta s - \int_a^t \frac{f(t,s)}{\rho(t)-t}\Delta s \\ &= \int_a^t \frac{f(\rho(t),s)-f(t,s)}{\rho(t)-t}\Delta s + (\rho(t)-t)\frac{f(\rho(t),\rho(t))}{\rho(t)-t} \\ &= f(\rho(t),\rho(t)) + \int_a^t f^\nabla(t,s)\Delta s. \end{aligned}$$

2. t is left-dense. Take arbitrary $\varepsilon > 0$. Since f and f^∇ are continuous on $[a,b]_{\mathbb{T}}$, they are uniformly continuous on $[a,b]_{\mathbb{T}}$. Then there is a neighborhood U of the point t such that

$$\begin{aligned} |f^\nabla(\tau_1,s) - f^\nabla(\xi_1,s)| &< \frac{\varepsilon}{2}, \\ |f(r,s_1) - f(t,t)| &< \frac{\varepsilon}{2} \end{aligned}$$

for all $\tau_1,\xi_1,r,s_1\in U$ and for any fixed $s\in[a,b]_{\mathbb{T}}$. On the other hand, for any fixed $s\in[a,b]_{\mathbb{T}}$ and $r\in U\backslash\{t\}$, by the mean value theorem it follows that there are $\tau,\xi\in U\backslash\{t\}$ such that

$$f^\nabla(\tau,s) \le \frac{f(t,s)-f(r,s)}{t-r} \le f^\nabla(\xi,s).$$

Hence we conclude that there is a neighborhood $U_1 \subset U \backslash \{t\}$ such that for all $r \in U_1$, we have

$$\left| \frac{f(t,s) - f(r,s)}{t-r} - f^{\nabla}(t,s) \right| < \frac{\varepsilon}{2(b-a)}.$$

Let now $r \in U_1$. Without loss of generality, suppose that $t > r$. Then we have

$$\begin{aligned}
&\left| \frac{g(t) - g(r)}{t-r} - \int_a^t f^{\nabla}(t,s)\Delta s - f(t,t) \right| \\
&= \left| \frac{1}{t-r} \int_a^t f(t,s)\Delta s - \frac{1}{t-r} \int_a^r f(r,s)\Delta s - \int_a^t f^{\nabla}(t,s)\Delta s - f(t,t) \right| \\
&= \left| \frac{1}{t-r} \int_a^t f(t,s)\Delta s - \frac{1}{t-r} \int_a^t f(r,s)\Delta s + \frac{1}{t-r} \int_r^t f(r,s)\Delta s - \int_a^t f^{\nabla}(t,s)\Delta s - f(t,t) \right| \\
&= \left| \frac{1}{t-r} \int_a^t (f(t,s) - f(r,s))\Delta s \quad \int_a^t f^{\nabla}(t,s)\Delta s + \frac{1}{t-r} \int_r^t (f(r,s) - f(t,t))\Delta s \right| \\
&\le \left| \frac{1}{t-r} \int_a^t (f(t,s) - f(r,s))\Delta s - \int_a^t f^{\nabla}(t,s)\Delta s \right| + \left| \frac{1}{t-r} \int_r^t (f(r,s) - f(t,t))\Delta s \right| \\
&\le \left| \int_a^t \left(\frac{f(t,s) - f(r,s)}{t-r} - f^{\nabla}(t,s) \right)\Delta s \right| + \int_r^t \frac{|f(r,s) - f(t,t)|}{t-r}\Delta s \\
&\le \int_a^t \left| \frac{f(t,s) - f(r,s)}{t-r} - f^{\nabla}(t,s) \right| \Delta s + \frac{\varepsilon}{2}\frac{1}{t-r}(t-r) \\
&< \frac{\varepsilon}{2(b-a)} \int_a^b \Delta s + \frac{\varepsilon}{2} \\
&= \frac{\varepsilon}{2} + \frac{\varepsilon}{2} \\
&= \varepsilon.
\end{aligned}$$

Example 3.130. Let f and f^{Δ} be continuous functions on $[a,b]_{\mathbb{T}}$. Then

$$\left(\int_a^t f(t,s)\nabla s \right)^{\Delta} = f(\sigma(t), \sigma(t)) + \int_a^t f_t^{\Delta}(t,s)\nabla s, \quad t \in [a,b]_{\mathbb{T}}.$$

Indeed, define the function

$$g(t) = \int_a^t f(t,s)\nabla s, \quad t \in \mathbb{T}.$$

Take arbitrary $t \in [a,b]_{\mathbb{T}}$. We will consider the following cases.

1. t is right-scattered. Then

$$\begin{aligned}
g^{\Delta}(t) &= \frac{g(\sigma(t)) - g(t)}{\sigma(t) - t} \\
&= \frac{\int_a^{\sigma(t)} f(\sigma(t),s)\nabla s - \int_a^t f(t,s)\nabla s}{\sigma(t) - t} \\
&= \int_a^{\sigma(t)} \frac{f(\sigma(t),s)}{\sigma(t)-t}\nabla s - \int_a^t \frac{f(t,s)}{\sigma(t)-t}\nabla s \\
&= \int_a^t \frac{f(\sigma(t),s)}{\sigma(t)-t}\nabla s + \int_t^{\sigma(t)} \frac{f(\sigma(t),s)}{\sigma(t)-t}\nabla s - \int_a^t \frac{f(t,s)}{\sigma(t)-t}\Delta s \\
&= \int_a^t \frac{f(\sigma(t),s) - f(t,s)}{\sigma(t)-t}\sigma s + (\sigma(t)-t)\frac{f(\sigma(t),\sigma(t))}{\sigma(t)-t} \\
&= f(\sigma(t),\sigma(t)) + \int_a^t f^{\Delta}(t,s)\nabla s.
\end{aligned}$$

2. t is right-dense. Take arbitrary $\varepsilon > 0$. Since f and f^{Δ} are continuous on $[a,b]_{\mathbb{T}}$, they are uniformly continuous on $[a,b]_{\mathbb{T}}$. Then there is a neighborhood U of the point t such that

$$\begin{aligned}
|f^{\Delta}(\tau_1,s) - f^{\Delta}(\xi_1,s)| &< \frac{\varepsilon}{2}, \\
|f(r,s_1) - f(t,t)| &< \frac{\varepsilon}{2}
\end{aligned}$$

for all $\tau_1, \xi_1, r, s_1 \in U$ and any fixed $s \in [a,b]_{\mathbb{T}}$. On the other hand, for any fixed $s \in [a,b]_{\mathbb{T}}$ and $r \in U\backslash\{t\}$, by the mean value theorem it follows that there are $\tau, \xi \in U\backslash\{t\}$ such that

$$f^{\Delta}(\tau,s) \le \frac{f(t,s) - f(r,s)}{t-r} \le f^{\Delta}(\xi,s).$$

Hence we conclude that there is a neighborhood $U_1 \subset U\backslash\{t\}$ such that for all $r \in U_1$, we have

$$\left|\frac{f(t,s) - f(r,s)}{t-r} - f^{\Delta}(t,s)\right| < \frac{\varepsilon}{2(b-a)}.$$

Let now $r \in U_1$. Without loss of generality, suppose that $t > r$. Then we have

$$
\begin{aligned}
&\left|\frac{g(t)-g(r)}{t-r} - \int_a^t f^{\Delta}(t,s)\nabla s - f(t,t)\right| \\
&= \left|\frac{1}{t-r}\int_a^t f(t,s)\nabla s - \frac{1}{t-r}\int_a^r f(r,s)\nabla s - \int_a^t f^{\Delta}(t,s)\nabla s - f(t,t)\right| \\
&= \left|\frac{1}{t-r}\int_a^t f(t,s)\nabla s - \frac{1}{t-r}\int_a^t f(r,s)\nabla s + \frac{1}{t-r}\int_r^t f(r,s)\nabla s - \int_a^t f^{\Delta}(t,s)\nabla s - f(t,t)\right| \\
&= \left|\frac{1}{t-r}\int_a^t (f(t,s)-f(r,s))\nabla s - \int_a^t f^{\Delta}(t,s)\nabla s + \frac{1}{t-r}\int_r^t (f(r,s)-f(t,t))\nabla s\right| \\
&\le \left|\frac{1}{t-r}\int_a^t (f(t,s)-f(r,s))\nabla s - \int_a^t f^{\Delta}(t,s)\nabla s\right| + \left|\frac{1}{t-r}\int_r^t (f(r,s)-f(t,t))\nabla s\right| \\
&\le \left|\int_a^t \left(\frac{f(t,s)-f(r,s)}{t-r} - f^{\Delta}(t,s)\right)\nabla s\right| + \int_r^t \frac{|f(r,s)-f(t,t)|}{t-r}\nabla s \\
&\le \int_a^t \left|\frac{f(t,s)-f(r,s)}{t-r} - f^{\Delta}(t,s)\right|\nabla s + \frac{\varepsilon}{2}\frac{1}{t-r}(t-r) \\
&< \frac{\varepsilon}{2(b-a)}\int_a^b \nabla s + \frac{\varepsilon}{2} \\
&= \frac{\varepsilon}{2} + \frac{\varepsilon}{2} \\
&= \varepsilon.
\end{aligned}
$$

Definition 3.27. Let f be the real-valued function on $[a,\infty)_{\mathbb{T}}$ such that it is nabla integrable from a to every point $A \in \mathbb{T}, A \ge a$. If the integral

$$
F(A) = \int_a^A f(t)\nabla t
$$

has a finite limit as $A \to \infty$, then we call that limit the improper nabla integral (shortly improper integral) of the first kind from a to ∞, and we write

$$
\int_a^\infty f(t)\nabla t = \lim_{A\to\infty}\left\{\int_a^A f(t)\nabla t\right\}. \tag{3.71}
$$

In such a case, we say that the improper integral

$$\int_a^\infty f(t)\nabla t \tag{3.72}$$

exists or that it converges. If the limit (3.71) does not exist, then we say that the improper integral (3.72) does not exist and that it diverges.

Definition 3.28. Let $\mathbb{T}$ be a time scale, let $a, b \in \mathbb{T}$, $a < b$, be fixed points, and let b be left-dense. Suppose that a function f is defined in the interval $[a, b)_{\mathbb{T}}$ and is integrable on every interval $[a, c]_{\mathbb{T}}$ with $c < b$ and is unbounded on $[a, b)_{\mathbb{T}}$. Then the ordinary Riemann integral of f on $[a, b)_{\mathbb{T}}$ cannot exist since a Riemann-integrable function from a to b must be bounded on $[a, b)_{\mathbb{T}}$. The formal expression

$$\int_a^b f(t)\nabla t \tag{3.73}$$

is called the improper integral of the second kind. We say that integral (3.73) is improper at $t = b$. Sometimes, we say that f has singularity at $t = b$. If the left-sided limit

$$\lim_{c \to b-} \int_a^c f(t)\nabla t \tag{3.74}$$

exists, then we call this limit the value of the improper integral (3.73), and we write

$$\int_a^b f(t)\nabla t = \lim_{c \to b-} \int_a^c f(t)\nabla t.$$

If, moreover, this limit is finite, then we say that the improper integral (3.73) converges. Otherwise, if the limit (3.74) is infinite or does not exist, then we say that the improper integral (3.73) diverges.

Example 3.131. Let

$$\mathbb{T} = \{t_k : k \in \mathbb{N}_0\} \quad \text{with } 0 < t_0 < t_0 < \cdots, \quad \lim_{k \to \infty} t_k = \infty. \tag{3.75}$$

If $f : [t_0, \infty)_{\mathbb{T}} \to \mathbb{R}$ is a nonincreasing function, then

$$\int_{t_0}^\infty f(t)\nabla t \le \int_{t_0}^\infty f(t)dt \le \int_{t_0}^\infty f(t)\Delta t.$$

Indeed, we have

$$\int_{t_0}^\infty f(t)\nabla t = \sum_{k=0}^\infty f(t_{k+1})(t_{k+1} - t_k)$$

and

$$\int_{t_0}^{\infty} f(t)\Delta t = \sum_{k=0}^{\infty} f(t_k)(t_{k+1} - t_k).$$

Because the function f is nonincreasing on $[t_0, \infty)_{\mathbb{T}}$, we get

$$\begin{aligned} f(t_{k+1})(t_{k+1} - t_k) &\leq \int_{t_k}^{t_{k+1}} f(t)dt \\ &\leq f(t_k)(t_{k+1} - t_k) \quad \text{for all } k \in \mathbb{N}_0. \end{aligned}$$

Hence

$$\begin{aligned} \int_{t_0}^{\infty} f(t)\nabla t &= \sum_{k=0}^{\infty} f(t_{k+1})(t_{k+1} - t_k) \\ &\leq \sum_{k=0}^{\infty} \int_{t_k}^{t_{k+1}} f(t)dt \\ &= \int_{t_0}^{\infty} f(t)dt \\ &\leq \sum_{k=0}^{\infty} f(t_k)(t_{k+1} - t_k) \\ &= \int_{t_0}^{\infty} f(t)\Delta t. \end{aligned}$$

Example 3.132. Let $\mathbb{T}$ be as in (3.75). Suppose that f is nonincreasing on $[t_0, \infty)_{\mathbb{T}}$ and

$$\int_{t_0}^{\infty} f(t)dt = \infty.$$

Then applying Example 3.131, we get that the integral

$$\int_{t_0}^{\infty} f(t)\Delta t$$

diverges.

Example 3.133. Let $\mathbb{T}$ be as in (3.75), and let f be nonincreasing on $[t_0, \infty)_{\mathbb{T}}$. Suppose that

$$\int_{t_0}^{\infty} f(t)dt < \infty.$$

Then applying Example 3.131, we conclude that the integral

$$\int_{t_0}^{\infty} f(t)\nabla t$$

converges.

Example 3.134. Let $\mathbb{T}$ be a time scale as in (3.75). Consider the integral

$$\int_{t_0}^{\infty} \frac{1}{t^p}\nabla t, \quad p > 1.$$

Let

$$f(t) = \frac{1}{t^p}, \quad t \in \mathbb{T}.$$

Then the function f is nonincreasing on $[t_0, \infty)_{\mathbb{T}}$. Note that

$$\begin{aligned}
\int_{t_0}^{\infty} \frac{1}{t^p} dt &= \lim_{A\to\infty} \int_{t_0}^{A} \frac{1}{t^p} dt \\
&= \frac{1}{1-p} \lim_{A\to\infty} \left(\frac{1}{t^{p-1}} \Big|_{t=t_0}^{t=A} \right) \\
&= \frac{1}{1-p} \lim_{A\to\infty} \left(\frac{1}{A^{p-1}} - \frac{1}{t_0^{p-1}} \right) \\
&= \frac{1}{(p-1)t_0^{p-1}},
\end{aligned}$$

i. e., the integral

$$\int_{t_0}^{\infty} \frac{1}{t^p} dt$$

converges. Now applying Example 3.131, we conclude that the integral

$$\int_{t_0}^{\infty} \frac{1}{t^p}\nabla t, \quad p > 1,$$

converges.

Example 3.135. Let $\mathbb{T}$ be a time scale as in (3.75). Consider the integral

$$\int_{t_0}^{\infty} \frac{1}{t} \nabla t.$$

We have

$$\int_{t_0}^{\infty} \frac{1}{t} \nabla t = \sum_{k=0}^{\infty} \frac{t_{k+1} - t_k}{t_{k+1}}.$$

Assume that

$$\sum_{k=0}^{\infty} \frac{t_{k+1} - t_k}{t_{k+1}} < \infty.$$

Then

$$\lim_{k \to \infty} \frac{t_{k+1} - t_k}{t_{k+1}} = 0.$$

Hence

$$\lim_{k \to \infty} \frac{t_{k+1}}{t_k} = 1.$$

Because $t_k < t_{k+1}$, $k \in \mathbb{N}_0$, there is $N \in \mathbb{N}$ such that

$$\frac{t_{k+1}}{t_k} < 2 \quad \text{for all } k > N.$$

Since

$$\int_{t_0}^{\infty} \frac{1}{t} dt = \infty,$$

applying Example 3.131, we get that

$$\int_{t_0}^{\infty} \frac{\Delta t}{t} = \infty.$$

Because

$$\int_{t_0}^{\infty} \frac{\Delta t}{t} = \sum_{k=0}^{\infty} \frac{t_{k+1} - t_k}{t_k},$$

we get

$$\sum_{k=0}^{\infty} \frac{t_{k+1} - t_k}{t_k} = \infty. \tag{3.76}$$

On the other hand, we have

$$\begin{aligned}
\sum_{k=0}^{\infty} \frac{t_{k+1} - t_k}{t_k} &= \sum_{k=0}^{N-1} \frac{t_{k+1} - t_k}{t_k} + \sum_{k=N}^{\infty} \frac{t_{k+1} - t_k}{t_k} \\
&= \sum_{k=0}^{N-1} \frac{t_{k+1} - t_k}{t_k} + \sum_{k=N}^{\infty} \frac{t_{k+1} - t_k}{t_k + 1} \frac{t_{k+1}}{t_k} \\
&< \sum_{k=0}^{N-1} \frac{t_{k+1} - t_k}{t_k} + 2 \sum_{k=N}^{\infty} \frac{t_{k+1} - t_k}{t_{k+1}} \\
&< \infty,
\end{aligned}$$

which contradicts with (3.76). Therefore

$$\sum_{k=0}^{\infty} \frac{t_{k+1} - t_k}{t_{k+1}} = \infty$$

and

$$\int_{t_0}^{\infty} \frac{1}{t} \nabla t = \infty.$$

Example 3.136. Let $\mathbb{T}$ be as in (3.75). Consider the integral

$$\int_{t_0}^{\infty} \frac{1}{t^p} \nabla t, \quad p \in [0, 1).$$

Since

$$\lim_{k \to \infty} t_k = \infty,$$

there exists $t_l \in \mathbb{T}$ such that $t_l > 1$. Therefore

$$\begin{aligned}
\int_{t_l}^{\infty} \frac{1}{t^p} \nabla t &\geq \int_{t_l}^{\infty} \frac{1}{t} \nabla t \\
&= \infty.
\end{aligned}$$

Hence

$$\int_{t_0}^{\infty} \frac{1}{t^p} \nabla t = \int_{t_0}^{t_l} \frac{1}{t^p} \nabla t + \int_{t_l}^{\infty} \frac{1}{t^p} \nabla t$$
$$= \infty.$$

Thus the considered integral diverges.

3.12 Advanced practical problems

Problem 3.1. Let $\mathbb{T} = (-5\mathbb{N}_0) \cup 4^{\mathbb{N}_0}$. Check if the function

$$f(t) = \frac{t^3 - 2t^2 + t - 1}{t^3 - t^2 + 3t - 3}, \quad t \in \mathbb{T},$$

is regulated.

Problem 3.2. Let $\mathbb{T} = [-1, 3] \cup [7, 9] \cup [10, \infty)$. Check if the function

$$f(t) = \frac{t^{101} - 101t + 100}{t^2 - 2t + 1}, \quad t \in \mathbb{T},$$

is regulated.

Problem 3.3. Let $\mathbb{T} = \{0\} \cup \{1 - \frac{1}{4^n}\}_{n \in \mathbb{N}_0} \cup 2^{\mathbb{N}_0}$. Check if the function

$$f(t) = \frac{3}{1 - t^3} + \frac{1}{t - 1}, \quad t \in \mathbb{T},$$

is regulated.

Problem 3.4. Let $\mathbb{T} = \{2\} \cup \{2 + \frac{1}{n}\}_{n \in \mathbb{N}} \cup [4, 9] \cup \{9 + \frac{1}{n^2}\}_{n \in \mathbb{N}}$. Check if the function

$$f(t) = \frac{2t^2 - 11t - 21}{t^2 - 9t + 14}, \quad t \in \mathbb{T},$$

is regulated.

Problem 3.5. Let $\mathbb{T} = [1, 3] \cup 5^{\mathbb{N}}$. Check if the function

$$f(t) = \frac{t^4 - 2t + 1}{t^8 - 2t + 1}, \quad t \in \mathbb{T},$$

is regulated.

Problem 3.6. Let $\mathbb{T} = [-1, 3] \cup [7, 9] \cup [10, \infty)$, and let

$$f(t) = \begin{cases} 3t & \text{if } t \in [-1, 3] \cup (7, 9] \cup [10, \infty), \\ \frac{t+1}{t+20} & \text{if } t \in \{7, 10\}. \end{cases}$$

Check if f is delta predifferentiable, and if it is, then find its region D.

Problem 3.7. Let $\mathbb{T} = \{2\} \cup \{2 + \frac{1}{n}\}_{n\in\mathbb{N}} \cup [4,9] \cup \{9 + \frac{1}{n^2}\}_{n\in\mathbb{N}}$, and let

$$f(t) = \begin{cases} t & \text{if } t \in \{2 + \frac{1}{n}\}_{n\in\mathbb{N}} \cup (4,9] \cup \{9 + \frac{1}{n^2}\}_{n\in\mathbb{N}}, \\ t^2 & \text{if } t \in \{2,4\}. \end{cases}$$

Check if f is delta predifferentiable, and if it is, then find its region D.

Problem 3.8. Let $\mathbb{T} = (-5\mathbb{N}_0) \cup 4^{\mathbb{N}_0}$, and let

$$f(t) = \begin{cases} 3t^2 + 15t + 25 & \text{if } t \in (-5\mathbb{N}), \\ 1 & \text{if } t = 0, \\ 21t^2 & \text{if } t \in 4^{\mathbb{N}_0}. \end{cases}$$

Prove that

$$\int f(t)\Delta t = t^3 + C, \quad t \in \mathbb{T},$$

where C is a constant.

Problem 3.9. Let $\mathbb{T} = [-1,3] \cup [7,9] \cup [10,\infty)$, and let

$$f(t) = \begin{cases} 4 + 18t & \text{if } t \in [-1,3) \cup [7,9) \cup [10,\infty), \\ 94 & \text{if } t = 3, \\ 175 & \text{if } t = 9. \end{cases}$$

Prove that

$$\int f(t)\Delta t = 4t + 9t^2 + C, \quad t \in \mathbb{T},$$

where C is a constant.

Problem 3.10. Let $\mathbb{T} = \{0\} \cup \{1 - \frac{1}{4^n}\}_{n\in\mathbb{N}_0} \cup 2^{\mathbb{N}_0}$, and let

$$f(t) = \begin{cases} \frac{4(3+5t)}{(2+t^2)(41+6t+t^2)} & \text{if } t \in \{1 - \frac{1}{4^n}\}_{n\in\mathbb{N}}, \\ \frac{6}{41} & \text{if } t = 0, \\ \frac{3t}{(2+t^2)(2+9t^2)} & \text{if } t \in 2^{\mathbb{N}_0}. \end{cases}$$

Prove that

$$\int f(t)\Delta t = \frac{1+t^2}{2+t^2} + C, \quad t \in \mathbb{T},$$

where C is a constant.

Problem 3.11. Let $\mathbb{T} = \{2\} \cup \{2 + \frac{1}{n}\}_{n \in \mathbb{N}} \cup [4, 9] \cup \{9 + \frac{1}{n^2}\}_{n \in \mathbb{N}}$, and let

$$f(t) = \begin{cases} -3 - 2t & \text{if } t \in \{2\} \cup [4, 9) \cup \{10\}, \\ \frac{13-t-t^2}{t-3} & \text{if } t \in \{2 + \frac{1}{n}\}_{n \in \mathbb{N}, n \geq 2}, \\ -10 & \text{if } t = 3, \\ -21 & \text{if } t = 9, \\ -12 - t - \frac{t-9}{(\sqrt{t-9}-1)^2} & \text{if } t \in \{9 + \frac{1}{n^2}\}_{n \in \mathbb{N}}. \end{cases}$$

Prove that

$$\int f(t)\Delta t = -3t - t^2 + C, \quad t \in \mathbb{T},$$

where C is a constant.

Problem 3.12. Let $\mathbb{T} = [1, 3] \cup 5^{\mathbb{N}}$, and let

$$f(t) = \begin{cases} 4t^3 - 6t & \text{if } t \in [1, 3), \\ 248 & \text{if } t = 3, \\ 156t^3 - 18t & \text{if } t \in 5^{\mathbb{N}}. \end{cases}$$

Prove that

$$\int f(t)\Delta t = t^4 - 3t^2 + C, \quad t \in \mathbb{T},$$

where C is a constant.

Problem 3.13. Let $\mathbb{T} = 2^{\mathbb{N}_0}$. Prove that

$$\int_t^{2t} (\tau^4 - 3\tau^2 + \tau)\Delta\tau = t^5 - 3t^3 + t^2, \quad t \in \mathbb{T}.$$

Problem 3.14. Let $\mathbb{T} = \mathbb{Z}$. Prove that

$$\int_{-1}^{2} (t^3 - t)\Delta t = 0.$$

Problem 3.15. Let $\mathbb{T} = (-5\mathbb{N}_0) \cup 4^{\mathbb{N}_0}$. Prove that

$$\int_{-10}^{16} \frac{2+t}{5-t}\Delta t = \frac{4094}{60}.$$

Problem 3.16. Let $\mathbb{T} = [-1,3] \cup [7,9] \cup [10,\infty)$. Prove that

$$\int_{-1}^{11} \frac{1}{2+t} \Delta t = \log \frac{715}{108} + \frac{49}{55}.$$

Problem 3.17. Let $\mathbb{T} = \{0\} \cup \{1 - \frac{1}{4^n}\}_{n\in\mathbb{N}_0} \cup 2^{\mathbb{N}_0}$. Prove that

$$\int_0^4 \frac{1}{1+2t} \Delta t = \sum_{n=0}^{\infty} \frac{3}{8(3 \cdot 2^{2n-1} - 1)} + \frac{11}{15}.$$

Problem 3.18. Investigate for convergence and divergence the following integrals:

1. $\int_1^\infty \frac{\sin\frac{t}{2}\cos\frac{3t}{2}}{t} \Delta t$, $\mathbb{T} = 2^{\mathbb{N}_0}$,
2. $\int_1^\infty \log\frac{t+1}{t} \Delta t$, $\mathbb{T} = \mathbb{Z}$,
3. $\int_1^\infty \frac{e^{3t}-e^t}{2t} \Delta t$, $\mathbb{T} = 3^{\mathbb{N}_0}$,
4. $\int_1^\infty \frac{1}{\sqrt{t}(\sqrt{t}+1)(\sqrt{t}+2)} \Delta t$, $\mathbb{T} = 4^{\mathbb{N}_0}$.

Problem 3.19. Investigate for convergence and divergence the following integrals:

1. $\int_1^\infty \frac{t^3+2t^2+3t+10}{t^7+t^4+11t+20} \Delta t$, $\mathbb{T} = \mathbb{Z}$,
2. $\int_1^\infty \cos(t + \frac{1}{2}) \Delta t$, $\mathbb{T} = \mathbb{Z}$,
3. $\int_2^\infty \frac{1}{t}(t^2 + t + 1) \Delta t$, $\mathbb{T} = 2^{\mathbb{N}_0}$,
4. $\int_1^\infty \frac{\sin t \sin(2t)}{2t(t^2+1)(t^4+10)} \Delta t$, $\mathbb{T} = 3^{\mathbb{N}_0}$,
5. $\int_1^\infty (\sin^2 t + 2\sin^3 t - \cos^4 t + e_1(t,0) + 10)(t^2 + t) \Delta t$, $\mathbb{T} = 2^{\mathbb{N}_0}$,
6. $\int_2^\infty \frac{1}{t^2+10t+20} \Delta t$, $\mathbb{T} = 4^{\mathbb{N}_0}$.

Problem 3.20. Let $\mathbb{T} = 3^{\mathbb{N}_0}$. Investigate for convergence and divergence the integral

$$\int_1^\infty \frac{1}{2t+3} \operatorname{arctanh} \frac{2t}{1-3t^2} \Delta t.$$

Problem 3.21. Let $\mathbb{T} = 2^{\mathbb{N}_0}$. Prove that the integral

$$\int_1^\infty \frac{\sin^2 t - 4\cos^2 t + 10}{(t^4 + 2t^2 + 11)(t^2 + 3t + 15)} \Delta t$$

converges.

Problem 3.22. Investigate for convergence and divergence the following integrals:

1. $\int_{-1}^7 \frac{e^{\frac{1}{t}}}{t^3} \Delta t$, $\mathbb{T} = [-1,0] \cup \{1,4,7\}$, where $[-1,0]$ is a real-valued interval.

2. $\int_0^4 \frac{\Delta t}{t \log^2 t}$, $\mathbb{T} = [0, \frac{1}{2}] \cup \{1, 3, 4\}$, where $[0, \frac{1}{2}]$ is a real-valued interval.
3. $\int_0^5 \frac{\Delta t}{t \log^2 t}$, $\mathbb{T} = [0, 1] \cup \{2, 3, 4, 5\}$, where $[0, 1]$ is a real-valued interval.
4. $\int_{-3}^1 \frac{\Delta t}{t \log t}$, $\mathbb{T} = \{-3, -2, -1\} \cup [0, 1]$, where $[0, 1]$ is a real-valued interval.
5. $\int_{-4}^1 \frac{\log t}{\sqrt{t}} \Delta t$, $\mathbb{T} = \{-4, -1\} \cup [0, 1]$, where $[0, 1]$ is a real-valued interval.
6. $\int_{-3}^{\pi} \tan t \Delta t$, $\mathbb{T} = \{-3, -2, -1\} \cup [0, \pi]$, where $[0, \pi]$ is a real-valued interval.

Problem 3.23. Let $\mathbb{T}$ be a time scale satisfying (3.58). Investigate for convergence and divergence the following integrals:
1. $\int_a^b \frac{(t-a)^{2\alpha+3}}{t^4+2t^2+7} \Delta t$,
2. $\int_a^b e_{t^2}(t, 1)(b-t)^{\beta+2} \Delta t$.
3. $\int_a^b (t-a)^{\alpha+1}(b-t)^{2\beta+7} \Delta t$.

Problem 3.24. Let $\mathbb{T} = (-4\mathbb{N}_0) \cup 2^{\mathbb{N}_0}$. Find

$$h_1(t, s), \quad h_2(t, s), \quad \text{and} \quad h_3(t, s), \quad s, t \in \mathbb{T}.$$

Problem 3.25. Let $\mathbb{T} = (-3\mathbb{N}_0) \cup \{2 + \frac{1}{n}\}_{n \in \mathbb{N}} \cup 3^{\mathbb{N}_0}$. Find

$$h_1(t, s), \quad h_2(t, s), \quad \text{and} \quad h_3(t, s), \quad s, t \in \mathbb{T}.$$

Problem 3.26. Let $\mathbb{T} = (-\mathbb{N}_0) \cup [1, 2] \cup 4^{\mathbb{N}}$, and let

$$f(t) = \begin{cases} 3t^2 - 3t + 2 & \text{if } t \in (-\mathbb{N}_0), \\ 2 & \text{if } t = 1, \\ 3t^2 + 1 & \text{if } t \in (1, 2], \\ 29 & \text{if } t = 4, \\ \frac{21}{16} t^2 + 1 & \text{if } t \in 4^{\mathbb{N}}, \quad t \geq 16. \end{cases}$$

Prove that

$$\int f(t) \nabla t = t^3 + t + C, \quad t \in \mathbb{T}_\kappa,$$

where C is a constant.

4 Delta elementary functions

4.1 Hilger's complex plane

Definition 4.1. Let $h > 0$.

1. The Hilger complex numbers are defined by

$$\mathbb{C}_h = \left\{ z \in \mathbb{C} : z \neq -\frac{1}{h} \right\}.$$

2. The Hilger real axis is defined as

$$\mathbb{R}_h = \left\{ z \in \mathbb{C} : z > -\frac{1}{h} \right\}.$$

3. The Hilger alternative axis is defined as

$$A_h = \left\{ z \in \mathbb{C} : z < -\frac{1}{h} \right\}.$$

4. The Hilger imaginary circle is defined by

$$\mathbb{I}_h = \left\{ z \in \mathbb{C} : \left| z + \frac{1}{h} \right| = \frac{1}{h} \right\}.$$

For $h = 0$, we set

$$\mathbb{C}_0 = \mathbb{C}, \quad \mathbb{R}_0 = \mathbb{R}, \quad \mathbb{A}_0 = \emptyset, \quad \mathbb{I}_0 = i\mathbb{R}.$$

Definition 4.2. Let $h > 0$ and $z \in \mathbb{C}_h$. We define the Hilger real part of z by

$$\operatorname{Re}_h(z) = \frac{|zh + 1| - 1}{h}$$

and the Hilger imaginary part of z by

$$\operatorname{Im}_h(z) = \frac{\operatorname{Arg}(zh + 1)}{h},$$

where $\operatorname{Arg}(z)$ denotes the principal argument of z, i. e.,

$$-\pi < \operatorname{Arg}(z) \leq \pi.$$

Note that

$$-\frac{1}{h} < \operatorname{Re}_h(z) < \infty \quad \text{and} \quad -\frac{\pi}{h} < \operatorname{Im}_h(z) < \frac{\pi}{h}.$$

In particular, $\operatorname{Re}_h(z) \in \mathbb{R}_h$.

https://doi.org/10.1515/9783112232088-004

Definition 4.3. Let $-\frac{\pi}{h} < w \le \frac{\pi}{h}$. We define the Hilger purely imaginary number $\mathring{\iota}$ by

$$\mathring{\iota} w = \frac{e^{iwh} - 1}{h}.$$

Example 4.1. Let $h = 2$ and $z = 1 - i$. Then

$$\begin{aligned} zh &= 2(1-i) \\ &= 2 - 2i, \\ zh + 1 &= 2 - 2i + 1 \\ &= 3 - 2i \\ &= \sqrt{13}\left(\frac{3}{\sqrt{13}} - \frac{2}{\sqrt{13}} i\right). \end{aligned}$$

Let $\alpha \in [-\pi, \pi]$ be such that

$$\begin{aligned} \cos\alpha &= \frac{3}{\sqrt{13}}, \\ \sin\alpha &= -\frac{2}{\sqrt{13}}. \end{aligned}$$

Then

$$\tan\alpha = -\frac{2}{3},$$

and

$$\alpha = \arctan\left(-\frac{2}{3}\right).$$

Hence

$$\begin{aligned} zh + 1 &= \sqrt{13}(\cos\alpha + i\sin\alpha) \\ &= \sqrt{13} e^{i\alpha}, \end{aligned}$$

and

$$\begin{aligned} \operatorname{Re}_h(z) &= \frac{|zh+1| - 1}{h} \\ &= \frac{\sqrt{13} - 1}{2}, \\ \operatorname{Im}_h(z) &= \frac{\operatorname{Arg}(zh+1)}{h} \\ &= \frac{\arctan(-\frac{2}{3})}{2}. \end{aligned}$$

Example 4.2. Let $h = 3$ and $z = \frac{-1-i}{3}$. Then

$$\begin{aligned} zh &= 3\left(\frac{-1-i}{3}\right) \\ &= -1-i, \\ zh+1 &= 1-1-i \\ &= -i \\ &= \cos\left(-\frac{\pi}{2}\right) + i\sin\left(-\frac{\pi}{2}\right). \end{aligned}$$

Hence

$$\begin{aligned} |zh+1| &= 1, \\ \operatorname{Arg}(zh+1) &= -\frac{\pi}{2}, \end{aligned}$$

and

$$\begin{aligned} \operatorname{Re}_h(z) &= \frac{|zh+1|-1}{h} \\ &= \frac{1-1}{3} \\ &= 0, \\ \operatorname{Im}_h(z) &= \frac{\operatorname{Arg}(zh+1)}{h} \\ &= \frac{-\frac{\pi}{2}}{3} \\ &= -\frac{\pi}{6}. \end{aligned}$$

Example 4.3. Let $h = 2$ and $z = \frac{-1-\sqrt{3}}{2} + \frac{1}{2}i$. Then

$$\begin{aligned} zh &= 2\left(\frac{-1-\sqrt{3}}{2} + \frac{1}{2}i\right) \\ &= -1-\sqrt{3}+i, \\ zh+1 &= 1-1-\sqrt{3}+i \\ &= -\sqrt{3}+i \\ &= 2\left(-\frac{\sqrt{3}}{2} + \frac{1}{2}i\right) \\ &= 2\left(\cos\left(\frac{2\pi}{3}\right) + i\sin\left(\frac{2\pi}{3}\right)\right). \end{aligned}$$

Hence

$$|zh+1| = 2,$$

$$\operatorname{Arg}(zh+1) = \frac{2\pi}{3},$$

and

$$\begin{aligned}\operatorname{Re}_h(z) &= \frac{|zh+1|-1}{h}\\ &= \frac{2-1}{2}\\ &= \frac{1}{2},\\ \operatorname{Im}_h(z) &= \frac{\operatorname{Arg}(zh+1)}{h}\\ &= \frac{\frac{2\pi}{3}}{2}\\ &= \frac{\pi}{3}.\end{aligned}$$

Example 4.4. Let $h = 4$ and $z = -1 - i$. Then

$$\begin{aligned}zh &= 4(-1-i)\\ &= -4-4i,\\ zh+1 &= 1-4-4i\\ &= -3-4i\\ &= 5\left(-\frac{3}{5}-\frac{4}{5}i\right).\end{aligned}$$

Let $\alpha \in [-\pi, \pi]$ be such that

$$\begin{aligned}\cos\alpha &= -\frac{3}{5},\\ \sin\alpha &= -\frac{4}{5}.\end{aligned}$$

Then

$$\tan\alpha = \frac{4}{3},$$

and

$$\alpha = \arctan\frac{4}{3}.$$

Hence

$$\begin{aligned}|zh+1| &= 5,\\ \operatorname{Arg}(zh+1) &= \arctan\frac{4}{3},\end{aligned}$$

and

$$\begin{aligned}\operatorname{Re}_h(z) &= \frac{|zh+1|-1}{h}\\ &= \frac{5-1}{4}\\ &= 1,\\ \operatorname{Im}_h(z) &= \frac{\operatorname{Arg}(zh+1)}{h}\\ &= \frac{1}{4}\arctan\frac{4}{3}.\end{aligned}$$

Example 4.5. Let $h = 3$ and $z = \frac{1}{6} - \frac{\sqrt{3}}{6}i$. Then

$$\begin{aligned}zh &= 3\left(\frac{1}{6} - \frac{\sqrt{3}}{6}i\right)\\ &= \frac{1}{2} - \frac{\sqrt{3}}{2}i,\\ zh+1 &= 1 + \frac{1}{2} - \frac{\sqrt{3}}{2}i\\ &= \frac{3}{2} - \frac{\sqrt{3}}{2}i\\ &= \sqrt{3}\left(\frac{\sqrt{3}}{2} - \frac{1}{2}i\right)\\ &= \sqrt{3}\left(\cos\left(-\frac{\pi}{6}\right) + i\sin\left(-\frac{\pi}{6}\right)\right).\end{aligned}$$

Hence

$$\begin{aligned}|zh+1| &= \sqrt{3},\\ \operatorname{Arg}(zh+1) &= -\frac{\pi}{6},\end{aligned}$$

and

$$\begin{aligned}\operatorname{Re}_h(z) &= \frac{|zh+1|-1}{h}\\ &= \frac{\sqrt{3}-1}{3},\\ \operatorname{Im}_h(z) &= \frac{\operatorname{Arg}(zh+1)}{h}\\ &= \frac{-\frac{\pi}{6}}{3}\\ &= -\frac{\pi}{18}.\end{aligned}$$

Exercise 4.1. Find $\mathrm{Re}_h(z)$ and $\mathrm{Im}_h(z)$, where
1. $h = 3, z = -\frac{1}{3} + \frac{1}{3}i$.
2. $h = 2, z = \frac{\sqrt{3}-2}{4} - \frac{1}{4}i$.
3. $h = 4, z = \frac{\sqrt{2+\sqrt{3}}-2}{8} + \frac{\sqrt{2-\sqrt{3}}}{8}i$.
4. $h = 5, z = -\frac{3}{10} - \frac{\sqrt{3}}{10}i$.
5. $h = 2, z = \frac{\sqrt{2}-1}{4} - \frac{\sqrt{2}}{4}i$.

Example 4.6. Let $z \in \mathbb{C}_h$. We will prove that $\mathring{\iota}\,\mathrm{Im}_h(z) \in \mathbb{I}_h$. Indeed, we have

$$\mathring{\iota}\,\mathrm{Im}_h(z) = \frac{e^{iw\,\mathrm{Im}_h(z)} - 1}{h}$$

and

$$\begin{aligned}\left|\mathring{\iota}\,\mathrm{Im}_h(z) + \frac{1}{h}\right| &= \left|\frac{e^{iw\,\mathrm{Im}_h(z)} - 1}{h} + \frac{1}{h}\right| \\ &= \frac{|e^{iw\,\mathrm{Im}_h(z)}|}{h} \\ &= \frac{1}{h},\end{aligned}$$

completing the proof.

Example 4.7. We will prove that

$$\lim_{h\to 0}[\mathrm{Re}_h(z) + \mathring{\iota}\,\mathrm{Im}_h(z)] = \mathrm{Re}(z) + i\,\mathrm{Im}(z).$$

Indeed, we have

$$\begin{aligned} z &= \mathrm{Re}(z) + \mathrm{Im}(z), \\ zh + 1 &= (\mathrm{Re}(z) + i\,\mathrm{Im}(z))h + 1 \\ &= h\,\mathrm{Re}(z) + 1 + ih\,\mathrm{Im}(z), \\ \mathrm{Arg}(zh+1) &= \arcsin\frac{h\,\mathrm{Im}(z)}{\sqrt{(h\,\mathrm{Re}(z)+1)^2 + h^2(\mathrm{Im}(z))^2}}, \\ \mathrm{Im}_h(z) &= \frac{\mathrm{Arg}(zh+1)}{h} \\ &= \frac{1}{h}\arcsin\frac{h\,\mathrm{Im}(z)}{\sqrt{(h\,\mathrm{Re}(z)+1)^2 + h^2(\mathrm{Im}(z))^2}}, \\ |zh+1| &= \sqrt{(h\,\mathrm{Re}(z)+1)^2 + h^2(\mathrm{Im}(z))^2}, \\ \mathrm{Re}_h(z) &= \frac{|zh+1| - 1}{h} \end{aligned}$$

$$= \frac{\sqrt{(h\operatorname{Re}(z)+1)^2 + h^2(\operatorname{Im}(z))^2} - 1}{h}.$$

Hence

$$\begin{aligned}
\lim_{h\to 0} \operatorname{Re}_h(z) &= \lim_{h\to 0} \frac{\sqrt{(h\operatorname{Re}(z)+1)^2 + h^2(\operatorname{Im}(z))^2} - 1}{h} \\
&= \lim_{h\to 0} \frac{(h\operatorname{Re}(z)+1)\operatorname{Re}(z) + h(\operatorname{Im}(z))^2}{\sqrt{(h\operatorname{Re}(z)+1)^2 + h^2(\operatorname{Im}(z))^2}} \\
&= \operatorname{Re}(z),
\end{aligned}$$

and

$$\begin{aligned}
\lim_{h\to 0} \operatorname{Im}_h(z) &= \lim_{h\to 0} \frac{1}{h} \arcsin \frac{h\operatorname{Im}(z)}{\sqrt{(h\operatorname{Re}(z)+1)^2 + h^2(\operatorname{Im}(z))^2}} \\
&= \lim_{h\to 0} \frac{1}{\sqrt{1 - \frac{h^2(\operatorname{Im}(z))^2}{(h\operatorname{Re}(z)+1)^2 + h^2(\operatorname{Im}(z))^2}}} \\
&\quad \times \frac{\operatorname{Im}(z)\sqrt{(h\operatorname{Re}(z)+1)^2 + h^2(\operatorname{Im}(z))^2} - h\operatorname{Im}(z)\frac{(h\operatorname{Re}(z)+1)\operatorname{Re}(z)+h(\operatorname{Im}(z))^2}{\sqrt{(h\operatorname{Re}(z)+1)^2+h^2(\operatorname{Im}(z))^2}}}{(h\operatorname{Re}(z)+1)^2 + h^2(\operatorname{Im}(z))^2} \\
&= \operatorname{Im}(z),
\end{aligned}$$

which completes the proof.

Example 4.8. Let $-\frac{\pi}{h} < w \le \frac{\pi}{h}$. We will prove that

$$|\mathring{\iota} w|^2 = \frac{4}{h^2}\left(\sin\frac{wh}{2}\right)^2.$$

Indeed, we have

$$\mathring{\iota} w = \frac{e^{iwh} - 1}{h} = \frac{\cos(wh) - 1 + i\sin(wh)}{h}.$$

Hence

$$\begin{aligned}
|\mathring{\iota} w|^2 &= \frac{(\cos(wh)-1)^2}{h^2} + \frac{(\sin(wh))^2}{h^2} \\
&= \frac{(\cos(wh))^2 - 2\cos(wh) + 1 + (\sin(wh))^2}{h^2} \\
&= \frac{2(1-\cos(wh))}{h^2} \\
&= \frac{4}{h^2}\left(\sin\frac{wh}{2}\right)^2,
\end{aligned}$$

completing the proof.

Definition 4.4. The circle plus addition $\oplus_h$ on $\mathbb{C}_h$ is defined by

$$z \oplus_h w = z + w + zwh.$$

Example 4.9. Let $z = 1 + i$, $w = 2 - i$, and $h = 2$. Then

$$\begin{aligned} z \oplus_h w &= (1 + i) + (2 - i) + 2(1 + i)(2 - i) \\ &= 3 + 2(2 - i + 2i + 1) \\ &= 3 + 2(3 + i) \\ &= 3 + 6 + 2i \\ &= 9 + 2i. \end{aligned}$$

Example 4.10. Let $z = -i$, $w = 2 + 3i$, and $h = 3$. Then

$$\begin{aligned} z \oplus_h w &= (-i) + (2 + 3i) + 3(-i)(2 + 3i) \\ &= 2 + 2i + 3(-2i + 3) \\ &= 2 + 2i - 6i + 9 \\ &= 11 - 4i. \end{aligned}$$

Example 4.11. Let $z = 2$, $w = 1 - 4i$, and $h = 2$. Then

$$\begin{aligned} z \oplus_h w &= 2 + (1 - 4i) + 2(2)(1 - 4i) \\ &= 3 - 4i + 4 - 16i \\ &= 7 - 20i. \end{aligned}$$

Example 4.12. Let $z = 2 - 3i$, $w = 1 + 4i$, and $h = 2$. Then

$$\begin{aligned} z \oplus_h w &= (2 - 3i) + (1 + 4i) + 2(2 - 3i)(1 + 4i) \\ &= 3 + i + 2(2 + 8i - 3i + 12) \\ &= 3 + i + 28 + 10i \\ &= 31 + 11i. \end{aligned}$$

Example 4.13. Let $z = 1 - 5i$, $w = 2 + 7i$, and $h = 3$. Then

$$\begin{aligned} z \oplus_h w &= (1 - 5i) + (2 + 7i) + 3(1 - 5i)(2 + 7i) \\ &= 3 + 2i + 3(2 + 7i - 10i + 35) \\ &= 3 + 2i + 74 - 9i \\ &= 77 - 7i. \end{aligned}$$

Exercise 4.2. Find $z \oplus_h w$, where
1. $z = i, w = 2 - 7i, h = 4$.
2. $z = 3 - i, w = 1 + i, h = 3$.
3. $z = i, w = 3 - 2i, h = 5$.
4. $z = 3, w = i, h = 2$.
5. $z = 1 + 7i, w = 3 - i, h = 3$.

Example 4.14. We will prove that $(\mathbb{C}_h, \oplus_h)$ is an Abelian group. For this aim, take $z, w \in \mathbb{C}_h$. Then $z, w \in \mathbb{C}$, and $z, w \neq -\frac{1}{h}$. Therefore $z \oplus_h w \in \mathbb{C}$. Since

$$\begin{aligned} h(z \oplus_h w) + 1 &= h(z + w + zwh) + 1 \\ &= 1 + hz + hw + zwh^2 \\ &= 1 + hz + hw(1 + hz) \\ &= (1 + hw)(1 + hz) \\ &\neq 0, \end{aligned}$$

we conclude that $z \oplus_h w \in \mathbb{C}_h$. Also,

$$0 \oplus_h z = z \oplus_h 0 = z,$$

i. e., 0 is the additive identity for $\oplus_h$. For $z \in \mathbb{C}_h$, we have

$$\begin{aligned} z \oplus_h \left(-\frac{z}{1 + zh}\right) &= z - \frac{z}{1 + zh} - z\frac{z}{1 + zh}h \\ &= \frac{z^2 h}{1 + zh} - \frac{z^2 h}{1 + zh} \\ &= 0, \end{aligned}$$

i. e., the additive inverse of z under the addition $\oplus_h$ is $-\frac{z}{1+zh}$. Note that

$$-\frac{z}{1 + zh} \in \mathbb{C}$$

and

$$1 - \frac{zh}{1 + zh} = \frac{1}{1 + zh} \neq 0,$$

i. e., $-\frac{z}{1+zh} \neq -\frac{1}{h}$. Therefore $-\frac{z}{1+zh} \in \mathbb{C}_h$. For $z, w, v \in \mathbb{C}_h$, we have

$$\begin{aligned} (z \oplus_h w) \oplus_h v &= (z + w + zwh) \oplus_h v \\ &= z + w + zwh + v + (z + w + zwh)vh \\ &= z + w + zwh + v + zvh + wvh + zwvh^2 \end{aligned}$$

and

$$\begin{aligned} z \oplus_h (w \oplus_h v) &= z + (w \oplus_h v) + z(w \oplus_h v)h \\ &= z + w + v + wvh + z(w + v + wvh)h \\ &= z + w + v + wvh + zwh + zvh + zwvh^2. \end{aligned}$$

Consequently,

$$z \oplus_h (w \oplus_h v) = (z \oplus_h w) \oplus_h v,$$

i. e., the associative law holds in $(\mathbb{C}_h, \oplus_h)$. For $z, w \in \mathbb{C}_h$, we have

$$\begin{aligned} z \oplus_h w &= z + w + zwh \\ &= w + z + wzh \\ &= w \oplus_h z, \end{aligned}$$

which completes the proof.

Example 4.15. Let $z \in \mathbb{C}_h$ and $w \in \mathbb{C}$ be such that $z + w \in \mathbb{C}_h$. We will simplify the expression

$$A = z \oplus_h \frac{w}{1 + hz}.$$

We have

$$\begin{aligned} A &= z + \frac{w}{1 + hz} + \frac{zw}{1 + hz}h \\ &= z + \frac{(1 + hz)w}{1 + hz} \\ &= z + w. \end{aligned}$$

Example 4.16. For $z \in \mathbb{C}_h$, we will prove that

$$z = \operatorname{Re}_h(z) \oplus_h \mathring{\iota} \operatorname{Im}_h(z).$$

We have

$$\begin{aligned} \operatorname{Re}_h(z) \oplus_h \mathring{\iota} \operatorname{Im}_h(z) &= \frac{|zh + 1| - 1}{h} \oplus_h \mathring{\iota} \frac{\operatorname{Arg}(zh + 1)}{h} \\ &= \frac{|zh + 1| - 1}{h} \oplus_h \frac{e^{i \operatorname{Arg}(zh+1)} - 1}{h} \\ &= \frac{|zh + 1| - 1}{h} + \frac{e^{i \operatorname{Arg}(zh+1)} - 1}{h} \\ &\quad + \frac{|zh + 1| - 1}{h} \frac{e^{i \operatorname{Arg}(zh+1)} - 1}{h} h \\ &= \frac{1}{h}(|zh + 1| - 1 + e^{i \operatorname{Arg}(zh+1)} - 1 \end{aligned}$$

$$
\begin{aligned}
&+ |zh+1|e^{i\operatorname{Arg}(zh+1)} \\
&- |zh+1| - e^{i\operatorname{Arg}(zh+1)} + 1) \\
&= \frac{1}{h}(|zh+1|e^{i\operatorname{Arg}(zh+1)} - 1) \\
&= \frac{1}{h}(zh+1-1) \\
&= z,
\end{aligned}
$$

completing the proof.

Definition 4.5. Let $n \in \mathbb{N}$ and $z \in \mathbb{C}_h$. We define the circle dot multiplication $\odot_h$ by

$$n \odot_h z = z \oplus_h z \oplus_h \cdots \oplus_h z.$$

Example 4.17. Let $z = 1 + i$ and $h = 2$. We will find $3 \odot_h z$. We have

$$3 \odot_h z = z \oplus_h z \oplus_h z.$$

Then

$$
\begin{aligned}
z \oplus_h z &= 1 + i + 1 + i + 2(1+i)^2 \\
&= 2 + 2i + 2(1 + 2i - 1) \\
&= 2 + 2i + 4i \\
&= 2 + 6i,
\end{aligned}
$$

and hence we arrive at

$$
\begin{aligned}
3 \odot_h z &= (z \oplus_h z) \oplus_h z \\
&= 3 + 7i + 2(2 + 6i + 2i - 6) \\
&= 3 + 7i + 2(-4 + 8i) \\
&= 3 + 7i - 8 + 16i \\
&= -5 + 23i.
\end{aligned}
$$

Example 4.18. Let $z = -i$ and $h = 2$. We will find $4 \odot_h z$. We have

$$4 \odot_h z = \big((z \oplus_h z) \oplus_h z\big) \oplus_h z.$$

Then

$$
\begin{aligned}
z \oplus_h z &= -i + (-i) + 2(-i)(-i) \\
&= -2i - 2, \\
(z \oplus_h z) \oplus_h z &= -2i - 2 + 2(-i)(-2i - 2)
\end{aligned}
$$

$$\begin{aligned} &= -2i - 2 + 2(2i - 2) \\ &= -2i - 2 + 4i - 4 \\ &= -6 + 2i, \end{aligned}$$

and

$$\begin{aligned} 4 \odot_h z &= ((z \oplus_h z) \oplus_h z) \oplus_h z \\ &= -6 + 2i + (-i) + 2(-i)(-6 + 2i) \\ &= -6 + i + 12i + 4 \\ &= -2 + 13i. \end{aligned}$$

Example 4.19. Let $z = 2$ and $h = 3$. We will find $4 \odot_h z$. We have

$$4 \odot_h z = ((z \oplus_h z) \oplus_h z) \oplus_h z.$$

Then

$$\begin{aligned} z \oplus_h z &= 2 + 2(2)(2) \\ &= 2 + 8 \\ &= 10, \\ (z \oplus_h z) \oplus_h z &= 10 + 2 + 2(2)(10) \\ &= 12 + 40 \\ &= 52, \end{aligned}$$

and

$$\begin{aligned} 4 \odot_h z &= ((z \oplus_h z) \oplus_h z) \oplus_h z \\ &= 52 + 2 + 2(2)(52) \\ &= 54 + 208 \\ &= 262. \end{aligned}$$

Example 4.20. Let $z = 2 - 3i$ and $h = 2$. We will find $3 \odot_h z$. We have

$$3 \odot_h z = (z \oplus_h z) \oplus_h z.$$

Then

$$\begin{aligned} z \oplus_h z &= 2 - 3i + 2 - 3i + 2(2 - 3i)^2 \\ &= 4 - 6i + 2(4 - 12i - 9) \\ &= 4 - 6i + 2(-5 - 12i) \end{aligned}$$

$$\begin{aligned} &= 4 - 6i - 10 - 24i \\ &= -6 - 30i, \end{aligned}$$

and

$$\begin{aligned} (z \oplus_h z) \oplus_h z &= -6 - 30i + 2 - 3i + 2(-6 - 30i)(2 - 3i) \\ &= -4 - 33i + 2(-12 + 18i - 60i - 90) \\ &= -4 - 33i + 2(-102 - 42i) \\ &= -4 - 33i - 204 - 84i \\ &= -208 - 117i. \end{aligned}$$

Example 4.21. Let $z = 1 - 5i$ and $h = 2$. We will find $4 \odot_h z$. We have

$$4 \odot_h z = \big((z \oplus_h z) \oplus_h z\big) \oplus_h z.$$

Then

$$\begin{aligned} z \oplus_h z &= 1 - 5i + 1 - 5i + 2(1 - 5i)^2 \\ &= 2 - 10i + 2(1 - 10i - 25) \\ &= 2 - 10i + 2(-24 - 10i) \\ &= 2 - 10i - 48 - 20i \\ &= -46 - 30i, \\ (z \oplus_h z) \oplus_h z &= -46 - 30i + 1 - 5i + 2(-46 - 30i)(1 - 5i) \\ &= -45 - 35i + 2(-46 + 230i - 30i - 150) \\ &= -45 - 35i + 2(-196 + 200i) \\ &= -45 - 35i - 392 + 400i \\ &= -437 + 365i, \end{aligned}$$

and

$$\begin{aligned} 4 \odot_h z &= \big((z \oplus_h z) \oplus_h z\big) \oplus_h z \\ &= -437 + 365i + 1 - 5i + 2(-437 + 365i)(1 - 5i) \\ &= -436 + 360i + 2(-437 + 2185i + 365i + 1825) \\ &= -436 + 360i + 2(1388 + 2550i) \\ &= -436 + 360i + 2776 + 5100i \\ &= 2340 + 5460i. \end{aligned}$$

Exercise 4.3. Find $n \odot_h z$, where
1. $n = 3, z = i, h = 4.$
2. $n = 4, z = 3 - i, h = 3.$
3. $n = 4, z = 3 - 2i, h = 2.$
4. $n = 4, z = 1 + 2i, h = 3.$
5. $n = 4, z = 1, h = 2.$

Example 4.22. Let $n \in \mathbb{N}$ and $z \in \mathbb{C}_h$. We will prove that

$$n \odot_h z = \frac{(zh+1)^n - 1}{h}. \tag{4.1}$$

For this aim, we will use induction.

1. Let $n = 2$. Then

$$\begin{aligned} 2 \odot_h z &= z \oplus_h z \\ &= z + z + z^2 h \\ &= 2z + zh \\ &= \frac{1}{h}(z^2h^2 + 2zh) \\ &= \frac{1}{h}(z^2h^2 + 2zh + 1 - 1) \\ &= \frac{(zh+1)^2 - 1}{h}. \end{aligned}$$

2. Assume that

$$n \odot_h z = \frac{(zh+1)^n - 1}{h}$$

   for some $n \in \mathbb{N}$.
3. We will prove that

$$(n+1) \odot_h z = \frac{(zh+1)^{n+1} - 1}{h}.$$

   Indeed,

$$\begin{aligned} (n+1) \odot_h z &= (n \odot_h z) \oplus_h z \\ &= \frac{(zh+1)^n - 1}{h} \oplus_h z \\ &= \frac{(zh+1)^n - 1}{h} + z + \frac{(zh+1)^n - 1}{h} zh \\ &= \frac{(zh+1)^n - 1 + zh + zh(zh+1)^n - zh}{h} \\ &= \frac{(zh+1)^{n+1} - 1}{h}. \end{aligned}$$

Hence we conclude that (4.1) holds for all $n \in \mathbb{N}$.

Definition 4.6. Let $z \in \mathbb{C}_h$. We define the circle minus $\ominus_h$ of z as

$$\ominus_h z = \frac{-z}{1 + zh}.$$

Example 4.23. Let $z = 2 - i$ and $h = 2$. Then

$$\begin{aligned}
\ominus_h z &= -\frac{z}{1 + hz} \\
&= -\frac{2 - i}{1 + 2(2 - i)} \\
&= -\frac{2 - i}{5 - 2i} \\
&= -\frac{(2 - i)(5 + 2i)}{(5 - 2i)(5 + 2i)} \\
&= -\frac{10 + 4i - 5i + 2}{25 + 4} \\
&= -\frac{12 - i}{29} \\
&= -\frac{12}{29} + \frac{1}{29} i.
\end{aligned}$$

Example 4.24. Let $z = 2 + 3i$ and $h = 3$. Then

$$\begin{aligned}
\ominus_h z &= -\frac{z}{1 + hz} \\
&= -\frac{2 + 3i}{1 + 3(2 + 3i)} \\
&= -\frac{2 + 3i}{7 + 9i} \\
&= -\frac{(2 + 3i)(7 - 9i)}{(7 + 9i)(7 - 9i)} \\
&= -\frac{14 - 18i + 21i + 27}{49 + 81} \\
&= -\frac{41 + 3i}{130} \\
&= -\frac{41}{130} - \frac{3}{130} i.
\end{aligned}$$

Example 4.25. Let $z = 1 - 4i$ and $h = 2$. Then

$$\begin{aligned}
\ominus_h z &= -\frac{z}{1 + hz} \\
&= -\frac{1 - 4i}{1 + 2(1 - 4i)}
\end{aligned}$$

$$
\begin{aligned}
&= -\frac{1-4i}{3-8i} \\
&= -\frac{(1-4i)(3+8i)}{(3-8i)(3+8i)} \\
&= -\frac{3+8i-12i+32}{9+64} \\
&= -\frac{35-4i}{73} \\
&= -\frac{35}{73} + \frac{4}{73}i.
\end{aligned}
$$

Example 4.26. Let $z = 1 + 4i$ and $h = 2$. Then

$$
\begin{aligned}
\ominus_h z &= -\frac{z}{1+hz} \\
&= -\frac{1+4i}{1+2(1+4i)} \\
&= -\frac{1+4i}{3+8i} \\
&= -\frac{(1+4i)(3-8i)}{(3+8i)(3-8i)} \\
&= -\frac{3-8i+12i+32}{9+64} \\
&= -\frac{35+4i}{73} \\
&= -\frac{35}{73} - \frac{4}{73}i.
\end{aligned}
$$

Example 4.27. Let $z = 2 + 7i$ and $h = 3$. Then

$$
\begin{aligned}
\ominus_h z &= -\frac{z}{1+hz} \\
&= -\frac{2+7i}{1+3(2+7i)} \\
&= -\frac{2+7i}{7+21i} \\
&= -\frac{(2+7i)(7-21i)}{(7+21i)(7-21i)} \\
&= -\frac{14-42i+49i+147}{49+441} \\
&= -\frac{161+7i}{490} \\
&= -\frac{161}{490} - \frac{1}{70}i.
\end{aligned}
$$

Exercise 4.4. Find $\ominus_h z$, where
1. $z = 2 - 7i, h = 4$.
2. $z = 1 + i, h = 3$.
3. $z = 3 - 2i, h = 5$.
4. $z = i, h = 2$.
5. $z = 3 - i, h = 3$.

Example 4.28. Let $z \in \mathbb{C}_h$. We will prove that $\ominus_h z$ is the additive inverse of z under the operation $\oplus_h$, i. e.,

$$\ominus_h(\ominus_h z) = z.$$

Indeed, we have

$$\begin{aligned}\ominus_h(\ominus_h z) &= -\frac{\ominus_h z}{1 + (\ominus_h z)h}\\ &= -\frac{\frac{-z}{1+zh}}{1 + \frac{-z}{1+zh}h}\\ &= \frac{\frac{z}{1+zh}}{\frac{1+zh-zh}{1+zh}}\\ &= z,\end{aligned}$$

completing the proof.

Definition 4.7. Let $z, w \in \mathbb{C}_h$. We define the circle minus subtraction by

$$z \ominus_h w = z \oplus_h (\ominus_h w).$$

Remark 4.1. For $z, w \in \mathbb{C}_h$, we have

$$\begin{aligned}z \ominus_h w &= z \oplus_h (\ominus_h w)\\ &= z + (\ominus_h w) + z(\ominus_h w)h\\ &= z - \frac{w}{1 + wh} - \frac{zwh}{1 + wh}\\ &= \frac{z + zwh - w - zwh}{1 + wh}\\ &= \frac{z - w}{1 + wh},\end{aligned}$$

i. e.,

$$z \ominus_h w = \frac{z - w}{1 + wh}. \tag{4.2}$$

Example 4.29. Let $z = 1 + i$, $w = 2 - i$, and $h = 2$. We will find $z \ominus_h w$. By the computations in Example 4.23 we find

$$
\begin{aligned}
z \ominus_h w &= z \oplus_h (\ominus_h w)\\
&= z + (\ominus_h w) + hz(\ominus_h w)\\
&= 1 + i + \left(-\frac{12}{29} + \frac{1}{29}i\right) + 2(1+i)\left(-\frac{12}{29} + \frac{1}{29}i\right)\\
&= 1 + i - \frac{12}{29} + \frac{1}{29}i - \frac{24}{29} + \frac{2}{29}i - \frac{24}{29}i - \frac{2}{29}\\
&= 1 + i - \frac{38}{29} - \frac{21}{29}i\\
&= -\frac{9}{29} + \frac{8}{29}i.
\end{aligned}
$$

Example 4.30. Let $z = -i$, $w = 2 + 3i$, and $h = 3$. We will find $z \ominus_h w$. By the computations in Example 4.24 we find

$$
\begin{aligned}
z \ominus_h w &= z \oplus_h (\ominus_h w)\\
&= z + (\ominus_h w) + hz(\ominus_h w)\\
&= -i + \left(-\frac{41}{130} - \frac{3}{130}i\right) + 3(-i)\left(-\frac{41}{130} - \frac{3}{130}i\right)\\
&= -i - \frac{41}{130} - \frac{3}{130}i + \frac{123}{130}i - \frac{9}{130}\\
&= -i - \frac{5}{13} + \frac{12}{13}i\\
&= -\frac{5}{13} - \frac{1}{13}i.
\end{aligned}
$$

Example 4.31. Let $z = 2$, $w = 1 - 4i$, and $h = 2$. We will find $z \ominus_h w$. By the computations in Example 4.25 we get

$$
\begin{aligned}
z \ominus_h w &= z \oplus_h (\ominus_h w)\\
&= z + (\ominus_h w) + hz(\ominus_h w)\\
&= 2 + \left(-\frac{35}{73} + \frac{4}{73}i\right) + 2(2)\left(-\frac{35}{73} + \frac{4}{73}i\right)\\
&= 2 - \frac{35}{73} + \frac{4}{73}i - \frac{140}{73} + \frac{16}{73}i\\
&= 2 - \frac{175}{73} + \frac{20}{73}i\\
&= -\frac{29}{73} + \frac{20}{73}i.
\end{aligned}
$$

Example 4.32. Let $z = 2-3i$, $w = 1+4i$, and $h = 2$. We will find $z \ominus_h w$. By the computations in Example 4.26 we find

$$
\begin{aligned}
z \ominus_h w &= z \oplus_h (\ominus_h w)\\
&= z + (\ominus_h w) + hz(\ominus_h w)
\end{aligned}
$$

$$
\begin{aligned}
&= 2 - 3i + \left(-\frac{35}{73} - \frac{4}{73}i\right) + 2(2 - 3i)\left(-\frac{35}{73} - \frac{4}{73}i\right) \\
&= 2 - 3i - \frac{35}{73} - \frac{4}{73}i - \frac{140}{73} - \frac{16}{73}i + \frac{210}{73}i - \frac{24}{73} \\
&= 2 - 3i - \frac{199}{73} + \frac{190}{73}i \\
&= -\frac{53}{73} - \frac{29}{73}i.
\end{aligned}
$$

Example 4.33. Let $z = 1-5i$, $w = 2+7i$, and $h = 3$. We will find $z\ominus_h w$. By the computations in Example 4.27 we get

$$
\begin{aligned}
z \ominus_h w &= z \oplus_h (\ominus_h w) \\
&= z + (\ominus_h w) + hz(\ominus_h w) \\
&= 1 - 5i + \left(-\frac{161}{490} - \frac{1}{70}i\right) + 3(2 + 7i)\left(-\frac{161}{490} - \frac{1}{70}i\right) \\
&= 1 - 5i + 7(1 + 3i)\left(-\frac{161}{490} - \frac{1}{70}i\right) \\
&= 1 - 5i + (1 + 3i)\left(-\frac{161}{70} - \frac{1}{10}i\right) \\
&= 1 - 5i - \frac{161}{70} - \frac{1}{10}i - \frac{483}{70}i + \frac{3}{10} \\
&= 1 - 5i - 2 - 7i \\
&= -1 - 12i.
\end{aligned}
$$

Exercise 4.5. Find $z \ominus_h w$, where z, w, and h are the same as in Exercise 4.2.

Example 4.34. Let $z \in \mathbb{C}_h$. We will prove that $\overline{z} = \ominus_h z$ iff $z \in \mathbb{I}_h$. Indeed, by (4.2) we have

$$\overline{z} = \ominus_h z = -\frac{z}{1 + zh}$$

iff

$$\overline{z} + \overline{z}zh = -z$$

iff

$$2\,\mathrm{Re}(z) + |z|^2 h = 0.$$

Also, $z \in \mathbb{I}_h$ iff

$$\left|z + \frac{1}{h}\right| = \frac{1}{h}$$

iff

$$\begin{aligned}\frac{1}{h^2} &= \left|z + \frac{1}{h}\right|^2 \\ &= \left(\mathrm{Re}(z) + \frac{1}{h}\right)^2 + (\mathrm{Im}(z))^2 \\ &= (\mathrm{Re}(z))^2 + \frac{2}{h}\mathrm{Re}(z) + \frac{1}{h^2} + (\mathrm{Im}(z))^2\end{aligned}$$

iff

$$|z|^2 + \frac{2}{h}\mathrm{Re}(z) = 0$$

iff

$$2\,\mathrm{Re}(z) + |z|^2 h = 0,$$

which completes the proof.

Example 4.35. Let $-\frac{\pi}{h} < w \le \frac{\pi}{h}$. We will prove that

$$\ominus_h(\mathring{\iota}w) = \overline{\mathring{\iota}w}.$$

Indeed, we have

$$\begin{aligned}\ominus_h(\mathring{\iota}w) &= -\frac{\mathring{\iota}w}{1 + \mathring{\iota}wh} \\ &= -\frac{\frac{e^{iwh}-1}{h}}{1 + \frac{e^{iwh}-1}{h}h} \\ &= -\frac{e^{iwh} - 1}{he^{iwh}} \\ &= \frac{e^{-iwh} - 1}{h} \\ &= \overline{\mathring{\iota}w},\end{aligned}$$

completing the proof.

Definition 4.8. Let $z \in \mathbb{C}_h$. The circle square $^{②}{}_h$ of z is defined by

$$z^{②_h} = (-z)(\ominus_h z).$$

Remark 4.2. We have

$$z^{②_h} = -z\frac{-z}{1 + zh} = \frac{z^2}{1 + zh}.$$

Example 4.36. Let $z = 2 - i$ and $h = 2$. By the computations in Example 4.23 we get

$$\begin{aligned} z^{②}{}_h &= (-z)(\ominus_h z) \\ &= (i-2)\left(-\frac{12}{29} + \frac{1}{29}i\right) \\ &= -\frac{12}{29}i - \frac{1}{29} + \frac{24}{29} - \frac{2}{29}i \\ &= \frac{23}{29} - \frac{14}{29}i. \end{aligned}$$

Example 4.37. Let $z = 2 + 3i$ and $h = 3$. By the computations in Example 4.24 we get

$$\begin{aligned} z^{②}{}_h &= (-z)(\ominus_h z) \\ &= (-2-3i)\left(-\frac{41}{130} - \frac{3}{130}i\right) \\ &= \frac{82}{130} + \frac{6}{130}i + \frac{123}{130}i - \frac{9}{130}i \\ &= \frac{73}{130} + \frac{129}{130}i. \end{aligned}$$

Example 4.38. Let $z = 1 - 4i$ and $h = 2$. By the computations in Example 4.25 we get

$$\begin{aligned} z^{②}{}_h &= (-z)(\ominus_h z) \\ &= (4i-1)\left(-\frac{35}{73} + \frac{4}{73}i\right) \\ &= -\frac{140}{73}i - \frac{16}{73} + \frac{35}{73} - \frac{4}{73}i \\ &= \frac{19}{73} - \frac{144}{73}i. \end{aligned}$$

Example 4.39. Let $z = 1 + 4i$ and $h = 2$. By the computations in Example 4.26 we get

$$\begin{aligned} z^{②}{}_h &= (-z)(\ominus_h z) \\ &= (-1-4i)\left(-\frac{35}{73} - \frac{4}{73}i\right) \\ &= \frac{35}{73} + \frac{4}{73}i + \frac{140}{73}i - \frac{16}{73} \\ &= \frac{19}{73} + \frac{144}{73}i. \end{aligned}$$

Example 4.40. Let $z = 2 + 7i$ and $h = 3$. By the computations in Example 4.27 we get

$$\begin{aligned} z^{②}{}_h &= (-z)(\ominus_h z) \\ &= (-2-7i)\left(-\frac{161}{490} - \frac{1}{70}i\right) \end{aligned}$$

$$= \frac{322}{490} + \frac{2}{70}i + \frac{1127}{490}i - \frac{7}{70}$$
$$= \frac{39}{70} + \frac{163}{70}i.$$

Exercise 4.6. Find $z^{\textcircled{2}}{}_h$, where z and h are the same as in Exercise 4.4.

Example 4.41. For $z \in \mathbb{C}_h$, we have

$$(\ominus_h z)^{\textcircled{2}}{}_h = z^{\textcircled{2}}{}_h.$$

Indeed,

$$\begin{aligned}(\ominus_h z)^{\textcircled{2}}{}_h &= -(\ominus_h z)(\ominus_h(\ominus_h z))\\ &= \frac{z}{1+zh}z\\ &= \frac{z^2}{1+zh}\\ &= z^{\textcircled{2}},\end{aligned}$$

completing the proof.

Example 4.42. For $z \in \mathbb{C}_h$, we have

$$1 + zh = \frac{z^2}{z^{\textcircled{2}}{}_h}.$$

Indeed,

$$\frac{z^2}{z^{\textcircled{2}}{}_h} = \frac{z^2}{\frac{z^2}{1+zh}} = 1 + zh,$$

completing the proof.

Example 4.43. For $z \in \mathbb{C}_h$, we have

$$z + (\ominus_h z) = z^{\textcircled{2}}{}_h h.$$

Indeed,

$$z^{\textcircled{2}}{}_h h = \frac{z^2}{1+zh}h,$$

and

$$z + (\ominus_h z) = z - \frac{z}{1+zh} = \frac{z^2 h}{1+zh},$$

which completes the proof.

Example 4.44. For $z \in \mathbb{C}_h$, we have

$$z \oplus_h z^{②}{}_h = z + z^2.$$

Indeed,

$$\begin{aligned} z \oplus_h z^{②}{}_h &= z + z^{②} + z z^{②} h \\ &= z + \frac{z^2}{1+zh} + \frac{z^3 h}{1+zh} \\ &= z + \frac{z^2(1+zh)}{1+zh} \\ &= z + z^2, \end{aligned}$$

completing the proof.

Example 4.45. Let $-\frac{\pi}{h} < w \le \frac{\pi}{h}$. Then

$$-(\mathring{\iota} w)^{②}{}_h = \frac{4}{h^2}\sin^2\left(\frac{wh}{2}\right).$$

Indeed,

$$\begin{aligned} -(\mathring{\iota} w)^{②}{}_h &= -(\mathring{\iota} w)(\ominus_h \mathring{\iota} w) \\ &= (\mathring{\iota} w)\overline{\mathring{\iota} w} \\ &= |\mathring{\iota} w|^2 \\ &= \frac{4}{h^2}\sin^2\left(\frac{wh}{2}\right), \end{aligned}$$

completing the proof.

Exercise 4.7. Let $z \in \mathbb{C}_h$. Prove that

$$z^{②} \in \mathbb{R} \quad \text{iff} \quad z \in \mathbb{R}_h \cup \mathbb{A}_h \cup \mathbb{I}_h.$$

4.2 Delta regressive functions

For $h > 0$, we define the strip

$$\mathbb{Z}_h = \left\{ z \in \mathbb{C} : -\frac{\pi}{h} < \operatorname{Im}(z) \le \frac{\pi}{h} \right\}.$$

For $h = 0$, we set $\mathbb{Z}_0 = \mathbb{C}$.

Definition 4.9. For $h > 0$, we define the cylinder transformation $\xi_h : \mathbb{C}_h \to \mathbb{Z}_h$ by

$$\xi_h(z) = \frac{1}{h}\operatorname{Log}(1 + zh),$$

where Log is the principal logarithm function. Moreover, we define $\xi_0(z) = z$ for all $z \in \mathbb{C}$.

Remark 4.3. Note that

$$\xi_h^{-1}(z) = \frac{e^{zh} - 1}{h}$$

for $z \in \mathbb{Z}_h$.

Definition 4.10. We say that a function $f : \mathbb{T} \to \mathbb{R}$ is delta regressive if

$$1 + \mu(t)p(t) \neq 0 \quad \text{for all } t \in \mathbb{T}^\kappa.$$

The set of all delta regressive and rd-continuous functions $f : \mathbb{T} \to \mathbb{R}$ is denoted by $\mathscr{R}_\mu$, $\mathscr{R}_\mu(\mathbb{T})$, or $\mathscr{R}_\mu(\mathbb{T}, \mathbb{R})$.

Example 4.46. Let $\mathbb{T} = \mathbb{Z}$, and let

$$f(t) = t^2 - 6t + 4, \quad t \in \mathbb{T}.$$

Here

$$\mu(t) = 1, \quad t \in \mathbb{T}.$$

Then f is a delta regressive function for $t \neq 1, 5$:

$$\begin{aligned} 1 + \mu(t)f(t) &= 1 + t^2 - 6t + 4 \\ &= t^2 - 6t + 5, \quad t \in \mathbb{T}. \end{aligned}$$

Example 4.47. Let $\mathbb{T} = 2^{\mathbb{N}_0} \cup (-\mathbb{N}_0)$, and let

$$f(t) = \begin{cases} t + 2 - \frac{25}{t} & \text{if } t \in 2^{\mathbb{N}_0}, \\ t^2 + 2t - 25 & \text{if } t \in -\mathbb{N}_0. \end{cases}$$

We have the following cases.

1. $t \in 2^{\mathbb{N}_0}$. Then

$$\mu(t) = t, \quad t \in \mathbb{T},$$

and

$$
\begin{aligned}
1+\mu(t)f(t) &= 1+t\left(t+2-\frac{25}{t}\right)\\
&= 1+t^2+2t-25\\
&= t^2+2t-24\\
&\neq 0 \quad \text{if and only if} \quad t\neq 4.
\end{aligned}
$$

2. t $t\in(-\mathbb{N}_0)$. Then

$$
\mu(t)=1, \quad t\in\mathbb{T}.
$$

We have

$$
\begin{aligned}
1+\mu(t)f(t) &= 1+t^2+2t-25\\
&= t^2+2t-24\\
&\neq 0 \quad \text{if and only if} \quad t\neq -6.
\end{aligned}
$$

Thus f is delta regressive if $t\neq -6, 4$.

Example 4.48. Let $\mathbb{T}=\{-3\}\cup\{-3+\frac{1}{n}\}_{n\in\mathbb{N}}\cup[-2,0]\cup 7^{\mathbb{N}_0}$, and let

$$
f(t)=\begin{cases}
3 & \text{if } t=-3,\\
-1 & \text{if } t\in\{-3+\frac{1}{n}\}_{n\in\mathbb{N}},\\
t^2-5 & \text{if } t\in[-2,0],\\
\frac{t}{6}-\frac{2}{3}-\frac{11}{3t} & \text{if } t\in 7^{\mathbb{N}_0}.
\end{cases}
$$

We have the following cases.

1. $t=-3$. Then

$$
\mu(-3)=0,
$$

and

$$
\begin{aligned}
1+\mu(t)f(t) &= 1\\
&\neq 0.
\end{aligned}
$$

2. $t=-3+\frac{1}{n}$ for some $n\in\mathbb{N}$, $n\geq 2$. Then

$$
\begin{aligned}
n &= \frac{1}{t+3},\\
\sigma(t) &= -3+\frac{1}{n-1}\\
&= -3+\frac{1}{\frac{1}{t+3}-1}
\end{aligned}
$$

$$\begin{aligned}
&= -3 - \frac{t+3}{t+2}\\
&= -\frac{3t+6+t+3}{t+2}\\
&= -\frac{4t+9}{t+2},
\end{aligned}$$

and

$$\begin{aligned}
\mu(t) &= \sigma(t) - t\\
&= -\frac{4t+9}{t+2} - t\\
&= -\frac{t^2+2t+4t+9}{t+2}\\
&= -\frac{t^2+6t+9}{t+2}\\
&= -\frac{(t+3)^2}{t+2}.
\end{aligned}$$

Hence

$$\begin{aligned}
1+\mu(t)f(t) &= 1 + \frac{(t+3)^2}{t+2}\\
&= \frac{t^2+6t+11}{t+2}\\
&\neq 0.
\end{aligned}$$

3. $t \in [-2, 0]$. Then

$$\mu(t) = 0,$$

and

$$\begin{aligned}
1+\mu(t)f(t) &= 1\\
&\neq 0.
\end{aligned}$$

4. $t \in 7^{\mathbb{N}_0}$. Then

$$\mu(t) = 6t,$$

and

$$\begin{aligned}
1+\mu(t)f(t) &= 1 + 6t\left(\frac{t}{6} - \frac{2}{3} - \frac{11}{3t}\right)\\
&= 1 + t^2 - 4t - 22\\
&= t^2 - 4t - 21\\
&\neq 0 \quad \text{if and only if} \quad t \neq 7.
\end{aligned}$$

Thus f is delta regressive if $t \neq 7$.

Example 4.49. Let $\mathbb{T} = \{(\frac{1}{2})^{2^n} : n \in \mathbb{N}_0\} \cup \{0, 1\}$, and let

$$f(t) = \begin{cases} -\frac{16}{3} & \text{if } t \in \{(\frac{1}{2})^{2^n} : n \in \mathbb{N}\}, \\ -2 & \text{if } t = \frac{1}{2}, \\ t^2 - \frac{16}{3} & \text{if } t = 0, \\ 37 & \text{if } t = 1. \end{cases}$$

We will use the computations in Chapter 1. We have the following cases.

1. $t \in \{(\frac{1}{2})^{2^n} : n \in \mathbb{N}\}$. Then

$$\mu(t) = \sqrt{t} - t,$$

and

$$\begin{aligned} 1 + \mu(t) f(t) &= 1 - \frac{16}{3}(\sqrt{t} - t) \\ &\neq 0 \quad \text{if and only if} \quad t \neq \frac{1}{16}. \end{aligned}$$

2. $t = \frac{1}{2}$. Then

$$\mu\left(\frac{1}{2}\right) = \frac{1}{2},$$

and

$$\begin{aligned} 1 + \mu\left(\frac{1}{2}\right) f\left(\frac{1}{2}\right) &= 1 + \frac{1}{2}(-2) \\ &= 1 - 1 \\ &= 0. \end{aligned}$$

3. $t = 0$. Then

$$\mu(0) = 0,$$

and

$$\begin{aligned} 1 + \mu(0) f(0) &= 1 \\ &\neq 0. \end{aligned}$$

4. $t = 1$. Then

$$\mu(1) = 0,$$

and

$$\begin{aligned}1+\mu(1)f(1) &= 1\\ &\neq 0.\end{aligned}$$

Thus f is delta regressive for $t \neq \frac{1}{2}, \frac{1}{16}$.

Example 4.50. Let $U = \{\frac{1}{2^n} : n \in \mathbb{N}\}$, let

$$\mathbb{T} = U \cup (1-U) \cup (1+U) \cup (2-U) \cup (2+U) \cup \{0,1,2\},$$

and let

$$f(t) = \begin{cases} 3 & \text{if } t = 0,\\ -4 & \text{if } t = \frac{1}{2},\\ 3 & \text{if } t = 1,\\ -4 & \text{if } t = \frac{3}{2},\\ 4 & \text{if } t = 2,\\ 10 & \text{if } t = \frac{5}{2},\\ -\frac{1}{t} & \text{if } t \in U\backslash\{\frac{1}{2}\},\\ 4t - \frac{39}{2} & \text{if } t \in (1-U)\backslash\{\frac{1}{2}\},\\ 3 & \text{if } t \in (1+U)\backslash\{\frac{3}{2}\},\\ -16 & \text{if } t \in (2-U)\backslash\{\frac{3}{2}\},\\ 5 & \text{if } t \in (2+U)\backslash\{\frac{5}{2}\}.\end{cases}$$

We will use the computations in Example 1.19. We have the following cases.

1. $t = 0$. Then

$$\mu(0) = 0,$$

and

$$\begin{aligned}1+\mu(0)f(0) &= 1\\ &\neq 0.\end{aligned}$$

2. $t = \frac{1}{2}$. Then

$$\mu\left(\frac{1}{2}\right) = \frac{1}{4}.$$

Hence

$$\begin{aligned} 1 + \mu\left(\frac{1}{2}\right)f\left(\frac{1}{2}\right) &= 1 + \frac{1}{4}(-4) \\ &= 1 - 1 \\ &= 0. \end{aligned}$$

3. $t = 1$. Then

$$\mu(1) = 0,$$

and

$$\begin{aligned} 1 + \mu(1)f(1) &= 1 \\ &\neq 0. \end{aligned}$$

4. $t = \frac{3}{2}$. Then

$$\mu\left(\frac{3}{2}\right) = \frac{1}{4},$$

and

$$\begin{aligned} 1 + \mu\left(\frac{3}{2}\right)f\left(\frac{3}{2}\right) &= 1 + \frac{1}{4}(-4) \\ &= 1 - 1 \\ &= 0. \end{aligned}$$

5. $t = 2$. Then

$$\mu(2) = 0,$$

and

$$\begin{aligned} 1 + \mu(2)f(2) &= 1 \\ &\neq 0. \end{aligned}$$

6. $t = \frac{5}{2}$. Then

$$\mu\left(\frac{5}{2}\right) = 0,$$

and

$$\begin{aligned} 1 + \mu\left(\frac{5}{2}\right)f\left(\frac{5}{2}\right) &= 1 \\ &\neq 0. \end{aligned}$$

7. $t \in U \backslash \{\frac{1}{2}\}$. Then

$$\mu(t) = t,$$

and

$$\begin{aligned} 1 + \mu(t)f(t) &= 1 + t\left(-\frac{1}{t}\right) \\ &= 1 - 1 \\ &= 0. \end{aligned}$$

8. $t \in (1 - U) \backslash \{\frac{1}{2}\}$. Then

$$\mu(t) = \frac{1-t}{2},$$

and

$$\begin{aligned} 1 + \mu(t)f(t) &= 1 + \frac{1-t}{2}\left(4t - \frac{39}{2}\right) \\ &\neq 0 \quad \text{if } t \neq \frac{7}{8}. \end{aligned}$$

9. $t \in (1 + U) \backslash \{\frac{3}{2}\}$. Then

$$\mu(t) = t - 1,$$

and

$$\begin{aligned} 1 + \mu(t)f(t) &= 1 + 3(t-1) \\ &= 3t - 2 \\ &\neq 0. \end{aligned}$$

10. $t \in (2 - U) \backslash \{\frac{3}{2}\}$. Then

$$\mu(t) = \frac{2-t}{2},$$

and

$$\begin{aligned} 1 + \mu(t)f(t) &= 1 + \frac{2-t}{2}(-16) \\ &= 8t - 15 \\ &\neq 0 \quad \text{if } t \neq \frac{15}{8}. \end{aligned}$$

11. $t \in (2+U)\backslash\{\frac{5}{2}\}$. Then

$$\mu(t) = t - 2,$$

and

$$\begin{aligned} 1 + \mu(t)f(t) &= 1 + 5(t-2) \\ &= 5t - 9 \\ &\neq 0. \end{aligned}$$

Therefore f is delta regressive for $t \neq \frac{1}{2}, \frac{3}{2}, \frac{7}{8}, \frac{15}{8}$ and $t \notin U\backslash\{\frac{1}{2}\}$.

Exercise 4.8. Let $\mathbb{T} = 2\mathbb{Z}$, and let

$$f(t) = t^2 + \frac{t}{2} - \frac{11}{2}, \quad t \in \mathbb{T}.$$

Check if f is delta regressive on $\mathbb{T}$.

Exercise 4.9. Let $\mathbb{T} = (-2\mathbb{N}_0) \cup 3^{\mathbb{N}_0}$, and let

$$f(t) = \begin{cases} \frac{t^2}{2} + \frac{3}{2}t - \frac{71}{2} & \text{if } t \in (-2\mathbb{N}), \\ -1 & \text{if } t = 0, \\ -\frac{t}{162} & \text{if } t \in 3^{\mathbb{N}_0}. \end{cases}$$

Check if f is delta regressive on $\mathbb{T}$.

Exercise 4.10. Let $\mathbb{T} = P_{3,7} \cup [4,6]$, and let

$$f(t) = \begin{cases} t^2 - t & \text{if } t \in \left(\bigcup_{k=0}^{\infty}[10k, 10k+3)\right) \cup [4,6], \\ -\frac{1}{7} & \text{if } t \in \bigcup_{k=0}^{\infty}\{10k+3\}. \end{cases}$$

Check if f is delta regressive on $\mathbb{T}$.

Exercise 4.11. Let $\mathbb{T} = \{0\} \cup 11^{\mathbb{N}_0}$, and let

$$f(t) = \begin{cases} -3 & \text{if } t = 0, \\ -\frac{t}{1210} & \text{if } t \in 11^{\mathbb{N}_0}. \end{cases}$$

Check if $\mathbb{T}$ is delta regressive on $\mathbb{T}$.

Exercise 4.12. Let $\mathbb{T} = [1,2] \cup [3,4] \cup [7,8] \cup 9^{\mathbb{N}}$, and let

$$f(t) = \begin{cases} 2 & \text{if } t \in [1,2) \cup [3,4) \cup [7,8), \\ -\frac{1}{72} & \text{if } t \in 9^{\mathbb{N}}, \\ -1 & \text{if } t = 2, \\ 16 & \text{if } t = 4, \\ 3 & \text{if } t = 8. \end{cases}$$

Check if f is delta regressive on $\mathbb{T}$.

Definition 4.11. In $\mathscr{R}_\mu$, we define the circle plus addition by

$$f \oplus_\mu g = f + g + \mu f g, \quad f, g \in \mathscr{R}_\mu.$$

Example 4.51. Let $\mathbb{T} = \mathbb{Z}$, let f be as in Example 4.46, and let $g(t) = t$, $t \in \mathbb{T}$. We will find $(f \oplus_\mu g)(t)$, $t \in \mathbb{T}$. By Example 4.46 we get that f is delta regressive for $t \neq 1, 5$. Next,

$$\begin{aligned} 1 + \mu(t)g(t) &= 1 + t \\ &\neq 0 \quad \text{if } t \neq -1. \end{aligned}$$

Thus g is delta regressive for $t \neq -1$. Hence

$$\begin{aligned} (f \oplus_\mu g)(t) &= f(t) + g(t) + \mu(t)f(t)g(t) \\ &= t^2 - 6t + 4 + t + t(t^2 - 6t + 4) \\ &= t^2 - 5t + 4 + t^3 - 6t^2 + 4t \\ &= t^3 - 5t^2 - t + 4, \quad t \in \mathbb{T}, \ t \neq -1, 1, 5. \end{aligned}$$

Example 4.52. Let $\mathbb{T} = 2^{\mathbb{N}_0} \cup (-\mathbb{N}_0)$, let f be as in Example 4.47, and let $g(t) = t$, $t \in \mathbb{T}$. We will find $(f \oplus_\mu g)(t)$, $t \in \mathbb{T}$. We have the following cases.

1. $t \in 2^{\mathbb{N}_0}$. By Example 4.47 we have that f is delta regressive for $t \neq 4$. Next,

$$\begin{aligned} 1 + \mu(t)g(t) &= 1 + t^2 \\ &\neq 0, \quad t \in \mathbb{T}. \end{aligned}$$

   Thus g is delta regressive on $\mathbb{T}$. Then

$$\begin{aligned} (f \oplus_\mu g)(t) &= f(t) + g(t) + \mu(t)f(t)g(t) \\ &= t + 2 - \frac{25}{t} + t + t^2\left(t + 2 - \frac{25}{t}\right) \\ &= 2t + 2 - \frac{25}{t} + t^3 + 2t^2 - 25t \\ &= t^3 + 2t^2 - 23t - \frac{25}{t} + 2, \quad t \neq 4. \end{aligned}$$

2. $t \in (-\mathbb{N}_0)$. By Example 4.47 we have that f is delta regressive for $t \neq -6$. Next,

$$\begin{aligned} 1 + \mu(t)g(t) &= 1 + t \\ &\neq 0 \quad \text{if } t \neq -1. \end{aligned}$$

Hence

$$\begin{aligned} (f \oplus_\mu g)(t) &= f(t) + g(t) + \mu(t)f(t)g(t) \\ &= t^2 + 2t - 25 + t + t(t^2 + 2t - 25) \\ &= t^2 + 3t - 25 + t^3 + 2t^2 - 25t \\ &= t^3 + 3t^2 - 22t - 25, \quad t \neq -1, -6. \end{aligned}$$

Example 4.53. Let $\mathbb{T} = \{-3\} \cup \{-3 + \frac{1}{n}\}_{n \in \mathbb{N}} \cup [-2, 0] \cup 7^{\mathbb{N}_0}$, let f be as in Example 4.48, and let $g(t) = t$, $t \in \mathbb{T}$. We will find $(f \oplus_\mu g)(t)$, $t \in \mathbb{T}$. We have the following cases.

1. $t = -3$. Then by Example 4.48 we have that f is delta regressive for $t = -3$. Next,

$$\begin{aligned} 1 + \mu(-3)g(-3) &= 1 \\ &\neq 0. \end{aligned}$$

Thus g is delta regressive for $t = -3$. Hence

$$\begin{aligned} (f \oplus_\mu g)(-3) &= f(-3) + g(-3) + \mu(-3)f(-3)g(-3) \\ &= 3 + (-3) \\ &= 0. \end{aligned}$$

2. $t = -3 + \frac{1}{n}$, $n \in \mathbb{N}$. By Example 4.48 it follows that f is delta regressive for t. Next,

$$\begin{aligned} 1 + \mu(t)g(t) &= 1 - \frac{t(t+3)^2}{t+2} \\ &\neq 0, \quad t \neq -2, \end{aligned}$$

i. e., g is delta regressive for $t \neq -2$. Hence

$$\begin{aligned} (f \oplus_\mu g)(t) &= f(t) + g(t) + \mu(t)f(t)g(t) \\ &= -1 + t - \frac{t(t+3)^2}{t+2} \\ &= \frac{(t+2)(t-1) - t(t^2 + 6t + 9)}{t+2} \\ &= \frac{t^2 + t - 2 - t^3 - 6t^2 - 9t}{t+2} \end{aligned}$$

$$= \frac{-t^3 - 5t^2 - 8t - 2}{t+2}, \quad t \neq -2.$$

3. $t \in [-2, 0]$. By Example 4.48 it follows that f is regressive for t. Next,

$$\begin{aligned} 1 + \mu(t)g(t) &= 1 \\ &\neq 0. \end{aligned}$$

Therefore g is delta regressive for t. Hence

$$\begin{aligned} (f \oplus_\mu g)(t) &= f(t) + g(t) + \mu(t)f(t)g(t) \\ &= t^2 + t - 5. \end{aligned}$$

4. $t \in 7^{\mathbb{N}_0}$. By Example 4.48 it follows that f is delta regressive for $t \neq 7$. Next,

$$\begin{aligned} 1 + \mu(t)g(t) &= 1 + 6t^2 \\ &\neq 0, \end{aligned}$$

i. e., g is delta regressive for t. Hence

$$\begin{aligned} (f \oplus_\mu g)(t) &= f(t) + g(t) + \mu(t)f(t)g(t) \\ &= \frac{t}{6} - \frac{2}{3} - \frac{11}{3t} + t + 6t\left(\frac{t}{6} - \frac{2}{3} - \frac{11}{3t}\right)t \\ &= \frac{7t}{6} - \frac{2}{3} - \frac{11}{3t} + t^3 - 4t^2 - 22t \\ &= t^3 - 4t^2 - \frac{125}{6}t - \frac{2}{3} - \frac{11}{3t}. \end{aligned}$$

Example 4.54. Let $\mathbb{T}$ and f be as in Example 4.49, and let $g(t) = t$, $t \in \mathbb{T}$. We will find $(f \oplus_\mu g)(t)$, $t \in \mathbb{T}$. We have the following cases.

1. $t \in \{(\frac{1}{2})^{2^n} : n \in \mathbb{N}\}$. By Example 4.49 we have that f is delta regressive for $t \neq \frac{1}{16}$. Next,

$$\begin{aligned} 1 + \mu(t)g(t) &= 1 + t(\sqrt{t} - t) \\ &\neq 0, \end{aligned}$$

i. e., g is delta regressive for t. Hence

$$\begin{aligned} (f \oplus_\mu g)(t) &= f(t) + g(t) + \mu(t)f(t)g(t) \\ &= -\frac{16}{3} + t + (\sqrt{t} - t)\left(-\frac{16}{3}\right)t \\ &= -\frac{16}{3} + t + \frac{16}{3}t^2 - \frac{16}{3}t^{\frac{3}{2}}. \end{aligned}$$

2. $t = \frac{1}{2}$. By Example 4.49 it follows that f is not delta regressive for $t = \frac{1}{2}$. Thus $(f \oplus_\mu g)(t)$ does not exist.
3. $t = 0$. By Example 4.49 it follows that f is delta regressive for $t = 0$. Next,

$$\begin{aligned} 1 + \mu(0)g(0) &= 1 \\ &\neq 0. \end{aligned}$$

Thus g is delta regressive for $t = 0$. Hence

$$\begin{aligned} (f \oplus_\mu g)(0) &= f(0) + g(0) + \mu(0)f(0)g(0) \\ &= -\frac{16}{3} + 0 \\ &= -\frac{16}{3}. \end{aligned}$$

4. $t = 1$. By Example 4.49 we have that f is delta regressive for $t = 1$. Next,

$$\begin{aligned} 1 + \mu(1)g(1) &= 1 \\ &\neq 0. \end{aligned}$$

Thus g is delta regressive for $t = 1$. Hence

$$\begin{aligned} (f \oplus_\mu g)(1) &= f(1) + g(1) + \mu(1)f(1)g(1) \\ &= 37 + 1 \\ &= 38. \end{aligned}$$

Example 4.55. Let $\mathbb{T}$ and f be as in Example 4.50, and let $g(t) = t$, $t \in \mathbb{T}$. We will find $(f \oplus_\mu g)(t)$, $t \in \mathbb{T}$. We have the following cases.

1. $t = 0$. By Example 4.50 it follows that f is delta regressive for $t = 0$. Next,

$$\begin{aligned} 1 + \mu(0)g(0) &= 1 \\ &\neq 0, \end{aligned}$$

i. e., g is delta regressive for $t = 0$. Hence

$$\begin{aligned} (f \oplus_\mu g)(0) &= f(0) + g(0) + \mu(0)f(0)g(0) \\ &= 3 + 0 \\ &= 3. \end{aligned}$$

2. $t = \frac{1}{2}$. By Example 4.50, we have that f is delta regressive for $t = \frac{1}{2}$. Next,

$$1 + \mu\left(\frac{1}{2}\right)g\left(\frac{1}{2}\right) = 1 + \left(\frac{1}{2}\right)\left(\frac{1}{4}\right)$$

$$\begin{aligned}&= 1 + \frac{1}{8}\\ &= \frac{9}{8}\\ &\neq 0,\end{aligned}$$

i. e., g is delta regressive for $t = \frac{1}{2}$. Hence

$$\begin{aligned}(f \oplus_\mu g)\left(\frac{1}{2}\right) &= f\left(\frac{1}{2}\right) + g\left(\frac{1}{2}\right) + \mu\left(\frac{1}{2}\right)f\left(\frac{1}{2}\right)g\left(\frac{1}{2}\right)\\ &= -4 + \frac{1}{2} + \frac{1}{4}(-4)\left(\frac{1}{2}\right)\\ &= -\frac{7}{2} - \frac{1}{2}\\ &= -4.\end{aligned}$$

3. $t = 1$. By Example 4.50 we have that f is delta regressive for $t = 1$. Next,

$$\begin{aligned}1 + \mu(1)g(1) &= 1\\ &\neq 0,\end{aligned}$$

i. e., g is delta regressive for $t = 1$. Hence

$$\begin{aligned}(f \oplus_\mu g)(1) &= f(1) + g(1) + \mu(1)f(1)g(1)\\ &= 3 + 1\\ &= 4.\end{aligned}$$

4. $t = \frac{3}{2}$. By Example 4.50 we have that f is delta regressive for $t = \frac{3}{2}$. Next,

$$\begin{aligned}1 + \mu\left(\frac{3}{2}\right)g\left(\frac{3}{2}\right) &= 1 + \frac{1}{4}\left(\frac{3}{2}\right)\\ &= 1 + \frac{3}{8}\\ &= \frac{11}{8}\\ &\neq 0,\end{aligned}$$

i. e., g is delta regressive for $t = \frac{3}{2}$. Hence

$$\begin{aligned}(f \oplus_\mu g)\left(\frac{3}{2}\right) &= f\left(\frac{3}{2}\right) + g\left(\frac{3}{2}\right) + \mu\left(\frac{3}{2}\right)f\left(\frac{3}{2}\right)g\left(\frac{3}{2}\right)\\ &= -4 + \frac{3}{2} + \frac{1}{4}(-4)\left(\frac{3}{2}\right)\end{aligned}$$

$$= -\frac{5}{2} - \frac{3}{2}$$
$$= -4.$$

5. $t = 2$. By Example 4.50 we have that f is delta regressive for $t = 2$. Next,

$$\begin{aligned} 1 + \mu(2)g(2) &= 1 \\ &\neq 0, \end{aligned}$$

i. e., g is delta regressive for $t = 2$. Hence

$$\begin{aligned} (f \oplus_\mu g)(2) &= f(2) + g(2) + \mu(2)f(2)g(2) \\ &= 4 + 2 \\ &= 6. \end{aligned}$$

6. $t = \frac{5}{2}$. By Example 4.50 we have that f is delta regressive for $t = \frac{5}{2}$. Next,

$$\begin{aligned} 1 + \mu\left(\frac{5}{2}\right)g\left(\frac{5}{2}\right) &= 1 \\ &\neq 0, \end{aligned}$$

i. e., g is delta regressive for $t = \frac{5}{2}$. Hence

$$\begin{aligned} (f \oplus_m g)\left(\frac{5}{2}\right) &= f\left(\frac{5}{2}\right) + g\left(\frac{5}{2}\right) + \mu\left(\frac{5}{2}\right)f\left(\frac{5}{2}\right)g\left(\frac{5}{2}\right) \\ &= 10 + \frac{5}{2} \\ &= \frac{25}{2}. \end{aligned}$$

7. $t \in U\backslash\{\frac{1}{2}\}$. By Example 4.50 we have that f is not delta regressive for t. Thus $(f \oplus_\mu g)(t)$ does not exist.
8. $t \in (1 - U)\backslash\{\frac{1}{2}\}$. By Example 4.50, we have that f is delta regressive for $t \neq \frac{7}{8}$. Next,

$$\begin{aligned} 1 + \mu(t)g(t) &= 1 + t\frac{1-t}{2} \\ &= \frac{t - t^2 + 2}{2} \\ &\neq 0, \end{aligned}$$

i. e., g is delta regressive for t. Hence

$$\begin{aligned} (f \oplus_\mu g)(t) &= f(t) + g(t) + \mu(t)f(t)g(t) \\ &= 4t - \frac{39}{2} + t + \frac{1-t}{2}\left(4t - \frac{39}{2}\right)t \end{aligned}$$

$$
\begin{aligned}
&= 5t - \frac{39}{2} + 2t^2 - \frac{39}{4}t - 2t^3 + \frac{39}{4}t^2 \\
&= -2t^3 + \frac{47}{4}t^2 - \frac{19}{4}t - \frac{39}{2}, \quad t \neq \frac{7}{8}.
\end{aligned}
$$

9. $t \in (1+U)\backslash\{\frac{3}{2}\}$. By Example 4.50, we have that f is delta regressive for t. Next,

$$
\begin{aligned}
1 + \mu(t)g(t) &= 1 + t(t-1) \\
&= t^2 - t + 1 \\
&\neq 0,
\end{aligned}
$$

i. e., g is delta regressive for t. Hence

$$
\begin{aligned}
(f \oplus_\mu g)(t) &= f(t) + g(t) + \mu(t)f(t)g(t) \\
&= 3 + t + 3t(t-1) \\
&= 3 + t + 3t^2 - 3t \\
&= 3t^2 - 2t + 3.
\end{aligned}
$$

10. $t \in (2-U)\backslash\{\frac{3}{2}\}$. By Example 4.50 we have that f is delta regressive for $t \neq \frac{15}{8}$. Next,

$$
\begin{aligned}
1 + \mu(t)g(t) &= 1 + t\frac{2-t}{2} \\
&= \frac{2 + 2t - t^2}{2} \\
&\neq 0,
\end{aligned}
$$

i. e., g is delta regressive for t. Hence

$$
\begin{aligned}
(f \oplus_\mu g)(t) &= f(t) + g(t) + \mu(t)f(t)g(t) \\
&= -16 + t - 16t\frac{2-t}{2} \\
&= -16 + t - 16t + 8t^2 \\
&= 8t^2 - 15t - 16, \quad t \neq \frac{15}{8}.
\end{aligned}
$$

11. $t \in (2+U)\backslash\{\frac{5}{2}\}$. By Example 4.50 we have that f is delta regressive for t. Next,

$$
\begin{aligned}
1 + \mu(t)g(t) &= 1 + t(t-2) \\
&= t^2 - 2t + 1 \\
&= (t-1)^2 \\
&\neq 0,
\end{aligned}
$$

i. e., g is delta regressive for t. Hence

$$\begin{aligned}(f \oplus_\mu g)(t) &= f(t) + g(t) + \mu(t)f(t)g(t)\\ &= 5 + t + 5t(t-2)\\ &= 5 + t + 5t^2 - 10t\\ &= 5t^2 - 9t + 5.\end{aligned}$$

Exercise 4.13. Let $\mathbb{T}$ and f be as in Exercise 4.8, and let $g(t) = t$, $t \in \mathbb{T}$. Find $(f \oplus_\mu g)(t)$, $t \in \mathbb{T}$.

Exercise 4.14. Let $\mathbb{T}$ and f be as in Exercise 4.9, and let $g(t) = t$, $t \in \mathbb{T}$. Find $(f \oplus_\mu g)(t)$, $t \in \mathbb{T}$.

Exercise 4.15. Let $\mathbb{T}$ and f be as in Exercise 4.10, and let $g(t) = t$, $t \in \mathbb{T}$. Find $(f \oplus_\mu g)(t)$, $t \in \mathbb{T}$.

Exercise 4.16. Let $\mathbb{T}$ and f be as in Exercise 4.11, and let $g(t) = t$, $t \in \mathbb{T}$. Find $(f \oplus_\mu g)(t)$, $t \in \mathbb{T}$.

Exercise 4.17. Let $\mathbb{T}$ and f be as in Exercise 4.12, and let $g(t) = t$, $t \in \mathbb{T}$. Find $(f \oplus_\mu g)(t)$, $t \in \mathbb{T}$.

Exercise 4.18. Prove that $(\mathscr{R}_\mu, \oplus_\mu)$ is an Abelian group.

Definition 4.12. The group $(\mathscr{R}_\mu, \oplus_\mu)$ is called the *regressive group*.

Definition 4.13. For $f \in \mathscr{R}_\mu$, we define the circle minus by

$$\ominus_\mu f = -\frac{f}{1+\mu f}.$$

Example 4.56. Let $\mathbb{T}$ and g be as in Example 4.51. Then

$$\ominus_\mu g(t) = -\frac{t}{1+t}, \quad t \in \mathbb{T}.$$

Example 4.57. Let $\mathbb{T}$ and g be as in Example 4.52.

1. If $t \in 2^{\mathbb{N}_0}$, then

$$\ominus_\mu g(t) = -\frac{t}{1+t^2}.$$

2. If $t \in (-\mathbb{N}_0)$, then

$$\ominus_\mu g(t) = -\frac{t}{1+t}.$$

Example 4.58. Let $\mathbb{T}$ and g be as in Example 4.53.

1. If $t = -3$, then
$$\ominus_\mu g(-3) = 3.$$
2. If $t = -3 + \frac{1}{n}$ for some $n \in \mathbb{N}$, then
$$\ominus_\mu g(t) = \frac{t(t+2)}{t^3 + 6t^2 + 8t - 2}.$$
3. If $t \in [-2, 0]$, then
$$\ominus_\mu g(t) = -t.$$
4. If $t \in 7^{\mathbb{N}_0}$, then
$$\ominus_\mu g(t) = -\frac{t}{1 + 6t^2}.$$

Example 4.59. Let $\mathbb{T}$ and g be as in Example 4.54.

1. If $t \in \{(\frac{1}{2})^{2^n} : n \in \mathbb{N}\}$, then
$$\ominus_\mu g(t) = -\frac{t}{1 + t(\sqrt{t} - t)}.$$
2. If $t = \frac{1}{2}$, then
$$\begin{aligned}\ominus_m g\left(\frac{1}{2}\right) &= -\frac{\frac{1}{2}}{\frac{5}{4}}\\ &= -\frac{2}{5}.\end{aligned}$$
3. If $t = 0$, then
$$\ominus_\mu g(0) = 0.$$
4. If $t = 1$, then
$$\ominus_\mu g(1) = -1.$$

Example 4.60. Let $\mathbb{T}$ and g be as in Example 4.55.

1. If $t = 0$, then
$$\ominus_\mu g(0) = 0.$$
2. If $t = \frac{1}{2}$, then
$$\ominus_\mu g\left(\frac{1}{2}\right) = -\frac{\frac{1}{2}}{\frac{9}{8}}$$

$$= -\frac{4}{9}.$$

3. If $t = 1$, then

$$\ominus_\mu g(1) = -1.$$

4. If $t = \frac{3}{2}$, then

$$\ominus_\mu g\left(\frac{3}{2}\right) = -\frac{\frac{3}{2}}{\frac{11}{8}} = -\frac{12}{11}.$$

5. If $t = 2$, then

$$\ominus_\mu g(2) = -2.$$

6. If $t = \frac{5}{2}$, then

$$\ominus_\mu g\left(\frac{5}{2}\right) = -\frac{5}{2}.$$

7. If $t \in U\backslash\{\frac{1}{2}\}$, then

$$\ominus_\mu g(t) = -\frac{t}{1+t^2}.$$

8. If $t \in (1-U)\backslash\{\frac{1}{2}\}$, then

$$\ominus_\mu g(t) = -\frac{2t}{-t^2+t+2}.$$

9. If $t \in (1+U)\backslash\{\frac{3}{2}\}$, then

$$\ominus_\mu g(t) = -\frac{t}{t^2-t+1}.$$

10. If $t \in (2-U)\backslash\{\frac{3}{2}\}$, then

$$\ominus_\mu g(t) = -\frac{2t}{2+2t-t^2}.$$

11. If $t \in (2+U)\backslash\{\frac{5}{2}\}$, then

$$\ominus_\mu g(t) = -\frac{t}{(t-1)^2}.$$

Exercise 4.19. Let $\mathbb{T}$ and g be as in Exercise 4.13. Find $\ominus_\mu g(t)$, $t \in \mathbb{T}$.

Exercise 4.20. Let $\mathbb{T}$ and g be as in Exercise 4.14. Find $\ominus_\mu g(t)$, $t \in \mathbb{T}$.

Exercise 4.21. Let $\mathbb{T}$ and g be as in Exercise 4.15. Find $\ominus_\mu g(t)$, $t \in \mathbb{T}$.

Exercise 4.22. Let $\mathbb{T}$ and g be as in Exercise 4.16. Find $\ominus_\mu g(t)$, $t \in \mathbb{T}$.

Exercise 4.23. Let $\mathbb{T}$ and g be as in Exercise 4.17. Find $\ominus_\mu g(t)$, $t \in \mathbb{T}$.

Exercise 4.24. Let $f \in \mathscr{R}_\mu$. Prove that $(\ominus_\mu f) \in \mathscr{R}$.

Definition 4.14. We define the *circle minus subtraction* $\ominus_\mu$ on $\mathscr{R}_\mu$ by

$$f \ominus_\mu g = f \oplus_\mu (\ominus_\mu g), \quad f, g \in \mathscr{R}_\mu.$$

Remark 4.4. For $f, g \in \mathscr{R}_\mu$, we have

$$\begin{aligned} f \ominus_\mu g &= f \oplus_\mu (\ominus_\mu g) \\ &= f \oplus_\mu \left(-\frac{g}{1+\mu g}\right) \\ &= f - \frac{g}{1+\mu g} - \frac{\mu f g}{1+\mu g} \\ &= \frac{f-g}{1+\mu g}. \end{aligned}$$

Example 4.61. Let $\mathbb{T}$, f, and g be as in Example 4.46. Then by the computations of Example 4.56 we have

$$\begin{aligned} f \ominus_\mu g(t) &= f(t) + (\ominus_\mu g(t)) + \mu(t) f(t)(\ominus_\mu g(t)) \\ &= t^2 - 6t + 4 - \frac{t}{1+t} - (t^2 - 6t + 4)\frac{t}{1+t} \\ &= \frac{(t^2 - 6t + 4)(t+1) - t - t(t^2 - 6t + 4)}{1+t} \\ &= \frac{t^2 - 6t + 4 - t}{1+t} \\ &= \frac{t^2 - 7t + 4}{1+t}, \quad t \in \mathbb{T}. \end{aligned}$$

Example 4.62. Let $\mathbb{T}$, f, and g be as in Example 4.47. Then we have the following cases.

1. $t \in 2^{\mathbb{N}_0}$. By the computations in Example 4.57 we find

$$f \ominus_\mu g(t) = f(t) + (\ominus_\mu g(t)) + \mu(t) f(t)(\ominus_\mu g(t))$$

$$
\begin{aligned}
&= t + 2 - \frac{25}{t} - \frac{t}{1+t^2} - t\left(t + 2 - \frac{25}{t}\right)\frac{t}{1+t^2} \\
&= \frac{(1+t^2)(t+2-\frac{25}{t}) - t - t^2(t+2-\frac{25}{t})}{1+t^2} \\
&= \frac{t+2-\frac{25}{t} - t}{1+t^2} \\
&= \frac{2t-25}{t(1+t^2)}.
\end{aligned}
$$

2. $t \in (-\mathbb{N}_0)$. Then by the computations in Example 4.47 we have

$$
\begin{aligned}
f \ominus_\mu g(t) &= f(t) + (\ominus_\mu g(t)) + \mu(t) f(t)(\ominus_\mu g(t)) \\
&= t^2 + 2t - 25 - \frac{t}{1+t} - (t^2 + 2t - 25)\frac{t}{1+t} \\
&= \frac{(t^2+2t-25)(1+t) - t - t(t^2+2t-25)}{1+t} \\
&= \frac{t^2+2t-25-t}{1+t} \\
&= \frac{t^2+t-25}{1+t}.
\end{aligned}
$$

Example 4.63. Let $\mathbb{T}, f$, and g be as in Example 4.48. Then we have the following cases.

1. $t = -3$. Then

$$
\begin{aligned}
(f \ominus_\mu g)(-3) &= f(-3) + (\ominus_\mu g(-3)) + \mu(-3) f(-3)(\ominus_\mu g(-3)) \\
&= 3 + 3 \\
&= 6.
\end{aligned}
$$

2. $t = -3 + \frac{1}{n}$ for some $n \in \mathbb{N}$. By the computations in Example 4.58 we have

$$
\begin{aligned}
f \ominus_\mu g(t) &= f(t) + (\ominus_\mu g(t)) + \mu(t) f(t)(\ominus_\mu g(t)) \\
&= -1 + \frac{t(t+2)}{t^3+6t^2+8t-2} - \frac{(t+3)^2}{t+2}(-1)\left(\frac{t(t+2)}{t^3+6t^2+8t-2}\right) \\
&= -1 + \frac{t(t+2)}{t^3+6t^2+8t-2} + \frac{t(t+3)^2}{t^3+6t^2+8t-2} \\
&= \frac{-t^3-6t^2-8t+2+t^2+2t+t^3+6t^2+9t}{t^3+6t^2+8t-2} \\
&= \frac{t^2+3t}{t^3+6t^2+8t-2}.
\end{aligned}
$$

3. $t \in [-2, 0]$. Then

$$
f \ominus_\mu g(t) = f(t) + (\ominus_\mu g(t)) + \mu(t) f(t)(\ominus_\mu g(t))
$$

$$= t^2 - t - 5.$$

4. $t \in 7^{\mathbb{N}_0}$. Then by the computations in Example 4.58 we have

$$\begin{aligned}
f \ominus_\mu g(t) &= f(t) + (\ominus_\mu g(t)) + \mu(t) f(t)(\ominus_\mu g(t)) \\
&= \frac{t}{6} - \frac{2}{3} - \frac{11}{3t} - \frac{t}{1+6t^2} + 6t\left(\frac{t}{6} - \frac{2}{3} - \frac{11}{3t}\right)\left(-\frac{t}{1+6t^2}\right) \\
&= \frac{(1+6t^2)(\frac{t}{6} - \frac{2}{3} - \frac{11}{3t}) - t - 6t^2(\frac{t}{6} - \frac{2}{3} - \frac{11}{3t})}{1+6t^2} \\
&= \frac{\frac{t}{6} - \frac{2}{3} - \frac{11}{3t} - t}{1+6t^2} \\
&= \frac{t^2 - 4t - 22 - 6t^2}{6t(1+6t^2)} \\
&= -\frac{5t^2 + 4t + 22}{6t(1+6t^2)}.
\end{aligned}$$

Example 4.64. Let t, f, and g be as in Example 4.49. Then we have the following cases.

1. $t \in \{(\frac{1}{2})^{2^n} : n \in \mathbb{N}\}$. By the computations in Example 4.49 we have

$$\begin{aligned}
f \ominus_\mu g(t) &= f(t) + (\ominus_\mu g(t)) + \mu(t) f(t)(\ominus_\mu g(t)) \\
&= -\frac{16}{3} - \frac{t}{1+t(\sqrt{t}-t)} + (\sqrt{t}-t)\left(-\frac{16}{3}\right)\left(-\frac{t}{1+t(\sqrt{t}-t)}\right) \\
&= -\frac{16}{3} - \frac{t}{1+t(\sqrt{t}-t)} + \frac{16t(\sqrt{t}-t)}{3(1+t(\sqrt{t}-t))} \\
&= \frac{-16(1+t(\sqrt{t}-t)) - 3t + 16t(\sqrt{t}-t)}{3(1+t(\sqrt{t}-t))} \\
&= -\frac{16+3t}{3(1+t(\sqrt{t}-t))}.
\end{aligned}$$

2. $t = \frac{1}{2}$. Then by the computations in Example 4.49 we have that f is not delta regressive for $t = \frac{1}{2}$. Thus $(f \ominus_\mu g)(\frac{1}{2})$ does not exist.
3. $t = 0$. Then by the computations in Example 4.49 we have

$$\begin{aligned}
(f \ominus_\mu g)(0) &= f(0) + (\ominus_\mu g(0)) + \mu(0) f(0)(\ominus_\mu g(0)) \\
&= -\frac{16}{3}.
\end{aligned}$$

4. $t = 1$. Then by the computations in Example 4.49 we have

$$\begin{aligned}
(f \ominus_\mu g)(1) &= f(1) + (\ominus_\mu g(1)) + \mu(1) f(1)(\ominus_\mu g(1)) \\
&= 37 + (-1) \\
&= 36.
\end{aligned}$$

Example 4.65. Let $\mathbb{T}, f$, and g be as in Example 4.50. Then we have the following cases.

1. $t = 0$. Then

$$\begin{aligned}(f \ominus_\mu g)(0) &= f(0) + (\ominus_\mu g(0)) + \mu(0) f(0)(\ominus_\mu g(0)) \\ &= 3 + 0 \\ &= 3.\end{aligned}$$

2. $t = \frac{1}{2}$. Then by the computations in Example 4.60 we get

$$\begin{aligned}(f \ominus g)\left(\frac{1}{2}\right) &= f\left(\frac{1}{2}\right) + \left(\ominus_\mu g\left(\frac{1}{2}\right)\right) + \mu\left(\frac{1}{2}\right) f\left(\frac{1}{2}\right)\left(\ominus_\mu g\left(\frac{1}{2}\right)\right) \\ &= -4 + \left(-\frac{4}{9}\right) + \frac{1}{4}(-4)\left(-\frac{4}{9}\right) \\ &= -4 - \frac{4}{9} + \frac{4}{9} \\ &= -4.\end{aligned}$$

3. $t = 1$. Then by the computations of Example 4.60 we get

$$\begin{aligned}(f \ominus_\mu g)(1) &= f(1) + (\ominus_\mu g(1)) + \mu(1) f(1)(\ominus_\mu g(1)) \\ &= 3 + (-1) \\ &= 2.\end{aligned}$$

4. $t = \frac{3}{2}$. Then by the computations in Example 4.60 we have

$$\begin{aligned}(f \ominus_\mu g)\left(\frac{3}{2}\right) &= f\left(\frac{3}{2}\right) + \left(\ominus_\mu g\left(\frac{3}{2}\right)\right) + \mu\left(\frac{3}{2}\right) f\left(\frac{3}{2}\right)\left(\ominus_\mu g\left(\frac{3}{2}\right)\right) \\ &= -4 + \left(-\frac{12}{11}\right) + \frac{1}{4}(-4)\left(-\frac{12}{11}\right) \\ &= -4 - \frac{12}{11} + \frac{12}{11} \\ &= -4.\end{aligned}$$

5. $t = 2$. Then by the computations of Example 4.60 we have

$$\begin{aligned}(f \ominus_\mu g)(2) &= f(2) + (\ominus_\mu g(2)) + \mu(2) f(2)(\ominus_\mu g(2)) \\ &= 4 + (-2) \\ &= 2.\end{aligned}$$

6. $t = \frac{5}{2}$. Then by the computations in Example 4.60 we have

$$\begin{aligned}(f \ominus_\mu g)\left(\frac{5}{2}\right) &= f\left(\frac{5}{2}\right) + \left(\ominus_\mu g\left(\frac{5}{2}\right)\right) + \mu\left(\frac{5}{2}\right)\left(\ominus_\mu g\left(\frac{5}{2}\right)\right) \\ &= 10 + \left(-\frac{5}{2}\right)\end{aligned}$$

$$= \frac{15}{2}.$$

7. $t \in U \setminus \{\frac{1}{2}\}$. Then by the computations in Example 4.60 we have

$$\begin{aligned}
f \ominus_\mu g(t) &= f(t) + (\ominus_\mu g(t)) + \mu(t) f(t)(\ominus_\mu g(t)) \\
&= -\frac{1}{t} - \frac{t}{1+t^2} + t\left(-\frac{1}{t}\right)\left(-\frac{t}{1+t^2}\right) \\
&= -\frac{1}{t} - \frac{t}{1+t^2} + \frac{t}{1+t^2} \\
&= -\frac{1}{t}.
\end{aligned}$$

8. $t \in (1-U) \setminus \{\frac{1}{2}\}$. By the computations in Example 4.50 we have that the function f is delta regressive for $t \neq \frac{7}{8}$. Hence, using the computations in Example 4.60 we get

$$\begin{aligned}
f \ominus_\mu g(t) &= f(t) + (\ominus_\mu g(t)) + \mu(t) f(t)(\ominus_\mu g(t)) \\
&= 4t - \frac{39}{2} - \frac{2t}{t-t^2+2} + \frac{1-t}{2}\left(4t - \frac{39}{2}\right)\left(-\frac{2t}{t-t^2+2}\right) \\
&= \frac{(4t - \frac{39}{2})(t - t^2 + 2) - 2t - (t-t^2)(4t - \frac{39}{2})}{t - t^2 + 2} \\
&= \frac{8t - 39 - 2t}{t - t^2 + 2} \\
&= \frac{6t - 39}{t - t^2 + 2}, \quad t \neq \frac{7}{8}.
\end{aligned}$$

9. $t \in (1+U) \setminus \{\frac{1}{2}\}$. Then by the computations in Example 4.60 we get

$$\begin{aligned}
f \ominus_\mu g(t) &= f(t) + (\ominus_\mu g(t)) + \mu(t) f(t)(\ominus_\mu g(t)) \\
&= 3 - \frac{t}{t^2 - t + 1} + (t-1)3\left(-\frac{t}{t^2 - t + 1}\right) \\
&= \frac{3(t^2 - t + 1) - t - 3(t^2 - t)}{t^2 - t + 1} \\
&= \frac{3 - t}{t^2 - t + 1}.
\end{aligned}$$

10. $t \in (2-U) \setminus \{\frac{3}{2}\}$. By Example 4.50 we have that f is delta regressive for $t \neq \frac{15}{8}$. Now using the computations in Example 4.60, we get

$$\begin{aligned}
f \ominus_\mu g(t) &= f(t) + (\ominus_\mu g(t)) + \mu(t) f(t)(\ominus_\mu g(t)) \\
&= -16 + \left(-\frac{2t}{2 + 2t - t^2}\right) + \frac{2-t}{2}(-16)\left(-\frac{2t}{2 + 2t - t^2}\right) \\
&= \frac{-16(2 + 2t - t^2) - 2t + 16(2t - t^2)}{2 + 2t - t^2} \\
&= -\frac{32 + 2t}{2 + 2t - t^2}, \quad t \neq \frac{15}{8}.
\end{aligned}$$

11. $t \in (2 + U)\backslash\{\frac{5}{2}\}$. Then by the computations in Example 4.60 we have

$$\begin{aligned} f \ominus_\mu g(t) &= f(t) + (\ominus_\mu g(t)) + \mu(t) f(t)(\ominus_\mu g(t)) \\ &= 5 + \left(-\frac{t}{(t-1)^2}\right) + (t-2)5\left(-\frac{t}{(t-1)^2}\right) \\ &= \frac{5(t^2 - 2t + 1) - t - 5(t^2 - 2t)}{(t-1)^2} \\ &= \frac{5-t}{(t-1)^2}. \end{aligned}$$

Exercise 4.25. Let $\mathbb{T}, f$, and g be as in Exercise 4.13. Find $(f \ominus_\mu g)(t)$, $t \in \mathbb{T}$.

Exercise 4.26. Let $\mathbb{T}, f$, and g be as in Exercise 4.14. Find $(f \ominus_\mu g)(t)$, $t \in \mathbb{T}$.

Exercise 4.27. Let $\mathbb{T}, f$, and g be as in Exercise 4.15. Find $(f \ominus_\mu g)(t)$, $t \in \mathbb{T}$.

Exercise 4.28. Let $\mathbb{T}, f$, and g be as in Exercise 4.16. Find $(f \ominus_\mu g)(t)$, $t \in \mathbb{T}$.

Exercise 4.29. Let $\mathbb{T}, f$, and g be as in Exercise 4.17. Find $(f \ominus_\mu g)(t)$, $t \in \mathbb{T}$.

Example 4.66. Let $f \in \mathscr{R}_\mu$. We will prove that

$$f \ominus_\mu f = 0.$$

Indeed, we have

$$\begin{aligned} f \ominus_\mu f &= f \oplus_\mu (\ominus_\mu f) \\ &= f \oplus_\mu \left(-\frac{f}{1+\mu f}\right) \\ &= f - \frac{f}{1+\mu f} - \frac{\mu f^2}{1+\mu f} \\ &= \frac{f + \mu f^2 - f - \mu f^2}{1+\mu f} \\ &= 0. \end{aligned}$$

Example 4.67. Let $f \in \mathscr{R}_\mu$. We will prove that

$$\ominus_\mu(\ominus_\mu f) = f.$$

We have

$$\begin{aligned}\ominus_\mu(\ominus_\mu f) &= \ominus_\mu\left(-\frac{f}{1+\mu f}\right)\\ &= \frac{\frac{f}{1+\mu f}}{1-\frac{\mu f}{1+\mu f}}\\ &= f.\end{aligned}$$

Example 4.68. Let $f, g \in \mathscr{R}_\mu$. We will prove that

$$f \ominus_\mu g \in \mathscr{R}.$$

We have

$$\begin{aligned}1+\mu(f\ominus_\mu g) &= 1+\frac{\mu f-\mu g}{1+\mu g}\\ &= \frac{1+\mu f}{1+\mu g} \neq 0.\end{aligned}$$

Note that $\frac{f-g}{1+\mu g}$ is rd-continuous. Therefore $f \ominus_\mu g \in \mathscr{R}$.

Example 4.69. Let $f, g \in \mathscr{R}_\mu$. We will prove that

$$\ominus_\mu(f\ominus_\mu g) = g\ominus_\mu f.$$

We have

$$\begin{aligned}\ominus_\mu(f\ominus_\mu g) &= \ominus_\mu\left(\frac{f-g}{1+\mu g}\right)\\ &= -\frac{\frac{f-g}{1+\mu g}}{1+\mu\frac{f-g}{1+\mu g}}\\ &= -\frac{f-g}{1+\mu f}\\ &= \frac{g-f}{1+\mu f}\\ &= g\ominus_\mu f.\end{aligned}$$

Example 4.70. Let $f, g \in \mathscr{R}_\mu]$. We will prove that

$$\ominus_\mu(f\oplus_\mu g) = (\ominus_\mu f)\oplus_\mu(\ominus_\mu g).$$

We have

$$\begin{aligned}\ominus_\mu(f\oplus_\mu g) &= \ominus_\mu(f+g+\mu fg)\\ &= -\frac{f+g+\mu fg}{1+\mu f+\mu g+\mu^2 fg}\end{aligned}$$

$$= -\frac{f + g + \mu fg}{(1 + \mu f)(1 + \mu g)}.$$

Since

$$\ominus_\mu f = -\frac{f}{1 + \mu f} \quad \text{and} \quad \ominus_\mu g = -\frac{g}{1 + \mu g},$$

we also have

$$\begin{aligned}
(\ominus_\mu f) \oplus_\mu (\ominus_\mu g) &= \ominus_\mu f + (\ominus_\mu g) + \mu(\ominus_\mu f)(\ominus_\mu g) \\
&= -\frac{f}{1 + \mu f} - \frac{g}{1 + \mu g} + \frac{\mu fg}{(1 + \mu f)(1 + \mu g)} \\
&= \frac{-f(1 + \mu g) - g(1 + \mu f) + \mu fg}{(1 + \mu f)(1 + \mu g)} \\
&= -\frac{f + g + \mu fg}{(1 + \mu f)(1 + \mu g)}.
\end{aligned}$$

Example 4.71. Let $f, g \in \mathscr{R}$. We will prove that

$$f \oplus_\mu \frac{g}{1 + \mu f} = f + g.$$

We have

$$\begin{aligned}
f \oplus_\mu \frac{g}{1 + \mu f} &= f + \frac{g}{1 + \mu f} + \frac{\mu fg}{1 + \mu f} \\
&= f + g.
\end{aligned}$$

4.3 The delta exponential function

Definition 4.15. For $f \in \mathscr{R}_\mu$, we define the generalized delta exponential function by

$$e_{f,\mu}(t, s) = e^{\int_s^t \xi_{\mu(\tau)}(f(\tau))\Delta\tau} \quad \text{for } s, t \in \mathbb{T}.$$

Remark 4.5. In fact, using the definition for the cylinder transformation, we have

$$e_{f,\mu}(t, s) = e^{\int_s^t \frac{1}{\mu(\tau)} \operatorname{Log}(1 + \mu(\tau) f(\tau))\Delta\tau} \quad \text{for } s, t \in \mathbb{T}.$$

Example 4.72. Let $\mathbb{T} = h\mathbb{Z}$ and $f \in \mathscr{R}_\mu$. Then

$$\int_s^t \frac{1}{\mu(\tau)} \log(1 + \mu(\tau) f(\tau))\Delta\tau = \int_s^t \frac{1}{h} \log(1 + hf(\tau))\Delta\tau$$

$$= \sum_{\tau=s}^{t-h} \log(1 + hf(\tau)), \quad s, t \in \mathbb{T}, \ s \leq t.$$

Hence

$$\begin{aligned} e_{f,\mu}(t,s) &= e^{\int_s^t \frac{1}{\mu(\tau)} \log(1+\mu(\tau)f(\tau))\Delta\tau} \\ &= e^{\sum_{\tau=s}^{t-h} \log(1+hf(\tau))} \\ &= \prod_{\tau=s}^{t-h} (1 + hf(\tau)). \end{aligned}$$

Example 4.73. Let $\mathbb{T} = q^{\mathbb{N}_0}$, $q > 1$, and $f \in \mathscr{R}_\mu$. Then

$$\begin{aligned} \int_s^t \frac{1}{\mu(\tau)} \log(1 + \mu(\tau)f(\tau))\Delta\tau &= \int_s^t \frac{1}{(q-1)\tau} \log(1 + (q-1)\tau f(\tau))\Delta\tau \\ &= \sum_{\tau=s}^{\frac{t}{q}} \log(1 + (q-1)\tau f(\tau)) \quad t, s \in \mathbb{T}, \ s \leq t. \end{aligned}$$

Hence

$$\begin{aligned} e_{f,\mu}(t,s) &= e^{\int_s^t \frac{1}{\mu(\tau)} \log(1+\mu(\tau)f(\tau))\Delta\tau} \\ &= e^{\sum_{\tau=s}^{\frac{t}{q}} \log(1+(q-1)\tau f(\tau))} \\ &= \prod_{\tau=s}^{\frac{t}{q}} (1 + (q-1)\tau f(\tau)). \end{aligned}$$

Example 4.74. Let $\mathbb{T} = (-2\mathbb{N}_0) \cup 2^{\mathbb{N}_0}$ and $f \in \mathscr{R}_\mu$. We will find $e_{f,\mu}(t,s)$, $t, s \in \mathbb{T}$, $s \leq t$. We have the following cases.

1. $s \in (-2\mathbb{N}_0)$, $t \in 2^{\mathbb{N}}$. Then

$$\begin{aligned} \int_s^t \frac{1}{\mu(\tau)} \log(1 + \mu(\tau)f(\tau))\Delta\tau &= \int_s^0 \frac{1}{\mu(\tau)} \log(1 + \mu(\tau)f(\tau))\Delta\tau \\ &\quad + \int_0^1 \frac{1}{\mu(\tau)} \log(1 + \mu(\tau)f(\tau))\Delta\tau \\ &\quad + \int_1^t \frac{1}{\mu(\tau)} \log(1 + \mu(\tau)f(\tau))\Delta\tau \\ &= \sum_{\tau=s}^{-2} \log(1 + 2f(\tau)) + \log(1 + f(0)) + \sum_{\tau=1}^{\frac{t}{2}} \log(1 + \tau f(\tau)). \end{aligned}$$

Hence

$$\begin{aligned}
e_{f,\mu}(t,s) &= e^{\int_s^t \frac{1}{\mu(\tau)}\log(1+\mu(\tau)f(\tau))\Delta\tau}\\
&= e^{\sum_{\tau=s}^{-2}\log(1+2f(\tau))+\log(1+f(0))+\sum_{\tau=1}^{\frac{t}{2}}\log(1+\tau f(\tau))}\\
&= e^{\sum_{\tau=s}^{-2}\log(1+2f(\tau))}e^{\log(1+f(0))}e^{\sum_{\tau=1}^{\frac{t}{2}}\log(1+\tau f(\tau))}\\
&= \prod_{\tau=s}^{-2}(1+2f(\tau))(1+f(0))\prod_{\tau=1}^{\frac{t}{2}}(1+\tau f(\tau)).
\end{aligned}$$

2. $s \in (-2\mathbb{N}_0)$, $t = 1$. Then

$$\begin{aligned}
\int_s^1 \frac{1}{\mu(\tau)}\log(1+\mu(\tau)f(\tau))\Delta\tau &= \int_s^0 \frac{1}{\mu(\tau)}\log(1+\mu(\tau)f(\tau))\Delta\tau\\
&\quad + \int_0^1 \frac{1}{\mu(\tau)}\log(1+\mu(\tau)f(\tau))\Delta\tau\\
&= \sum_{\tau=s}^{-2}\log(1+2f(\tau)) + \log(1+f(0))
\end{aligned}$$

   Hence

$$\begin{aligned}
e_{f,\mu}(t,s) &= e^{\int_s^t \frac{1}{\mu(\tau)}\log(1+\mu(\tau)f(\tau))\Delta\tau}\\
&= e^{\sum_{\tau=s}^{-2}\log(1+2f(\tau))+\log(1+f(0))}\\
&= e^{\sum_{\tau=s}^{-2}\log(1+2f(\tau))}e^{\log(1+f(0))}\\
&= \prod_{\tau=s}^{-2}(1+2f(\tau))(1+f(0)).
\end{aligned}$$

3. $s, t \in (-2\mathbb{N}_0)$. Then

$$\int_s^t \frac{1}{\mu(\tau)}\log(1+\mu(\tau)f(\tau))\Delta\tau = \sum_{\tau=s}^{t-2}\log(1+2f(\tau)),$$

   and

$$\begin{aligned}
e_{f,\mu}(t,s) &= e^{\int_s^t \frac{1}{\mu(\tau)}\log(1+\mu(\tau)f(\tau))\Delta\tau}\\
&= e^{\sum_{\tau=s}^{t-2}\log(1+2f(\tau))}\\
&= \prod_{\tau=s'}^{t-2}(1+2f(\tau)).
\end{aligned}$$

4. $s = 0, t \in 2^{\mathbb{N}}$. Then

$$\begin{aligned}\int_s^t \frac{1}{\mu(\tau)}\log(1+\mu(\tau)f(\tau))\Delta\tau &= \int_0^1 \frac{1}{\mu(\tau)}\log(1+\mu(\tau)f(\tau))\Delta\tau \\ &\quad + \int_1^t \frac{1}{\mu(\tau)}\log(1+\mu(\tau)f(\tau))\Delta\tau \\ &= \log(1+f(0)) + \sum_{\tau=1}^{\frac{t}{2}} \log(1+\tau f(\tau)).\end{aligned}$$

Hence

$$\begin{aligned}e_{f,\mu}(t,s) &= e^{\int_0^t \frac{1}{\mu(\tau)}\log(1+\mu(\tau)f(\tau))\Delta\tau} \\ &= e^{\log(1+f(0))+\sum_{\tau=1}^{\frac{t}{2}}\log(1+\tau f(\tau))} \\ &= e^{\log(1+f(0))}e^{\sum_{\tau=1}^{\frac{t}{2}}\log(1+\tau f(\tau))} \\ &= (1+f(0))\prod_{\tau=1}^{\frac{t}{2}}(1+\tau f(\tau)).\end{aligned}$$

5. $s, t \in 2^{\mathbb{N}_0}$. Then

$$\int_s^t \frac{1}{\mu(\tau)}\log(1+\mu(\tau)f(\tau))\Delta\tau = \sum_{\tau=s}^{\frac{t}{2}} \log(1+\tau f(\tau)).$$

Hence

$$\begin{aligned}e_{f,\mu}(t,s) &= e^{\int_s^t \frac{1}{\mu(\tau)}\log(1+\mu(\tau)f(\tau))\Delta\tau} \\ &= e^{\sum_{\tau=s}^{\frac{t}{2}}\log(1+\tau f(\tau))} \\ &= \prod_{\tau=s}^{\frac{t}{2}}(1+\tau f(\tau)).\end{aligned}$$

Example 4.75. Let $\mathbb{T} = \{(\frac{1}{2})^{2^n} : n \in \mathbb{N}_0\} \cup \{0,1\}$. We will find $e_{f,\mu}(t,s)$, $s,t \in \mathbb{T}$, $s \le t$. We have the following cases.

1. $s = 0$ and $t = 1$. Then

$$\begin{aligned}\int_0^1 \frac{1}{\mu(\tau)}\log(1+\mu(\tau)f(\tau))\Delta\tau &= \log\left(1+\frac{1}{2}f\left(\frac{1}{2}\right)\right) \\ &\quad + \sum_{\tau=1}^{\infty}\log\left(1+\left(\left(\frac{1}{2}\right)^{2^{\tau-1}} - \left(\frac{1}{2}\right)^{2^\tau}\right)f\left(\left(\frac{1}{2}\right)^{2^\tau}\right)\right),\end{aligned}$$

and

$$\begin{aligned}
e_{f,\mu}(1,0) &= e^{\int_0^1 \frac{1}{\mu(\tau)} \log(1+\mu(\tau)f(\tau))\Delta\tau} \\
&= e^{\log(1+\frac{1}{2}f(\frac{1}{2}))+\sum_{\tau=1}^{\infty}\log(1+((\frac{1}{2})^{2^{\tau-1}}-(\frac{1}{2})^{2^\tau})f((\frac{1}{2})^{2^\tau}))} \\
&= \left(1+\frac{1}{2}f\left(\frac{1}{2}\right)\right)\prod_{\tau=1}^{\infty}\log\left(1+\left(\left(\frac{1}{2}\right)^{2^{\tau-1}}-\left(\frac{1}{2}\right)^{2^\tau}\right)f\left(\left(\frac{1}{2}\right)^{2^\tau}\right)\right).
\end{aligned}$$

2. $s=0$ and $t=(\frac{1}{2})^{2^n}$ for some $n\in\mathbb{N}_0$. Then

$$\int_0^t \frac{1}{\mu(\tau)}\log(1+\mu(\tau)f(\tau))\Delta\tau = \sum_{s=n}^{\infty}\log\left(1+\left(\left(\frac{1}{2}\right)^{2^{s-1}}-\left(\frac{1}{2}\right)^{2^s}\right)f\left(\left(\frac{1}{2}\right)^{2^s}\right)\right).$$

Hence

$$\begin{aligned}
e_{f,\mu}(t,0) &= e^{\int_0^t \frac{1}{\mu(\tau)}\log(1+\mu(\tau)f(\tau))\Delta\tau} \\
&= e^{\sum_{s=n}^{\infty}\log(1+((\frac{1}{2})^{2^{s-1}}-(\frac{1}{2})^{2^s})f((\frac{1}{2})^{2^s}))} \\
&= \prod_{s=n}^{\infty}\log\left(1+\left(\left(\frac{1}{2}\right)^{2^{s-1}}-\left(\frac{1}{2}\right)^{2^s}\right)f\left(\left(\frac{1}{2}\right)^{2^s}\right)\right).
\end{aligned}$$

3. $s=\frac{1}{2}$ and $t=1$. Then

$$\int_{\frac{1}{2}}^1 \frac{1}{\mu(\tau)}\log(1+\mu(\tau)f(\tau))\Delta\tau = \log\left(1+\frac{1}{2}f\left(\frac{1}{2}\right)\right),$$

and

$$\begin{aligned}
e_{f,\mu}\left(1,\frac{1}{2}\right) &= e^{\int_{\frac{1}{2}}^1 \frac{1}{\mu(\tau)}\log(1+\mu(\tau)f(\tau))\Delta\tau} \\
&= e^{\log(1+\frac{1}{2}f(\frac{1}{2}))} \\
&= 1+\frac{1}{2}f\left(\frac{1}{2}\right).
\end{aligned}$$

4. $s=(\frac{1}{2})^{2^n}$ for some $n\in\mathbb{N}$, $t=1$. Then

$$\begin{aligned}
\int_s^1 \frac{1}{\mu(\tau)}\log(1+\mu(\tau)f(\tau))\Delta\tau &= \int_s^{\frac{1}{2}} \frac{1}{\mu(\tau)}\log(1+\mu(\tau)f(\tau))\Delta\tau \\
&\quad + \int_{\frac{1}{2}}^1 \frac{1}{\mu(\tau)}\log(1+\mu(\tau)f(\tau))\Delta\tau
\end{aligned}$$

$$= \sum_{s=1}^{n} \log\left(1 + \left(\left(\frac{1}{2}\right)^{2^{s-1}} - \left(\frac{1}{2}\right)^{2^s}\right) f\left(\left(\frac{1}{2}\right)^{2^s}\right)\right) + \log\left(1 + \frac{1}{2} f\left(\frac{1}{2}\right)\right),$$

and

$$\begin{aligned} e_{f,\mu}(1,s) &= e^{\int_s^1 \frac{1}{\mu(\tau)} \log(1+\mu(\tau)f(\tau))\Delta\tau} \\ &= e^{\sum_{s=1}^n \log(1+((\frac{1}{2})^{2^{s-1}} - (\frac{1}{2})^{2^s}) f((\frac{1}{2})^{2^s})) + \log(1+\frac{1}{2}f(\frac{1}{2}))} \\ &= \prod_{s=1}^{n} \left(1 + \left(\left(\frac{1}{2}\right)^{2^{s-1}} - \left(\frac{1}{2}\right)^{2^s}\right) f\left(\left(\frac{1}{2}\right)^{2^s}\right)\right)\left(1 + \frac{1}{2} f\left(\frac{1}{2}\right)\right). \end{aligned}$$

5. $s = (\frac{1}{2})^{2^n}$ and $t = (\frac{1}{2})^{2^m}$ for some $m, n \in \mathbb{N}$, $m < n$. Then

$$\int_s^t \frac{1}{\mu(\tau)} \log(1 + \mu(\tau)f(\tau))\Delta\tau = \sum_{s=m+1}^{n} \log\left(1 + \left(\left(\frac{1}{2}\right)^{2^{s-1}} - \left(\frac{1}{2}\right)^{2^s}\right) f\left(\left(\frac{1}{2}\right)^{2^s}\right)\right),$$

and

$$\begin{aligned} e_{f,\mu}(t,s) &= e^{\int_s^t \frac{1}{\mu(\tau)} \log(1+\mu(\tau)f(\tau))\Delta\tau} \\ &= e^{\sum_{s=m+1}^n \log(1+((\frac{1}{2})^{2^{s-1}} - (\frac{1}{2})^{2^s}) f((\frac{1}{2})^{2^s}))} \\ &= \prod_{s=m+1}^{n} \left(1 + \left(\left(\frac{1}{2}\right)^{2^{s-1}} - \left(\frac{1}{2}\right)^{2^s}\right) f\left(\left(\frac{1}{2}\right)^{2^s}\right)\right). \end{aligned}$$

Example 4.76. Let $\mathbb{T} = \{1 - \frac{1}{2^n}\}_{n \in \mathbb{N}_0} \cup [1,2] \cup 3^{\mathbb{N}}$, and let $f \in \mathcal{R}_\mu$. We will find $e_{f,\mu}(t,s)$ for $s, t \in \mathbb{T}$, $s \le t$. We have the following cases.

1. $s = 1 - \frac{1}{2^n}$ for some $n \in \mathbb{N}_0$, $t \in 3^{\mathbb{N}}$, $t \ge 9$. Then

$$\begin{aligned} \int_s^t \frac{1}{\mu(\tau)} \log(1 + \mu(\tau)f(\tau))\Delta\tau &= \int_s^1 \frac{1}{\mu(\tau)} \log(1 + \mu(\tau)f(\tau))\Delta\tau \\ &+ \int_1^2 \frac{1}{\mu(\tau)} \log(1 + \mu(\tau)f(\tau))\Delta\tau \\ &+ \int_2^3 \frac{1}{\mu(\tau)} \log(1 + \mu(\tau)f(\tau))\Delta\tau \\ &+ \int_3^t \frac{1}{\mu(\tau)} \log(1 + \mu(\tau)f(\tau))\Delta\tau \end{aligned}$$

$$= \sum_{\tau=n}^{\infty} \log\left(1 + \frac{1}{2^{\tau+1}} f\left(1 - \frac{1}{2^{\tau}}\right)\right) + \int_1^2 f(\tau)\Delta\tau$$

$$+ \log(1 + f(2)) + \sum_{\tau=3}^{\frac{t}{3}} \log(1 + 2\tau f(\tau)).$$

Hence

$$\begin{aligned}
e_{f,\mu}(t,s) &= e^{\int_s^t \frac{1}{\mu(\tau)} \log(1+\mu(\tau)f(\tau))\Delta\tau} \\
&= e^{\sum_{\tau=n}^{\infty} \log(1+\frac{1}{2^{\tau+1}} f(1-\frac{1}{2^{\tau}})) + \int_1^2 f(\tau)\Delta\tau + \log(1+f(2)) + \sum_{\tau=3}^{\frac{t}{3}} \log(1+2\tau f(\tau))} \\
&= e^{\sum_{\tau=n}^{\infty} \log(1+\frac{1}{2^{\tau+1}} f(1-\frac{1}{2^{\tau}}))} e^{\int_1^2 f(\tau)\Delta\tau} \\
&\quad \times e^{\log(1+f(2))} e^{\sum_{\tau=3}^{\frac{t}{3}} \log(1+2\tau f(\tau))} \\
&= \prod_{\tau=n}^{\infty} \left(1 + \frac{1}{2^{\tau+1}} f\left(1 - \frac{1}{2^{\tau}}\right)\right) e^{\int_1^2 f(\tau)\Delta\tau} (1 + f(2)) \prod_{\tau=3}^{\frac{t}{3}} (1 + 2\tau f(\tau)).
\end{aligned}$$

2. $s = 1 - \frac{1}{2^n}$ for some $n \in \mathbb{N}_0$, $t = 3$. Then

$$\begin{aligned}
\int_s^t \frac{1}{\mu(\tau)} \log(1 + \mu(\tau)f(\tau))\Delta\tau &= \int_s^1 \frac{1}{\mu(\tau)} \log(1 + \mu(\tau)f(\tau))\Delta\tau \\
&\quad + \int_1^2 \frac{1}{\mu(\tau)} \log(1 + \mu(\tau)f(\tau))\Delta\tau \\
&\quad + \int_2^3 \frac{1}{\mu(\tau)} \log(1 + \mu(\tau)f(\tau))\Delta\tau \\
&= \sum_{\tau=n}^{\infty} \log\left(1 + \frac{1}{2^{\tau+1}} f\left(1 - \frac{1}{2^{\tau}}\right)\right) + \int_1^2 f(\tau)\Delta\tau \\
&\quad + \log(1 + f(2)).
\end{aligned}$$

Hence

$$\begin{aligned}
e_{f,\mu}(t,s) &= e^{\int_s^t \frac{1}{\mu(\tau)} \log(1+\mu(\tau)f(\tau))\Delta\tau} \\
&= e^{\sum_{\tau=n}^{\infty} \log(1+\frac{1}{2^{\tau+1}} f(1-\frac{1}{2^{\tau}})) + \int_1^2 f(\tau)\Delta\tau + \log(1+f(2))} \\
&= e^{\sum_{\tau=n}^{\infty} \log(1+\frac{1}{2^{\tau+1}} f(1-\frac{1}{2^{\tau}}))} e^{\int_1^2 f(\tau)\Delta\tau} \\
&\quad \times e^{\log(1+f(2))} \\
&= \prod_{\tau=n}^{\infty} \left(1 + \frac{1}{2^{\tau+1}} f\left(1 - \frac{1}{2^{\tau}}\right)\right) e^{\int_1^2 f(\tau)\Delta\tau} (1 + f(2)) \prod_{\tau=3}^{\frac{t}{3}} .
\end{aligned}$$

3. $s = 1 - \frac{1}{2^n}$ for some $n \in \mathbb{N}_0$, $t \in [1,2]$. Then

$$\begin{aligned}\int_s^t \frac{1}{\mu(\tau)} \log(1+\mu(\tau)f(\tau))\Delta\tau &= \int_s^1 \frac{1}{\mu(\tau)} \log(1+\mu(\tau)f(\tau))\Delta\tau \\ &\quad + \int_1^2 \frac{1}{\mu(\tau)} \log(1+\mu(\tau)f(\tau))\Delta\tau \\ &= \sum_{\tau=n}^{\infty} \log\left(1+\frac{1}{2^{\tau+1}} f\left(1-\frac{1}{2^\tau}\right)\right) + \int_1^2 f(\tau)\Delta\tau.\end{aligned}$$

Hence

$$\begin{aligned}e_{f,\mu}(t,s) &= e^{\int_s^t \frac{1}{\mu(\tau)} \log(1+\mu(\tau)f(\tau))\Delta\tau} \\ &= e^{\sum_{\tau=n}^{\infty} \log(1+\frac{1}{2^{\tau+1}} f(1-\frac{1}{2^\tau})) + \int_1^2 f(\tau)\Delta\tau} \\ &= e^{\sum_{\tau=n}^{\infty} \log(1+\frac{1}{2^{\tau+1}} f(1-\frac{1}{2^\tau}))} e^{\int_1^2 f(\tau)\Delta\tau} \\ &= \prod_{\tau=n}^{\infty} \left(1+\frac{1}{2^{\tau+1}} f\left(1-\frac{1}{2^\tau}\right)\right) e^{\int_1^2 f(\tau)\Delta\tau}.\end{aligned}$$

4. $s = 1 - \frac{1}{2^n}$ and $t = 1 - \frac{1}{2^m}$ for some $m, n \in \mathbb{N}_0$, $m > n$. Then

$$\int_s^t \frac{1}{\mu(\tau)} \log(1+\mu(\tau)f(\tau))\Delta\tau = \sum_{\tau=n}^{m} \log\left(1+\frac{1}{2^{\tau+1}} f\left(1-\frac{1}{2^\tau}\right)\right).$$

Hence

$$\begin{aligned}e_{f,\mu}(t,s) &= e^{\int_s^t \frac{1}{\mu(\tau)} \log(1+\mu(\tau)f(\tau))\Delta\tau} \\ &= e^{\sum_{\tau=n}^{m} \log(1+\frac{1}{2^{\tau+1}} f(1-\frac{1}{2^\tau}))} \\ &= \prod_{\tau=n}^{m} \left(1+\frac{1}{2^{\tau+1}} f\left(1-\frac{1}{2^\tau}\right)\right).\end{aligned}$$

5. $s \in [1,2]$ and $t \in 3^{\mathbb{N}}$, $t \geq 9$. Then

$$\begin{aligned}\int_s^t \frac{1}{\mu(\tau)} \log(1+\mu(\tau)f(\tau))\Delta\tau &= \int_s^2 \frac{1}{\mu(\tau)} \log(1+\mu(\tau)f(\tau))\Delta\tau \\ &\quad + \int_2^3 \frac{1}{\mu(\tau)} \log(1+\mu(\tau)f(\tau))\Delta\tau \\ &\quad + \int_3^t \frac{1}{\mu(\tau)} \log(1+\mu(\tau)f(\tau))\Delta\tau\end{aligned}$$

$$= \int_s^2 f(\tau)\Delta\tau + \log(1+f(2)) + \sum_{\tau=3}^{\frac{t}{3}} \log(1+2\tau f(\tau)).$$

Hence

$$\begin{aligned} e_{f,\mu}(t,s) &= e^{\int_s^t \frac{1}{\mu(\tau)}\log(1+\mu(\tau)f(\tau))\Delta\tau} \\ &= e^{\int_s^2 f(\tau)\Delta\tau + \log(1+f(2)) + \sum_{\tau=3}^{\frac{t}{3}} \log(1+2\tau f(\tau))} \\ &= e^{\int_s^2 f(\tau)\Delta\tau} e^{\log(1+f(2))} e^{\sum_{\tau=3}^{\frac{t}{3}} \log(1+2\tau f(\tau))} \\ &= e^{\int_s^2 f(\tau)\Delta\tau}(1+f(2)) \prod_{\tau=3}^{\frac{t}{3}} (1+2\tau f(\tau)). \end{aligned}$$

6. $s, t \in [1,2]$. Then

$$\int_s^t \frac{1}{\mu(\tau)} \log(1+\mu(\tau)f(\tau))\Delta\tau = \int_s^t f(\tau)\Delta\tau.$$

Hence

$$\begin{aligned} e_{f,\mu}(t,s) &= e^{\int_s^t \frac{1}{\mu(\tau)}\log(1+\mu(\tau)f(\tau))\Delta\tau} \\ &= e^{\int_s^t f(\tau)\Delta\tau}. \end{aligned}$$

7. $s \in [1,2]$ and $t = 3$. Then

$$\begin{aligned} \int_s^3 \frac{1}{\mu(\tau)} \log(1+\mu(\tau)f(\tau))\Delta\tau &= \int_s^2 \frac{1}{\mu(\tau)} \log(1+\mu(\tau)f(\tau))\Delta\tau \\ &\quad + \int_2^3 \frac{1}{\mu(\tau)} \log(1+\mu(\tau)f(\tau))\Delta\tau \\ &= \int_s^2 f(\tau)\Delta\tau + \log(1+f(2)). \end{aligned}$$

Hence,

$$\begin{aligned} e_{f,\mu}(3,s) &= e^{\int_s^3 \frac{1}{\mu(\tau)}\log(1+\mu(\tau)f(\tau))\Delta\tau} \\ &= e^{\int_s^2 f(\tau)\Delta\tau + \log(1+f(2))} \\ &= e^{\int_s^2 f(\tau)\Delta\tau} e^{\log(1+f(2))} \\ &= e^{\int_s^2 f(\tau)\Delta\tau}(1+f(2)). \end{aligned}$$

8. Let $s, t \in 3^{\mathbb{N}}$, $s \le t$. Then

$$\int_s^t \frac{1}{\mu(\tau)} \log(1+\mu(\tau)f(\tau))\Delta\tau = \sum_{\tau=s}^{\frac{t}{3}} \log(1+2\tau f(\tau))$$

and

$$\begin{aligned} e_{f,\mu}(t,s) &= e^{\int_s^t \frac{1}{\mu(\tau)} \log(1+\mu(\tau)f(\tau))\Delta\tau} \\ &= e^{\sum_{\tau=s}^{\frac{t}{3}} \log(1+2\tau f(\tau))} \\ &= \prod_{\tau=s}^{\frac{t}{3}} (1+2\tau f(\tau)). \end{aligned}$$

Exercise 4.30. Let $\mathbb{T} = \{2 - \frac{1}{2^n}\}_{n \in \mathbb{N}_0} \cup [2,3] \cup 4^{\mathbb{N}}$ and $f \in \mathscr{R}_\mu$. Prove that

$$e_{f,\mu}(t,s) = \begin{cases} \prod_{\tau=n}^{\infty}\left(1+\frac{1}{2^{\tau+1}}f\left(2-\frac{1}{2^\tau}\right)\right)e^{\int_2^3 f(\tau)\Delta\tau}(1+f(3))\prod_{\tau=4}^{\frac{t}{4}}\left(1+3\tau f(\tau)\right) \\ \quad \text{if } s = 2-\frac{1}{2^n} \text{ for some } n \in \mathbb{N}_0 \text{ and } t \in 4^{\mathbb{N}},\ t \ge 16 \\ \prod_{\tau=n}^{\infty}\left(1+\frac{1}{2^{\tau+1}}f\left(2-\frac{1}{2^\tau}\right)\right)e^{\int_2^3 f(\tau)\Delta\tau}\left(1+f(3)\right) \\ \quad \text{if } s = 2-\frac{1}{2^n} \text{ for some } n \in \mathbb{N}_0 \text{ and } t = 4 \\ \prod_{\tau=n}^{\infty}\left(1+\frac{1}{2^{\tau+1}}f\left(2-\frac{1}{2^\tau}\right)\right)e^{\int_2^t f(\tau)\Delta\tau} \\ \quad \text{if } s = 2-\frac{1}{2^n} \text{ for some } n \in \mathbb{N}_0 \text{ and } t \in [2,3] \\ \prod_{\tau=n}^{m}\left(1+\frac{1}{2^{\tau+1}}f\left(2-\frac{1}{2^\tau}\right)\right) \\ \quad \text{if } s = 2-\frac{1}{2^n} \text{ and } t = 2-\frac{1}{2^m} \text{ for some } m,n \in \mathbb{N},\ m > n, \\ e^{\int_s^3 f(\tau)\Delta\tau}\left(1+f(3)\right)\prod_{\tau=4}^{\frac{t}{4}}\left(1+3\tau f(\tau)\right) \\ \quad \text{if } s \in [2,3] \text{ and } t \in 4^{\mathbb{N}},\ t \ge 16, \\ e^{\int_s^t f(\tau)\Delta\tau} \quad \text{if } s,t \in [2,3], \\ e^{\int_s^3 f(\tau)\Delta\tau}(1+f(3)) \quad \text{if } s \in [2,3],\ t = 4, \\ \prod_{\tau=4}^{\frac{t}{4}}(1+3\tau f(\tau)) \quad \text{if } s,t \in 4^{\mathbb{N}},\ s \le t. \end{cases}$$

Example 4.77 (Semigroup property). If $f \in \mathscr{R}_\mu$, we will prove that

$$e_{f,\mu}(t,r)e_{f,\mu}(r,s) = e_{f,\mu}(t,s) \quad \text{for all } t,r,s \in \mathbb{T}.$$

We have

$$\begin{aligned} e_{f,\mu}(t,r)e_{f,\mu}(r,s) &= e^{\int_r^t \xi_{\mu(\tau)}(f(\tau))\Delta\tau} e^{\int_s^r \xi_{\mu(\tau)}(f(\tau))\Delta\tau} \\ &= e^{\int_r^t \xi_{\mu(\tau)}(f(\tau))\Delta\tau + \int_s^r \xi_{\mu(\tau)}(f(\tau))\Delta\tau} \\ &= e^{\int_s^t \xi_{\mu(\tau)}(f(\tau))\Delta\tau} \\ &= e_{f,\mu}(t,s), \quad t,r,s \in \mathbb{T}, \end{aligned}$$

completing the proof.

Exercise 4.31. Let $f \in \mathscr{R}_\mu$. Prove that

$$e_{0,\mu}(t,s) = 1 \quad \text{and} \quad e_{f,\mu}(t,t) = 1, \quad t,s \in \mathbb{T}.$$

Example 4.78. Let $f \in \mathscr{R}_\mu$ and fix $t_0 \in \mathbb{T}$. We will prove that

$$e^{\Delta}_{f,\mu}(t,t_0) = f(t)e_{f,\mu}(t,t_0), \quad t \in \mathbb{T}.$$

Really, we have the following cases.

1. If $\sigma(t) > t$, then

$$\begin{aligned}
e^{\Delta}_{f,\mu}(t,t_0) &= \frac{e_{f,\mu}(\sigma(t),t_0) - e_{f,\mu}(t,t_0)}{\mu(t)} \\
&= \frac{e^{\int_{t_0}^{\sigma(t)} \xi_{\mu(\tau)}(f(\tau))\Delta\tau} - e^{\int_{t_0}^{t} \xi_{\mu(\tau)}(f(\tau))\Delta\tau}}{\mu(t)} \\
&= \frac{e^{\int_{t_0}^{t} \xi_{\mu(\tau)}(f(\tau))\Delta\tau + \int_{t}^{\sigma(t)} \xi_{\mu(\tau)}(f(\tau))\Delta\tau} - e^{\int_{t_0}^{t} \xi_{\mu(\tau)}(f(\tau))\Delta\tau}}{\mu(t)} \\
&= \frac{e^{\int_{t}^{\sigma(t)} \xi_{\mu(\tau)}(f(\tau))\Delta\tau} - 1}{\mu(t)} e^{\int_{t_0}^{t} \xi_{\mu(\tau)}(f(\tau))\Delta\tau} \\
&= \frac{e^{\mu(t)\xi_{\mu(t)}(f(t))} - 1}{\mu(t)} e_{f,\mu}(t,t_0) \\
&= f(t)e_{f,\mu}(t,t_0).
\end{aligned}$$

2. If $\sigma(t) = t$, then

$$\begin{aligned}
&\left|e_{f,\mu}(t,t_0) - e_{f,\mu}(s,t_0) - f(t)e_{f,\mu}(t,t_0)(t-s)\right| \\
&\quad = \left|e_{f,\mu}(t,t_0) - e_{f,\mu}(t,t_0)e_{f,\mu}(s,t) - f(t)e_{f,\mu}(t,t_0)(t-s)\right| \\
&\quad = \left|e_{f,\mu}(t,t_0)\right|\left|1 - e_{f,\mu}(s,t) - f(t)(t-s)\right| \\
&\quad = \left|e_{f,\mu}(t,t_0)\right|\Bigg|1 - \int_s^t \xi_{\mu(\tau)}(f(\tau))\Delta\tau - e_{f,\mu}(s,t) \\
&\qquad + \int_s^t \xi_{\mu(\tau)}(f(\tau))\Delta\tau - f(t)(t-s)\Bigg| \\
&\quad \le \left|e_{f,\mu}(t,t_0)\right|\Bigg(\left|1 - \int_s^t \xi_{\mu(\tau)}(f(\tau))\Delta\tau - e_{f,\mu}(s,t)\right| \\
&\qquad + \left|\int_s^t \xi_{\mu(\tau)}(f(\tau))\Delta\tau - f(t)(t-s)\right|\Bigg)
\end{aligned}$$

$$\le |e_{f,\mu}(t,t_0)|\left(\left|1-\int_s^t \xi_{\mu(\tau)}(f(\tau))\Delta\tau - e_{f,\mu}(s,t)\right| + \left|\int_s^t (\xi_{\mu(\tau)}(f(\tau)) - \xi_0(f(t)))\Delta\tau\right|\right),$$

i. e.,

$$\begin{aligned}|e_{f,\mu}(t,t_0) - e_{f,\mu}(s,t_0) - f(t)e_{f,\mu}(t,t_0)(t-s)| &\le |e_{f,\mu}(t,t_0)|\left|1-\int_s^t \xi_{\mu(\tau)}(f(\tau))\Delta\tau - e_{f,\mu}(s,t)\right| \\ &\quad + |e_{f,\mu}(t,t_0)|\left|\int_s^t (\xi_{\mu(\tau)}(f(\tau)) - \xi_0(f(t)))\Delta\tau\right|.\end{aligned} \tag{4.3}$$

Since $\sigma(t) = t$ and $f \in \mathrm{C_{rd}}$, we get

$$\lim_{r\to t} \xi_{\mu(r)}(f(r)) = \xi_0(f(t)).$$

Therefore there exists a neighborhood U_1 of t such that

$$|\xi_{\mu(\tau)}(f(\tau)) - \xi_0(f(t))| < \frac{\varepsilon}{3|e_{f,\mu}(t,t_0)|} \quad \text{for all } \tau \in U_1.$$

Let $s \in U_1$. Then

$$|e_{f,\mu}(t,t_0)|\left|\int_s^t (\xi_{\mu(\tau)}(f(\tau)) - \xi_0(f(t)))\Delta\tau\right| \le \frac{\varepsilon}{3}|t-s|. \tag{4.4}$$

Also, using that

$$\lim_{z\to 0} \frac{1-z-e^{-z}}{z} = 0,$$

we conclude that there exists a neighborhood U_2 of t such that if $s \in U_2$ and $s < t$, then

$$\left|\frac{1-\int_s^t \xi_{\mu(\tau)}(f(\tau))\Delta\tau - e_{f,\mu}(s,t)}{\int_s^t \xi_{\mu(\tau)}(f(\tau))\Delta\tau}\right| < \varepsilon^*,$$

where

$$\varepsilon^* = \min\left\{1, \frac{\varepsilon}{1+3|f(t)||e_{f,\mu}(t,t_0)|}\right\}.$$

Let $s \in U = U_1 \cap U_2$, $s \neq t$. Then

$$
\begin{aligned}
&|e_{f,\mu}(t,t_0)|\left|1 - \int_s^t \xi_{\mu(\tau)}(f(\tau))\Delta\tau - e_{f,\mu}(s,t)\right| \\
&= |e_{f,\mu}(t,t_0)|\frac{|1 - \int_s^t \xi_{\mu(\tau)}(f(\tau))\Delta\tau - e_{f,\mu}(s,t)|}{|\int_s^t \xi_{\mu(\tau)}(f(\tau))\Delta\tau|}\left|\int_s^t \xi_{\mu(\tau)}(f(\tau))\Delta\tau\right| \\
&\leq |e_{f,\mu}(t,t_0)|\varepsilon^*\left|\int_s^t \xi_{\mu(\tau)}(f(\tau))\Delta\tau\right| \\
&\leq |e_{f,\mu}(t,t_0)|\varepsilon^*\left\{\left|\int_s^t (\xi_{\mu(\tau)}(f(\tau)) - \xi_0(f(t)))\Delta\tau\right| + |f(t)||t-s|\right\} \\
&\leq |e_{f,\mu}(t,t_0)|\left|\int_s^t (\xi_{\mu(\tau)}(f(\tau)) - \xi_0(f(t)))\Delta\tau\right| + |e_{f,\mu}(t,t_0)|\varepsilon^*|f(t)||t-s| \\
&\leq \frac{\varepsilon}{3}|t-s| + \frac{\varepsilon}{3}|t-s| \\
&= \frac{2\varepsilon}{3}|t-s|.
\end{aligned}
$$

From the last inequality and from (4.3) and (4.4) we conclude that

$$
|e_{f,\mu}(t,t_0) - e_{f,\mu}(s,t_0) - f(t)e_{f,\mu}(t,t_0)(t-s)| \leq \frac{2\varepsilon}{3}|t-s| + \frac{\varepsilon}{3}|t-s| = \varepsilon|t-s|,
$$

which completes the proof.

Example 4.79. Let $f \in \mathscr{R}_\mu$ and fix $t_0 \in \mathbb{T}$. Then $e_{f,\mu}(\cdot,t_0)$ is a solution to the Cauchy problem

$$
y^\Delta(t) = f(t)y(t), \quad t \in \mathbb{T}, \quad y(t_0) = 1. \tag{4.5}
$$

Example 4.80. Let $f \in \mathscr{R}_\mu$ and fix $t_0 \in \mathbb{T}$. Then $e_{f,\mu}(\cdot,t_0)$ is a unique solution to problem (4.5).

Indeed, let y be a solution to problem (4.5). Then

$$
\begin{aligned}
\left(\frac{y}{e_{f,\mu}(\cdot,t_0)}\right)^\Delta(t) &= \frac{y^\Delta(t)e_{f,\mu}(t,t_0) - y(t)e_{f,\mu}^\Delta(t,t_0)}{e_{f,\mu}(\sigma(t),t_0)e_{f,\mu}(t,t_0)} \\
&= \frac{f(t)y(t)e_{f,\mu}(t,t_0) - y(t)f(t)e_{f,\mu}(t,t_0)}{e_{f,\mu}(\sigma(t),t_0)e_{f,\mu}(t,t_0)} \\
&= 0, \quad t \in \mathbb{T}^\kappa.
\end{aligned}
$$

Consequently, $y = ce_{f,\mu}(\cdot,t_0)$, where c is a constant. Thus

$$1 = y(t_0) = c e_{f,\mu}(t_0, t_0) = c.$$

Therefore $y = e_{f,\mu}(\cdot, t_0)$.

Example 4.81. If $f \in \mathscr{R}_\mu$, then

$$e_{f,\mu}(\sigma(t), s) = (1 + \mu(t) f(t)) e_{f,\mu}(t, s), \quad t, s \in \mathbb{T}.$$

We have

$$\begin{aligned} e_{f,\mu}(\sigma(t), s) - e_{f,\mu}(t, s) &= \mu(t) e_{f,\mu}^{\Delta}(t, s) \\ &= \mu(t) f(t) e_{f,\mu}(t, s), \quad t, s \in \mathbb{T}, \end{aligned}$$

which completes the proof.

Example 4.82. If $f \in \mathscr{R}_\mu$, then

$$e_{f,\mu}(t, s) = \frac{1}{e_{f,\mu}(s, t)} = e_{\ominus_\mu f}(s, t), \quad t, s \in \mathbb{T}.$$

We have

$$\begin{aligned} e_{f,\mu}(t, s) &= e^{\int_s^t \xi_{\mu(\tau)}(f(\tau)) \Delta\tau} \\ &= e^{-\int_t^s \xi_{\mu(\tau)}(f(\tau)) \Delta\tau} \\ &= \frac{1}{e^{\int_t^s \xi_{\mu(\tau)}(f(\tau)) \Delta\tau}} \\ &= \frac{1}{e_{f,\mu}(s, t)}, \quad t, s \in \mathbb{T}. \end{aligned}$$

Now we fix $t_0 \in \mathbb{T}$ and consider the problem

$$y^{\Delta}(t) = (\ominus_\mu f)(t) y(t), \quad t \in \mathbb{T}, \quad y(t_0) = 1.$$

Its solution is $e_{\ominus_\mu f}(t, s)$. Also, using the quotient rule, we have

$$\begin{aligned} \left(\frac{1}{e_{f,\mu}(\cdot, s)} \right)^{\Delta} (t) &= -\frac{e_{f,\mu}^{\Delta}(t, s)}{e_{f,\mu}(\sigma(t), s) e_{f,\mu}(t, s)} \\ &= -\frac{f(t) e_{f,\mu}(t, s)}{(1 + \mu(t) f(t)) e_{f,\mu}(t, s) e_{f,\mu}(t, s)} \\ &= -\frac{f(t)}{(1 + \mu(t) f(t)) e_{f,\mu}(t, s)} \\ &= (\ominus_\mu f)(t) \frac{1}{e_{f,\mu}(t, s)}, \quad t, s \in \mathbb{T}. \end{aligned}$$

Therefore

$$\frac{1}{e_{f,\mu}(t,s)} = e_{\ominus_\mu f,\mu}(t,s), \quad t,s \in \mathbb{T},$$

completing the proof.

Example 4.83. If $f, g \in \mathscr{R}_\mu$, then

$$e_{f,\mu}(t,s)e_{g,\mu}(t,s) = e_{f\oplus_\mu g,\mu}(t,s), \quad t,s \in \mathbb{T}.$$

We have

$$\begin{aligned}
e_{f,\mu}(t,s)e_{g,\mu}(t,s) &= e^{\int_s^t \xi_{\mu(\tau)}(f(\tau))\Delta\tau} e^{\int_s^t \xi_{\mu(\tau)}(g(\tau))\Delta\tau} \\
&= e^{\int_s^t (\xi_{\mu(\tau)}(f(\tau))+\xi_{\mu(\tau)}(g(\tau)))\Delta\tau} \\
&= e^{\int_s^t \frac{1}{\mu(\tau)}(\mathrm{Log}(1+\mu(\tau)f(\tau))+\mathrm{Log}(1+\mu(\tau)g(\tau)))\Delta\tau} \\
&= e^{\int_s^t \frac{1}{\mu(\tau)}\mathrm{Log}((1+\mu(\tau)f(\tau))(1+\mu(\tau)g(\tau)))\Delta\tau} \\
&= e^{\int_s^t \frac{1}{\mu(\tau)}\mathrm{Log}(1+\mu(\tau)(f(\tau)+g(\tau)+\mu(\tau)f(\tau)g(\tau)))\Delta\tau} \\
&= e^{\int_s^t \xi_{\mu(\tau)}((f\oplus_\mu g)(\tau))\Delta\tau} \\
&= e_{f\oplus_\mu g,\mu}(t,s), \quad t,s \in \mathbb{T},
\end{aligned}$$

completing the proof.

Example 4.84. If $f, g \in \mathscr{R}_\mu$, then

$$\frac{e_{f,\mu}(t,s)}{e_{g,\mu}(t,s)} = e_{f\ominus_\mu g,\mu}(t,s), \quad t,s \in \mathbb{T}.$$

We have

$$\begin{aligned}
\frac{e_{f,\mu}(t,s)}{e_{g,\mu}(t,s)} &= e_{f,\mu}(t,s)e_{\ominus_\mu g,\mu}(t,s) \\
&= e_{f\oplus_\mu(\ominus_\mu g),\mu}(t,s) \\
&= e_{f\ominus_\mu g,\mu}(t,s), \quad t,s \in \mathbb{T},
\end{aligned}$$

completing the proof.

Example 4.85. If $f \in \mathscr{R}_\mu$, then

$$e_{f,\mu}(t,\sigma(s))e_{f,\mu}(s,r) = \frac{1}{1+\mu(s)f(s)}e_{f,\mu}(t,r), \quad t,r,s \in \mathbb{T}.$$

We have

$$e_{f,\mu}(t,\sigma(s))e_{f,\mu}(s,r) = e^{\int_{\sigma(s)}^t \xi_{\mu(\tau)}(f(\tau))\Delta\tau} e^{\int_r^s \xi_{\mu(\tau)}(f(\tau))\Delta\tau}$$

$$
\begin{aligned}
&= e^{\int_{\sigma(s)}^{s} \xi_{\mu(\tau)}(f(\tau))\Delta\tau + \int_s^t \xi_{\mu(\tau)}(f(\tau))\Delta\tau + \int_r^s \xi_{\mu(\tau)}(f(\tau))\Delta\tau} \\
&= e^{-\xi_{\mu(s)}(f(s))\mu(s) + \int_r^t \xi_{\mu(\tau)}(f(\tau))\Delta\tau} \\
&= e^{-\operatorname{Log}(1+f(s)\mu(s))} e^{\int_r^t \xi_{\mu(\tau)}(f(\tau))\Delta\tau} \\
&= \frac{1}{1+\mu(s)f(s)} e_{f,\mu}(t,r), \quad t,r,s \in \mathbb{T},
\end{aligned}
$$

completing the proof.

Example 4.86. If $f, g \in \mathscr{R}_\mu$, then

$$
e^{\Delta}_{f\ominus_\mu g,\mu}(t,t_0) = \frac{(f(t)-g(t))e_{f,\mu}(t,t_0)}{e_{g,\mu}(\sigma(t),t_0)}, \quad t,t_0 \in \mathbb{T}.
$$

We have

$$
\begin{aligned}
e^{\Delta}_{f\ominus_\mu g,\mu}(t,t_0) &= \left(\frac{e_{f,\mu}(\cdot,t_0)}{e_{g,\mu}(\cdot,t_0)}\right)^{\Delta}(t) \\
&= \frac{e^{\Delta}_{f,\mu}(t,t_0)e_{g,\mu}(t,t_0) - e_{f,\mu}(t,t_0)e^{\Delta}_{g,\mu}(t,t_0)}{e_{g,\mu}(t,t_0)e_{g,\mu}(\sigma(t),t_0)} \\
&= \frac{f(t)e_{f,\mu}(t,t_0)e_{g,\mu}(t,t_0) - g(t)e_{f,\mu}(t,t_0)e_{g,\mu}(t,t_0)}{e_{g,\mu}(t,t_0)e_{g,\mu}(\sigma(t),t_0)} \\
&= \frac{(f(t)-g(t))e_{f,\mu}(t,t_0)}{e_{g,\mu}(\sigma(t),t_0)}, \quad t,t_0 \in \mathbb{T},
\end{aligned}
$$

completing the proof.

Example 4.87. Let $f \in \mathscr{R}_\mu$ and $a, b, c \in \mathbb{T}$. Then

$$
(e_{f,\mu}(c,\cdot))^{\Delta}(t) = -f(t)e_{f,\mu}(c,\sigma(t)), \quad t \in \mathbb{T},
$$

and

$$
\int_a^b f(t)e_{f,\mu}(c,\sigma(t))\Delta t = e_{f,\mu}(c,a) - e_{f,\mu}(c,b).
$$

We have

$$
\begin{aligned}
f(t)e_{f,\mu}(c,\sigma(t)) &= f(t)e_{\ominus_\mu f,\mu}(\sigma(t),c) \\
&= f(t)(1+\mu(t)\ominus_\mu f(t))e_{\ominus_\mu f,\mu}(t,c) \\
&= f(t)\left(1 - \frac{\mu(t)f(t)}{1+\mu(t)f(t)}\right)e_{\ominus_\mu f,\mu}(t,c) \\
&= \frac{f(t)}{1+\mu(t)f(t)} e_{\ominus_\mu f,\mu}(t,c)
\end{aligned}
$$

$$\begin{aligned}&= -(\ominus_\mu f)(t)e_{\ominus_\mu f,\mu}(t,c)\\ &= -e^{\Delta}_{\ominus_\mu f,\mu}(t,c), \quad t \in \mathbb{T}.\end{aligned}$$

Hence

$$\begin{aligned}\int_a^b f(t)e_{f,\mu}(c,\sigma(t))\Delta t &= -\int_a^b e^{\Delta}_{\ominus_\mu f}(t,c)\Delta t\\ &= e_{\ominus_\mu f}(a,c) - e_{\ominus_\mu f}(b,c)\\ &= e_{f,\mu}(c,a) - e_{f,\mu}(c,b),\end{aligned}$$

completing the proof.

Example 4.88. Let $1+\mu(t)\frac{2}{t} \neq 0$ and $1+\mu(t)\frac{5}{t} \neq 0$ for all $t \in \mathbb{T}\cap(0,\infty)$. Let $t_0 \in \mathbb{T}\cap(0,\infty)$. We will evaluate the integral

$$I = \int_{t_0}^{t} \frac{e_{f,\mu}(s,t_0)}{se^{\sigma}_g(s,t_0)}\Delta s, \quad \text{where } f(t) = \frac{5}{t},\ g(t) = \frac{2}{t}, \quad t \in \mathbb{T}.$$

We have

$$e^{\Delta}_{f\ominus_\mu g,\mu}(t,t_0) = (f(t)-g(t))\frac{e_{f,\mu}(t,t_0)}{e_{g,\mu}(\sigma(t),t_0)} = 3\frac{e_{f,\mu}(t,t_0)}{e_{g,\mu}(\sigma(t),t_0)}, \quad t \in \mathbb{T},$$

so that

$$I = \frac{1}{3}\int_{t_0}^{t} e^{\Delta}_{f\ominus_\mu g,\mu}(s,t_0)\Delta s = \frac{1}{3}(e_{f\ominus_\mu g,\mu}(t,t_0)-1) = \frac{1}{3}e_{f\ominus_\mu g,\mu}(t,t_0) - \frac{1}{3}, \quad t \in \mathbb{T},$$

with

$$h(t) = (f \ominus_\mu g)(t) = \frac{\frac{3}{t}}{1+\mu(t)\frac{2}{t}} = \frac{3}{t+2\mu(t)}, \quad t \in \mathbb{T}.$$

Example 4.89. Let $\alpha \in \mathbb{R}$. Suppose that the exponentials

$$e_{f,\mu}(t,t_0) \quad \text{and} \quad e_{g,\mu}(t,t_0) \quad \text{with} \quad f(t) = \frac{\alpha^2}{t} - \frac{(\alpha-1)^2}{\sigma(t)} \quad \text{and} \quad g(t) = \frac{\alpha}{t}$$

exist for all $t, t_0 \in \mathbb{T}\cap(0,\infty)$. We will prove that

1. $\frac{e_{f,\mu}(t,t_0)}{e_{g,\mu}(t,t_0)} = e_{\frac{\alpha-1}{\sigma},\mu}(t,t_0)$ and
2. $\frac{e_{\frac{\alpha-1}{\sigma},\mu}(t,t_0)}{e_{g,\mu}(t,t_0)} = e_{-\frac{1}{\sigma},\mu}(t,t_0) = \frac{t_0}{t}$

for all $t, t_0 \in \mathbb{T} \cap (0, \infty)$.

1. We have

$$
\begin{aligned}
(f \ominus_\mu g)(t) &= \frac{f(t) - g(t)}{1 + \mu(t)g(t)} \\
&= \frac{\frac{\alpha^2}{t} - \frac{(\alpha-1)^2}{\sigma(t)} - \frac{\alpha}{t}}{1 + \mu(t)\frac{\alpha}{t}} \\
&= \frac{1}{\sigma(t)} \cdot \frac{\alpha(\alpha-1)\sigma(t) - (\alpha-1)^2 t}{t + \alpha\mu(t)} \\
&= \frac{\alpha - 1}{\sigma(t)} \cdot \frac{\alpha\sigma(t) - (\alpha-1)t}{t + \alpha\mu(t)} \\
&= \frac{\alpha - 1}{\sigma(t)}, \quad t \in \mathbb{T}.
\end{aligned}
$$

Hence

$$
\frac{e_{f,\mu}(t, t_0)}{e_{g,\mu}(t, t_0)} = e_{f \ominus_\mu g, \mu}(t, t_0) = e_{\frac{\alpha-1}{\sigma}, \mu}(t, t_0), \quad t, t_0 \in \mathbb{T}.
$$

2. We have

$$
\begin{aligned}
\left(\frac{\alpha - 1}{\sigma} \ominus_\mu g\right)(t) &= \frac{\frac{\alpha-1}{\sigma(t)} - \frac{\alpha}{t}}{1 + \mu(t)\frac{\alpha}{t}} \\
&= \frac{1}{\sigma(t)} \cdot \frac{(\alpha-1)t - \alpha\sigma(t)}{t + \alpha\mu(t)} \\
&= -\frac{1}{\sigma(t)}, \quad t, t_0 \in \mathbb{T}.
\end{aligned}
$$

Hence

$$
\frac{e_{\frac{\alpha-1}{\sigma}, \mu}(t, t_0)}{e_{g,\mu}(t, t_0)} = e_{-\frac{1}{\sigma}, \mu}(t, t_0), \quad t, t_0 \in \mathbb{T}.
$$

Now we show that

$$
e_{-\frac{1}{\sigma}, \mu}(t, t_0) = \frac{t_0}{t}, \quad t, t_0 \in \mathbb{T}. \tag{4.6}
$$

Define $p(t) = \frac{t_0}{t}$, $t, t_0 \in \mathbb{T}$. Then $p(t_0) = 1$, and

$$
p^\Delta(t) = -\frac{t_0}{t\sigma(t)} = -\frac{1}{\sigma(t)} p(t), \quad t, t_0 \in \mathbb{T}.
$$

and thus (4.6) holds.

Exercise 4.32. Let $\mathbb{T} = \mathbb{Z}$. Prove that

$$e_a(t, t_0) = \prod_{s\in[t_0,t)} (1 + a(s)), \quad t, t_0 \in \mathbb{T}.$$

Exercise 4.33. Let $\mathbb{T} = h\mathbb{Z}$, $h > 0$. Prove that

$$e_a(t, t_0) = \prod_{s\in[t_0,t)} (1 + ha(s)), \quad t, t_0 \in \mathbb{T}.$$

4.4 Delta hyperbolic functions

In this and the following section, we often "skip" the arguments; e. g., we write $e_{f,\mu}$ for $e_{f,\mu}(t, s)$.

Definition 4.16. Let $f \in C_{rd}$ and $-\mu f^2 \in \mathscr{R}_\mu$. Define the delta hyperbolic functions $\cosh_{f,\mu}$ and $\sinh_{f,\mu}$ as

$$\cosh_{f,\mu} = \frac{e_{f,\mu} + e_{-f,\mu}}{2} \quad \text{and} \quad \sinh_{f,\mu} = \frac{e_{f,\mu} - e_{-f,\mu}}{2}.$$

Example 4.90. Let $f \in C_{rd}$ and $-\mu f^2 \in \mathscr{R}_\mu$. WE will prove that

$$\cosh^\Delta_{f,\mu} = f \sinh_{f,\mu}, \quad \sinh^\Delta_{f,\mu} = f \cosh_{f,\mu}, \quad \text{and} \quad \cosh^2_{f,\mu} - \sinh^2_{f,\mu} = e_{-\mu f^2,\mu}.$$

We have

$$\begin{aligned}
\cosh^\Delta_{f,\mu} &= \left(\frac{e_{f,\mu} + e_{-f,\mu}}{2}\right)^\Delta \\
&= \frac{e^\Delta_{f,\mu} + e^\Delta_{-f,\mu}}{2} \\
&= \frac{fe_{f,\mu} - fe_{-f,\mu}}{2} \\
&= f\frac{e_{f,\mu} - e_{-f,\mu}}{2} \\
&= f \sinh_{f,\mu}, \\
\sinh^\Delta_{f,\mu} &= \left(\frac{e_{f,\mu} - e_{-f,\mu}}{2}\right)^2 \\
&= \frac{e^\Delta_{f,\mu} - e^\Delta_{-f,\mu}}{2} \\
&= \frac{fe_{f,\mu} + fe_{-f,\mu}}{2} \\
&= f\frac{e_{f,\mu} + e_{-f,\mu}}{2}
\end{aligned}$$

$$= f\cosh_{f,\mu},$$

and

$$\begin{aligned}\cosh^2_{f,\mu} - \sinh^2_{f,\mu} &= \left(\frac{e_{f,\mu} + e_{-f,\mu}}{2}\right)^2 - \left(\frac{e_{f,\mu} - e_{-f,\mu}}{2}\right)^2 \\ &= \frac{e^2_{f,\mu} + 2e_{f,\mu}e_{-f,\mu} + e^2_{-f,\mu} - e^2_{f,\mu} + 2e_{f,\mu}e_{-f,\mu} - e^2_{-f,\mu}}{4} \\ &= e_{f,\mu}e_{-f,\mu} \\ &= e_{f\oplus_\mu(-f),\mu} \\ &= e_{-\mu f^2,\mu}.\end{aligned}$$

The proof is complete.

Example 4.91. Let $\mathbb{T} = \mathbb{Z}$. We will compute $\cosh_{\frac{1}{2},\mu}(t,t_0)$ and $\sinh_{\frac{1}{2},\mu}(t,t_0)$ for $t, t_0 \in \mathbb{T}$. Here $\mu(t) = 1$, $t \in \mathbb{T}$. Let $\alpha = \frac{1}{2}$. Then

$$1 - \alpha^2\mu^2(t) = 1 - \frac{1}{4} = \frac{3}{4} \neq 0, \quad t \in \mathbb{T}.$$

Therefore $\cosh_{\frac{1}{2},\mu}(t,t_0)$ and $\sinh_{\frac{1}{2},\mu}(t,t_0)$, $t, t_0 \in \mathbb{T}$, are well defined. We have

$$e_{\frac{1}{2},\mu}(t,t_0) = \left(\frac{3}{2}\right)^{t-t_0} \quad\text{and}\quad e_{-\frac{1}{2},\mu}(t,t_0) = \left(\frac{1}{2}\right)^{t-t_0}, \quad t, t_0 \in \mathbb{T}.$$

Hence

$$\begin{aligned}\cosh_{\frac{1}{2},\mu}(t,t_0) &= \frac{e_{\frac{1}{2},\mu}(t,t_0) + e_{-\frac{1}{2},\mu}(t,t_0)}{2} \\ &= \frac{\frac{3^{t-t_0}}{2^{t-t_0}} + \frac{1}{2^{t-t_0}}}{2} \\ &= \frac{3^{t-t_0}+1}{2^{1+t-t_0}}, \\ \sinh_{\frac{1}{2},\mu}(t,t_0) &= \frac{e_{\frac{1}{2},\mu}(t,t_0) - e_{-\frac{1}{2},\mu}(t,t_0)}{2} \\ &= \frac{3^{t-t_0}-1}{2^{1+t-t_0}}, \quad t, t_0 \in \mathbb{T}.\end{aligned}$$

Example 4.92. Let $\mathbb{T} = 2^{\mathbb{N}_0}$. We will compute $\cosh_{2,\mu}(t,t_0)$ and $\sinh_{2,\mu}(t,t_0)$ for $t, t_0 \in \mathbb{T}$. Here $\mu(t) = t$, $t \in \mathbb{T}$. Let $\alpha = 2$. Then

$$1 - \alpha^2\mu^2(t) = 1 - 4t^2 \neq 0 \quad \text{for all } t \in \mathbb{T}.$$

Therefore $\cosh_{2,\mu}(t,t_0)$ and $\sinh_{2,\mu}(t,t_0)$, $t, t_0 \in \mathbb{T}$, are well defined. We have

$$e_{2,\mu}(t,t_0) = \prod_{s\in[t_0,t)} (1+2s), \quad e_{-2,\mu}(t,t_0) = \prod_{s\in[t_0,t)} (1-2s), \quad t,t_0 \in \mathbb{T}.$$

Hence

$$\begin{aligned}\cosh_{2,\mu}(t,t_0) &= \frac{e_{2,\mu}(t,t_0)+e_{-2,\mu}(t,t_0)}{2}\\ &= \frac{1}{2}\Big(\prod_{s\in[t_0,t)}(1+2s) + \prod_{s\in[t_0,t)}(1-2s)\Big),\\ \sinh_2(t,t_0) &= \frac{e_{2,\mu}(t,t_0)-e_{-2,\mu}(t,t_0)}{2}\\ &= \frac{1}{2}\Big(\prod_{s\in[t_0,t)}(1+2s) - \prod_{s\in[t_0,t)}(1-2s)\Big), \quad t,t_0\in\mathbb{T}.\end{aligned}$$

Example 4.93. Let $\mathbb{T} = \mathbb{N}_0^2$. We compute $\cosh_{2,\mu}(t,t_0)$ and $\sinh_{2,\mu}(t,t_0)$ for $t,t_0 \in \mathbb{T}$. Here

$$\sigma(t) = (1+\sqrt{t})^2, \quad \mu(t) = 1+2\sqrt{t}, \quad t\in\mathbb{T}.$$

Let $\alpha = 2$. Then

$$1-\alpha^2\mu^2(t) = 1-4(1+2\sqrt{t})^2 \neq 0 \quad \text{for all } t\in\mathbb{T}.$$

Therefore $\cosh_{2,\mu}(t,t_0)$ and $\sinh_{2,\mu}(t,t_0)$ are well defined. Note that

$$\begin{aligned}e_{2,\mu}(t,t_0) &= \prod_{s\in[t_0,t)}(1+2(1+2\sqrt{s}))\\ &= \prod_{s\in[t_0,t)}(3+4\sqrt{s}),\\ e_{-2,\mu}(t,t_0) &= \prod_{s\in[t_0,t)}(1-2(1+2\sqrt{s}))\\ &= \prod_{s\in[t_0,t)}(-1-4\sqrt{s}), \quad t,t_0\in\mathbb{T}.\end{aligned}$$

Hence

$$\begin{aligned}\cosh_{2,\mu}(t,t_0) &= \frac{1}{2}(e_{2,\mu}(t,t_0)+e_{-2,\mu}(t,t_0))\\ &= \frac{1}{2}\Big(\prod_{s\in[t_0,t)}(3+4\sqrt{s}) + \prod_{s\in[t_0,t)}(-1-4\sqrt{s})\Big),\\ \sinh_2(t,t_0) &= \frac{1}{2}(e_2(t,t_0)-e_{-2}(t,t_0))\\ &= \frac{1}{2}\Big(\prod_{s\in[t_0,t)}(3+4\sqrt{s}) - \prod_{s\in[t_0,t)}(-1-4\sqrt{s})\Big), \quad t,t_0\in\mathbb{T}.\end{aligned}$$

Example 4.94. Let $\mathbb{T} = (-2\mathbb{N}_0) \cup 2^{\mathbb{N}_0}$ and $f \in \mathscr{R}_\mu$. We will find $\cosh_{f,\mu}(t,s)$ and $\sinh_{f,\mu}(t,s)$, $t, s \in \mathbb{T}$, $s \le t$, for $f \in \mathrm{C}_{\mathrm{rd}}$ and $-\mu f^2 \in \mathscr{R}_\mu$. We have the following cases.

1. $s \in (-2\mathbb{N}_0)$, $t \in 2^{\mathbb{N}}$. By Example 4.74 we have

$$e_{f,\mu}(t,s) = \prod_{\tau=s}^{-2}(1+2f(\tau))(1+f(0))\prod_{\tau=1}^{\frac{t}{2}}(1+\tau f(\tau))$$

and

$$e_{-f,\mu}(t,s) = \prod_{\tau=s}^{-2}(1-2f(\tau))(1-f(0))\prod_{\tau=1}^{\frac{t}{2}}(1-\tau f(\tau)).$$

Therefore

$$\begin{aligned}\cosh_{f,\mu}(t,s) = \frac{1}{2}\Bigg(&\prod_{\tau=s}^{-2}(1+2f(\tau))(1+f(0))\prod_{\tau=1}^{\frac{t}{2}}(1+\tau f(\tau))\\ &+\prod_{\tau=s}^{-2}(1-2f(\tau))(1-f(0))\prod_{\tau=1}^{\frac{t}{2}}(1-\tau f(\tau))\Bigg),\end{aligned}$$

and

$$\begin{aligned}\sinh_{f,\mu}(t,s) = \frac{1}{2}\Bigg(&\prod_{\tau=s}^{-2}(1+2f(\tau))(1+f(0))\prod_{\tau=1}^{\frac{t}{2}}(1+\tau f(\tau))\\ &-\prod_{\tau=s}^{-2}(1-2f(\tau))(1-f(0))\prod_{\tau=1}^{\frac{t}{2}}(1-\tau f(\tau))\Bigg).\end{aligned}$$

2. $s \in (-2\mathbb{N}_0)$, $t = 1$. By Example 4.74 we have

$$e_{f,\mu}(t,s) = \prod_{\tau=s}^{-2}(1+2f(\tau))(1+f(0))$$

and

$$e_{-f,\mu}(t,s) = \prod_{\tau=s}^{-2}(1-2f(\tau))(1-f(0)).$$

Therefore

$$\cosh_{f,\mu}(t,s) = \frac{1}{2}\left(\prod_{\tau=s}^{-2}(1+2f(\tau))(1+f(0)) + \prod_{\tau=s}^{-2}(1-2f(\tau))(1-f(0))\right),$$

and

$$\sinh_{f,\mu}(t,s) = \frac{1}{2}\left(\prod_{\tau=s}^{-2}(1+2f(\tau))(1+f(0)) - \prod_{\tau=s}^{-2}(1-2f(\tau))(1-f(0))\right).$$

3. $s, t \in (-2\mathbb{N}_0)$. By Example 4.74 we have

$$e_{f,\mu}(t,s) = \prod_{\tau=s'}^{t-2}(1+2f(\tau))$$

and

$$e_{-f,\mu}(t,s) = \prod_{\tau=s'}^{t-2}(1-2f(\tau)).$$

Therefore

$$\cosh_{f,\mu}(t,s) = \frac{1}{2}\left(\prod_{\tau=s'}^{t-2}(1+2f(\tau)) + \prod_{\tau=s'}^{t-2}(1-2f(\tau))\right),$$

and

$$\sinh_{f,\mu}(t,s) = \frac{1}{2}\left(\prod_{\tau=s'}^{t-2}(1+2f(\tau)) - \prod_{\tau=s'}^{t-2}(1-2f(\tau))\right).$$

4. $s = 0$, $t \in 2^{\mathbb{N}}$. By Example 4.74 we have

$$e_{f,\mu}(t,s) = (1+f(0))\prod_{\tau=1}^{\frac{t}{2}}(1+\tau f(\tau))$$

and

$$e_{-f,\mu}(t,s) = (1-f(0))\prod_{\tau=1}^{\frac{t}{2}}(1-\tau f(\tau)).$$

Therefore

$$\cosh_{f,\mu}(t,s) = \frac{1}{2}\left((1+f(0))\prod_{\tau=1}^{\frac{t}{2}}(1+\tau f(\tau)) + (1-f(0))\prod_{\tau=1}^{\frac{t}{2}}(1-\tau f(\tau))\right),$$

and

$$\sinh_{f,\mu}(t,s) = \frac{1}{2}\left((1+f(0))\prod_{\tau=1}^{\frac{t}{2}}(1+\tau f(\tau)) - (1-f(0))\prod_{\tau=1}^{\frac{t}{2}}(1-\tau f(\tau))\right).$$

5. $s, t \in 2^{\mathbb{N}_0}$. By Example 4.74 we have

$$e_{f,\mu}(t,s) = \prod_{\tau=s}^{\frac{t}{2}} (1 + \tau f(\tau))$$

and

$$e_{-f,\mu}(t,s) = \prod_{\tau=s}^{\frac{t}{2}} (1 - \tau f(\tau)).$$

Therefore

$$\cosh_{f,\mu}(t,s) = \frac{1}{2}\left(\prod_{\tau=s}^{\frac{t}{2}} (1 + \tau f(\tau)) + \prod_{\tau=s}^{\frac{t}{2}} (1 - \tau f(\tau))\right),$$

and

$$\sinh_{f,\mu}(t,s) = \frac{1}{2}\left(\prod_{\tau=s}^{\frac{t}{2}} (1 + \tau f(\tau)) - \prod_{\tau=s}^{\frac{t}{2}} (1 - \tau f(\tau))\right).$$

Example 4.95. Let $\mathbb{T} = \{(\frac{1}{2})^{2^n} : n \in \mathbb{N}_0\} \cup \{0, 1\}$. We will find $\cosh_{f,\mu}(t,s)$ and $\sinh_{f,\mu}(t,s)$, $s, t \in \mathbb{T}$, $s \le t$, for $f \in \mathrm{C_{rd}}$ and $-\mu f^2 \in \mathscr{R}_\mu$. We have the following cases.

1. $s = 0$ and $t = 1$. By Example 4.75 we have

$$e_{f,\mu}(1,0) = \left(1 + \frac{1}{2} f\left(\frac{1}{2}\right)\right) \prod_{\tau=1}^{\infty} \log\left(1 + \left(\left(\frac{1}{2}\right)^{2^{\tau-1}} - \left(\frac{1}{2}\right)^{2^\tau}\right) f\left(\left(\frac{1}{2}\right)^{2^\tau}\right)\right)$$

and

$$e_{-f,\mu}(1,0) = \left(1 - \frac{1}{2} f\left(\frac{1}{2}\right)\right) \prod_{\tau=1}^{\infty} \left(1 - \left(\left(\frac{1}{2}\right)^{2^{\tau-1}} - \left(\frac{1}{2}\right)^{2^\tau}\right) f\left(\left(\frac{1}{2}\right)^{2^\tau}\right)\right).$$

Therefore

$$\begin{aligned}\cosh_{f,\mu}(t,s) = \frac{1}{2}\Bigg(&\left(1 + \frac{1}{2} f\left(\frac{1}{2}\right)\right) \prod_{\tau=1}^{\infty} \log\left(1 + \left(\left(\frac{1}{2}\right)^{2^{\tau-1}} - \left(\frac{1}{2}\right)^{2^\tau}\right) f\left(\left(\frac{1}{2}\right)^{2^\tau}\right)\right)\\ &+ \left(1 - \frac{1}{2} f\left(\frac{1}{2}\right)\right) \prod_{\tau=1}^{\infty} \left(1 - \left(\left(\frac{1}{2}\right)^{2^{\tau-1}} - \left(\frac{1}{2}\right)^{2^\tau}\right) f\left(\left(\frac{1}{2}\right)^{2^\tau}\right)\right)\Bigg),\end{aligned}$$

and

$$\sinh_{f,\mu}(t,s) = \frac{1}{2}\Bigg(\left(1 + \frac{1}{2} f\left(\frac{1}{2}\right)\right) \prod_{\tau=1}^{\infty} \log\left(1 + \left(\left(\frac{1}{2}\right)^{2^{\tau-1}} - \left(\frac{1}{2}\right)^{2^\tau}\right) f\left(\left(\frac{1}{2}\right)^{2^\tau}\right)\right)$$

$$-\left(1-\frac{1}{2}f\left(\frac{1}{2}\right)\right)\prod_{\tau=1}^{\infty}\left(1-\left(\left(\frac{1}{2}\right)^{2^{\tau-1}}-\left(\frac{1}{2}\right)^{2^{\tau}}\right)f\left(\left(\frac{1}{2}\right)^{2^{\tau}}\right)\right)\right).$$

2. $s = 0$ and $t = (\frac{1}{2})^{2^n}$ for some $n \in \mathbb{N}_0$. By Example 4.75 we have

$$e_{f,\mu}(t,0) = \prod_{s=n}^{\infty}\log\left(1+\left(\left(\frac{1}{2}\right)^{2^{s-1}}-\left(\frac{1}{2}\right)^{2^{s}}\right)f\left(\left(\frac{1}{2}\right)^{2^{s}}\right)\right)$$

and

$$e_{-f,\mu}(t,0) = \prod_{s=n}^{\infty}\log\left(1-\left(\left(\frac{1}{2}\right)^{2^{s-1}}-\left(\frac{1}{2}\right)^{2^{s}}\right)f\left(\left(\frac{1}{2}\right)^{2^{s}}\right)\right).$$

Therefore

$$\begin{aligned}\cosh_{f,\mu}(t,0) = \frac{1}{2}\Bigg(&\prod_{s=n}^{\infty}\log\left(1+\left(\left(\frac{1}{2}\right)^{2^{s-1}}-\left(\frac{1}{2}\right)^{2^{s}}\right)f\left(\left(\frac{1}{2}\right)^{2^{s}}\right)\right)\\&+\prod_{s=n}^{\infty}\log\left(1-\left(\left(\frac{1}{2}\right)^{2^{s-1}}-\left(\frac{1}{2}\right)^{2^{s}}\right)f\left(\left(\frac{1}{2}\right)^{2^{s}}\right)\right)\Bigg),\end{aligned}$$

and

$$\begin{aligned}\sinh_{f,\mu}(t,0) = \frac{1}{2}\Bigg(&\prod_{s=n}^{\infty}\log\left(1+\left(\left(\frac{1}{2}\right)^{2^{s-1}}-\left(\frac{1}{2}\right)^{2^{s}}\right)f\left(\left(\frac{1}{2}\right)^{2^{s}}\right)\right)\\&-\prod_{s=n}^{\infty}\log\left(1-\left(\left(\frac{1}{2}\right)^{2^{s-1}}-\left(\frac{1}{2}\right)^{2^{s}}\right)f\left(\left(\frac{1}{2}\right)^{2^{s}}\right)\right)\Bigg).\end{aligned}$$

3. $s = \frac{1}{2}$ and $t = 1$. By Example 4.75 we have

$$e_{f,\mu}\left(1,\frac{1}{2}\right) = 1+\frac{1}{2}f\left(\frac{1}{2}\right)$$

and

$$e_{-f,\mu}\left(1,\frac{1}{2}\right) = 1-\frac{1}{2}f\left(\frac{1}{2}\right).$$

Therefore

$$\cosh_{f,\mu}\left(1,\frac{1}{2}\right) = 1,$$

and

$$\sinh_{f,\mu}\left(1,\frac{1}{2}\right)=\frac{1}{2}f\left(\frac{1}{2}\right).$$

4. $s=(\frac{1}{2})^{2^n}$ for some $n\in\mathbb{N}$, $t=1$. By Example 4.75 we have

$$e_{f,\mu}(1,s)=\prod_{s=1}^{n}\left(1+\left(\left(\frac{1}{2}\right)^{2^{s-1}}-\left(\frac{1}{2}\right)^{2^s}\right)f\left(\left(\frac{1}{2}\right)^{2^s}\right)\right)\left(1+\frac{1}{2}f\left(\frac{1}{2}\right)\right)$$

and

$$e_{-f,\mu}(1,s)=\prod_{s=1}^{n}\left(1-\left(\left(\frac{1}{2}\right)^{2^{s-1}}-\left(\frac{1}{2}\right)^{2^s}\right)f\left(\left(\frac{1}{2}\right)^{2^s}\right)\right)\left(1-\frac{1}{2}f\left(\frac{1}{2}\right)\right).$$

Therefore

$$\begin{aligned}\cosh_{f,\mu}(1,s)=\frac{1}{2}\Bigg(&\prod_{s=1}^{n}\left(1+\left(\left(\frac{1}{2}\right)^{2^{s-1}}-\left(\frac{1}{2}\right)^{2^s}\right)f\left(\left(\frac{1}{2}\right)^{2^s}\right)\right)\left(1+\frac{1}{2}f\left(\frac{1}{2}\right)\right)\\&+\prod_{s=1}^{n}\left(1-\left(\left(\frac{1}{2}\right)^{2^{s-1}}-\left(\frac{1}{2}\right)^{2^s}\right)f\left(\left(\frac{1}{2}\right)^{2^s}\right)\right)\left(1-\frac{1}{2}f\left(\frac{1}{2}\right)\right)\Bigg),\end{aligned}$$

and

$$\begin{aligned}\sinh_{f,\mu}(1,s)=\frac{1}{2}\Bigg(&\prod_{s=1}^{n}\left(1+\left(\left(\frac{1}{2}\right)^{2^{s-1}}-\left(\frac{1}{2}\right)^{2^s}\right)f\left(\left(\frac{1}{2}\right)^{2^s}\right)\right)\left(1+\frac{1}{2}f\left(\frac{1}{2}\right)\right)\\&-\prod_{s=1}^{n}\left(1-\left(\left(\frac{1}{2}\right)^{2^{s-1}}-\left(\frac{1}{2}\right)^{2^s}\right)f\left(\left(\frac{1}{2}\right)^{2^s}\right)\right)\left(1-\frac{1}{2}f\left(\frac{1}{2}\right)\right)\Bigg).\end{aligned}$$

5. $s=(\frac{1}{2})^{2^n}$ and $t=(\frac{1}{2})^{2^m}$ for some $m,n\in\mathbb{N}$, $m<n$. By Example 4.75 we have

$$e_{f,\mu}(t,s)=\prod_{s=m+1}^{n}\left(1+\left(\left(\frac{1}{2}\right)^{2^{s-1}}-\left(\frac{1}{2}\right)^{2^s}\right)f\left(\left(\frac{1}{2}\right)^{2^s}\right)\right)$$

and

$$e_{-f,\mu}(t,s)=\prod_{s=m+1}^{n}\left(1-\left(\left(\frac{1}{2}\right)^{2^{s-1}}-\left(\frac{1}{2}\right)^{2^s}\right)f\left(\left(\frac{1}{2}\right)^{2^s}\right)\right).$$

Therefore

$$\cosh_{f,\mu}(t,s)=\frac{1}{2}\Bigg(\prod_{s=m+1}^{n}\left(1+\left(\left(\frac{1}{2}\right)^{2^{s-1}}-\left(\frac{1}{2}\right)^{2^s}\right)f\left(\left(\frac{1}{2}\right)^{2^s}\right)\right)$$

$$+\prod_{s=m+1}^{n}\left(1-\left(\left(\frac{1}{2}\right)^{2^{s-1}}-\left(\frac{1}{2}\right)^{2^s}\right)f\left(\left(\frac{1}{2}\right)^{2^s}\right)\right)\right),$$

and

$$\sinh_{f,\mu}(t,s)=\frac{1}{2}\left(\prod_{s=m+1}^{n}\left(1+\left(\left(\frac{1}{2}\right)^{2^{s-1}}-\left(\frac{1}{2}\right)^{2^s}\right)f\left(\left(\frac{1}{2}\right)^{2^s}\right)\right)\right.$$
$$\left.-\prod_{s=m+1}^{n}\left(1-\left(\left(\frac{1}{2}\right)^{2^{s-1}}-\left(\frac{1}{2}\right)^{2^s}\right)f\left(\left(\frac{1}{2}\right)^{2^s}\right)\right)\right).$$

Example 4.96. Let $\mathbb{T}=\{1-\frac{1}{2^n}\}_{n\in\mathbb{N}_0}\cup[1,2]\cup 3^{\mathbb{N}}$ and $f\in\mathscr{R}_\mu$. We will find $\cosh_{f,\mu}(t,s)$ and $\sinh_{f,\mu}(t,s)$ for $s,t\in\mathbb{T}$, $s\le t$, and $f\in C_{rd}$, $-\mu f^2\in\mathscr{R}_\mu$. We have the following cases.

1. $s=1-\frac{1}{2^n}$ for some $n\in\mathbb{N}_0$, and $t\in 3^{\mathbb{N}}$, $t\ge 9$. By Example 4.76 we have

$$e_{f,\mu}(t,s)=\prod_{\tau=n}^{\infty}\left(1+\frac{1}{2^{\tau+1}}f\left(1-\frac{1}{2^\tau}\right)\right)e^{\int_1^2 f(\tau)\Delta\tau}(1+f(2))\prod_{\tau=3}^{\frac{t}{3}}(1+2\tau f(\tau))$$

and

$$e_{-f,\mu}(t,s)=\prod_{\tau=n}^{\infty}\left(1-\frac{1}{2^{\tau+1}}f\left(1-\frac{1}{2^\tau}\right)\right)e^{-\int_1^2 f(\tau)\Delta\tau}(1-f(2))\prod_{\tau=3}^{\frac{t}{3}}(1-2\tau f(\tau)).$$

Consequently,

$$\cosh_{f,\mu}(t,s)=\frac{1}{2}\left(\prod_{\tau=n}^{\infty}\left(1+\frac{1}{2^{\tau+1}}f\left(1-\frac{1}{2^\tau}\right)\right)e^{\int_1^2 f(\tau)\Delta\tau}(1+f(2))\prod_{\tau=3}^{\frac{t}{3}}(1+2\tau f(\tau))\right.$$
$$\left.+\prod_{\tau=n}^{\infty}\left(1-\frac{1}{2^{\tau+1}}f\left(1-\frac{1}{2^\tau}\right)\right)e^{-\int_1^2 f(\tau)\Delta\tau}(1-f(2))\prod_{\tau=3}^{\frac{t}{3}}(1-2\tau f(\tau))\right),$$

and

$$\sinh_{f,\mu}(t,s)=\frac{1}{2}\left(\prod_{\tau=n}^{\infty}\left(1+\frac{1}{2^{\tau+1}}f\left(1-\frac{1}{2^\tau}\right)\right)e^{\int_1^2 f(\tau)\Delta\tau}(1+f(2))\prod_{\tau=3}^{\frac{t}{3}}(1+2\tau f(\tau))\right.$$
$$\left.-\prod_{\tau=n}^{\infty}\left(1-\frac{1}{2^{\tau+1}}f\left(1-\frac{1}{2^\tau}\right)\right)e^{-\int_1^2 f(\tau)\Delta\tau}(1-f(2))\prod_{\tau=3}^{\frac{t}{3}}(1-2\tau f(\tau))\right).$$

2. $s=1-\frac{1}{2^n}$ for some $n\in\mathbb{N}_0$, and $t=3$. By Example 4.76 we have

$$e_{f,\mu}(t,s)=\prod_{\tau=n}^{\infty}\left(1+\frac{1}{2^{\tau+1}}f\left(1-\frac{1}{2^\tau}\right)\right)e^{\int_1^2 f(\tau)\Delta\tau}(1+f(2))$$

and

$$e_{-f,\mu}(t,s) = \prod_{\tau=n}^{\infty}\left(1 - \frac{1}{2^{\tau+1}}f\left(1 - \frac{1}{2^{\tau}}\right)\right)e^{-\int_1^2 f(\tau)\Delta\tau}(1 - f(2)).$$

Consequently,

$$\begin{aligned}\cosh_{f,\mu}(t,s) &= \frac{1}{2}\left(\prod_{\tau=n}^{\infty}\left(1 + \frac{1}{2^{\tau+1}}f\left(1 - \frac{1}{2^{\tau}}\right)\right)e^{\int_1^2 f(\tau)\Delta\tau}(1 + f(2))\right.\\ &\quad\left. + \prod_{\tau=n}^{\infty}\left(1 - \frac{1}{2}\left(2 - \frac{1}{2^{\tau+1}}f\left(1 - \frac{1}{2^{\tau}}\right)\right)\right)e^{-\int_1^2 f(\tau)\Delta\tau}(1 - f(2))\right),\end{aligned}$$

and

$$\begin{aligned}\sinh_{f,\mu}(t,s) &= \frac{1}{2}\left(\prod_{\tau=n}^{\infty}\left(1 + \frac{1}{2^{\tau+1}}f\left(1 - \frac{1}{2^{\tau}}\right)\right)e^{\int_1^2 f(\tau)\Delta\tau}(1 + f(2))\right.\\ &\quad\left. - \prod_{\tau=n}^{\infty}\left(1 - \frac{1}{2}\left(2 - \frac{1}{2^{\tau+1}}f\left(1 - \frac{1}{2^{\tau}}\right)\right)\right)e^{-\int_1^2 f(\tau)\Delta\tau}(1 - f(2))\right).\end{aligned}$$

3. $s = 1 - \frac{1}{2^n}$ for some $n \in \mathbb{N}_0$, and $t \in [1,2]$. By Example 4.76 we have

$$e_{f,\mu}(t,s) = \prod_{\tau=n}^{\infty}\left(1 + \frac{1}{2^{\tau+1}}f\left(1 - \frac{1}{2^{\tau}}\right)\right)e^{\int_1^2 f(\tau)\Delta\tau}$$

and

$$e_{-f,\mu}(t,s) = \prod_{\tau=n}^{\infty}\left(1 - \frac{1}{2^{\tau+1}}f\left(1 - \frac{1}{2^{\tau}}\right)\right)e^{-\int_1^2 f(\tau)\Delta\tau}.$$

Consequently,

$$\begin{aligned}\cosh_{f,\mu}(t,s) &= \frac{1}{2}\left(\prod_{\tau=n}^{\infty}\left(1 + \frac{1}{2^{\tau+1}}f\left(1 - \frac{1}{2^{\tau}}\right)\right)e^{\int_1^2 f(\tau)\Delta\tau}\right.\\ &\quad\left. + \prod_{\tau=n}^{\infty}\left(1 - \frac{1}{2^{\tau+1}}f\left(1 - \frac{1}{2^{\tau}}\right)\right)e^{-\int_1^2 f(\tau)\Delta\tau}\right),\end{aligned}$$

and

$$\begin{aligned}\sinh_{f,\mu}(t,s) &= \frac{1}{2}\left(\prod_{\tau=n}^{\infty}\left(1 + \frac{1}{2^{\tau+1}}f\left(1 - \frac{1}{2^{\tau}}\right)\right)e^{\int_1^2 f(\tau)\Delta\tau}\right.\\ &\quad\left. - \prod_{\tau=n}^{\infty}\left(1 - \frac{1}{2^{\tau+1}}f\left(1 - \frac{1}{2^{\tau}}\right)\right)e^{-\int_1^2 f(\tau)\Delta\tau}\right).\end{aligned}$$

4. $s = 1 - \frac{1}{2^n}$ and $t = 1 - \frac{1}{2^m}$ for some $m, n \in \mathbb{N}_0$, $m > n$. By Example 4.76 we have

$$e_{f,\mu}(t,s) = \prod_{\tau=n}^{m} \left(1 + \frac{1}{2^{\tau+1}} f\left(1 - \frac{1}{2^\tau}\right)\right)$$

and

$$e_{-f,\mu}(t,s) = \prod_{\tau=n}^{m} \left(1 - \frac{1}{2^{\tau+1}} f\left(1 - \frac{1}{2^\tau}\right)\right).$$

Consequently,

$$\begin{aligned}\cosh_{f,\mu}(t,s) = \frac{1}{2}\Bigg(&\prod_{\tau=n}^{m} \left(1 + \frac{1}{2^{\tau+1}} f\left(1 - \frac{1}{2^\tau}\right)\right) \\ &+ \prod_{\tau=n}^{m} \left(1 - \frac{1}{2^{\tau+1}} f\left(1 - \frac{1}{2^\tau}\right)\right)\Bigg),\end{aligned}$$

and

$$\begin{aligned}\sinh_{f,\mu}(t,s) = \frac{1}{2}\Bigg(&\prod_{\tau=n}^{m} \left(1 + \frac{1}{2^{\tau+1}} f\left(1 - \frac{1}{2^\tau}\right)\right) \\ &- \prod_{\tau=n}^{m} \left(1 - \frac{1}{2^{\tau+1}} f\left(1 - \frac{1}{2^\tau}\right)\right)\Bigg).\end{aligned}$$

5. $s \in [1,2]$, $t \in 3^{\mathbb{N}}$, $t \geq 9$. By Example 4.76 we have

$$e_{f,\mu}(t,s) = e^{\int_s^2 f(\tau)\Delta\tau}(1 + f(2)) \prod_{\tau=3}^{\frac{t}{3}} (1 + 2\tau f(\tau))$$

and

$$e_{-f,\mu}(t,s) = e^{-\int_s^2 f(\tau)\Delta\tau}(1 - f(2)) \prod_{\tau=3}^{\frac{t}{3}} (1 - 2\tau f(\tau)).$$

Consequently,

$$\begin{aligned}\cosh_{f,\mu}(t,s) = \frac{1}{2}\Bigg(&e^{\int_s^2 f(\tau)\Delta\tau}(1 + f(2)) \prod_{\tau=3}^{\frac{t}{3}} (1 + 2\tau f(\tau)) \\ &+ e^{-\int_s^2 f(\tau)\Delta\tau}(1 - f(2)) \prod_{\tau=3}^{\frac{t}{3}} (1 - 2\tau f(\tau))\Bigg),\end{aligned}$$

and

$$\sinh_{f,\mu}(t,s) = \frac{1}{2}\Bigg(e^{\int_s^2 f(\tau)\Delta\tau}(1+f(2))\prod_{\tau=3}^{\frac{t}{3}}(1+2\tau f(\tau)) - e^{-\int_s^2 f(\tau)\Delta\tau}(1-f(2))\prod_{\tau=3}^{\frac{t}{3}}(1-2\tau f(\tau))\Bigg).$$

6. $s, t \in [1,2]$. By Example 4.76 we have

$$e_{f,\mu}(t,s) = e^{\int_s^t f(\tau)\Delta\tau}$$

and

$$e_{-f,\mu}(t,s) = e^{-\int_s^t f(\tau)\Delta\tau}.$$

Consequently,

$$\cosh_{f,\mu}(t,s) = \frac{1}{2}(e^{\int_s^t f(\tau)\Delta\tau} + e^{-\int_s^t f(\tau)\Delta\tau}),$$

and

$$\sinh_{f,\mu}(t,s) = \frac{1}{2}(e^{\int_s^t f(\tau)\Delta\tau} - e^{-\int_s^t f(\tau)\Delta\tau}).$$

7. $s \in [1,2]$, $t = 3$. By Example 4.76 we have

$$e_{f,\mu}(3,s) = e^{\int_s^2 f(\tau)\Delta\tau}(1+f(2))$$

and

$$e_{-f,\mu}(3,s) = e^{-\int_s^2 f(\tau)\Delta\tau}(1-f(2)).$$

Consequently,

$$\cosh_{f,\mu}(3,s) = \frac{1}{2}(e^{\int_s^2 f(\tau)\Delta\tau}(1+f(2)) + e^{-\int_s^2 f(\tau)\Delta\tau}(1-f(2))),$$

and

$$\sinh_{f,\mu}(3,s) = \frac{1}{2}(e^{\int_s^2 f(\tau)\Delta\tau}(1+f(2)) - e^{-\int_s^2 f(\tau)\Delta\tau}(1-f(2))).$$

8. $s, t \in 3^{\mathbb{N}}$, $s \le t$. By Example 4.76 we have

$$e_{f,\mu}(t,s) = \prod_{\tau=s}^{\frac{t}{3}}(1+2\tau f(\tau))$$

and

$$e_{-f,\mu}(t,s) = \prod_{\tau=s}^{\frac{t}{3}} (1 - 2\tau f(\tau)).$$

Consequently,

$$\cosh_{f,\mu}(t,s) = \frac{1}{2}\left(\prod_{\tau=s}^{\frac{t}{3}} (1 + 2\tau f(\tau)) + \prod_{\tau=s}^{\frac{t}{3}} (1 - 2\tau f(\tau)) \right),$$

and

$$\sinh_{f,\mu}(t,s) = \frac{1}{2}\left(\prod_{\tau=s}^{\frac{t}{3}} (1 + 2\tau f(\tau)) - \prod_{\tau=s}^{\frac{t}{3}} (1 - 2\tau f(\tau)) \right).$$

Exercise 4.34. Let $\mathbb{T} = 3^{\mathbb{N}_0}$. Compute $\cosh_2(t, t_0)$ and $\sinh_2(t, t_0)$ for $t, t_0 \in \mathbb{T}$.

Example 4.97. Let $f \in \mathrm{C}_{\mathrm{rd}}$ and $-\mu f^2 \in \mathscr{R}_\mu$. Then

$$(\cosh_{f,\mu}(c,\cdot))^{\Delta}(t) = -f(t)\sinh_{f,\mu}(c,\sigma(t)), \quad t, c \in \mathbb{T}.$$

Using Example 4.87, we have

$$\begin{aligned}
(\cosh_{f,\mu}(c,\cdot))^{\Delta}(t) &= \left(\frac{e_{f,\mu}(c,\cdot) + e_{-f,\mu}(c,\cdot)}{2} \right)^{\Delta}(t) \\
&= \frac{-f(t)e_{f,\mu}(c,\sigma(t)) + f(t)e_{-f,\mu}(c,\sigma(t))}{2} \\
&= -f(t)\frac{e_{f,\mu}(c,\sigma(t)) - e_{-f,\mu}(c,\sigma(t))}{2} \\
&= -f(t)\sinh_{f,\mu}(c,\sigma(t)), \quad t, c \in \mathbb{T},
\end{aligned}$$

completing the proof.

Exercise 4.35. Let $f \in \mathrm{C}_{\mathrm{rd}}$ and $-\mu f^2 \in \mathscr{R}_\mu$. Prove that

$$(\sinh_{f,\mu}(c,\cdot))^{\Delta}(t) = -f(t)\cosh_{f,\mu}(c,\sigma(t)). \quad t, c \in \mathbb{T}.$$

Example 4.98. Let $f \in \mathrm{C}_{\mathrm{rd}}$ and $-\mu f^2 \in \mathscr{R}_\mu$. Then

$$\int_a^b f(t)\sinh_{f,\mu}(c,\sigma(t))\Delta t = \cosh_{f,\mu}(c,a) - \cosh_{f,\mu}(c,b), \quad a, b, c \in \mathbb{T}.$$

By Example 4.97 we have

$$f(t)\sinh_{f,\mu}(c,\sigma(t)) = -(\cosh_{f,\mu}(c,\cdot))^{\Delta}(t), \quad c,t \in \mathbb{T}.$$

Hence

$$\begin{aligned}\int_a^b f(t)\sinh_{f,\mu}(c,\sigma(t)) &= -\int_a^b (\cosh_{f,\mu}(c,\cdot))^{\Delta}(t)\Delta t\\ &= \cosh_{f,\mu}(c,a) - \cosh_{f,\mu}(c,b), \quad a,b,c \in \mathbb{T}.\end{aligned}$$

completing the proof.

Exercise 4.36. Let $f \in \mathrm{C}_{\mathrm{rd}}$ and $-\mu f^2 \in \mathscr{R}_\mu$. Prove that

$$\int_a^b f(t)\cosh_{f,\mu}\big(c,\sigma(t)\big)\Delta t = \sinh_{f,\mu}(c,a) - \sinh_{f,\mu}(c,b), \quad a,b,c \in \mathbb{T}.$$

Example 4.99. Let $f \in \mathrm{C}_{\mathrm{rd}}$ and $-\mu f^2 \in \mathscr{R}_\mu$. We will simplify

1. $A = \frac{\cosh_{f,\mu}(s,t_0)\cosh_{f,\mu}(t,t_0) - \sinh_{f,\mu}(s,t_0)\sinh_{f,\mu}(t,t_0)}{\cosh^2_{f,\mu}(s,t_0) - \sinh^2_{f,\mu}(s,t_0)}$,
2. $B = \frac{\cosh_{f,\mu}(s,t_0)\sinh_{f,\mu}(t,t_0) - \sinh_{f,\mu}(s,t_0)\cosh_{f,\mu}(t,t_0)}{\cosh^2_{f,\mu}(s,t_0) - \sinh^2_{f,\mu}(s,t_0)}$, $s,t,t_0 \in \mathbb{T}$.

Consider the following cases.

1. We have

$$\begin{aligned}&\cosh_{f,\mu}(s,t_0)\cosh_{f,\mu}(t,t_0) - \sinh_{f,\mu}(s,t_0)\sinh_{f,\mu}(t,t_0)\\ &\quad= \frac{e_{f,\mu}(s,t_0) + e_{-f,\mu}(s,t_0)}{2}\,\frac{e_{f,\mu}(t,t_0) + e_{-f,\mu}(t,t_0)}{2}\\ &\qquad - \frac{e_{f,\mu}(s,t_0) - e_{-f,\mu}(s,t_0)}{2}\,\frac{e_{f,\mu}(t,t_0) - e_{-f,\mu}(t,t_0)}{2}\\ &\quad= \frac{1}{4}\big(e_{f,\mu}(s,t_0)e_{f,\mu}(t,t_0) + e_{f,\mu}(s,t_0)e_{-f,\mu}(t,t_0)\\ &\qquad + e_{-f,\mu}(s,t_0)e_{f,\mu}(t,t_0) + e_{-f,\mu}(s,t_0)e_{-f,\mu}(t,t_0)\\ &\quad - e_{f,\mu}(s,t_0)e_{f,\mu}(t,t_0) + e_{f,\mu}(s,t_0)e_{-f,\mu}(t,t_0)\\ &\quad + e_{-f,\mu}(s,t_0)e_{f,\mu}(t,t_0) - e_{-f,\mu}(s,t_0)e_{-f,\mu}(t,t_0)\big)\\ &\quad= \frac{e_{f,\mu}(s,t_0)e_{-f,\mu}(t,t_0) + e_{-f,\mu}(s,t_0)e_{f,\mu}(t,t_0)}{2}, \quad s,t,t_0 \in \mathbb{T}.\end{aligned}$$

Also,

$$\begin{aligned}\cosh^2_{f,\mu}(s,t_0) - \sinh^2_{f,\mu}(s,t_0) &= \left(\frac{e_{f,\mu}(s,t_0) + e_{-f,\mu}(s,t_0)}{2}\right)^2\\ &\quad - \left(\frac{e_{f,\mu}(s,t_0) - e_{-f,\mu}(s,t_0)}{2}\right)^2\end{aligned}$$

$$= \frac{e^2_{f,\mu}(s,t_0) + 2e_{f,\mu}(s,t_0)e_{-f,\mu}(s,t_0) + e^2_{-f,\mu}(s,t_0)}{4} - \frac{e^2_{f,\mu}(s,t_0) - 2e_{f,\mu}(s,t_0)e_{-f,\mu}(s,t_0) + e^2_{-f,\mu}(s,t_0)}{4}$$
$$= e_{f,\mu}(s,t_0)e_{-f,\mu}(s,t_0), \quad s,t)_0 \in \mathbb{T}.$$

Therefore

$$\begin{aligned} A &= \frac{1}{2}\frac{e_{f,\mu}(s,t_0)e_{-f,\mu}(t,t_0) + e_{-f,\mu}(s,t_0)e_{f,\mu}(t,t_0)}{e_{f,\mu}(s,t_0)e_{-f,\mu}(s,t_0)} \\ &= \frac{1}{2}\left(\frac{e_{-f,\mu}(t,t_0)}{e_{-f,\mu}(s,t_0)} + \frac{e_{f,\mu}(t,t_0)}{e_{f,\mu}(s,t_0)}\right) \\ &= \frac{1}{2}(e_{f,\mu}(t,s) + e_{-f,\mu}(t,s)) \\ &= \cosh_{f,\mu}(t,s), \quad s,t,t_0 \in \mathbb{T}. \end{aligned}$$

2. We have

$$\begin{aligned} &\cosh_{f,\mu}(s,t_0)\sinh_{f,\mu}(t,t_0) - \sinh_{f,\mu}(s,t_0)\cosh_{f,\mu}(t,t_0) \\ &= \frac{e_{f,\mu}(s,t_0) + e_{-f,\mu}(s,t_0)}{2}\frac{e_{f,\mu}(t,t_0) - e_{-f,\mu}(t,t_0)}{2} \\ &\quad - \frac{e_{f,\mu}(s,t_0) - e_{-f,\mu}(s,t_0)}{2}\frac{e_{f,\mu}(t,t_0) + e_{-f,\mu}(t,t_0)}{2} \\ &= \frac{1}{4}(e_{f,\mu}(s,t_0)e_{f,\mu}(t,t_0) - e_{f,\mu}(s,t_0)e_{-f,\mu}(t,t_0) \\ &\quad + e_{-f,\mu}(s,t_0)e_{f,\mu}(t,t_0) - e_{-f,\mu}(s,t_0)e_{-f,\mu}(t,t_0)) \\ &\quad - \frac{1}{4}(e_{f,\mu}(s,t_0)e_{f,\mu}(t,t_0) + e_{f,\mu}(s,t_0)e_{-f,\mu}(t,t_0) \\ &\quad - e_{-f,\mu}(s,t_0)e_{f,\mu}(t,t_0) - e_{-f,\mu}(s,t_0)e_{-f,\mu}(t,t_0)) \\ &= \frac{-e_{f,\mu}(s,t_0)e_{-f,\mu}(t,t_0) + e_{-f,\mu}(s,t_0)e_{f,\mu}(t,t_0)}{2}, \quad s,t,t_0 \in \mathbb{T}. \end{aligned}$$

Hence

$$\begin{aligned} B &= \frac{1}{2}\frac{-e_{f,\mu}(s,t_0)e_{-f,\mu}(t,t_0) + e_{-f,\mu}(s,t_0)e_{f,\mu}(t,t_0)}{e_{f,\mu}(s,t_0)e_{-f,\mu}(s,t_0)} \\ &= \frac{1}{2}\left(-\frac{e_{-f,\mu}(t,t_0)}{e_{-f,\mu}(s,t_0)} + \frac{e_{f,\mu}(t,t_0)}{e_{f,\mu}(s,t_0)}\right) \\ &= \frac{1}{2}(e_{f,\mu}(t,s) - e_{-f,\mu}(t,s)) \\ &= \sinh_{f,\mu}(t,s), \quad s,t,t_0 \in \mathbb{T}. \end{aligned}$$

Definition 4.17. Let $f, g \in \mathcal{C}_{rd}$ and suppose that

$$2f + \mu(f^2 - g^2) = 0. \tag{4.7}$$

We define the delta hyperbolic functions $\mathrm{ch}_{fg,\mu}$ and $\mathrm{sh}_{fg,\mu}$ as

$$\mathrm{ch}_{fg,\mu} = \frac{e_{f+g,\mu} + e_{f-g,\mu}}{2} \quad \text{and} \quad \mathrm{sh}_{fg,\mu} = \frac{e_{f+g,\mu} - e_{f-g,\mu}}{2}.$$

Example 4.100. Let $\mathbb{T} = \{(\frac{1}{2})^{2^n} : n \in \mathbb{N}_0\} \cup \{0, 1\}$. We will find $\mathrm{ch}_{fg,\mu}(t,s)$ and $\mathrm{sh}_{fg,\mu}(t,s)$, $s, t \in \mathbb{T}$, $s \le t$, for $f, g \in \mathcal{C}_{rd}$ that satisfy (4.7). We have the following cases.

1. $s = 0$ and $t = 1$. By Example 4.75 we have

$$e_{f+g,\mu}(1,0) = \left(1 + \frac{1}{2}(f+g)\left(\frac{1}{2}\right)\right) \prod_{\tau=1}^{\infty} \log\left(1 + \left(\left(\frac{1}{2}\right)^{2^{\tau-1}} - \left(\frac{1}{2}\right)^{2^\tau}\right)(f+g)\left(\left(\frac{1}{2}\right)^{2^\tau}\right)\right)$$

and

$$e_{f-g,\mu}(1,0) = \left(1 - \frac{1}{2}(g-f)\left(\frac{1}{2}\right)\right) \prod_{\tau=1}^{\infty} \left(1 - \left(\left(\frac{1}{2}\right)^{2^{\tau-1}} - \left(\frac{1}{2}\right)^{2^\tau}\right)(g-f)\left(\left(\frac{1}{2}\right)^{2^\tau}\right)\right).$$

Therefore

$$\begin{aligned}\mathrm{ch}_{f,\mu}(t,s) = \frac{1}{2}\Bigg(&\left(1 + \frac{1}{2}(f+g)\left(\frac{1}{2}\right)\right) \prod_{\tau=1}^{\infty} \log\left(1 + \left(\left(\frac{1}{2}\right)^{2^{\tau-1}} - \left(\frac{1}{2}\right)^{2^\tau}\right) f\left(\left(\frac{1}{2}\right)^{2^\tau}\right)\right)\\ &+ \left(1 - \frac{1}{2}(g-f)\left(\frac{1}{2}\right)\right) \prod_{\tau=1}^{\infty} \left(1 - \left(\left(\frac{1}{2}\right)^{2^{\tau-1}} - \left(\frac{1}{2}\right)^{2^\tau}\right)(g-f)\left(\left(\frac{1}{2}\right)^{2^\tau}\right)\right)\Bigg),\end{aligned}$$

and

$$\begin{aligned}\mathrm{sh}_{f,\mu}(t,s) = \frac{1}{2}\Bigg(&\left(1 + \frac{1}{2}(f+g)\left(\frac{1}{2}\right)\right) \prod_{\tau=1}^{\infty} \log\left(1 + \left(\left(\frac{1}{2}\right)^{2^{\tau-1}} - \left(\frac{1}{2}\right)^{2^\tau}\right)(f+g)\left(\left(\frac{1}{2}\right)^{2^\tau}\right)\right)\\ &- \left(1 - \frac{1}{2}(g-f)\left(\frac{1}{2}\right)\right) \prod_{\tau=1}^{\infty} \left(1 - \left(\left(\frac{1}{2}\right)^{2^{\tau-1}} - \left(\frac{1}{2}\right)^{2^\tau}\right)(g-f)\left(\left(\frac{1}{2}\right)^{2^\tau}\right)\right)\Bigg).\end{aligned}$$

2. $s = 0$ and $t = (\frac{1}{2})^{2^n}$ for some $n \in \mathbb{N}_0$. By Example 4.75 we have

$$e_{f+g,\mu}(t,0) = \prod_{s=n}^{\infty} \log\left(1 + \left(\left(\frac{1}{2}\right)^{2^{s-1}} - \left(\frac{1}{2}\right)^{2^s}\right)(f+g)\left(\left(\frac{1}{2}\right)^{2^s}\right)\right)$$

and

$$e_{f-g,\mu}(t,0)=\prod_{s=n}^{\infty}\log\left(1-\left(\left(\frac{1}{2}\right)^{2^{s-1}}-\left(\frac{1}{2}\right)^{2^s}\right)(g-f)\left(\left(\frac{1}{2}\right)^{2^s}\right)\right).$$

Therefore

$$\begin{aligned}\operatorname{ch}_{fg,\mu}(t,0)=\frac{1}{2}\Bigg(&\prod_{s=n}^{\infty}\log\left(1+\left(\left(\frac{1}{2}\right)^{2^{s-1}}-\left(\frac{1}{2}\right)^{2^s}\right)(f+g)\left(\left(\frac{1}{2}\right)^{2^s}\right)\right)\\&+\prod_{s=n}^{\infty}\left(1-\left(\left(\frac{1}{2}\right)^{2^{s-1}}-\left(\frac{1}{2}\right)^{2^s}\right)(g-f)\left(\left(\frac{1}{2}\right)^{2^s}\right)\right)\Bigg),\end{aligned}$$

and

$$\begin{aligned}\operatorname{sh}_{fg,\mu}(t,0)=\frac{1}{2}\Bigg(&\prod_{s=n}^{\infty}\log\left(1+\left(\left(\frac{1}{2}\right)^{2^{s-1}}-\left(\frac{1}{2}\right)^{2^s}\right)(f+g)\left(\left(\frac{1}{2}\right)^{2^s}\right)\right)\\&-\prod_{s=n}^{\infty}\log\left(1-\left(\left(\frac{1}{2}\right)^{2^{s-1}}-\left(\frac{1}{2}\right)^{2^s}\right)(g-f)\left(\left(\frac{1}{2}\right)^{2^s}\right)\right)\Bigg).\end{aligned}$$

3. $s=\frac{1}{2}$ and $t=1$. By Example 4.75 we have

$$e_{f+g,\mu}\left(1,\frac{1}{2}\right)=1+\frac{1}{2}(f+g)\left(\frac{1}{2}\right)$$

and

$$e_{f-g,\mu}\left(1,\frac{1}{2}\right)=1-\frac{1}{2}(g-f)\left(\frac{1}{2}\right).$$

Therefore

$$\operatorname{ch}_{fg,\mu}\left(1,\frac{1}{2}\right)=1+\frac{1}{2}f\left(\frac{1}{2}\right),$$

and

$$\operatorname{sh}_{fg,\mu}\left(1,\frac{1}{2}\right)=\frac{1}{2}f\left(\frac{1}{2}\right).$$

4. $s=(\frac{1}{2})^{2^n}$ for some $n\in\mathbb{N}$, and $t=1$. By Example 4.75 we have

$$e_{f+g,\mu}(1,s)=\prod_{s=1}^{n}\left(1+\left(\left(\frac{1}{2}\right)^{2^{s-1}}-\left(\frac{1}{2}\right)^{2^s}\right)(f+g)\left(\left(\frac{1}{2}\right)^{2^s}\right)\right)\left(1+\frac{1}{2}(f+g)\left(\frac{1}{2}\right)\right)$$

and

$$e_{f-g,\mu}(1,s)=\prod_{s=1}^{n}\left(1-\left(\left(\frac{1}{2}\right)^{2^{s-1}}-\left(\frac{1}{2}\right)^{2^{s}}\right)(g-f)\left(\left(\frac{1}{2}\right)^{2^{s}}\right)\right)\left(1-\frac{1}{2}(g-f)\left(\frac{1}{2}\right)\right).$$

Therefore

$$\begin{aligned}\operatorname{ch}_{fg,\mu}(1,s)=\frac{1}{2}\Bigg(&\prod_{s=1}^{n}\left(1+\left(\left(\frac{1}{2}\right)^{2^{s-1}}-\left(\frac{1}{2}\right)^{2^{s}}\right)(f+g)\left(\left(\frac{1}{2}\right)^{2^{s}}\right)\right)\left(1+\frac{1}{2}(f+g)\left(\frac{1}{2}\right)\right)\\&+\prod_{s=1}^{n}\left(1-\left(\left(\frac{1}{2}\right)^{2^{s-1}}-\left(\frac{1}{2}\right)^{2^{s}}\right)(g-f)\left(\left(\frac{1}{2}\right)^{2^{s}}\right)\right)\left(1-\frac{1}{2}(g-f)\left(\frac{1}{2}\right)\right)\Bigg),\end{aligned}$$

and

$$\begin{aligned}\operatorname{sh}_{fg,\mu}(1,s)=\frac{1}{2}\Bigg(&\prod_{s=1}^{n}\left(1+\left(\left(\frac{1}{2}\right)^{2^{s-1}}-\left(\frac{1}{2}\right)^{2^{s}}\right)(f+g)\left(\left(\frac{1}{2}\right)^{2^{s}}\right)\right)\left(1+\frac{1}{2}(f+g)\left(\frac{1}{2}\right)\right)\\&-\prod_{s=1}^{n}\left(1-\left(\left(\frac{1}{2}\right)^{2^{s-1}}-\left(\frac{1}{2}\right)^{2^{s}}\right)(g-f)\left(\left(\frac{1}{2}\right)^{2^{s}}\right)\right)\left(1-\frac{1}{2}(g-f)\left(\frac{1}{2}\right)\right)\Bigg).\end{aligned}$$

5. $s=(\frac{1}{2})^{2^n}$ and $t=(\frac{1}{2})^{2^m}$ for some $m,n\in\mathbb{N}$, $m<n$. By Example 4.75 we have

$$e_{f+g,\mu}(t,s)=\prod_{s=m+1}^{n}\left(1+\left(\left(\frac{1}{2}\right)^{2^{s-1}}-\left(\frac{1}{2}\right)^{2^{s}}\right)(f+g)\left(\left(\frac{1}{2}\right)^{2^{s}}\right)\right)$$

and

$$e_{f-g,\mu}(t,s)=\prod_{s=m+1}^{n}\left(1-\left(\left(\frac{1}{2}\right)^{2^{s-1}}-\left(\frac{1}{2}\right)^{2^{s}}\right)(g-f)\left(\left(\frac{1}{2}\right)^{2^{s}}\right)\right).$$

Therefore

$$\begin{aligned}\operatorname{ch}_{fg,\mu}(t,s)=\frac{1}{2}\Bigg(&\prod_{s=m+1}^{n}\left(1+\left(\left(\frac{1}{2}\right)^{2^{s-1}}-\left(\frac{1}{2}\right)^{2^{s}}\right)(f+g)\left(\left(\frac{1}{2}\right)^{2^{s}}\right)\right)\\&+\prod_{s=m+1}^{n}\left(1-\left(\left(\frac{1}{2}\right)^{2^{s-1}}-\left(\frac{1}{2}\right)^{2^{s}}\right)(g-f)\left(\left(\frac{1}{2}\right)^{2^{s}}\right)\right)\Bigg),\end{aligned}$$

and

$$\begin{aligned}\operatorname{sh}_{fg,\mu}(t,s)=\frac{1}{2}\Bigg(&\prod_{s=m+1}^{n}\left(1+\left(\left(\frac{1}{2}\right)^{2^{s-1}}-\left(\frac{1}{2}\right)^{2^{s}}\right)(f+g)\left(\left(\frac{1}{2}\right)^{2^{s}}\right)\right)\\&-\prod_{s=m+1}^{n}\left(1-\left(\left(\frac{1}{2}\right)^{2^{s-1}}-\left(\frac{1}{2}\right)^{2^{s}}\right)(g-f)\left(\left(\frac{1}{2}\right)^{2^{s}}\right)\right)\Bigg).\end{aligned}$$

Example 4.101. Let $\mathbb{T} = \{1 - \frac{1}{2^n}\}_{n\in\mathbb{N}_0} \cup [1,2] \cup 3^{\mathbb{N}}$ and $f \in \mathscr{R}_\mu$. We will find $\operatorname{ch}_{fg,\mu}(t,s)$ and $\operatorname{sh}_{fg,\mu}(t,s)$ for $s,t \in \mathbb{T}$, $s \le t$, and $f,g \in \mathrm{C_{rd}}$ that satisfy (4.7). We have the following cases.

1. $s = 1 - \frac{1}{2^n}$ for some $n \in \mathbb{N}_0$, and $t \in 3^{\mathbb{N}}$, $t \ge 9$. By Example 4.76 we have

$$
\begin{aligned}
e_{f+g,\mu}(t,s) &= \prod_{\tau=n}^{\infty}\left(1 + \frac{1}{2^{\tau+1}}(f+g)\left(1 - \frac{1}{2^\tau}\right)\right) \\
&\quad \times e^{\int_1^2 (f+g)(\tau)\Delta\tau}(1 + f(2) + g(2)) \prod_{\tau=3}^{\frac{t}{3}}(1 + 2\tau(f+g)(\tau))
\end{aligned}
$$

and

$$
\begin{aligned}
e_{f-g,\mu}(t,s) &= \prod_{\tau=n}^{\infty}\left(1 - \frac{1}{2^{\tau+1}}(g-f)\left(1 - \frac{1}{2^\tau}\right)\right) \\
&\quad \times e^{-\int_1^2 (g-f)(\tau)\Delta\tau}(1 + f(2) - g(2)) \prod_{\tau=3}^{\frac{t}{3}}(1 - 2\tau(g-f)(\tau)).
\end{aligned}
$$

Consequently,

$$
\begin{aligned}
\operatorname{ch}_{fg,\mu}(t,s) &= \frac{1}{2}\Bigg(\prod_{\tau=n}^{\infty}\left(1 + \frac{1}{2^{\tau+1}}(f+g)\left(1 - \frac{1}{2^\tau}\right)\right) \\
&\quad \times e^{\int_1^2 (f+g)(\tau)\Delta\tau}(1 + f(2) + g(2)) \prod_{\tau=3}^{\frac{t}{3}}(1 + 2\tau(f+g)(\tau)) \\
&\quad + \prod_{\tau=n}^{\infty}\left(1 - \frac{1}{2^{\tau+1}}(g-f)\left(1 - \frac{1}{2^\tau}\right)\right) \\
&\quad \times e^{-\int_1^2 (g-f)(\tau)\Delta\tau}(1 + f(2) - g(2)) \prod_{\tau=3}^{\frac{t}{3}}(1 - 2\tau(g-f)(\tau))\Bigg),
\end{aligned}
$$

and

$$
\begin{aligned}
\operatorname{sh}_{fg,\mu}(t,s) &= \frac{1}{2}\Bigg(\prod_{\tau=n}^{\infty}\left(1 + \frac{1}{2^{\tau+1}}(f+g)\left(1 - \frac{1}{2^\tau}\right)\right) \\
&\quad \times e^{\int_1^2 (f+g)(\tau)\Delta\tau}(1 + f(2) + g(2)) \prod_{\tau=3}^{\frac{t}{3}}(1 + 2\tau(f+g)(\tau)) \\
&\quad - \prod_{\tau=n}^{\infty}\left(1 - \frac{1}{2^{\tau+1}}(g-f)\left(1 - \frac{1}{2^\tau}\right)\right) \\
&\quad \times e^{-\int_1^2 (g-f)(\tau)\Delta\tau}(1 + f(2) - g(2)) \prod_{\tau=3}^{\frac{t}{3}}(1 - 2\tau(g-f)(\tau))\Bigg).
\end{aligned}
$$

2. $s = 1 - \frac{1}{2^n}$ for some $n \in \mathbb{N}_0$, and $t = 3$. By Example 4.76 we have

$$e_{f+g,\mu}(t,s) = \prod_{\tau=n}^{\infty}\left(1 + \frac{1}{2^{\tau+1}}(f+g)\left(1 - \frac{1}{2^\tau}\right)\right)$$
$$\times e^{\int_1^2 (f+g)(\tau)\Delta\tau}(1 + f(2) + g(2))\prod_{\tau=3}^{\frac{t}{3}}(1 + 2\tau(f+g)(\tau))$$

and

$$e_{f-g,\mu}(t,s) = \prod_{\tau=n}^{\infty}\left(1 - \frac{1}{2^{\tau+1}}(g-f)\left(1 - \frac{1}{2^\tau}\right)\right)$$
$$\times e^{-\int_1^2 f(\tau)\Delta\tau}(1 + f(2) - g(2))\prod_{\tau=3}^{\frac{t}{3}}(1 - 2\tau(f-g)(\tau)).$$

Consequently,

$$\mathrm{ch}_{fg,\mu}(t,s) = \frac{1}{2}\left(\prod_{\tau=n}^{\infty}\left(1 + \frac{1}{2^{\tau+1}}(f+g)\left(1 - \frac{1}{2^\tau}\right)\right)\right.$$
$$\times e^{\int_1^2 (f+g)(\tau)\Delta\tau}(1 + f(2) + g(2))\prod_{\tau=3}^{\frac{t}{3}}(1 + 2\tau(f+g)(\tau))$$
$$+ \prod_{\tau=n}^{\infty}\left(1 - \frac{1}{2^{\tau+1}}(g-f)\left(1 - \frac{1}{2^\tau}\right)\right)$$
$$\left.\times e^{-\int_1^2 f(\tau)\Delta\tau}(1 + f(2) - g(2))\prod_{\tau=3}^{\frac{t}{3}}\right)(1 - 2\tau(g-f)(\tau)),$$

and

$$\mathrm{sh}_{fg,\mu}(t,s) = \frac{1}{2}\left(\prod_{\tau=n}^{\infty}\left(1 + \frac{1}{2^{\tau+1}}(f+g)\left(1 - \frac{1}{2^\tau}\right)\right)\right.$$
$$\times e^{\int_1^2 (f+g)(\tau)\Delta\tau}(1 + f(2) + g(2))\prod_{\tau=3}^{\frac{t}{3}}(1 + 2\,\tau(f+g)(\tau))$$
$$- \prod_{\tau=n}^{\infty}\left(1 - \frac{1}{2^{\tau+1}}(g-f)\left(1 - \frac{1}{2^\tau}\right)\right)$$
$$\left.\times e^{-\int_1^2 (g-f)(\tau)\Delta\tau}(1 + f(2) - g(2))\prod_{\tau=3}^{\frac{t}{3}}\right)(1 - 2\tau(g-f)(\tau)).$$

3. $s = 1 - \frac{1}{2^n}$ for some $n \in \mathbb{N}_0$, and $t \in [1, 2]$. By Example 4.76 we have

$$e_{f+g,\mu}(t,s) = \prod_{\tau=n}^{\infty}\left(1 + \frac{1}{2^{\tau+1}}(f+g)\left(1 - \frac{1}{2^\tau}\right)\right)e^{\int_1^2 (f+g)(\tau)\Delta\tau}$$

and

$$e_{f-g,\mu}(t,s) = \prod_{\tau=n}^{\infty}\left(1 - \frac{1}{2^{\tau+1}}(g-f)\left(1 - \frac{1}{2^\tau}\right)\right)e^{-\int_1^2 (g-f)(\tau)\Delta\tau}.$$

Consequently,

$$\begin{aligned}\mathrm{ch}_{fg,\mu}(t,s) = \frac{1}{2}\Bigg(&\prod_{\tau=n}^{\infty}\left(1 + \frac{1}{2^{\tau+1}}(f+g)\left(1 - \frac{1}{2^\tau}\right)\right)e^{\int_1^2 (f+g)(\tau)\Delta\tau}\\ &+\prod_{\tau=n}^{\infty}\left(1 - \frac{1}{2^{\tau+1}}(g-f)\left(1 - \frac{1}{2^\tau}\right)\right)e^{-\int_1^2 (g-f)(\tau)\Delta\tau}\Bigg),\end{aligned}$$

and

$$\begin{aligned}\mathrm{sh}_{f,\mu}(t,s) = \frac{1}{2}\Bigg(&\prod_{\tau=n}^{\infty}\left(1 + \frac{1}{2^{\tau+1}}(f+g)\left(1 - \frac{1}{2^\tau}\right)\right)e^{\int_1^2 (f+g)(\tau)\Delta\tau}\\ &-\prod_{\tau=n}^{\infty}\left(1 - \frac{1}{2^{\tau+1}}(g-f)\left(1 - \frac{1}{2^\tau}\right)\right)e^{-\int_1^2 (g-f)(\tau)\Delta\tau}\Bigg).\end{aligned}$$

4. $s = 1 - \frac{1}{2^n}$ and $t = 1 - \frac{1}{2^m}$ for some $m, n \in \mathbb{N}_0$, $m > n$. By Example 4.76 we have

$$e_{f+g,\mu}(t,s) = \prod_{\tau=n}^{m}\left(1 + \frac{1}{2^{\tau+1}}(f+g)\left(1 - \frac{1}{2^\tau}\right)\right)$$

and

$$e_{f-g,\mu}(t,s) = \prod_{\tau=n}^{m}\left(1 - \frac{1}{2^{\tau+1}}(g-f)\left(1 - \frac{1}{2^\tau}\right)\right).$$

Consequently,

$$\begin{aligned}\mathrm{ch}_{fg,\mu}(t,s) = \frac{1}{2}\Bigg(&\prod_{\tau=n}^{m}\left(1 + \frac{1}{2^{\tau+1}}(f+g)\left(1 - \frac{1}{2^\tau}\right)\right)\\ &+\prod_{\tau=n}^{m}\left(1 - \frac{1}{2^{\tau+1}}(g-f)\left(1 - \frac{1}{2^\tau}\right)\right)\Bigg),\end{aligned}$$

and

$$\mathrm{sh}_{fg,\mu}(t,s) = \frac{1}{2}\left(\prod_{\tau=n}^{m}\left(1+\frac{1}{2^{\tau+1}}(f+g)\left(1-\frac{1}{2^{\tau}}\right)\right)\right.$$
$$\left.-\prod_{\tau=n}^{m}\left(1-\frac{1}{2^{\tau+1}}(g-f)\left(1-\frac{1}{2^{\tau}}\right)\right)\right).$$

5. $s \in [1,2]$, and $t \in 3^{\mathbb{N}}$, $t \geq 9$. By Example 4.76 we have

$$e_{f+g,\mu}(t,s) = e^{\int_s^2 (f+g)(\tau)\Delta\tau}(1+f(2)+g(2))\prod_{\tau=3}^{\frac{t}{3}}(1+2\tau(f+g)(\tau))$$

and

$$e_{f-g,\mu}(t,s) = e^{-\int_s^2 (g-f)(\tau)\Delta\tau}(1+f(2)-g(2))\prod_{\tau=3}^{\frac{t}{3}}(1-2\tau(g-f)(\tau)).$$

Consequently,

$$\mathrm{ch}_{fg,\mu}(t,s) = \frac{1}{2}\left(e^{\int_s^2 (f+g)(\tau)\Delta\tau}(1+f(2)+g(2))\prod_{\tau=3}^{\frac{t}{3}}(1+2\tau(f+g)(\tau))\right.$$
$$\left.+e^{-\int_s^2 (g-f)(\tau)\Delta\tau}(1+f(2)-g(2))\prod_{\tau=3}^{\frac{t}{3}}(1-2\tau(g-f)(\tau))\right),$$

and

$$\mathrm{sh}_{f,\mu}(t,s) = \frac{1}{2}\left(e^{\int_s^2 (f+g)(\tau)\Delta\tau}(1+f(2)+g(2))\prod_{\tau=3}^{\frac{t}{3}}(1+2\tau(f+g)(\tau))\right.$$
$$\left.-e^{-\int_s^2 (g-f)(\tau)\Delta\tau}(1+f(2)-g(2))\prod_{\tau=3}^{\frac{t}{3}}(1-2\tau(g-f)(\tau))\right).$$

6. $s,t \in [1,2]$. By Example 4.76 we have

$$e_{f+g,\mu}(t,s) = e^{\int_s^t (f+g)(\tau)\Delta\tau}$$

and

$$e_{f-g,\mu}(t,s) = e^{-\int_s^t (g-f)(\tau)\Delta\tau}.$$

Consequently,

$$\mathrm{ch}_{fg,\mu}(t,s) = \frac{1}{2}\left(e^{\int_s^t (f+g)(\tau)\Delta\tau} + e^{-\int_s^t (g-f)(\tau)\Delta\tau}\right),$$

and

$$\mathrm{sh}_{fg,\mu}(t,s) = \frac{1}{2}\left(e^{\int_s^t (f+g)(\tau)\Delta\tau} - e^{-\int_s^t (g-f)(\tau)\Delta\tau}\right).$$

7. $s \in [1,2]$, and $t = 3$. By Example 4.76 we have

$$e_{f+g,\mu}(3,s) = e^{\int_s^2 (f+g)(\tau)\Delta\tau}(1 + f(2) + g(2))$$

and

$$e_{f-g,\mu}(3,s) = e^{-\int_s^2 (g-f)(\tau)\Delta\tau}(1 + f(2) - g(2)).$$

Consequently,

$$\mathrm{ch}_{fg,\mu}(3,s) = \frac{1}{2}\left(e^{\int_s^2 (f+g)(\tau)\Delta\tau}(1 + f(2) + g(2)) + e^{-\int_s^2 (g-f)(\tau)\Delta\tau}(1 + f(2) - g(2))\right),$$

and

$$\mathrm{sh}_{f,\mu}(3,s) = \frac{1}{2}\left(e^{\int_s^2 (f+g)(\tau)\Delta\tau}(1 + f(2) + g(2)) - e^{-\int_s^2 (g-f)(\tau)\Delta\tau}(1 + f(2) - g(2))\right).$$

8. $s,t \in 3^{\mathbb{N}}$, $s \le t$. By Example 4.76 we have

$$e_{f+g,\mu}(t,s) = \prod_{\tau=s}^{\frac{t}{3}} (1 + 2\tau(f+g)(\tau))$$

and

$$e_{f-g,\mu}(t,s) = \prod_{\tau=s}^{\frac{t}{3}} (1 - 2\tau(g-f)(\tau)).$$

Consequently,

$$\mathrm{ch}_{fg,\mu}(t,s) = \frac{1}{2}\left(\prod_{\tau=s}^{\frac{t}{3}} (1 + 2\tau(f+g)(\tau)) + \prod_{\tau=s}^{\frac{t}{3}} (1 - 2\tau(g-f)(\tau))\right),$$

and

$$\mathrm{sh}_{f,\mu}(t,s) = \frac{1}{2}\left(\prod_{\tau=s}^{\frac{t}{3}} (1 + 2\tau(f+g)(\tau)) - \prod_{\tau=s}^{\frac{t}{3}} (1 - 2\tau(g-f)(\tau))\right).$$

Example 4.102. If $f, g \in \mathcal{C}_{rd}$ satisfy (4.7), then

$$\mathrm{ch}^{\Delta}_{fg,\mu} = f\,\mathrm{ch}_{fg,\mu} + g\,\mathrm{sh}_{fg,\mu}, \quad \mathrm{sh}^{\Delta}_{fg,\mu} = g\,\mathrm{ch}_{fg,\mu} + f\,\mathrm{sh}_{fg,\mu},$$

and

$$\mathrm{ch}^2_{fg,\mu} - \mathrm{sh}^2_{fg,\mu} = 1.$$

We have

$$\begin{aligned}
\mathrm{ch}^{\Delta}_{fg,\mu} &= \left(\frac{e_{f+g,\mu} + e_{f-g,\mu}}{2}\right)^{\Delta} \\
&= \frac{e^{\Delta}_{f+g,\mu} + e^{\Delta}_{f-g,\mu}}{2} \\
&= \frac{(f+g)e_{f+g,\mu} + (f-g)e_{f-g,\mu}}{2} \\
&= f\frac{e_{f+g,\mu} + e_{f-g,\mu}}{2} + g\frac{e_{f+g,\mu} - e_{f-g,\mu}}{2} \\
&= f\,\mathrm{ch}_{fg,\mu} + g\,\mathrm{sh}_{fg,\mu}
\end{aligned}$$

and

$$\begin{aligned}
\mathrm{sh}^{\Delta}_{fg,\mu} &= \left(\frac{e_{f+g,\mu} - e_{f-g,\mu}}{2}\right)^{\Delta} \\
&= \frac{e^{\Delta}_{f+g,\mu} - pe^{\Delta}_{f-g,\mu}}{2} \\
&= \frac{(f+g)e_{f+g,\mu} - (f-g)e_{f-g,\mu}}{2} \\
&= f\frac{e_{f+g,\mu} - e_{f-g,\mu}}{2} + g\frac{e_{f+g,\mu} + e_{f-g,\mu}}{2} \\
&= f\,\mathrm{sh}_{fg,\mu} + g\,\mathrm{ch}_{fg,\mu}\,.
\end{aligned}$$

Next,

$$\begin{aligned}
\mathrm{ch}^2_{fg,\mu} - \mathrm{sh}^2_{fg,\mu} &= \left(\frac{e_{f+g,\mu} + e_{f-g,\mu}}{2}\right)^2 - \left(\frac{e_{f+g,\mu} - e_{f-g,\mu}}{2}\right)^2 \\
&= \frac{e^2_{f+g,\mu} + 2e_{f+g,\mu}e_{f-g,\mu} + e^2_{f-g,\mu}}{4} \\
&\quad - \frac{e^2_{f+g,\mu} - 2e_{f+g,\mu}e_{f-g,\mu} + e^2_{f-g,\mu}}{4} \\
&= e_{f+g,\mu}e_{f-g,\mu} \\
&= e_{(f+g)\oplus_\mu(f-g),\mu}.
\end{aligned}$$

Note that

$$(f+g)\oplus_\mu(f-g) = f + g + f - g + \mu(f+g)(f-g)$$

$$= 2f + \mu(f^2 - g^2)$$
$$= 0.$$

Therefore

$$\mathrm{ch}^2_{fg,\mu} - \mathrm{sh}^2_{fg,\mu} = 1,$$

completing the proof.

Exercise 4.37. Let $f, g \in \mathrm{C_{rd}}$ satisfy (4.7). Prove that

$$e_{f+g,\mu} = \mathrm{ch}_{fg,\mu} + \mathrm{sh}_{fg,\mu}$$

and

$$e_{f-g,\mu} = \mathrm{ch}_{fg,\mu} - \mathrm{sh}_{fg,\mu}.$$

Example 4.103. Let $f, g \in \mathrm{C_{rd}}$ satisfy (4.7). We will prove that

1. $\mathrm{sh}_{fg,\mu}(t,s) = -\mathrm{sh}_{fg,\mu}(s,t)$,
2. $\mathrm{sh}_{fg,\mu}(t,s) = \mathrm{sh}_{fg,\mu}(t,r)\,\mathrm{ch}_{fg,\mu}(r,s) - \mathrm{ch}_{fg,\mu}(t,r)\,\mathrm{sh}_{fg,\mu}(s,r)$,
3. $\mathrm{sh}_{fg,\mu}(t,r) = \mathrm{sh}_{fg,\mu}(t,s)\,\mathrm{ch}_{fg,\mu}(s,r) + \mathrm{ch}_{fg,\mu}(t,s)\,\mathrm{sh}_{fg,\mu}(s,r)$, $t, s, r \in \mathbb{T}$.

1. We have

$$\begin{aligned}
\mathrm{sh}_{fg,\mu}(t,s) &= \frac{e_{f+g,\mu}(t,s) - e_{f-g,\mu}(t,s)}{2} \\
&= \frac{\frac{1}{e_{f+g,\mu}(s,t)} - \frac{1}{e_{f-g,\mu}(s,t)}}{2} \\
&= \frac{e_{f-g,\mu}(s,t) - e_{f+g,\mu}(s,t)}{2e_{f+g,\mu}(s,t)e_{f-g,\mu}(s,t)} \\
&= -\frac{e_{f+g,\mu}(s,t) - e_{f-g,\mu}(s,t)}{2} \\
&= -\mathrm{sh}_{fg,\mu}(s,t), \quad s,t \in \mathbb{T}.
\end{aligned}$$

2. We have

$$\begin{aligned}
&\mathrm{sh}_{fg,\mu}(t,r)\,\mathrm{ch}_{fg,\mu}(r,s) - \mathrm{ch}_{fg,\mu}(t,r)\,\mathrm{sh}_{fg,\mu}(s,r) \\
&\quad = \mathrm{sh}_{fg,\mu}(t,r)\,\mathrm{ch}_{fg,\mu}(r,s) + \mathrm{ch}_{fg,\mu}(t,r)\,\mathrm{sh}_{fg,\mu}(r,s) \\
&\quad = \frac{e_{f+g,\mu}(t,r) - e_{f-g,\mu}(t,r)}{2}\,\frac{e_{f+g,\mu}(r,s) + e_{f-g,\mu}(r,s)}{2} \\
&\qquad + \frac{e_{f+g,\mu}(t,r) + e_{f-g,\mu}(t,r)}{2}\,\frac{e_{f+g,\mu}(r,s) + e_{f-g,\mu}(r,s)}{2} \\
&\quad = \frac{1}{4}\big(e_{f+g,\mu}(t,r)e_{f+g,\mu}(r,s) + e_{f+g,\mu}(t,r)e_{f-g,\mu}(r,s)
\end{aligned}$$

$$
\begin{aligned}
&- e_{f-g,\mu}(t,r)e_{f+g,\mu}(r,s) - e_{f-g,\mu}(t,r)e_{f-g,\mu}(r,s)) \\
&+ \frac{1}{4}(e_{f+g,\mu}(t,r)e_{f+g,\mu}(r,s) - e_{f+g,\mu}(t,r)e_{f-g,\mu}(r,s) \\
&- e_{f-g,\mu}(t,r)e_{f+g,\mu}(r,s) - e_{f-g,\mu}(t,r)e_{f-g,\mu}(r,s)) \\
= {}& \frac{e_{f+g,\mu}(t,s) - e_{f-g,\mu}(t,s)}{2} \\
= {}& \mathrm{sh}_{fg,\mu}(t,s), \quad t,s,r \in \mathbb{T}.
\end{aligned}
$$

3. We have

$$
\begin{aligned}
&\mathrm{sh}_{fg,\mu}(t,s)\,\mathrm{ch}_{fg,\mu}(s,r) + \mathrm{ch}_{fg,\mu}(t,s)\,\mathrm{sh}_{fg,\mu}(s,r) \\
&= \frac{e_{f+g,\mu}(t,s) - e_{f-g,\mu}(t,s)}{2}\,\frac{e_{f+g,\mu}(s,r) + e_{f-g,\mu}(s,r)}{2} \\
&\quad + \frac{e_{f+g,\mu}(t,s) + e_{f-g,\mu}(t,s)}{2}\,\frac{e_{f+g,\mu}(s,r) - e_{f-g,\mu}(s,r)}{2} \\
&= \frac{1}{4}(e_{f+g,\mu}(t,s)e_{f+g,\mu}(s,r) + e_{f+g,\mu}(t,s)e_{f-g,\mu}(s,r) \\
&\quad - e_{f-g,\mu}(t,s)e_{f+g,\mu}(s,r) - e_{f-g,\mu}(t,s)e_{f-g,\mu}(s,r)) \\
&\quad + \frac{1}{4}(e_{f+g,\mu}(t,s)e_{f+g,\mu}(s,r) - e_{f+g,\mu}(t,s)e_{f-g,\mu}(s,r) \\
&\quad + e_{f-g,\mu}(t,s)e_{f+g,\mu}(s,r) - e_{f-g,\mu}(t,s)e_{f-g,\mu}(s,r)) \\
&= \frac{e_{f+g,\mu}(t,r) - e_{f-g,\mu}(t,r)}{2} \\
&= shvtex_{fg,\mu}(t,r), \quad t,s,r \in \mathbb{T}.
\end{aligned}
$$

Exercise 4.38. Let $f, g \in C_{rd}$ satisfy (4.7). Prove that
1. $\mathrm{ch}_{fg}(t,s) = \mathrm{ch}_{fg}(s,t)$,
2. $\mathrm{ch}_{fg}(t,s) = \mathrm{ch}_{fg}(t,r)\,\mathrm{ch}_{fg}(s,r) - \mathrm{sh}_{fg}(t,r)\,\mathrm{sh}_{fg}(s,r)$,
3. $\mathrm{ch}_{fg}(t,r) = \mathrm{ch}_{fg}(t,s)\,\mathrm{ch}_{fg}(s,r) + \mathrm{sh}_{fg}(t,s)\,\mathrm{sh}_{fg}(s,r)$, $s,t,r \in \mathbb{T}$.

4.5 Delta trigonometric functions

Definition 4.18. Let $f \in C_{rd}$ and $\mu f^2 \in \mathscr{R}_\mu$. Define the delta trigonometric functions $\cos_{f,\mu}$ and $\sin_{f,\mu}$ as

$$
\cos_{f,\mu} = \frac{e_{if,\mu} + e_{-if,\mu}}{2} \quad \text{and} \quad \sin_{f,\mu} = \frac{e_{if,\mu} - e_{-if,\mu}}{2i}.
$$

Example 4.104. Let $\mathbb{T} = \{(\frac{1}{2})^{2^n} : n \in \mathbb{N}_0\} \cup \{0,1\}$. We will find $\cos_{f,\mu}(t,s)$ and $\sin_{f,\mu}(t,s)$, $s,t \in \mathbb{T}$, $s \le t$, for $f \in C_{rd}$ and $\mu f^2 \in \mathscr{R}_\mu$. We have the following cases.

1. $s = 0$ and $t = 1$. By Example 4.75 we have

$$e_{if,\mu}(1,0) = \left(1 + \frac{i}{2}f\left(\frac{1}{2}\right)\right)\prod_{\tau=1}^{\infty}\left(1 + i\left(\left(\frac{1}{2}\right)^{2^{\tau-1}} - \left(\frac{1}{2}\right)^{2^{\tau}}\right)f\left(\left(\frac{1}{2}\right)^{2^{\tau}}\right)\right)$$

and

$$e_{-if,\mu}(1,0) = \left(1 - \frac{i}{2}f\left(\frac{1}{2}\right)\right)\prod_{\tau=1}^{\infty}\left(1 - i\left(\left(\frac{1}{2}\right)^{2^{\tau-1}} - \left(\frac{1}{2}\right)^{2^{\tau}}\right)f\left(\left(\frac{1}{2}\right)^{2^{\tau}}\right)\right).$$

Therefore

$$\begin{aligned}\cos_{f,\mu}(t,s) = \frac{1}{2}\Bigg(&\left(1 + \frac{i}{2}f\left(\frac{1}{2}\right)\right)\prod_{\tau=1}^{\infty}\log\left(1 + i\left(\left(\frac{1}{2}\right)^{2^{\tau-1}} - \left(\frac{1}{2}\right)^{2^{\tau}}\right)f\left(\left(\frac{1}{2}\right)^{2^{\tau}}\right)\right)\\ &+ \left(1 - \frac{i}{2}f\left(\frac{1}{2}\right)\right)\prod_{\tau=1}^{\infty}\left(1 - i\left(\left(\frac{1}{2}\right)^{2^{\tau-1}} - \left(\frac{1}{2}\right)^{2^{\tau}}\right)f\left(\left(\frac{1}{2}\right)^{2^{\tau}}\right)\right)\Bigg),\end{aligned}$$

and

$$\begin{aligned}\sin_{f,\mu}(t,s) = \frac{1}{2i}\Bigg(&\left(1 + \frac{i}{2}f\left(\frac{1}{2}\right)\right)\prod_{\tau=1}^{\infty}\log\left(1 + i\left(\left(\frac{1}{2}\right)^{2^{\tau-1}} - \left(\frac{1}{2}\right)^{2^{\tau}}\right)f\left(\left(\frac{1}{2}\right)^{2^{\tau}}\right)\right)\\ &- \left(1 - \frac{i}{2}f\left(\frac{1}{2}\right)\right)\prod_{\tau=1}^{\infty}\left(1 - i\left(\left(\frac{1}{2}\right)^{2^{\tau-1}} - \left(\frac{1}{2}\right)^{2^{\tau}}\right)f\left(\left(\frac{1}{2}\right)^{2^{\tau}}\right)\right)\Bigg).\end{aligned}$$

2. $s = 0$, and $t = (\frac{1}{2})^{2^n}$ for some $n \in \mathbb{N}_0$. By Example 4.75 we have

$$e_{if,\mu}(t,0) = \prod_{s=n}^{\infty}\log\left(1 + i\left(\left(\frac{1}{2}\right)^{2^{s-1}} - \left(\frac{1}{2}\right)^{2^{s}}\right)f\left(\left(\frac{1}{2}\right)^{2^{s}}\right)\right)$$

and

$$e_{-if,\mu}(t,0) = \prod_{s=n}^{\infty}\log\left(1 - i\left(\left(\frac{1}{2}\right)^{2^{s-1}} - \left(\frac{1}{2}\right)^{2^{s}}\right)f\left(\left(\frac{1}{2}\right)^{2^{s}}\right)\right).$$

Therefore

$$\begin{aligned}\cos_{f,\mu}(t,0) = \frac{1}{2}\Bigg(&\prod_{s=n}^{\infty}\log\left(1 + i\left(\left(\frac{1}{2}\right)^{2^{s-1}} - \left(\frac{1}{2}\right)^{2^{s}}\right)f\left(\left(\frac{1}{2}\right)^{2^{s}}\right)\right)\\ &+ \prod_{s=n}^{\infty}\log\left(1 - i\left(\left(\frac{1}{2}\right)^{2^{s-1}} - \left(\frac{1}{2}\right)^{2^{s}}\right)f\left(\left(\frac{1}{2}\right)^{2^{s}}\right)\right)\Bigg),\end{aligned}$$

and

$$\sin_{f,\mu}(t,0) = \frac{1}{2i}\left(\prod_{s=n}^{\infty}\left(1 + i\left(\left(\frac{1}{2}\right)^{2^{s-1}} - \left(\frac{1}{2}\right)^{2^s}\right)f\left(\left(\frac{1}{2}\right)^{2^s}\right)\right)\right.$$
$$\left. - \prod_{s=n}^{\infty}\log\left(1 - i\left(\left(\frac{1}{2}\right)^{2^{s-1}} - \left(\frac{1}{2}\right)^{2^s}\right)f\left(\left(\frac{1}{2}\right)^{2^s}\right)\right)\right)$$

3. $s = \frac{1}{2}$ and $t = 1$. By Example 4.75 we have

$$e_{if,\mu}\left(1, \frac{1}{2}\right) = 1 + \frac{i}{2}f\left(\frac{1}{2}\right)$$

and

$$e_{-if,\mu}\left(1, \frac{1}{2}\right) = 1 - \frac{i}{2}f\left(\frac{1}{2}\right).$$

Therefore

$$\cos_{f,\mu}\left(1, \frac{1}{2}\right) = 1,$$

and

$$\sin_{f,\mu}\left(1, \frac{1}{2}\right) = \frac{1}{2}f\left(\frac{1}{2}\right).$$

4. $s = (\frac{1}{2})^{2^n}$ for some $n \in \mathbb{N}$, and $t = 1$. By Example 4.75 we have

$$e_{if,\mu}(1,s) = \prod_{s=1}^{n}\left(1 + i\left(\left(\frac{1}{2}\right)^{2^{s-1}} - \left(\frac{1}{2}\right)^{2^s}\right)f\left(\left(\frac{1}{2}\right)^{2^s}\right)\right)\left(1 + \frac{i}{2}f\left(\frac{1}{2}\right)\right)$$

and

$$e_{-if,\mu}(1,s) = \prod_{s=1}^{n}\left(1 - i\left(\left(\frac{1}{2}\right)^{2^{s-1}} - \left(\frac{1}{2}\right)^{2^s}\right)f\left(\left(\frac{1}{2}\right)^{2^s}\right)\right)\left(1 - \frac{i}{2}f\left(\frac{1}{2}\right)\right).$$

Therefore

$$\cos_{f,\mu}(1,s) = \frac{1}{2}\left(\prod_{s=1}^{n}\left(1 + i\left(\left(\frac{1}{2}\right)^{2^{s-1}} - \left(\frac{1}{2}\right)^{2^s}\right)f\left(\left(\frac{1}{2}\right)^{2^s}\right)\right)\left(1 + \frac{i}{2}f\left(\frac{1}{2}\right)\right)\right.$$
$$\left. + \prod_{s=1}^{n}\left(1 - i\left(\left(\frac{1}{2}\right)^{2^{s-1}} - \left(\frac{1}{2}\right)^{2^s}\right)f\left(\left(\frac{1}{2}\right)^{2^s}\right)\right)\left(1 - \frac{i}{2}f\left(\frac{1}{2}\right)\right)\right),$$

and

$$\sin_{f,\mu}(1,s) = \frac{1}{2i}\left(\prod_{s=1}^{n}\left(1 + i\left(\left(\frac{1}{2}\right)^{2^{s-1}} - \left(\frac{1}{2}\right)^{2^s}\right)f\left(\left(\frac{1}{2}\right)^{2^s}\right)\right)\left(1 + \frac{i}{2}f\left(\frac{1}{2}\right)\right)\right.$$
$$\left. - \prod_{s=1}^{n}\left(1 - i\left(\left(\frac{1}{2}\right)^{2^{s-1}} - \left(\frac{1}{2}\right)^{2^s}\right)f\left(\left(\frac{1}{2}\right)^{2^s}\right)\right)\left(1 - \frac{i}{2}f\left(\frac{1}{2}\right)\right)\right).$$

5. $s = (\frac{1}{2})^{2^n}$ and $t = (\frac{1}{2})^{2^m}$ for some $m, n \in \mathbb{N}$, $m < n$. By Example 4.75 we have

$$e_{if,\mu}(t,s) = \prod_{s=m+1}^{n}\left(1 + i\left(\left(\frac{1}{2}\right)^{2^{s-1}} - \left(\frac{1}{2}\right)^{2^s}\right)f\left(\left(\frac{1}{2}\right)^{2^s}\right)\right)$$

and

$$e_{-if,\mu}(t,s) = \prod_{s=m+1}^{n}\left(1 - i\left(\left(\frac{1}{2}\right)^{2^{s-1}} - \left(\frac{1}{2}\right)^{2^s}\right)f\left(\left(\frac{1}{2}\right)^{2^s}\right)\right).$$

Therefore

$$\cos_{f,\mu}(t,s) = \frac{1}{2}\left(\prod_{s=m+1}^{n}\left(1 + i\left(\left(\frac{1}{2}\right)^{2^{s-1}} - \left(\frac{1}{2}\right)^{2^s}\right)f\left(\left(\frac{1}{2}\right)^{2^s}\right)\right)\right.$$
$$\left. + \prod_{s=m+1}^{n}\left(1 - i\left(\left(\frac{1}{2}\right)^{2^{s-1}} - \left(\frac{1}{2}\right)^{2^s}\right)f\left(\left(\frac{1}{2}\right)^{2^s}\right)\right)\right),$$

and

$$\sin_{f,\mu}(t,s) = \frac{1}{2i}\left(\prod_{s=m+1}^{n}\left(1 + i\left(\left(\frac{1}{2}\right)^{2^{s-1}} - \left(\frac{1}{2}\right)^{2^s}\right)f\left(\left(\frac{1}{2}\right)^{2^s}\right)\right)\right.$$
$$\left. - \prod_{s=m+1}^{n}\left(1 - i\left(\left(\frac{1}{2}\right)^{2^{s-1}} - \left(\frac{1}{2}\right)^{2^s}\right)f\left(\left(\frac{1}{2}\right)^{2^s}\right)\right)\right).$$

Example 4.105. Let $\mathbb{T} = \{1 - \frac{1}{2^n}\}_{n\in\mathbb{N}_0} \cup [1,2] \cup 3^{\mathbb{N}}$ and $f \in \mathscr{R}_\mu$. We will find $\cos_{f,\mu}(t,s)$ and $\sin_{f,\mu}(t,s)$ for $s, t \in \mathbb{T}$, $s \le t$, and $f \in \mathrm{C_{rd}}$, $\mu f^2 \in \mathscr{R}_\mu$. We have the following cases.

1. $s = 1 - \frac{1}{2^n}$ for some $n \in \mathbb{N}_0$, and $t \in 3^{\mathbb{N}}$, $t \ge 9$. By Example 4.76 we have

$$e_{if,\mu}(t,s) = \prod_{\tau=n}^{\infty}\left(1 + \frac{i}{2^{\tau+1}}f\left(1 - \frac{1}{2^\tau}\right)\right)e^{i\int_1^2 f(\tau)\Delta\tau}(1 + if(2))\prod_{\tau=3}^{\frac{t}{3}}(1 + 2i\tau f(\tau))$$

and

$$e_{-if,\mu}(t,s) = \prod_{\tau=n}^{\infty}\left(1 - \frac{i}{2^{\tau+1}}f\left(1 - \frac{1}{2^\tau}\right)\right)e^{-i\int_1^2 f(\tau)\Delta\tau}(1 - f(2))\prod_{\tau=3}^{\frac{t}{3}}(1 - 2i\tau f(\tau)).$$

Consequently,

$$\cos_{f,\mu}(t,s) = \frac{1}{2}\Bigg(\prod_{\tau=n}^{\infty}\left(1+\frac{i}{2^{\tau+1}}f\left(1-\frac{1}{2^{\tau}}\right)\right)e^{i\int_1^2 f(\tau)\Delta\tau}(1+if(2))\prod_{\tau=3}^{\frac{t}{3}}(1+2i\tau f(\tau))$$
$$+\prod_{\tau=n}^{\infty}\left(1-\frac{i}{2^{\tau+1}}f\left(1-\frac{1}{2^{\tau}}\right)\right)e^{-i\int_1^2 f(\tau)\Delta\tau}(1-if(2))\prod_{\tau=3}^{\frac{t}{3}}(1-2i\tau f(\tau))\Bigg),$$

and

$$\sin_{f,\mu}(t,s) = \frac{1}{2i}\Bigg(\prod_{\tau=n}^{\infty}\left(1+\frac{i}{2^{\tau+1}}f\left(1-\frac{1}{2^{\tau}}\right)\right)e^{i\int_1^2 f(\tau)\Delta\tau}(1+if(2))\prod_{\tau=3}^{\frac{t}{3}}(1+2i\tau f(\tau))$$
$$-\prod_{\tau=n}^{\infty}\left(1-\frac{i}{2^{\tau+1}}f\left(1-\frac{1}{2^{\tau}}\right)\right)e^{-i\int_1^2 f(\tau)\Delta\tau}(1-if(2))\prod_{\tau=3}^{\frac{t}{3}}(1-2i\tau f(\tau))\Bigg).$$

2. $s = 1-\frac{1}{2^n}$ for some $n \in \mathbb{N}_0$, and $t = 3$. By Example 4.76 we have

$$e_{if,\mu}(t,s) = \prod_{\tau=n}^{\infty}\left(1+\frac{i}{2^{\tau+1}}f\left(1-\frac{1}{2^{\tau}}\right)\right)e^{i\int_1^2 f(\tau)\Delta\tau}(1+if(2))$$

and

$$e_{-if,\mu}(t,s) = \prod_{\tau=n}^{\infty}\left(1-\frac{i}{2^{\tau+1}}f\left(1-\frac{1}{2^{\tau}}\right)\right)e^{-i\int_1^2 f(\tau)\Delta\tau}(1-if(2))\prod_{\tau=3}^{\frac{t}{3}}(1\_2i\tau f(\tau)).$$

Consequently,

$$\cos_{f,\mu}(t,s) = \frac{1}{2}\Bigg(\prod_{\tau=n}^{\infty}\left(1+\frac{i}{2^{\tau+1}}f\left(1-\frac{1}{2^{\tau}}\right)\right)e^{i\int_1^2 f(\tau)\Delta\tau}(1+if(2))\prod_{\tau=3}^{\frac{t}{3}}(1+2i\tau f(\tau))$$
$$+\prod_{\tau=n}^{\infty}\left(1-\frac{i}{2^{\tau+1}}f\left(1-\frac{1}{2^{\tau}}\right)\right)e^{-i\int_1^2 f(\tau)\Delta\tau}(1-if(2))\prod_{\tau=3}^{\frac{t}{3}}\Bigg)(1-2i\tau f(\tau)),$$

and

$$\sin_{f,\mu}(t,s) = \frac{1}{2i}\Bigg(\prod_{\tau=n}^{\infty}\left(1+\frac{i}{2^{\tau+1}}f\left(1-\frac{1}{2^{\tau}}\right)\right)e^{i\int_1^2 f(\tau)\Delta\tau}(1+if(2))\prod_{\tau=3}^{\frac{t}{3}}$$
$$-\prod_{\tau=n}^{\infty}\left(1-\frac{i}{2^{\tau+1}}f\left(1-\frac{1}{2^{\tau}}\right)\right)e^{-i\int_1^2 f(\tau)\Delta\tau}(1-if(2))\prod_{\tau=3}^{\frac{t}{3}}\Bigg).$$

3. $s = 1 - \frac{1}{2^n}$ for some $n \in \mathbb{N}_0$, and $t \in [1,2]$. By Example 4.76 we have

$$e_{if,\mu}(t,s) = \prod_{\tau=n}^{\infty}\left(1 + \frac{i}{2^{\tau+1}} f\left(1 - \frac{1}{2^\tau}\right)\right) e^{i\int_1^2 f(\tau)\Delta\tau}$$

and

$$e_{-if,\mu}(t,s) = \prod_{\tau=n}^{\infty}\left(1 - \frac{i}{2^{\tau+1}} f\left(1 - \frac{1}{2^\tau}\right)\right) e^{-i\int_1^2 f(\tau)\Delta\tau}.$$

Consequently,

$$\begin{aligned}\cos_{f,\mu}(t,s) = \frac{1}{2}\Bigg(&\prod_{\tau=n}^{\infty}\left(1 + \frac{i}{2^{\tau+1}} f\left(1 - \frac{1}{2^\tau}\right)\right) e^{i\int_1^2 f(\tau)\Delta\tau}\\ &+ \prod_{\tau=n}^{\infty}\left(1 - \frac{i}{2^{\tau+1}} f\left(1 - \frac{1}{2^\tau}\right)\right) e^{-i\int_1^2 f(\tau)\Delta\tau}\Bigg),\end{aligned}$$

and

$$\begin{aligned}\sin_{f,\mu}(t,s) = \frac{1}{2i}\Bigg(&\prod_{\tau=n}^{\infty}\left(1 + \frac{i}{2^{\tau+1}} f\left(1 - \frac{1}{2^\tau}\right)\right) e^{i\int_1^2 f(\tau)\Delta\tau}\\ &- \prod_{\tau=n}^{\infty}\left(1 - \frac{i}{2^{\tau+1}} f\left(1 - \frac{1}{2^\tau}\right)\right) e^{-i\int_1^2 f(\tau)\Delta\tau}\Bigg).\end{aligned}$$

4. $s = 1 - \frac{1}{2^n}$ and $t = 1 - \frac{1}{2^m}$ for some $m, n \in \mathbb{N}_0$, $m > n$. By Example 4.76 we have

$$e_{if,\mu}(t,s) = \prod_{\tau=n}^{m}\left(1 + \frac{i}{2^{\tau+1}} f\left(1 - \frac{1}{2^\tau}\right)\right)$$

and

$$e_{-if,\mu}(t,s) = \prod_{\tau=n}^{m}\left(1 - \frac{i}{2^{\tau+1}} f\left(1 - \frac{1}{2^\tau}\right)\right).$$

Consequently,

$$\begin{aligned}\cos_{f,\mu}(t,s) = \frac{1}{2}\Bigg(&\prod_{\tau=n}^{m}\left(1 + \frac{i}{2^{\tau+1}} f\left(1 - \frac{1}{2^\tau}\right)\right)\\ &+ \prod_{\tau=n}^{m}\left(1 - \frac{i}{2^{\tau+1}} f\left(1 - \frac{1}{2^\tau}\right)\right)\Bigg),\end{aligned}$$

and

$$\sin_{f,\mu}(t,s) = \frac{1}{2i}\left(\prod_{\tau=n}^{m}\left(1+\frac{i}{2^{\tau+1}}f\left(1-\frac{1}{2^{\tau}}\right)\right)\right.$$
$$\left.-\prod_{\tau=n}^{m}\left(1-\frac{i}{2^{\tau+1}}f\left(1-\frac{1}{2^{\tau}}\right)\right)\right).$$

5. $s \in [1,2]$ and $t \in 3^{\mathbb{N}}$, $t \geq 9$. By Example 4.76 we have

$$e_{if,\mu}(t,s) = e^{i\int_s^2 f(\tau)\Delta\tau}(1+if(2))\prod_{\tau=3}^{\frac{t}{3}}(1+2i\tau f(\tau))$$

and

$$e_{-if,\mu}(t,s) = e^{-i\int_s^2 f(\tau)\Delta\tau}(1-if(2))\prod_{\tau=3}^{\frac{t}{3}}(1-2i\tau f(\tau)).$$

Consequently,

$$\cos_{f,\mu}(t,s) = \frac{1}{2}\left(e^{i\int_s^2 f(\tau)\Delta\tau}(1+if(2))\prod_{\tau=3}^{\frac{t}{3}}(1+2i\tau f(\tau))\right.$$
$$\left.+e^{-i\int_s^2 f(\tau)\Delta\tau}(1-if(2))\prod_{\tau=3}^{\frac{t}{3}}(1-2i\tau f(\tau))\right),$$

and

$$\sin_{f,\mu}(t,s) = \frac{1}{2i}\left(e^{i\int_s^2 f(\tau)\Delta\tau}(1+if(2))\prod_{\tau=3}^{\frac{t}{3}}(1+2i\tau f(\tau))\right.$$
$$\left.-e^{-i\int_s^2 f(\tau)\Delta\tau}(1-if(2))\prod_{\tau=3}^{\frac{t}{3}}(1-2i\tau f(\tau))\right).$$

6. $s,t \in [1,2]$. By Example 4.76 we have

$$e_{if,\mu}(t,s) = e^{i\int_s^t f(\tau)\Delta\tau}$$

and

$$e_{-if,\mu}(t,s) = e^{-i\int_s^t f(\tau)\Delta\tau}.$$

Consequently,

$$\cos_{f,\mu}(t,s) = \frac{1}{2}\left(e^{i\int_s^t f(\tau)\Delta\tau}+e^{-i\int_s^t f(\tau)\Delta\tau}\right),$$

and

$$\sin_{f,\mu}(t,s) = \frac{1}{2i}\left(e^{i\int_s^t f(\tau)\Delta\tau} - e^{-i\int_s^t f(\tau)\Delta\tau}\right).$$

7. $s \in [1,2]$ and $t = 3$. By Example 4.76 we have

$$e_{if,\mu}(3,s) = e^{i\int_s^2 f(\tau)\Delta\tau}(1 + if(2))$$

and

$$e_{-if,\mu}(3,s) = e^{-i\int_s^2 f(\tau)\Delta\tau}(1 - if(2)).$$

Consequently,

$$\cos_{f,\mu}(3,s) = \frac{1}{2}\left(e^{i\int_s^2 f(\tau)\Delta\tau}(1 + if(2)) + e^{-i\int_s^2 f(\tau)\Delta\tau}(1 - if(2))\right),$$

and

$$\sin_{f,\mu}(3,s) = \frac{1}{2i}\left(e^{i\int_s^2 f(\tau)\Delta\tau}(1 + if(2)) - e^{-i\int_s^2 f(\tau)\Delta\tau}(1 - if(2))\right).$$

8. $s, t \in 3^{\mathbb{N}}$, $s \le t$. By Example 4.76 we have

$$e_{if,\mu}(t,s) = \prod_{\tau=s}^{\frac{t}{3}}(1 + 2i\tau f(\tau))$$

and

$$e_{-if,\mu}(t,s) = \prod_{\tau=s}^{\frac{t}{3}}(1 - 2i\tau f(\tau)).$$

Consequently,

$$\cos_{f,\mu}(t,s) = \frac{1}{2}\left(\prod_{\tau=s}^{\frac{t}{3}}(1 + 2i\tau f(\tau)) + \prod_{\tau=s}^{\frac{t}{3}}(1 - 2i\tau f(\tau))\right),$$

and

$$\sin_{f,\mu}(t,s) = \frac{1}{2i}\left(\prod_{\tau=s}^{\frac{t}{3}}(1 + 2i\tau f(\tau)) - \prod_{\tau=s}^{\frac{t}{3}}(1 - 2i\tau f(\tau))\right).$$

Example 4.106. Let $f \in \mathrm{C}_{\mathrm{rd}}$ and $\mu f^2 \in \mathscr{R}_\mu$. Then

$$\cos_{f,\mu}^{\Delta} = -f\sin_{f,\mu}, \quad \sin_{f,\mu}^{\Delta} = f\cos_{f,\mu},$$

and

$$\cos_{f,\mu}^2 + \sin_{f,\mu}^2 = e_{\mu f^2}.$$

We have

$$\begin{aligned}
\cos_{f,\mu}^{\Delta} &= \left(\frac{e_{if,\mu} + e_{-if,\mu}}{2}\right)^{\Delta} \\
&= \frac{e_{if,\mu}^{\Delta} + e_{-if,\mu}^{\Delta}}{2} \\
&= \frac{ife_{if,\mu} - ife_{-if,\mu}}{2} \\
&= -f\frac{e_{if,\mu} - e_{-if,\mu}}{2i} \\
&= -f \sin_{f,\mu}, \\
\sin_{f,\mu}^{\Delta} &= \left(\frac{e_{if,\mu} - e_{-if,\mu}}{2i}\right)^{\Delta} \\
&= \frac{e_{if,\mu}^{\Delta} - e_{-if,\mu}^{\Delta}}{2i} \\
&= \frac{ife_{if,\mu} + ife_{-if,\mu}}{2i} \\
&= f\frac{e_{if,\mu} + e_{-if,\mu}}{2} \\
&= f \cos_{f,\mu}, \\
\cos_f^2 + \sin_f^2 &= \left(\frac{e_{if,\mu} + e_{-if,\mu}}{2}\right)^2 + \left(\frac{e_{if,\mu} - e_{-if,\mu}}{2}\right)^2 \\
&= \frac{e_{if,\mu}^2 + 2e_{if,\mu}e_{-if,\mu} + e_{-if,\mu}^2}{4} - \frac{e_{if,\mu} - 2e_{if,\mu}e_{-if,\mu} + e_{-if,\mu}^2}{4} \\
&= e_{if,\mu}e_{-if,\mu} \\
&= e_{(if)\oplus_\mu(-if)}.
\end{aligned}$$

Since

$$(if) \oplus_\mu (-if) = (if) + (-if) + \mu(if)(-if) = \mu f^2,$$

we get the desired result.

Example 4.107. Let $f \in C_{rd}$ and $\mu f^2 \in \mathscr{R}_\mu$. We simplify

$$A = \frac{\cos_{f,\mu}(s,t_0)\cos_{f,\mu}(t,t_0) + \sin_{f,\mu}(s,t_0)\sin_{f,\mu}(t,t_0)}{\cos_f^2(s,t_0) + \sin_{f,\mu}^2(s,t_0)}, \quad s,t,t_0 \in \mathbb{T}.$$

We have

$$\begin{aligned}
&\cos_{f,\mu}(s,t_0)\cos_{f,\mu}(t,t_0)+\sin_{f,\mu}(s,t_0)\sin_{f,\mu}(t,t_0)\\
&\quad=\frac{e_{if,\mu}(s,t_0)+e_{-if,\mu}(s,t_0)}{2}\,\frac{e_{if,\mu}(t,t_0)+e_{-if,\mu}(t,t_0)}{2}\\
&\qquad+\frac{e_{if,\mu}(s,t_0)-e_{-if,\mu}(s,t_0)}{2i}\,\frac{e_{if,\mu}(t,t_0)-e_{-if,\mu}(t,t_0)}{2i}\\
&\quad=\frac{1}{4}\big(e_{if,\mu}(s,t_0)e_{if,\mu}(t,t_0)+e_{-if,\mu}(s,t_0)e_{if,\mu}(t,t_0)\\
&\qquad+e_{if,\mu}(s,t_0)e_{-if,\mu}(t,t_0)+e_{-if,\mu}(s,t_0)e_{-if,\mu}(t,t_0)\big)\\
&\qquad-\frac{1}{4}\big(e_{if,\mu}(s,t_0)e_{if,\mu}(t,t_0)-e_{if,\mu}(s,t_0)e_{-if,\mu}(t,t_0)\\
&\qquad-e_{-if,\mu}(s,t_0)e_{if,\mu}(t,t_0)+e_{-if,\mu}(s,t_0)e_{-if,\mu}(t,t_0)\big)\\
&\quad=\frac{e_{-if,\mu}(s,t_0)e_{if,\mu}(t,t_0)+e_{if,\mu}(s,t_0)e_{-if,\mu}(t,t_0)}{2},\\
&\cos_f^2(s,t_0)+\sin_{f,\mu}^2(s,t_0)\\
&\quad=\left(\frac{e_{if,\mu}(s,t_0)+e_{-if,\mu}(s,t_0)}{2}\right)^2+\left(\frac{e_{if,\mu}(s,t_0)-e_{-if,\mu}(s,t_0)}{2i}\right)^2\\
&\quad=\frac{e_{if,\mu}^2(s,t_0)+2e_{if,\mu}(s,t_0)e_{-if,\mu}(s,t_0)+e_{-if,\mu}^2(s,t_0)}{4}\\
&\qquad-\frac{e_{if,\mu}^2(s,t_0)-2e_{if,\mu}(s,t_0)e_{-if,\mu}(s,t_0)+e_{-if,\mu}^2(s,t_0)}{4}\\
&\quad=e_{if,\mu}(s,t_0)e_{-if,\mu}(s,t_0),\quad s,t,t_0\in\mathbb{T}.
\end{aligned}$$

Hence

$$\begin{aligned}
A&=\frac{1}{2}\,\frac{e_{-if,\mu}(s,t_0)e_{if,\mu}(t,t_0)+e_{if,\mu}(s,t_0)e_{-if,\mu}(t,t_0)}{e_{if,\mu}(s,t_0)e_{-if,\mu}(s,t_0)}\\
&=\frac{1}{2}\left(\frac{e_{if,\mu}(t,t_0)}{e_{if,\mu}(s,t_0)}+\frac{e_{-if,\mu}(t,t_0)}{e_{-if,\mu}(s,t_0)}\right)\\
&=\frac{1}{2}\big(e_{if,\mu}(t,s)+e_{-if,\mu}(t,s)\big)\\
&=\cos_{f,\mu}(t,s),\quad s,t,t_0\in\mathbb{T}.
\end{aligned}$$

Exercise 4.39. Let $f\in C_{rd}$ and $\mu f^2\in\mathscr{R}_\mu$. We simplify

$$B=\frac{\cos_{f,\mu}(s,t_0)\sin_{f,\mu}(t,t_0)-\sin_{f,\mu}(s,t_0)\cos_{f,\mu}(t,t_0)}{\cos_f^2(s,t_0)+\sin_{f,\mu}^2(s,t_0)},\quad s,t,t_0\in\mathbb{T}.$$

Definition 4.19. Let $f,g\in C_{rd}$ and assume that

$$2f+\mu(f^2+g^2)=0. \tag{4.8}$$

Define the delta trigonometric functions $\mathrm{c}_{fg,\mu}$ and $\mathrm{s}_{fg,\mu}$ as

$$\mathrm{c}_{fg,\mu} = \frac{e_{f+ig,\mu} + e_{f-ig,\mu}}{2} \quad \text{and} \quad \mathrm{s}_{fg,\mu} = \frac{e_{f+ig,\mu} - e_{f-ig,\mu}}{2i}.$$

Example 4.108. Let $\mathbb{T} = \{(\frac{1}{2})^{2^n} : n \in \mathbb{N}_0\} \cup \{0, 1\}$. We will find $\mathrm{c}_{fg,\mu}(t,s)$ and $\mathrm{s}_{fg,\mu}(t,s)$, $s, t \in \mathbb{T}$, $s \leq t$, for $f, g \in \mathcal{C}_{\mathrm{rd}}$ that satisfy (4.8). We have the following cases.

1. $s = 0$ and $t = 1$. By Example 4.75 we have

$$e_{f+ig,\mu}(1,0) = \left(1 + \frac{1}{2}(f+ig)\left(\frac{1}{2}\right)\right)\prod_{\tau=1}^{\infty}\log\left(1 + \left(\left(\frac{1}{2}\right)^{2^{\tau-1}} - \left(\frac{1}{2}\right)^{2^{\tau}}\right)(f+ig)\left(\left(\frac{1}{2}\right)^{2^{\tau}}\right)\right)$$

and

$$e_{f-ig,\mu}(1,0) = \left(1 - \frac{1}{2}(ig-f)\left(\frac{1}{2}\right)\right)\prod_{\tau=1}^{\infty}\left(1 - \left(\left(\frac{1}{2}\right)^{2^{\tau-1}} - \left(\frac{1}{2}\right)^{2^{\tau}}\right)(ig-f)\left(\left(\frac{1}{2}\right)^{2^{\tau}}\right)\right).$$

Therefore

$$\begin{aligned}\mathrm{c}_{f,\mu}(t,s) = \frac{1}{2}\Bigg(&\left(1 + \frac{1}{2}(f+ig)\left(\frac{1}{2}\right)\right)\prod_{\tau=1}^{\infty}\log\left(1 + \left(\left(\frac{1}{2}\right)^{2^{\tau-1}} - \left(\frac{1}{2}\right)^{2^{\tau}}\right)(f+ig)\left(\left(\frac{1}{2}\right)^{2^{\tau}}\right)\right)\\ &+ \left(1 - \frac{1}{2}(ig-f)\left(\frac{1}{2}\right)\right)\prod_{\tau=1}^{\infty}\left(1 - \left(\left(\frac{1}{2}\right)^{2^{\tau-1}} - \left(\frac{1}{2}\right)^{2^{\tau}}\right)(ig-f)\left(\left(\frac{1}{2}\right)^{2^{\tau}}\right)\right)\Bigg),\end{aligned}$$

and

$$\begin{aligned}\mathrm{s}_{f,\mu}(t,s) = \frac{1}{2i}\Bigg(&\left(1 + \frac{1}{2}(f+ig)\left(\frac{1}{2}\right)\right)\prod_{\tau=1}^{\infty}\log\left(1 + \left(\left(\frac{1}{2}\right)^{2^{\tau-1}} - \left(\frac{1}{2}\right)^{2^{\tau}}\right)(f+ig)\left(\left(\frac{1}{2}\right)^{2^{\tau}}\right)\right)\\ &- \left(1 - \frac{1}{2}(ig-f)\left(\frac{1}{2}\right)\right)\prod_{\tau=1}^{\infty}\left(1 - \left(\left(\frac{1}{2}\right)^{2^{\tau-1}} - \left(\frac{1}{2}\right)^{2^{\tau}}\right)(ig-f)\left(\left(\frac{1}{2}\right)^{2^{\tau}}\right)\right)\Bigg).\end{aligned}$$

2. $s = 0$, and $t = (\frac{1}{2})^{2^n}$ for some $n \in \mathbb{N}_0$. By Example 4.75 we have

$$e_{f+ig,\mu}(t,0) = \prod_{s=n}^{\infty}\log\left(1 + \left(\left(\frac{1}{2}\right)^{2^{s-1}} - \left(\frac{1}{2}\right)^{2^{s}}\right)(f+ig)\left(\left(\frac{1}{2}\right)^{2^{s}}\right)\right)$$

and

$$e_{f-ig,\mu}(t,0) = \prod_{s=n}^{\infty}\log\left(1 - \left(\left(\frac{1}{2}\right)^{2^{s-1}} - \left(\frac{1}{2}\right)^{2^{s}}\right)(ig-f)\left(\left(\frac{1}{2}\right)^{2^{s}}\right)\right).$$

Therefore

$$c_{fg,\mu}(t,0) = \frac{1}{2}\left(\prod_{s=n}^{\infty}\log\left(1+\left(\left(\frac{1}{2}\right)^{2^{s-1}}-\left(\frac{1}{2}\right)^{2^s}\right)(f+ig)\left(\left(\frac{1}{2}\right)^{2^s}\right)\right)\right.$$
$$\left.+\prod_{s=n}^{\infty}\left(1-\left(\left(\frac{1}{2}\right)^{2^{s-1}}-\left(\frac{1}{2}\right)^{2^s}\right)(ig-f)\left(\left(\frac{1}{2}\right)^{2^s}\right)\right)\right),$$

and

$$s_{fg,\mu}(t,0) = \frac{1}{2i}\left(\prod_{s=n}^{\infty}\log\left(1+\left(\left(\frac{1}{2}\right)^{2^{s-1}}-\left(\frac{1}{2}\right)^{2^s}\right)(f+ig)\left(\left(\frac{1}{2}\right)^{2^s}\right)\right)\right.$$
$$\left.-\prod_{s=n}^{\infty}\log\left(1-\left(\left(\frac{1}{2}\right)^{2^{s-1}}-\left(\frac{1}{2}\right)^{2^s}\right)(ig-f)\left(\left(\frac{1}{2}\right)^{2^s}\right)\right)\right).$$

3. $s = \frac{1}{2}$ and $t = 1$. By Example 4.75 we have

$$e_{f+ig,\mu}\left(1,\frac{1}{2}\right) = 1+\frac{1}{2}(f+ig)\left(\frac{1}{2}\right)$$

and

$$e_{f-ig,\mu}\left(1,\frac{1}{2}\right) = 1-\frac{1}{2}(ig-f)\left(\frac{1}{2}\right).$$

Therefore

$$c_{f,\mu}\left(1,\frac{1}{2}\right) = 1+\frac{1}{2}f\left(\frac{1}{2}\right),$$

and

$$s_{f,\mu}\left(1,\frac{1}{2}\right) = \frac{1}{2}g\left(\frac{1}{2}\right).$$

4. $s = (\frac{1}{2})^{2^n}$ for some $n \in \mathbb{N}$, and $t = 1$. By Example 4.75 we have

$$e_{f+ig,\mu}(1,s) = \prod_{s=1}^{n}\left(1+\left(\left(\frac{1}{2}\right)^{2^{s-1}}-\left(\frac{1}{2}\right)^{2^s}\right)(f+ig)\left(\left(\frac{1}{2}\right)^{2^s}\right)\right)\left(1+\frac{1}{2}(f+ig)\left(\frac{1}{2}\right)\right)$$

and

$$e_{f-ig,\mu}(1,s) = \prod_{s=1}^{n}\left(1-\left(\left(\frac{1}{2}\right)^{2^{s-1}}-\left(\frac{1}{2}\right)^{2^s}\right)(ig-f)\left(\left(\frac{1}{2}\right)^{2^s}\right)\right)\left(1-\frac{1}{2}(ig-f)\left(\frac{1}{2}\right)\right).$$

Therefore

$$\mathrm{c}_{fg,\mu}(1,s)=\frac{1}{2}\Bigg(\prod_{s=1}^{n}\Bigg(1+\bigg(\Big(\frac{1}{2}\Big)^{2^{s-1}}-\Big(\frac{1}{2}\Big)^{2^s}\bigg)(f+ig)\bigg(\Big(\frac{1}{2}\Big)^{2^s}\bigg)\Bigg)\Big(1+\frac{1}{2}(f+ig)\Big(\frac{1}{2}\Big)\Big)$$
$$+\prod_{s=1}^{n}\Bigg(1-\bigg(\Big(\frac{1}{2}\Big)^{2^{s-1}}-\Big(\frac{1}{2}\Big)^{2^s}\bigg)(ig-f)\bigg(\Big(\frac{1}{2}\Big)^{2^s}\bigg)\Bigg)\Big(1-\frac{1}{2}(ig-f)\Big(\frac{1}{2}\Big)\Big)\Bigg),$$

and

$$\mathrm{s}_{fg,\mu}(1,s)=\frac{1}{2i}\Bigg(\prod_{s=1}^{n}\Bigg(1+\bigg(\Big(\frac{1}{2}\Big)^{2^{s-1}}-\Big(\frac{1}{2}\Big)^{2^s}\bigg)(f+ig)\bigg(\Big(\frac{1}{2}\Big)^{2^s}\bigg)\Bigg)\Big(1+\frac{1}{2}(f+ig)\Big(\frac{1}{2}\Big)\Big)$$
$$-\prod_{s=1}^{n}\Bigg(1-\bigg(\Big(\frac{1}{2}\Big)^{2^{s-1}}-\Big(\frac{1}{2}\Big)^{2^s}\bigg)(ig-f)\bigg(\Big(\frac{1}{2}\Big)^{2^s}\bigg)\Bigg)\Big(1-\frac{1}{2}(ig-f)\Big(\frac{1}{2}\Big)\Big)\Bigg).$$

5. $s=(\frac{1}{2})^{2^n}$ and $t=(\frac{1}{2})^{2^m}$ for some $m,n\in\mathbb{N}$, $m<n$. By Example 4.75 we have

$$e_{f+ig,\mu}(t,s)=\prod_{s=m+1}^{n}\Bigg(1+\bigg(\Big(\frac{1}{2}\Big)^{2^{s-1}}-\Big(\frac{1}{2}\Big)^{2^s}\bigg)(f+ig)\bigg(\Big(\frac{1}{2}\Big)^{2^s}\bigg)\Bigg)$$

and

$$e_{f-ig,\mu}(t,s)=\prod_{s=m+1}^{n}\Bigg(1-\bigg(\Big(\frac{1}{2}\Big)^{2^{s-1}}-\Big(\frac{1}{2}\Big)^{2^s}\bigg)(ig-f)\bigg(\Big(\frac{1}{2}\Big)^{2^s}\bigg)\Bigg).$$

Therefore

$$\mathrm{c}_{fg,\mu}(t,s)=\frac{1}{2}\Bigg(\prod_{s=m+1}^{n}\Bigg(1+\bigg(\Big(\frac{1}{2}\Big)^{2^{s-1}}-\Big(\frac{1}{2}\Big)^{2^s}\bigg)(f+ig)\bigg(\Big(\frac{1}{2}\Big)^{2^s}\bigg)\Bigg)$$
$$+\prod_{s=m+1}^{n}\Bigg(1-\bigg(\Big(\frac{1}{2}\Big)^{2^{s-1}}-\Big(\frac{1}{2}\Big)^{2^s}\bigg)(ig-f)\bigg(\Big(\frac{1}{2}\Big)^{2^s}\bigg)\Bigg)\Bigg),$$

and

$$\mathrm{s}_{fg,\mu}(t,s)=\frac{1}{2i}\Bigg(\prod_{s=m+1}^{n}\Bigg(1+\bigg(\Big(\frac{1}{2}\Big)^{2^{s-1}}-\Big(\frac{1}{2}\Big)^{2^s}\bigg)(f+ig)\bigg(\Big(\frac{1}{2}\Big)^{2^s}\bigg)\Bigg)$$
$$-\prod_{s=m+1}^{n}\Bigg(1-\bigg(\Big(\frac{1}{2}\Big)^{2^{s-1}}-\Big(\frac{1}{2}\Big)^{2^s}\bigg)(ig-f)\bigg(\Big(\frac{1}{2}\Big)^{2^s}\bigg)\Bigg)\Bigg).$$

Example 4.109. Let $\mathbb{T}=\{1-\frac{1}{2^n}\}_{n\in\mathbb{N}_0}\cup[1,2]\cup 3^{\mathbb{N}}$ and $f\in\mathscr{R}_\mu$. We will find $\mathrm{c}_{fg,\mu}(t,s)$ and $\mathrm{s}_{fg,\mu}(t,s)$ for $s,t\in\mathbb{T}$, $s\le t$, and $f,g\in\mathrm{C}_{\mathrm{rd}}$ that satisfy (4.8). We have the following cases.

1. $s = 1 - \frac{1}{2^n}$ for some $n \in \mathbb{N}_0$, and $t \in 3^{\mathbb{N}}$, $t \geq 9$. By Example 4.76 we have

$$
\begin{aligned}
e_{f+ig,\mu}(t,s) &= \prod_{\tau=n}^{\infty}\left(1+\frac{1}{2^{\tau+1}}(f+ig)\left(1-\frac{1}{2^{\tau}}\right)\right)\\
&\quad\times e^{\int_1^2 (f+ig)(\tau)\Delta\tau}(1+f(2)+ig(2))\prod_{\tau=3}^{\frac{t}{3}}(1+2\tau(f+ig)(\tau))
\end{aligned}
$$

and

$$
\begin{aligned}
e_{f-ig,\mu}(t,s) &= \prod_{\tau=n}^{\infty}\left(1-\frac{1}{2^{\tau+1}}(ig-f)\left(1-\frac{1}{2^{\tau}}\right)\right)\\
&\quad\times e^{-\int_1^2 (g-f)(\tau)\Delta\tau}(1+f(2)-ig(2))\prod_{\tau=3}^{\frac{t}{3}}(1-2\tau(ig-f)(\tau)).
\end{aligned}
$$

Consequently,

$$
\begin{aligned}
c_{fg,\mu}(t,s) &= \frac{1}{2}\left(\prod_{\tau=n}^{\infty}\left(1+\frac{1}{2^{\tau+1}}(f+ig)\left(1-\frac{1}{2^{\tau}}\right)\right)\right.\\
&\quad\times e^{\int_1^2 (f+ig)(\tau)\Delta\tau}(1+f(2)+ig(2))\prod_{\tau=3}^{\frac{t}{3}}(1+2\tau(f+ig)(\tau))\\
&\quad+\prod_{\tau=n}^{\infty}\left(1-\frac{1}{2^{\tau+1}}(ig-f)\left(1-\frac{1}{2^{\tau}}\right)\right)\\
&\quad\left.\times e^{-\int_1^2 (ig-f)(\tau)\Delta\tau}(1+f(2)-ig(2))\prod_{\tau=3}^{\frac{t}{3}}(1-2\tau(ig-f)(\tau))\right),
\end{aligned}
$$

and

$$
\begin{aligned}
s_{fg,\mu}(t,s) &= \frac{1}{2i}\left(\prod_{\tau=n}^{\infty}\left(1+\frac{1}{2^{\tau+1}}(f+ig)\left(1-\frac{1}{2^{\tau}}\right)\right)\right.\\
&\quad\times e^{\int_1^2 (f+ig)(\tau)\Delta\tau}(1+f(2)+ig(2))\prod_{\tau=3}^{\frac{t}{3}}(1+2\tau(f+ig)(\tau))\\
&\quad-\prod_{\tau=n}^{\infty}\left(1-\frac{1}{2^{\tau+1}}(ig-f)\left(1-\frac{1}{2^{\tau}}\right)\right)\\
&\quad\left.\times e^{-\int_1^2 (ig-f)(\tau)\Delta\tau}(1+f(2)-ig(2))\prod_{\tau=3}^{\frac{t}{3}}(1-2\tau(ig-f)(\tau))\right).
\end{aligned}
$$

2. $s = 1 - \frac{1}{2^n}$ for some $n \in \mathbb{N}_0$, and $t = 3$. By Example 4.76 we have

$$
e_{f+ig,\mu}(t,s) = \prod_{\tau=n}^{\infty}\left(1+\frac{1}{2^{\tau+1}}(f+ig)\left(1-\frac{1}{2^{\tau}}\right)\right)
$$

$$\times e^{\int_1^2 (f+ig)(\tau)\Delta\tau}(1+f(2)+ig(2))\prod_{\tau=3}^{\frac{t}{3}}(1+2\tau(f+ig)(\tau))$$

and

$$e_{f-ig,\mu}(t,s)=\prod_{\tau=n}^{\infty}\left(1-\frac{1}{2^{\tau+1}}(ig-f)\left(1-\frac{1}{2^{\tau}}\right)\right)$$
$$\times e^{-\int_1^2 (ig-f)(\tau)\Delta\tau}(1+f(2)-ig(2))\prod_{\tau=3}^{\frac{t}{3}}(1-2\tau(ig-f)(\tau)).$$

Consequently,

$$\begin{aligned}c_{fg,\mu}(t,s)=\frac{1}{2}\Bigg(&\prod_{\tau=n}^{\infty}\left(1+\frac{1}{2^{\tau+1}}(f+ig)\left(1-\frac{1}{2^{\tau}}\right)\right)\\&\times e^{\int_1^2 (f+ig)(\tau)\Delta\tau}(1+f(2)+ig(2))\prod_{\tau=3}^{\frac{t}{3}}(1+2\tau(f+ig)(\tau))\\&+\prod_{\tau=n}^{\infty}\left(1-\frac{1}{2^{\tau+1}}(ig-f)\left(1-\frac{1}{2^{\tau}}\right)\right)\\&\times e^{-\int_1^2 (ig-f)(\tau)\Delta\tau}(1+f(2)-ig(2))\prod_{\tau=3}^{\frac{t}{3}}\Bigg)(1-2\tau(ig-f)(\tau)),\end{aligned}$$

and

$$\begin{aligned}s_{fg,\mu}(t,s)=\frac{1}{2i}\Bigg(&\prod_{\tau=n}^{\infty}\left(1+\frac{1}{2^{\tau+1}}(f+ig)\left(1-\frac{1}{2^{\tau}}\right)\right)\\&\times e^{\int_1^2 (f+ig)(\tau)\Delta\tau}(1+f(2)+ig(2))\prod_{\tau=3}^{\frac{t}{3}}(1+2\tau(f+ig)(\tau))\\&-\prod_{\tau=n}^{\infty}\left(1-\frac{1}{2^{\tau+1}}(ig-f)\left(1-\frac{1}{2^{\tau}}\right)\right)\\&\times e^{-\int_1^2 (ig-f)(\tau)\Delta\tau}(1+f(2)-ig(2))\prod_{\tau=3}^{\frac{t}{3}}\Bigg)(1-2\tau(ig-f)(\tau)).\end{aligned}$$

3. $s=1-\frac{1}{2^n}$ for some $n\in\mathbb{N}_0$, and $t\in[1,2]$. By Example 4.76 we have

$$e_{f+ig,\mu}(t,s)=\prod_{\tau=n}^{\infty}\left(1+\frac{1}{2^{\tau+1}}(f+ig)\left(1-\frac{1}{2^{\tau}}\right)\right)e^{\int_1^2 (f+ig)(\tau)\Delta\tau}$$

and

$$e_{f-ig,\mu}(t,s)=\prod_{\tau=n}^{\infty}\left(1-\frac{1}{2^{\tau+1}}(ig-f)\left(1-\frac{1}{2^{\tau}}\right)\right)e^{-\int_1^2 (ig-f)(\tau)\Delta\tau}.$$

Consequently,

$$\begin{aligned}\mathrm{c}_{fg,\mu}(t,s) &= \frac{1}{2}\Bigg(\prod_{\tau=n}^{\infty}\left(1+\frac{1}{2^{\tau+1}}(f+ig)\left(1-\frac{1}{2^{\tau}}\right)\right)e^{\int_1^2 (f+ig)(\tau)\Delta\tau}\\&\quad+\prod_{\tau=n}^{\infty}\left(1-\frac{1}{2^{\tau+1}}(ig-f)\left(1-\frac{1}{2^{\tau}}\right)\right)e^{-\int_1^2 (ig-f)(\tau)\Delta\tau}\Bigg),\end{aligned}$$

and

$$\begin{aligned}\mathrm{s}_{f,\mu}(t,s) &= \frac{1}{2i}\Bigg(\prod_{\tau=n}^{\infty}\left(1+\frac{1}{2^{\tau+1}}(f+ig)\left(1-\frac{1}{2^{\tau}}\right)\right)e^{\int_1^2 (f+ig)(\tau)\Delta\tau}\\&\quad-\prod_{\tau=n}^{\infty}\left(1-\frac{1}{2^{\tau+1}}(ig-f)\left(1-\frac{1}{2^{\tau}}\right)\right)e^{-\int_1^2 (ig-f)(\tau)\Delta\tau}\Bigg).\end{aligned}$$

4. $s = 1-\frac{1}{2^n}$ and $t = 1-\frac{1}{2^m}$ for some $m, n \in \mathbb{N}_0$, $m > n$. By Example 4.76 we have

$$e_{f+ig,\mu}(t,s) = \prod_{\tau=n}^{m}\left(1+\frac{1}{2^{\tau+1}}(f+ig)\left(1-\frac{1}{2^{\tau}}\right)\right)$$

and

$$e_{f-ig,\mu}(t,s) = \prod_{\tau=n}^{m}\left(1-\frac{1}{2^{\tau+1}}(ig-f)\left(1-\frac{1}{2^{\tau}}\right)\right).$$

Consequently,

$$\begin{aligned}\mathrm{c}_{fg,\mu}(t,s) &= \frac{1}{2}\Bigg(\prod_{\tau=n}^{m}\left(1+\frac{1}{2^{\tau+1}}(f+ig)\left(1-\frac{1}{2^{\tau}}\right)\right)\\&\quad+\prod_{\tau=n}^{m}\left(1-\frac{1}{2^{\tau+1}}(ig-f)\left(1-\frac{1}{2^{\tau}}\right)\right)\Bigg),\end{aligned}$$

and

$$\begin{aligned}\mathrm{s}_{fg,\mu}(t,s) &= \frac{1}{2i}\Bigg(\prod_{\tau=n}^{m}\left(1+\frac{1}{2^{\tau+1}}(f+ig)\left(1-\frac{1}{2^{\tau}}\right)\right)\\&\quad-\prod_{\tau=n}^{m}\left(1-\frac{1}{2^{\tau+1}}(ig-f)\left(1-\frac{1}{2^{\tau}}\right)\right)\Bigg).\end{aligned}$$

5. $s \in [1,2]$, and $t \in 3^{\mathbb{N}}$, $t \geq 9$. By Example 4.76 we have

$$e_{f+ig,\mu}(t,s) = e^{\int_s^2 (f+ig)(\tau)\Delta\tau}(1+f(2)+ig(2))\prod_{\tau=3}^{\frac{t}{3}}(1+2\tau(f+ig)(\tau))$$

and

$$e_{f-ig,\mu}(t,s) = e^{-\int_s^2 (ig-f)(\tau)\Delta\tau}(1+f(2)-ig(2))\prod_{\tau=3}^{\frac{t}{3}}(1-2\tau(ig-f)(\tau)).$$

Consequently,

$$\begin{aligned} \mathrm{c}_{fg,\mu}(t,s) = \frac{1}{2}\Bigg(& e^{\int_s^2 (f+ig)(\tau)\Delta\tau}(1+f(2)+ig(2))\prod_{\tau=3}^{\frac{t}{3}}(1+2\tau(f+ig)(\tau)) \\ & + e^{-\int_s^2 (ig-f)(\tau)\Delta\tau}(1+f(2)-ig(2))\prod_{\tau=3}^{\frac{t}{3}}(1-2\tau(ig-f)(\tau))\Bigg), \end{aligned}$$

and

$$\begin{aligned} \mathrm{s}_{f,\mu}(t,s) = \frac{1}{2i}\Bigg(& e^{\int_s^2 (f+ig)(\tau)\Delta\tau}(1+f(2)+ig(2))\prod_{\tau=3}^{\frac{t}{3}}(1+2\tau(f+ig)(\tau)) \\ & - e^{-\int_s^2 (ig-f)(\tau)\Delta\tau}(1+f(2)-ig(2))\prod_{\tau=3}^{\frac{t}{3}}(1-2\tau(ig-f)(\tau))\Bigg). \end{aligned}$$

6. $s, t \in [1,2]$. By Example 4.76 we have

$$e_{f+ig,\mu}(t,s) = e^{\int_s^t (f+ig)(\tau)\Delta\tau}$$

and

$$e_{f-ig,\mu}(t,s) = e^{-\int_s^t (ig-f)(\tau)\Delta\tau}.$$

Consequently,

$$\mathrm{c}_{fg,\mu}(t,s) = \frac{1}{2}(e^{\int_s^t (f+ig)(\tau)\Delta\tau} + e^{-\int_s^t (ig-f)(\tau)\Delta\tau}),$$

and

$$\mathrm{s}_{fg,\mu}(t,s) = \frac{1}{2i}(e^{\int_s^t (f+ig)(\tau)\Delta\tau} - e^{-\int_s^t (ig-f)(\tau)\Delta\tau}).$$

7. $s \in [1,2]$ and $t = 3$. By Example 4.76 we have

$$e_{f+ig,\mu}(3,s) = e^{\int_s^2 (f+ig)(\tau)\Delta\tau}(1+f(2)+ig(2))$$

and

$$e_{f-ig,\mu}(3,s) = e^{-\int_s^2 (ig-f)(\tau)\Delta\tau}(1+f(2)-ig(2)).$$

Consequently,

$$c_{fg,\mu}(3,s)=\frac{1}{2}(e^{\int_s^2(f+ig)(\tau)\Delta\tau}(1+f(2)+ig(2))+e^{-\int_s^2(ig-f)(\tau)\Delta\tau}(1+f(2)-ig(2))),$$

and

$$s_{fg,\mu}(3,s)=\frac{1}{2i}(e^{\int_s^2(f+ig)(\tau)\Delta\tau}(1+f(2)+ig(2))-e^{-\int_s^2(ig-f)(\tau)\Delta\tau}(1+f(2)-ig(2))).$$

8. $s,t\in 3^{\mathbb{N}}$, $s\le t$. By Example 4.76 we have

$$e_{f+ig,\mu}(t,s)=\prod_{\tau=s}^{\frac{t}{3}}(1+2\tau(f+ig)(\tau))$$

and

$$e_{f-ig,\mu}(t,s)=\prod_{\tau=s}^{\frac{t}{3}}(1-2\tau(ig-f)(\tau)).$$

Consequently,

$$c_{fg,\mu}(t,s)=\frac{1}{2}\left(\prod_{\tau=s}^{\frac{t}{3}}(1+2\tau(f+ig)(\tau))+\prod_{\tau=s}^{\frac{t}{3}}(1-2\tau(ig-f)(\tau))\right),$$

and

$$s_{f,\mu}(t,s)=\frac{1}{2i}\left(\prod_{\tau=s}^{\frac{t}{3}}(1+2\tau(f+ig)(\tau))-\prod_{\tau=s}^{\frac{t}{3}}(1-2\tau(ig-f)(\tau))\right).$$

Example 4.110. Let $f,g\in C_{rd}$ satisfy (4.8). Then

$$c_{fg,\mu}^{\Delta}=f\,c_{fg,\mu}-g\,s_{fg,\mu},\quad s_{fg,\mu}^{\Delta}=g\,c_{fg,\mu}+fs_{fg},$$

and

$$c_{fg,\mu}^2+s_{fg,\mu}^2=1.$$

We have

$$\begin{aligned}c_{fg,\mu}^{\Delta}&=\left(\frac{e_{f+ig,\mu}+e_{f-ig,\mu}}{2}\right)^{\Delta}\\&=\frac{e_{f+ig,\mu}^{\Delta}+e_{f-ig,\mu}^{\Delta}}{2}\\&=\frac{(f+ig)e_{f+ig,\mu}+(f-ig)e_{f-ig,\mu}}{2}\end{aligned}$$

$$
\begin{aligned}
&= f\frac{e_{f+ig,\mu}+e_{f-ig,\mu}}{2} + ig\frac{e_{f+ig,\mu}-e_{f-ig,\mu}}{2}\\
&= f\,\mathrm{c}_{fg,\mu} - g\frac{e_{f+ig,\mu}-e_{f-ig,\mu}}{2i}\\
&= f\,\mathrm{c}_{fg,\mu} - g\,\mathrm{s}_{fg,\mu},
\end{aligned}
$$

$$
\begin{aligned}
\mathrm{s}_{fg,\mu}^{\Delta} &= \left(\frac{e_{f+ig,\mu}-e_{f-ig,\mu}}{2i}\right)^{\Delta}\\
&= \frac{e_{f+ig,\mu}^{\Delta}-e_{f-ig,\mu}^{\Delta}}{2i}\\
&= \frac{(f+ig)e_{f+ig,\mu}-(f-ig)e_{f-ig,\mu}}{2i}\\
&= f\frac{e_{f+ig,\mu}-e_{f-ig,\mu}}{2i} + g\frac{e_{f+ig,\mu}+e_{f-ig,\mu}}{2}\\
&= f\,\mathrm{s}_{fg,\mu} + g\,\mathrm{c}_{fg,\mu},
\end{aligned}
$$

and

$$
\begin{aligned}
\mathrm{c}_{fg,\mu}^2+\mathrm{s}_{fg,\mu}^2 &= \left(\frac{e_{f+ig,\mu}+e_{f-ig,\mu}}{2}\right)^2 + \left(\frac{e_{f+ig,\mu}-e_{f-ig,\mu}}{2i}\right)^2\\
&= \frac{e_{f+ig,\mu}^2+2e_{f+ig,\mu}e_{f-ig,\mu}+e_{f-ig,\mu}^2}{4}\\
&\quad - \frac{e_{f+ig,\mu}^2-2e_{f+ig,\mu}e_{f-ig,\mu}+e_{f-ig,\mu}^2}{4}\\
&= e_{(f+ig)\oplus_\mu(f-ig)}.
\end{aligned}
$$

Since

$$
\begin{aligned}
(f+ig)\oplus_\mu(f-ig) &= f+ig+f-ig+\mu(f+ig)(f-ig)\\
&= 2f+\mu(f^2+g^2)\\
&= 0,
\end{aligned}
$$

we get

$$
\mathrm{c}_{fg,\mu}^2+\mathrm{s}_{fg,\mu}^2 = 1,
$$

completing the proof.

Example 4.111. Let $f, g \in \mathrm{C}_{\mathrm{rd}}$ satisfy (4.8). We will prove that
1. $\mathrm{c}_{fg,\mu}(s,t) = \mathrm{c}_{fg,\mu}(t,s)$,
2. $\mathrm{c}_{fg,\mu}(t,s) = \mathrm{c}_{fg,\mu}(t,r)\,\mathrm{c}_{fg,\mu}(s,r) + \mathrm{s}_{fg,\mu}(t,r)\,\mathrm{s}_{fg,\mu}(s,r)$,
3. $\mathrm{c}_{fg,\mu}(t,r) = \mathrm{c}_{fg,\mu}(t,s)\,\mathrm{c}_{fg,\mu}(s,r) - \mathrm{s}_{fg,\mu}(t,s)\,\mathrm{s}_{fg,\mu}(s,r)$, $t,s,r \in \mathbb{T}$.

1. We have

$$\begin{aligned}
\mathrm{c}_{fg,\mu}(t,s) &= \frac{e_{f+ig,\mu}(t,s)+e_{f-ig,\mu}(t,s)}{2} \\
&= \frac{\frac{1}{e_{f+ig,\mu}(s,t)}+\frac{1}{e_{f-ig,\mu}(s,t)}}{2} \\
&= \frac{e_{f+ig,\mu}(s,t)+e_{f-ig,\mu}(s,t)}{2e_{f+ig,\mu}(s,t)e_{f-ig,\mu}(s,t)} \\
&= \frac{e_{f+ig,\mu}(s,t)+e_{f-ig,\mu}(s,t)}{2} \\
&= \mathrm{c}_{fg,\mu}(s,t), \quad t,s\in\mathbb{T}.
\end{aligned}$$

2. We have

$$\begin{aligned}
\mathrm{c}_{fg,\mu}(t,r)\,\mathrm{c}_{fg,\mu}(s,r) &= \frac{e_{f+ig,\mu}(t,r)+e_{f-ig,\mu}(t,r)}{2}\,\frac{e_{f+ig,\mu}(s,r)+e_{f-ig,\mu}(s,r)}{2} \\
&= \frac{1}{4}\big(e_{f+ig,\mu}(t,r)e_{f+ig,\mu}(s,r)+e_{f+ig,\mu}(t,r)e_{f-ig,\mu}(s,r) \\
&\quad + e_{f-ig,\mu}(t,r)e_{f+ig,\mu}(s,r)+e_{f-ig,\mu}(t,r)e_{f-ig,\mu}(s,r)\big), \\
\mathrm{s}_{fg,\mu}(t,r)\,\mathrm{s}_{fg,\mu}(s,r) &= \frac{e_{f+ig,\mu}(t,r)-e_{f-ig,\mu}(t,r)}{2i}\,\frac{e_{f+ig,\mu}(s,r)-e_{f-ig,\mu}(s,r)}{2i} \\
&= -\frac{1}{4}\big(e_{f+ig,\mu}(t,r)e_{f+ig,\mu}(s,r)-e_{f+ig,\mu}(t,r)e_{f-ig,\mu}(s,r) \\
&\quad - e_{f-ig,\mu}(t,r)e_{f+ig,\mu}(s,r)+e_{f-ig,\mu}(t,r)e_{f-ig,\mu}(s,r)\big), \quad t,s,r\in\mathbb{T}.
\end{aligned}$$

Hence

$$\begin{aligned}
&\mathrm{c}_{fg,\mu}(t,r)\,\mathrm{c}_{fg,\mu}(s,r)+\mathrm{s}_{fg,\mu}(t,r)\,\mathrm{s}_{fg,\mu}(s,r) \\
&\quad = \frac{1}{2}\big(e_{f+ig,\mu}(t,r)e_{f-ig,\mu}(s,r)+e_{f-ig,\mu}(t,r)e_{f+ig,\mu}(s,r)\big) \\
&\quad = \frac{1}{2}\big(e_{f+ig,\mu}(t,r)e_{f+ig,\mu}(r,s)+e_{f-ig,\mu}(t,r)e_{f-ig,\mu}(r,s)\big) \\
&\quad = \frac{1}{2}\big(e_{f+ig,\mu}(t,s)+e_{f-ig,\mu}(t,s)\big) \\
&\quad = \mathrm{c}_{fg,\mu}(t,s), \quad t,s,r\in\mathbb{T}.
\end{aligned}$$

3. We have

$$\begin{aligned}
\mathrm{c}_{fg,\mu}(t,s)\,\mathrm{c}_{fg,\mu}(s,r) &= \frac{e_{f+ig,\mu}(t,s)+e_{f-ig,\mu}(t,s)}{2}\,\frac{e_{f+ig,\mu}(s,r)+e_{f-ig,\mu}(s,r)}{2} \\
&= \frac{1}{4}\big(e_{f+ig,\mu}(t,r)+e_{f-ig,\mu}(t,r)+e_{f+ig,\mu}(t,s)e_{f-ig,\mu}(s,r) \\
&\quad + e_{f-ig,\mu}(t,s)e_{f+ig,\mu}(s,r)\big),
\end{aligned}$$

$$
\begin{aligned}
s_{fg,\mu}(t,s)\, s_{fg,\mu}(s,r) &= \frac{e_{f+ig,\mu}(t,s)-e_{f-ig,\mu}(t,s)}{2i}\,\frac{e_{f+ig,\mu}(s,r)-e_{f-ig,\mu}(s,r)}{2i} \\
&= -\frac{1}{4}\big(e_{f+ig,\mu}(t,r)+e_{f-ig,\mu}(t,r)-e_{f+ig,\mu}(t,s)e_{f-ig,\mu}(s,r) \\
&\quad - e_{f-ig,\mu}(t,s)e_{f+ig,\mu}(s,r)\big), \quad t,s,r\in\mathbb{T}.
\end{aligned}
$$

Hence

$$
\begin{aligned}
c_{fg,\mu}(t,s)\, c_{fg,\mu}(s,r) - s_{fg,\mu}(t,s)\, s_{fg,\mu}(s,r) &= \frac{1}{2}\big(e_{f+ig,\mu}(t,r)+e_{f-ig,\mu}(t,r)\big) \\
&= c_{fg,\mu}(t,r), \quad t,s,r\in\mathbb{T}.
\end{aligned}
$$

Exercise 4.40. Let $f,g\in C_{rd}$ satisfy (4.8). Prove that
1. $s_{fg,\mu}(s,t) = -s_{fg,\mu}(t,s)$,
2. $s_{fg,\mu}(t,s) = s_{fg,\mu}(t,r)\, c_{fg,\mu}(s,r) - c_{fg,\mu}(t,r)\, s_{fg,\mu}(s,r)$,
3. $s_{fg,\mu}(t,r) = s_{fg,\mu}(t,s)\, c_{fg,\mu}(s,r) + c_{fg,\mu}(t,s)\, s_{fg,\mu}(s,r)$, $t,s,r\in\mathbb{T}$.

4.6 Advanced practical problems

Problem 4.1. Find $\mathrm{Re}_h(z)$ and $\mathrm{Im}_h(z)$, where
1. $h=3$, $z=\frac{1}{3}$.
2. $h=2$, $z=\frac{\sqrt{2+\sqrt{2+\sqrt{2}}-2}}{4}-\frac{\sqrt{2-\sqrt{2+\sqrt{2}}}}{4}i$.
3. $h=3$, $z=-\frac{\sqrt{2}+2}{6}+\frac{\sqrt{2}}{6}i$.
4. $h=5$, $z=\frac{2}{5}-i$.
5. $h=3$, $z=\frac{1}{3}-\frac{4}{3}i$.

Problem 4.2. Find $z\oplus_h w$, where
1. $z=-2i$, $w=7-i$, $h=2$.
2. $z=2+i$, $w=1+7i$, $h=3$.
3. $z=3+2i$, $w=1-i$, $h=4$.
4. $z=5$, $w=2+3i$, $h=2$.
5. $z=1-8i$, $w=1+10i$, $h=5$.

Problem 4.3. Find $n\odot_h z$, where
1. $n=3$, $z=2i$, $h=2$.
2. $n=4$, $z=2+i$, $h=3$.
3. $n=3$, $z=3+2i$, $h=2$.
4. $n=4$, $z=5$, $h=2$.
5. $n=3$, $z=1-i$, $h=4$.

Problem 4.4. Find $\ominus_h z$, where
1. $z = 7 - i, h = 2$.
2. $z = 1 + 7i, h = 3$.
3. $z = 1 - i, h = 4$.
4. $z = 2 + 3i, h = 2$.
5. $z = 1 - 10i, h = 5$.

Problem 4.5. Find $z \ominus_h w$, where z, w, and h are as in Problem 4.2.

Problem 4.6. Find $z^{②}$, where z and h are as in Problem 4.4.

Problem 4.7. Let $\mathbb{T} = (-5\mathbb{N}_0) \cup 4^{\mathbb{N}_0}$, and let

$$f(t) = \begin{cases} -\frac{1}{5} & \text{if } t \in (-5\mathbb{N}), \\ 2 & \text{if } t = 0, \\ -\frac{1}{48} & \text{if } t \in 4^{\mathbb{N}_0}. \end{cases}$$

Check if f is delta regressive on $\mathbb{T}$.

Problem 4.8. Let $\mathbb{T} = [-1, 3] \cup [7, 9] \cup [10, \infty)$, and let

$$f(t) = \begin{cases} t^2 + 3t & \text{if } t \in [-1, 3) \cup [7, 9) \cup [10, \infty), \\ -\frac{1}{4} & \text{if } t = 3, \\ -1 & \text{if } t = 9. \end{cases}$$

Check if f is delta regressive on $\mathbb{T}$.

Problem 4.9. Let $\mathbb{T} = -\{0\} \cup \{1 - \frac{1}{4^n}\}_{n\in\mathbb{N}_0} \cup 2^{\mathbb{N}_0}$, and let

$$f(t) = \begin{cases} 16 & \text{if } t \in \{1 - \frac{1}{4^n}\}_{n\in\mathbb{N}_0}, \\ -\frac{4}{3} & \text{if } t = 0, \\ t^2 + t & \text{if } t \in 2^{\mathbb{N}_0}. \end{cases}$$

Check if f is delta regressive on $\mathbb{T}$.

Problem 4.10. Let $\mathbb{T} = [1, 3] \cup 5^{\mathbb{N}}$, and let

$$f(t) = t, \quad t \in \mathbb{T}.$$

Check if f is delta regressive on $\mathbb{T}$.

Problem 4.11. Let $\mathbb{T} = \{2\} \cup \{2 + \frac{1}{n}\}_{n\in\mathbb{N}} \cup [4, 9] \cup \{9 + \frac{1}{n^2}\}_{n\in\mathbb{N}}$, and let

$$f(t) = \begin{cases} \frac{1+t}{1+2t} & \text{if } t \in \{2\} \cup [4,9) \cup \{10\}, \\ t^2 & \text{if } t \in \{2 + \frac{1}{n}\}_{n\in\mathbb{N}, n\geq 2}, \\ -1 & \text{if } t = 3, 9, \\ 7 & \text{if } t \in \{9 + \frac{1}{n^2}\}_{n\in\mathbb{N}}. \end{cases}$$

Check if f is delta regressive on $\mathbb{T}$.

Problem 4.12. Let $\mathbb{T}$ and f be as in Problem 4.7, and let $g(t) = t$, $t \in \mathbb{T}$. Find $(f \oplus_\mu g)(t)$, $t \in \mathbb{T}$.

Problem 4.13. Let $\mathbb{T}$ and f be as in Problem 4.8, and let $g(t) = t$, $t \in \mathbb{T}$. Find $(f \oplus_\mu g)(t)$, $t \in \mathbb{T}$.

Problem 4.14. Let $\mathbb{T}$ and f be as in Problem 4.9, and let $g(t) = t$, $t \in \mathbb{T}$. Find $(f \oplus_\mu g)(t)$, $t \in \mathbb{T}$.

Problem 4.15. Let $\mathbb{T}$ and f be as in Problem 4.10, and let $g(t) = t$, $t \in \mathbb{T}$. Find $(f \oplus_\mu g)(t)$, $t \in \mathbb{T}$.

Problem 4.16. Let $\mathbb{T}$ and f be as in Problem 4.11, and let $g(t) = t$, $t \in \mathbb{T}$. Find $(f \oplus_\mu g)(t)$, $t \in \mathbb{T}$.

Problem 4.17. Let $\mathbb{T}$ and g be as in Problem 4.12. Find $\ominus_\mu g(t)$, $t \in \mathbb{T}$.

Problem 4.18. Let $\mathbb{T}$ and g be as in Problem 4.13. Find $\ominus_\mu g(t)$, $t \in \mathbb{T}$.

Problem 4.19. Let $\mathbb{T}$ and g be as in Problem 4.14. Find $\ominus_\mu g(t)$, $t \in \mathbb{T}$.

Problem 4.20. Let $\mathbb{T}$ and g be as in Problem 4.15. Find $\ominus_\mu g(t)$, $t \in \mathbb{T}$.

Problem 4.21. Let $\mathbb{T}$ and g be as in Problem 4.16. Find $\ominus_\mu g(t)$, $t \in \mathbb{T}$.

Problem 4.22. Let $\mathbb{T}, f$, and g be as in Problem 4.12. Find $(f \ominus_\mu g)(t)$, $t \in \mathbb{T}$.

Problem 4.23. Let $\mathbb{T}, f$ and g be as in Problem 4.13. Find $(f \ominus_\mu g)(t)$, $t \in \mathbb{T}$.

Problem 4.24. Let $\mathbb{T}, f$, and g be as in Problem 4.14. Find $(f \ominus_\mu g)(t)$, $t \in \mathbb{T}$.

Problem 4.25. Let $\mathbb{T}, f$, and g be as in Problem 4.15. Find $(f \ominus_\mu g)(t)$, $t \in \mathbb{T}$.

Problem 4.26. Let $\mathbb{T}, f$, and g be as in Problem 4.16. Find $(f \ominus_\mu g)(t)$, $t \in \mathbb{T}$.

Problem 4.27. Let $\mathbb{T} = \{3 - \frac{1}{2^n}\}_{n\in\mathbb{N}_0} \cup [3,4] \cup 5^{\mathbb{N}}$ and $f \in \mathscr{R}_\mu$. Prove that

$$e_{f,\mu}(t,s) = \begin{cases} \prod_{\tau=n}^{\infty}(1 + \frac{1}{2^{\tau+1}} f(3 - \frac{1}{2^\tau})) e^{\int_3^4 f(\tau)\Delta\tau}(1 + f(4)) \prod_{\tau=5}^{\frac{t}{5}}(1 + 4\tau f(\tau)) \\ \quad \text{if } s = 3 - \frac{1}{2^n} \text{ for some } n \in \mathbb{N}_0 \text{ and } t \in 5^{\mathbb{N}},\ t \geq 25, \\ \prod_{\tau=n}^{\infty}(1 + \frac{1}{2^{\tau+1}} f(3 - \frac{1}{2^\tau})) e^{\int_3^4 f(\tau)\Delta\tau}(1 + f(4)) \\ \quad \text{if } s = 3 - \frac{1}{2^n} \text{ for some } n \in \mathbb{N}_0 \text{ and } t = 5, \\ \prod_{\tau=n}^{\infty}(1 + \frac{1}{2^{\tau+1}} f(3 - \frac{1}{2^\tau})) e^{\int_3^t f(\tau)\Delta\tau} \\ \quad \text{if } s = 3 - \frac{1}{2^n} \text{ for some } n \in \mathbb{N}_0 \text{ and } t \in [3,4], \\ \prod_{\tau=n}^{m}(1 + \frac{1}{2^{\tau+1}} f(3 - \frac{1}{2^\tau})) \\ \quad \text{if } s = 3 - \frac{1}{2^n} \text{ for some } n \in \mathbb{N}_0 \text{ and } t = 3 - \frac{1}{2^m} \text{ for some } m \in \mathbb{N},\ m > n, \\ e^{\int_s^4 f(\tau)\Delta\tau}(1 + f(4)) \prod_{\tau=4}^{\frac{t}{4}}(1 + 4\tau f(\tau)) \\ \quad \text{if } s \in [3,4] \quad and t \in 5^{\mathbb{N}},\ t \geq 25, \\ e^{\int_s^t f(\tau)\Delta\tau} \quad \text{if } s, t \in [3,4], \\ e^{\int_s^4 f(\tau)\Delta\tau}(1 + f(4)) \quad \text{if } s \in [3,4] \text{ and } t = 5, \\ \prod_{\tau=5}^{\frac{t}{5}}(1 + 4\tau f(\tau)) \quad \text{if } s, t \in 5^{\mathbb{N}},\ s \leq t. \end{cases}$$

Solutions, hints, and answers to exercises

Chapter 1

Exercise 1.1. Answer:

1. Yes.
2. No.
3. Yes.
4. No.
5. Yes.

Exercise 1.2. Answer:

$$\sigma(t) = \begin{cases} t & \text{if } t \in \bigcup_{k=0}^{\infty}[k(a+b), k(a+b)+a), \\ t+b & \text{if } t \in \bigcup_{k=0}^{\infty}\{k(a+b)+a\}. \end{cases}$$

Exercise 1.3. Answer:

1. $\sigma(t) = t + h, t \in \mathbb{T}$.
2. $$\sigma(t) = \begin{cases} t+2 & \text{if } t \in (-2\mathbb{N}), \\ 1 & \text{if } t = 0, \quad 3t \quad \text{if } t \in 3^{\mathbb{N}_0}. \end{cases}$$
3. $$\sigma(t) = \begin{cases} t & \text{if } t \in (\bigcup_{k=0}^{\infty}[10k, 10k+3)) \cup [4,6), \\ t+7 & \text{if } t = 10k+3,\ k \in \mathbb{N}; \quad 4 \quad \text{if } t = 3; \quad 10 \quad \text{if } t = 6. \end{cases}$$
4. $$\sigma(t) = \begin{cases} 11t & \text{if } t \in 11^{\mathbb{N}_0}, \\ 1 & \text{if } t = 0. \end{cases}$$
5. $$\sigma(t) = \begin{cases} t & \text{if } t \in [1,2) \cup [3,4) \cup [7,8), \\ 9t & \text{if } r \in 9^{\mathbb{N}}, \\ 3 & \text{if } t = 2, \\ 7 & \text{if } t = 4, \\ 9 & \text{if } t = 8. \end{cases}$$

Exercise 1.4. Answer:

1. $\mu(t) = h, t \in \mathbb{T}$.
2. $$\mu(t) = \begin{cases} 2 & \text{if } t \in (-2\mathbb{N}), \\ 1 & \text{if } t = 0; \quad 2t \quad \text{if } t \in 3^{\mathbb{N}_0}. \end{cases}$$

https://doi.org/10.1515/9783112232088-005

3.

$$\mu(t)=\begin{cases}0 & \text{if } t\in(\bigcup_{k=0}^{\infty}[10k,10k+3))\cup[4,6),\\ 7 & \text{if } t=10k+3,\ k\in\mathbb{N}_0;\quad 1\quad \text{if } t=3;\quad 4\quad \text{if } t=6.\end{cases}$$

4.

$$\mu(t)=\begin{cases}10t & \text{if } t\in 11^{\mathbb{N}_0},\\ 1 & \text{if } t=0.\end{cases}$$

5.

$$\mu(t)=\begin{cases}0 & \text{if } t\in[1,2)\cup[3,4)\cup[7,8),\\ 8t & \text{if } r\in 9^{\mathbb{N}},\\ 1 & \text{if } t\in\{2,8\},\\ 3 & \text{if } t=4.\end{cases}$$

Exercise 1.5. Answer:

1. $\rho(t)=t-h,\ t\in\mathbb{T}$.
2.

$$\rho(t)=\begin{cases}t-2 & \text{if } t\in(-2\mathbb{N}_0),\\ 0 & \text{if } t=1,\\ \frac{t}{3} & \text{if } t\in 3^{\mathbb{N}}.\end{cases}$$

3.

$$\rho(t)=\begin{cases}0 & \text{if } t=0,\\ t & \text{if } t\in(0,3]\cup\bigcup_{k=1}^{\infty}(10k,10k+3]\cup(4,6],\\ t-7 & \text{if } t\in\bigcup_{k=2}^{\infty}\{10k\},\\ 3 & \text{if } t=4,\\ 6 & \text{if } t=10.\end{cases}$$

4.

$$\rho(t)=\begin{cases}0 & \text{if } t\in\{0,1\},\\ \frac{t}{11} & \text{if } t\in 11^{\mathbb{N}}.\end{cases}$$

5.

$$\rho(t)=\begin{cases}t & \text{if } t\in(1,2]\cup(3,4]\cup(7,8],\\ 1 & \text{if } t=1,\\ 2 & \text{if } t=3,\\ 4 & \text{if } t=7,\\ 8 & \text{if } t=9,\\ \frac{t}{9} & \text{if } t\in 9^{\mathbb{N}}\backslash\{9\}.\end{cases}$$

Exercise 1.6. Answer:

1. $\nu(t)=h,\ t\in\mathbb{T}$.

2.

$$v(t) = \begin{cases} 2 & \text{if } t \in 2\mathbb{N}_0, \\ 1 & \text{if } t = 1, \\ \frac{2t}{3} & \text{if } t \in 3^{\mathbb{N}}. \end{cases}$$

3.

$$v(t) = \begin{cases} 0 & \text{if } t \in [0,3] \cup \bigcup_{k=1}^{\infty}(10k, 10k+3] \cup (4,6], \\ 7 & \text{if } t \in \bigcup_{k=2}^{\infty}\{10k\}, \\ 1 & \text{if } t = 4, \\ 4 & \text{if } t = 10. \end{cases}$$

4.

$$v(t) = \begin{cases} 0 & \text{if } t = 0, \quad 1 \quad \text{if } t = 1 \\ \frac{10t}{11} & \text{if } t \in 11^{\mathbb{N}}. \end{cases}$$

5.

$$v(t) = \begin{cases} 0 & \text{if } t \in [1,2] \cup (3,4] \cup (7,8], \\ 1 & \text{if } t \in \{3,9\}, \\ 3 & \text{if } t = 7, \\ \frac{8t}{9} & \text{if } t \in 9^{\mathbb{N}} \backslash \{9\}. \end{cases}$$

Exercise 1.7. Answer:

1. Any point is isolated.
2. Any point is isolated.
3. $t \in \bigcup_{k=0}^{\infty}(10k, 10k+3) \cup (4,6)$ is dense, $t \in \bigcup_{k=0}^{\infty}\{10k+3\}$ is right-scattered and left-dense, $t \in \bigcup_{k=1}^{\infty}\{10k\}$ is right-dense and left-scattered, $t = 0$ is right-dense, $t = 4$ is right-dense and left-scattered, $t = 6$ is right-scattered and left-dense.
4. $t = 0$ is right-scattered, $t \in 11^{\mathbb{N}_0}$ is isolated.
5. $t \in (1,2) \cup (3,4) \cup (7,8)$ is dense, $t = 1$ is right-dense, $t = 2$ is right-scattered and left-dense, $t = 3$ is right-dense and left-scattered, $t = 4$ is left-dense and right-scattered, $t = 7$ is right-dense and left-scattered, $t = 8$ is left-dense and right-scattered, $t \in 9^{\mathbb{N}}$ is isolated.

Exercise 1.8. Answer:

1.

$$\begin{aligned} [-3,9]_{\mathbb{T}} &= \{-3,1,5,9\}, \\ [-3,9)_{\mathbb{T}} &= \{-3,1,5\}, \\ (-3,9]_{\mathbb{T}} &= \{1,5,9\}, \\ (-3,9)_{\mathbb{T}} &= \{1,5\}. \end{aligned}$$

2.

$$\begin{aligned} [-4,9]_{\mathbb{T}} &= \{-4,-2,0,1,3,9\}, \\ [-4,9)_{\mathbb{T}} &= \{-4,-2,0,1,3\}, \\ (-4,9]_{\mathbb{T}} &= \{-2,0,1,3,9\}, \\ (-4,9)_{\mathbb{T}} &= \{-2,0,1,3\}. \end{aligned}$$

3.

$$[0,12]_{\mathbb{T}} = [0,3] \cup [4,6] \cup [10,12],$$
$$[0,12)_{\mathbb{T}} = [0,3] \cup [4,6] \cup [10,12),$$
$$(0,12]_{\mathbb{T}} = (0,3] \cup [4,6] \cup [10,12],$$
$$(0,12)_{\mathbb{T}} = (0,3] \cup [4,6] \cup [10,12).$$

4.

$$[0,121]_{\mathbb{T}} = \{0,1,11,121\},$$
$$[0,121)_{\mathbb{T}} = \{0,1,11\},$$
$$(0,121]_{\mathbb{T}} = \{1,11,121\},$$
$$(0,121)_{\mathbb{T}} = \{1,11\}.$$

5.

$$[2,9]_{\mathbb{T}} = \{2\} \cup [3,4] \cup [7,8] \cup \{9\},$$
$$[2,9)_{\mathbb{T}} = \{2\} \cup [3,4] \cup [7,8],$$
$$(2,9]_{\mathbb{T}} = [3,4] \cup [7,8] \cup \{9\},$$
$$(2,9)_{\mathbb{T}} = [3,4] \cup [7,8].$$

Exercise 1.9. Answer:

1.

$$\mathbb{T}^{\kappa} = \mathbb{T},$$
$$\mathbb{T}_{\kappa} = \mathbb{T}.$$

2.

$$\mathbb{T}^{\kappa} = \mathbb{T},$$
$$\mathbb{T}_{\kappa} = \mathbb{T}.$$

3.

$$\mathbb{T}^{\kappa} = \mathbb{T},$$
$$\mathbb{T}_{\kappa} = \mathbb{T}.$$

4.

$$\mathbb{T}^{\kappa} = \mathbb{T},$$
$$\mathbb{T}_{\kappa} = 11^{\mathbb{N}_0}.$$

5.

$$\mathbb{T}^{\kappa} = \mathbb{T},$$
$$\mathbb{T}_{\kappa} = \mathbb{T}.$$

Exercise 1.10. Answer:

1.

$$f^{\sigma}(t) = 2 - 4h + h^2 + 2(h-2)t + t^2,$$
$$f^{\sigma^2}(t) = 2 - 8h + 4h^2 + 4(h-1)t + t^2, \quad t \in \mathbb{T}.$$

2.

$$f^{\sigma}(t) = \begin{cases} t^2 - 2 & \text{if } t \in (-2\mathbb{N}), \\ -1 & \text{if } t = 0, \\ 2 - 12t + 9t^2 & \text{if } t \in 3^{\mathbb{N}_0}, \end{cases}$$

$$f^{\sigma^2}(t) = \begin{cases} 2 + 4t + t^2 & \text{if } t \in (-2\mathbb{N}), \\ -1 & \text{if } t = 0, \\ 2 - 36t + 81t^2 & \text{if } t \in 3^{\mathbb{N}_0}. \end{cases}$$

3.

$$f^{\sigma}(t) = \begin{cases} 2 - 4t + t^2 & \text{if } t \in \bigcup_{k=0}^{\infty}[10k, 10k+3) \cup [4,6], \\ 23 + 10t + t^2 & \text{if } t \in \bigcup_{k=0}^{\infty}\{10k+3\}, \quad 2 \quad \text{if } t = 3, \quad 62 \quad \text{if } t = 6, \end{cases}$$

$$f^{\sigma^2}(t) = \begin{cases} 2 - 4t + t^2 & \text{if } t \in \bigcup_{k=0}^{\infty}[10k, 10k+3) \cup [4,6], \\ 142 + 24t + t^2 & \text{if } t \in \bigcup_{k=0}^{\infty}\{10k+3\}, \quad 2 \quad \text{if } t = 3, \quad 62 \quad \text{if } t = 6. \end{cases}$$

4.

$$f^{\sigma}(t) = \begin{cases} -1 & \text{if } t = 0, \\ 2 - 44t + 121t^2 & \text{if } t \in 11^{\mathbb{N}_0}, \end{cases}$$

$$f^{\sigma^2}(t) = \begin{cases} 79 & \text{if } t = 0, \\ 2 - 484t + 14641t^2 & \text{if } t \in 11^{\mathbb{N}_0}. \end{cases}$$

5.

$$f^{\sigma}(t) = \begin{cases} 2 - 4t + t^2 & \text{if } t \in [1,2) \cup [3,4) \cup [7,8), \\ 2 - 36t + 81t^2 & \text{if } t \in 9^{\mathbb{N}}, \\ -1 & \text{if } t = 2, \\ 23 & \text{if } t = 4, \\ 47 & \text{if } t = 8, \end{cases}$$

$$f^{\sigma^2}(t) = \begin{cases} 2 - 4t + t^2 & \text{if } t \in [1,2) \cup [3,4) \cup [7,8), \\ 2 - 324t + 6561t^2 & \text{if } t \in 9^{\mathbb{N}}, \\ -1 & \text{if } t = 2, \\ 23 & \text{if } t = 4, \\ 6239 & \text{if } t = 8. \end{cases}$$

Exercise 1.11. Answer:

1.

$$f^{\rho^2}(t) = \sqrt[3]{\frac{1 - 2h + t}{2 - 14h + 7t}},$$

$$f^{\sigma^3\rho^2}(t) = \sqrt[3]{\frac{1 + h + t}{2 + 7h + 7t}},$$

$$f^{\rho^2\sigma^3}(t) = \sqrt[3]{\frac{1 + h + t}{2 + 7h + 7t}}, \quad t \in \mathbb{T}.$$

2.

$$f^{\rho^2}(t) = \begin{cases} \sqrt[3]{\frac{t-3}{7t-26}} & \text{if } t \in (-2\mathbb{N}), \\ \sqrt[3]{\frac{3}{26}} & \text{if } t = 0, \\ \sqrt[3]{\frac{1}{12}} & \text{if } t = 1, \quad \sqrt[3]{\frac{1}{2}} \text{ if } t = 3, \quad \sqrt[3]{\frac{2}{9}} \text{ if } t = 3^2, \\ \sqrt[3]{\frac{9+t}{18+7t}} & \text{if } t \in 3^{\mathbb{N}}, \quad t \geq 3^3, \end{cases}$$

$$f^{\sigma^3\rho^2}(t) = \begin{cases} \sqrt[3]{\frac{3+t}{16+7t}} & \text{if } t \in (-2\mathbb{N}), \\ \sqrt[3]{\frac{2}{9}} & \text{if } t = 0, \\ \sqrt[3]{\frac{2}{15}} & \text{if } t = 1, \quad \sqrt[3]{\frac{2}{13}} \text{ if } t = 3, \quad \sqrt[3]{\frac{28}{191}} \text{ if } t = 3^2, \\ \sqrt[3]{\frac{1+3t}{2+21t}} & \text{if } t \in 3^{\mathbb{N}}, \quad t \geq 3^3, \end{cases}$$

$$f^{\rho^2\sigma^3}(t) = \begin{cases} \sqrt[3]{\frac{3+t}{16+7t}} & \text{if } t \in (-2\mathbb{N}), \\ \sqrt[3]{\frac{2}{9}} & \text{if } t = 0, \\ \sqrt[3]{\frac{2}{15}} & \text{if } t = 1, \quad \sqrt[3]{\frac{2}{13}} \text{ if } t = 3, \quad \sqrt[3]{\frac{28}{191}} \text{ if } t = 3^2, \\ \sqrt[3]{\frac{1+3t}{2+21t}} & \text{if } t \in 3^{\mathbb{N}}. \end{cases}$$

3.

$$f^{\rho^2}(t) = \begin{cases} \sqrt[3]{\frac{1+t}{2+7t}} & \text{if } t \in \bigcup_{k=0}^{\infty}(10k, 10k+3) \cup (4,6), \\ \sqrt[3]{\frac{1+t}{2+7t}} & \text{if } t \in \bigcup_{k=0}^{\infty}\{10k+3\}, \\ \sqrt[3]{\frac{-13+t}{-96+7t}} & \text{if } t \in \bigcup_{k=0}^{\infty}\{10k\}, \\ \sqrt[3]{\frac{4}{23}} & \text{if } t = 4, \\ \sqrt[3]{\frac{7}{44}} & \text{if } t = 6, \end{cases}$$

$$f^{\sigma^3\rho^2}(t) = \begin{cases} \sqrt[3]{\frac{1+t}{2+7t}} & \text{if } t \in \bigcup_{k=0}^{\infty}(10k, 10k+3) \cup (4,6), \\ \sqrt[3]{\frac{22+t}{149+7t}} & \text{if } t \in \bigcup_{k=0}^{\infty}\{10k+3\}, \\ \sqrt[3]{\frac{-13+t}{-96+7t}} & \text{if } t \in \bigcup_{k=0}^{\infty}\{10k\}, \\ \sqrt[3]{\frac{4}{23}} & \text{if } t = 4, \\ \sqrt[3]{\frac{11}{72}} & \text{if } t = 6, \end{cases}$$

$$f^{\rho^2\sigma^3}(t) = \begin{cases} \sqrt[3]{\frac{1+t}{2+7t}} & \text{if } t \in \bigcup_{k=0}^{\infty}(10k, 10k+3) \cup (4,6), \\ \sqrt[3]{\frac{22+t}{149+7t}} & \text{if } t \in \bigcup_{k=0}^{\infty}\{10k+3\}, \\ \sqrt[3]{\frac{-13+t}{-96+7t}} & \text{if } t \in \bigcup_{k=0}^{\infty}\{10k\}, \\ \sqrt[3]{\frac{4}{23}} & \text{if } t = 4, \\ \sqrt[3]{\frac{7}{44}} & \text{if } t = 6. \end{cases}$$

4.

$$f^{\rho^2}(t)=\begin{cases}\sqrt[3]{\frac{1}{2}} & \text{if } t=0,\\ \sqrt[3]{\frac{1}{2}} & \text{if } t=1,\quad \text{if } \sqrt[3]{\frac{1}{2}} \quad \text{if } t=3,\quad \sqrt[3]{\frac{2}{9}} \quad \text{if } t=11^2,\\ \sqrt[3]{\frac{121+t}{242+7t}} & \text{if } t\in 11^{\mathbb{N}},\quad t\geq 11^3,\end{cases}$$

$$f^{\sigma^3\rho^2}(t)=\begin{cases}\sqrt[3]{\frac{2}{9}} & \text{if } t=0,\\ \sqrt[3]{\frac{12}{79}} & \text{if } t=1,\quad \text{if } \sqrt[3]{\frac{122}{849}} \quad \text{if } t=11,\quad \sqrt[3]{\frac{1332}{9319}} \quad \text{if } t=11^2,\\ \sqrt[3]{\frac{1+11t}{2+77t}} & \text{if } t\in 11^{\mathbb{N}},\quad t\geq 11^3,\end{cases}$$

$$f^{\rho^2\sigma^3}(t)=\begin{cases}\sqrt[3]{\frac{122}{849}} & \text{if } t=0,\\ \sqrt[3]{\frac{122}{849}} & \text{if } t=1,11;\quad \sqrt[3]{\frac{1332}{9319}} \quad \text{if } t=11^2,\\ \sqrt[3]{\frac{1+11t}{2+77t}} & \text{if } t\in 11^{\mathbb{N}},\quad t\geq 11^3.\end{cases}$$

5.

$$f^{\rho^2}(t)=\begin{cases}\sqrt[3]{\frac{1+t}{2+7t}} & \text{if } t\in(1,2)\cup(3,4)\cup(7,8),\\ \sqrt[3]{\frac{2}{9}} & \text{if } t=1,\\ \sqrt[3]{\frac{3}{16}} & \text{if } t=2,\\ \sqrt[3]{\frac{3}{16}} & \text{if } t=3,\\ \sqrt[3]{\frac{1}{6}} & \text{if } t=4,\\ \sqrt[3]{\frac{1}{6}} & \text{if } t=7,\\ \sqrt[3]{\frac{9}{58}} & \text{if } t=8,\\ \sqrt[3]{\frac{9}{58}} & \text{if } t=9,\quad \sqrt[3]{\frac{9}{58}} \quad \text{if } t=9^2,\\ \sqrt[3]{\frac{81+t}{162+7t}} & \text{if } t\in 9^{\mathbb{N}},\ t\geq 9^3.\end{cases}$$

$$f^{\sigma^3\rho^2}(t)=\begin{cases}\sqrt[3]{\frac{1+t}{2+7t}} & \text{if } t\in(1,2)\cup(3,4)\cup(7,8),\\ \sqrt[3]{\frac{2}{9}} & \text{if } t=1,\\ \sqrt[3]{\frac{3}{16}} & \text{if } t=2,\\ \sqrt[3]{\frac{3}{16}} & \text{if } t=3,\\ \sqrt[3]{\frac{1}{6}} & \text{if } t=4,\\ \sqrt[3]{\frac{1}{6}} & \text{if } t=7,\\ \sqrt[3]{\frac{2}{13}} & \text{if } t=8,\\ \sqrt[3]{\frac{82}{569}} & \text{if } t=9,\quad \sqrt[3]{\frac{146}{1021}} \quad \text{if } t=9^2,\\ \sqrt[3]{\frac{1+9t}{2+63t}} & \text{if } t\in 9^{\mathbb{N}},\ t\geq 9^3,\end{cases}$$

$$f^{\rho^2\sigma^3}(t)=\begin{cases}\sqrt[3]{\frac{1+t}{2+7t}} & \text{if } t\in(1,2)\cup(3,4)\cup(7,8),\\ \sqrt[3]{\frac{2}{9}} & \text{if } t=1,\\ \sqrt[3]{\frac{3}{16}} & \text{if } t=2,\\ \sqrt[3]{\frac{4}{23}} & \text{if } t=3,\\ \sqrt[3]{\frac{8}{51}} & \text{if } t=4,\\ \sqrt[3]{\frac{8}{51}} & \text{if } t=7,\\ \sqrt[3]{\frac{730}{1021}} & \text{if } t=8,\\ \sqrt[3]{\frac{730}{1021}} & \text{if } t=9,9^2\\ \sqrt[3]{\frac{1+9t}{2+63t}} & \text{if } t\in 9^{\mathbb{N}},\ t\ge 9^3.\end{cases}$$

Exercise 1.12. Answer:

1. 1.
2. $-2\sigma(t)-4\rho(t)$.
3. -1.
4. -1.
5. $\frac{\sigma(t)}{4(\sigma(t)+2)}$.

Chapter 2

Exercise 2.6. *Solution.* Note that all points of $\mathbb{T}$ are right-scattered and $\sigma(t)=t+k$, $\mu(t)=k$, $t\in\mathbb{T}$. Then

$$\begin{aligned}\frac{1}{2}f_2^{\Delta}(t)-\frac{k}{2}&=\frac{1}{2}\frac{(\sigma(t))^2-t^2}{\sigma(t)-t}-\frac{k}{2}\\&=\frac{1}{2}\frac{(\sigma(t)-t)(\sigma(t)+t)}{\sigma(t)-t}-\frac{k}{2}\\&=\frac{1}{2}(\sigma(t)+t)-\frac{k}{2}\\&=\frac{1}{2}(2t+k)-\frac{k}{2}\\&=t+\frac{k}{2}-\frac{k}{2}\\&=t,\\ \frac{1}{3}f_3^{\Delta}(t)-\frac{k}{2}f_2^{\Delta}(t)+\frac{1}{6}k^2&=\frac{1}{3}\frac{(\sigma(t))^3-t^3}{\sigma(t)-t}-\frac{k}{2}\frac{(\sigma(t))^2-t^2}{\sigma(t)-t}+\frac{1}{6}k^2\\&=\frac{1}{3}\frac{(\sigma(t)-t)((\sigma(t))^2+t\sigma(t)+t^2)}{\sigma(t)-t}\\&\quad-\frac{k}{2}\frac{(\sigma(t)-t)(\sigma(t)+t)}{\sigma(t)-t}+\frac{1}{6}k^2\end{aligned}$$

$$
\begin{aligned}
&= \frac{1}{3}\left((\sigma(t))^2 + t\sigma(t) + t^2\right) - \frac{k}{2}(\sigma(t) + t) + \frac{1}{6}k^2\\
&= \frac{1}{3}\left((t+k)^2 + t(t+k) + t^2\right) - \frac{k}{2}(t + k + t) + \frac{1}{6}k^2\\
&= \frac{1}{3}(t^2 + 2kt + k^2 + t^2 + kt + t^2) - \frac{k}{2}(2t + k) + \frac{1}{6}k^2\\
&= \frac{1}{3}(3t^2 + 3kt + k^2) - kt - \frac{k^2}{2} + \frac{1}{6}k^2\\
&= t^2 + kt + \frac{k^2}{3} - kt - \frac{k^2}{3}\\
&= t^2,
\end{aligned}
$$

$$
\begin{aligned}
&\frac{1}{4}f_4^{\Delta}(t) - \frac{1}{2}kf_3^{\Delta}(t) + \frac{1}{4}k^2f_2^{\Delta}(t)\\
&\quad= \frac{1}{4}\frac{(\sigma(t))^4 - t^4}{\sigma(t) - t} - \frac{1}{2}k\frac{(\sigma(t))^3 - t^3}{\sigma(t) - t} + \frac{1}{4}k^2\frac{(\sigma(t))^2 - t^2}{\sigma(t) - t}\\
&\quad= \frac{1}{4}\frac{(\sigma(t) - t)((\sigma(t))^3 + t(\sigma(t))^2 + t^2\sigma(t) + t^3)}{\sigma(t) - t}\\
&\qquad - \frac{1}{2}k\frac{(\sigma(t) - t)((\sigma(t))^2 + t\sigma(t) + t^2)}{\sigma(t) - t}\\
&\qquad + \frac{1}{4}k^2\frac{(\sigma(t) - t)(\sigma(t) + t)}{\sigma(t) - t}\\
&\quad= \frac{1}{4}\left((\sigma(t))^3 + t(\sigma(t))^2 + t^2\sigma(t) + t^3\right) - \frac{1}{2}k\left((\sigma(t))^2 + t\sigma(t) + t^2\right)\\
&\qquad + \frac{1}{4}k^2(\sigma(t) + t)\\
&\quad= \frac{1}{4}\left((t+k)^3 + t(t+k)^2 + t^2(t+k) + t^3\right) - \frac{1}{2}k\left((t+k)^2 + t(t+k) + t^2\right)\\
&\qquad + \frac{1}{4}k^2(t + k + t)\\
&\quad= \frac{1}{4}(t^3 + 3t^2k + 3tk^2 + k^3 + t(t^2 + 2tk + k^2) + t^3 + kt^2 + t^3)\\
&\qquad - \frac{1}{2}k(t^2 + 2kt + k^2 + t^2 + kt + t^2) + \frac{1}{4}k^2(2t + k)\\
&\quad= \frac{1}{4}(3t^3 + 4t^2k + 3tk^2 + k^3 + t^3 + 2t^2k + tk^2)\\
&\qquad - \frac{1}{2}k(3t^2 + 3kt + k^2) + \frac{1}{2}tk^2 + \frac{1}{4}k^2\\
&\quad= \frac{1}{4}(4t^3 + 6t^2k + 4tk^2 + k^3) - \frac{3}{2}kt^2 - \frac{3}{2}k^2t - \frac{1}{2}k^3 + \frac{1}{2}k^2t + \frac{1}{4}k^3\\
&\quad= t^3 + \frac{3}{2}kt^2 + k^2t + \frac{1}{4}k^3 - \frac{3}{2}kt^2 - \frac{3}{2}k^2t - \frac{1}{2}k^3 + \frac{1}{2}k^2t + \frac{1}{4}k^3\\
&\quad= t^3,
\end{aligned}
$$

$$\begin{aligned}
&\frac{1}{5}f_5^\Delta - \frac{k}{2}f_4^\Delta + \frac{1}{3}k^2 f_3^\Delta - \frac{1}{30}k^4 \\
&= \frac{1}{5}\frac{(\sigma(t))^5 - t^5}{\sigma(t)-t} - \frac{k}{2}\frac{(\sigma(t))^4 - t^4}{\sigma(t)-t} + \frac{1}{3}k^2\frac{(\sigma(t))^3 - t^3}{\sigma(t)-t} - \frac{1}{30}k^4 \\
&= \frac{1}{5}\frac{(\sigma(t)-t)((\sigma(t))^4 + t(\sigma(t))^3 + t^2(\sigma(t))^2 + t^3\sigma(t) + t^4)}{\sigma(t)-t} \\
&\quad - \frac{k}{2}\frac{(\sigma(t)-t)((\sigma(t))^3 + t(\sigma(t))^2 + t^2\sigma(t) + t^3)}{\sigma(t)-t} \\
&\quad + \frac{1}{3}k^2\frac{(\sigma(t)-t)((\sigma(t))^2 + t\sigma(t) + t^2)}{\sigma(t)-t} - \frac{1}{30}k^4 \\
&= \frac{1}{5}((t+k)^4 + t(t+k)^3 + t^2(t+k)^2 + t^3(t+k)_+ t^4) \\
&\quad - \frac{k}{2}((t+k)^3 + t(t+k)^2 + t^2(t+k) + t^3) \\
&\quad + \frac{1}{3}k^2((t+k)^2 + t(t+k) + t^2) - \frac{1}{30}k^4 \\
&= \frac{1}{5}(t^4 + 4t^2k^2 + k^4 + 4t^3k + 2t^2k^2 + 4tk^3 + t(t^3 + 3t^2k + 3tk^2 + k^3) \\
&\quad + t^2(t^2 + 2tk + k^2) + t^4 + kt^3 + t^4) \\
&\quad - \frac{k}{2}(t^3 + 3t^2k + 3tk^2 + k^3 + t(t^2 + 2tk + k^2) + t^3 + kt^2 + t^3) \\
&\quad + \frac{1}{3}k^2(t^2 + 2tk + k^2 + t^2 + kt + t^2) - \frac{1}{30}k^4 \\
&= \frac{1}{5}(t^4 + 6t^2k^2 + k^4 + 4t^3k + 4tk^3 + t^4 + 3t^3k + 3t^2k^2 + tk^3 \\
&\quad + t^4 + 2t^3k + t^2k^2 + 2t^4 + kt^3) - \frac{k}{2}(3t^3 + 4t^2k + 3tk^2 + k^3 + t^3 + 2t^2k + tk^2) \\
&\quad + \frac{1}{3}k^2(3t^2 + 3kt + k^2) - \frac{1}{30}k^4 \\
&= \frac{1}{5}(5t^4 + 10t^3k + 10t^2k^2 + 5tk^3 + k^4) - \frac{k}{2}(4t^3 + 6t^2k + 4tk^2 + k^3) \\
&\quad + k^2t^2 + k^3t + \frac{1}{3}k^4 - \frac{1}{30}k^4 \\
&= t^4 + 2t^3k + 2t^2k^2 + tk^3 + \frac{1}{5}k^4 - 2kt^3 - 3t^2k^2 - 2tk^3 - \frac{k^4}{2} \\
&\quad + k^2t^2 + k^3t + \frac{1}{3}k^4 - \frac{1}{30}k^4 \\
&= t^4, \quad t \in \mathbb{T}.
\end{aligned}$$

We leave the last equality to the reader as an exercise.

Exercise 2.7. *Solution.* We have

$$(2\sin t + t^2 - 3t^3)^\Delta = 2(\sin t)^\Delta + (t^2)^\Delta - 3\frac{(\sigma(t))^3 - t^3}{\mu(t)}$$

$$= 2\frac{\sin\sigma(t) - \sin t}{\mu(t)} + \frac{(\sigma(t))^2 - t^2}{\mu(t)}$$
$$- 3\frac{(\sigma(t) - t)((\sigma(t))^2 + t\sigma(t) + t^2)}{\mu(t)}$$
$$= 2\frac{\sin\frac{\sigma(t)-t}{2}\cos\frac{\sigma(t)+t}{2}}{\mu(t)} + \frac{(\sigma(t) - t)(\sigma(t) + t)}{\mu(t)}$$
$$- 3(\sigma(t))^2 - 3t\sigma(t) - 3t^2$$
$$= 2\frac{\sin\frac{\mu(t)}{2}\cos\frac{\sigma(t)+t}{2}}{\mu(t)} + \sigma(t) + t - 3(\sigma(t))^2 - 3t\sigma(t) - 3t^2$$
$$= 2\frac{\sin\frac{\mu(t)}{2}\cos\frac{\sigma(t)+t}{2}}{\mu(t)} - 3(\sigma(t))^2 + \sigma(t)(1 - 3t) + t - 3t^2, \quad t \in \mathbb{T}^\kappa.$$

Exercise 2.8. *Solution.* Here $\sigma(t) = 2t$, $\mu(t) = t$, $t \in \mathbb{T}$. Then

$$f^\Delta(t) = \frac{1}{(2t^2 + 3t + 1)(2(\sigma(t))^2 + 3\sigma(t) + 1)}((t^3 + t^2 - 2t)^\Delta(2t^2 + 3t + 1)$$
$$- (t^3 + t^2 - 2t)(2t^2 + 3t + 1)^\Delta)$$
$$= \frac{1}{(2t^2 + 3t + 1)(8t^2 + 6t + 1)}\left(\left(\frac{(\sigma(t))^3 - t^3}{\sigma(t) - t}\right.\right.$$
$$\left. + \frac{(\sigma(t))^2 - t^2}{\sigma(t) - t} - 2\right)(2t^2 + 3t + 1)$$
$$\left.- (t^3 + t^2 - 2t)\left(2\frac{(\sigma(t))^2 - t^2}{\sigma(t) - t} + 3\right)\right)$$
$$= \frac{1}{(2t^2 + 3t + 1)(8t^2 + 6t + 1)}(((\sigma(t))^2 + t\sigma(t) + t^2$$
$$+ \sigma(t) + t - 2)(2t^2 + 3t + 1)$$
$$- (t^3 + t^2 - 2t)(2\sigma(t) + 2t + 1))$$
$$= \frac{1}{(2t^2 + 3t + 1)(8t^2 + 6t + 1)}((4t^2 + 2t^2 + t^2 + 2t + t - 2)(2t^2 + 3t + 1)$$
$$- (t^3 + t^2 - 2t)(4t + 2t + 3))$$
$$= \frac{1}{(2t^2 + 3t + 1)(8t^2 + 6t + 1)}((7t^2 + 3t - 2)(2t^2 + 3t + 1)$$
$$- (t^3 + t^2 - 2t)(6t + 3))$$
$$= \frac{1}{(2t^2 + 3t + 1)(8t^2 + 6t + 1)}(14t^4 + 21t^3 + 7t^2 + 6t^3 + 9t^2 + 3t - 4t^2$$
$$- 6t - 2 - 6t^4 - 3t^3 - 6t^3 - 3t^2 + 12t^2 + 6t)$$
$$= \frac{8t^4 + 18t^3 + 21t^2 + 3t - 2}{(2t^2 + 3t + 1)(8t^2 + 6t + 1)}, \quad t \in \mathbb{T}.$$

Exercise 2.9. *Solution.* Here $\sigma(t) = 3t$, $\mu(t) = 2t$, $t \in \mathbb{T}$. Then

$$\begin{aligned} f^{\Delta}(t) &= \frac{1}{4}\frac{(\sigma(t))^4 - t^4}{\sigma(t) - t} + \frac{1}{3}\frac{(\sigma(t))^3 - t^3}{\sigma(t) - t} - \frac{(\sigma(t))^2 - t^2}{\sigma(t) - t} \\ &= \frac{1}{4}\left((\sigma(t))^3 + t(\sigma(t))^2 + t^2\sigma(t) + t^3\right) \\ &\quad + \frac{1}{3}\left((\sigma(t))^2 + t\sigma(t) + t^2\right) - (\sigma(t) + t) \\ &= \frac{1}{4}(27t^3 + 9t^3 + 3t^3 + t^3) + \frac{1}{3}(9t^2 + 3t^2 + t^2) - 4t \\ &= 10t^3 + \frac{13}{3}t^2 - 4t, \quad t \in \mathbb{T}. \end{aligned}$$

Exercise 2.10. Answer: $t^2 - 4t - \frac{37}{6}$.

Exercise 2.11. *Solution.* Let $t \in \mathbb{T}$, $t = \sqrt[4]{n}$. Then $n = t^4$, and

$$\begin{aligned} \sigma(t) &= \inf\{\sqrt[4]{l} : \sqrt[4]{l} > \sqrt[4]{n} : l \in \mathbb{N}\} \\ &= \sqrt[4]{n+1} \\ &= \sqrt[4]{t^4 + 1}. \end{aligned}$$

Hence

$$\begin{aligned} f^{\Delta}(t) &= \frac{1}{3}\frac{(\sigma(t))^3 - t^3}{\sigma(t) - t} - 2\frac{(\sigma(t))^2 - t^2}{\sigma(t) - t} - 3 \\ &= \frac{1}{3}\left((\sigma(t))^2 + t\sigma(t) + t^2\right) - 2(\sigma(t) + t) - 3 \\ &= \frac{1}{3}\left(\sqrt{t^4 + 1} + t\sqrt[4]{t^4 + 1} + t^2\right) - 2\left(\sqrt[4]{t^4 + 1} + t\right) - 3 \\ &= \left(\frac{1}{3}t - 2\right)\sqrt[4]{t^4 + 1} + \frac{1}{3}\sqrt{t^4 + 1} + \frac{1}{3}t^2 - 2t - 3. \end{aligned}$$

Exercise 2.12. Answer:

$$f^{\Delta}(t) = \begin{cases} -\frac{17}{(3+7t)^2} & \text{if } t \in [1,2) \cup [3,4) \cup [7,8), \\ -\frac{17}{3(3+7t)(1+21t)} & \text{if } t \in 9^{\mathbb{N}}, \\ -\frac{1}{24} & \text{if } t = 2, \\ -\frac{17}{1612} & \text{if } t = 4, \\ -\frac{17}{3894} & \text{if } t = 8. \end{cases}$$

Exercise 2.13. Answer: 2.

Exercise 2.14. Answer: $52t + 12$.

Exercise 2.15. Answer: $\frac{6}{((1+t)(1+2t)(1+4t)}$.

Exercise 2.16. Answer: $-\frac{4}{9(t+3)(3t^2+4t+1)}$.

Exercise 2.17. Answer: $\frac{2}{(t+1)(t+2)(t+3)}$.
Exercise 2.18. Answer:

$$f^{\Delta^2}(t) = \begin{cases} \frac{4}{(t-4)^3} & \text{if } t \in \left(\bigcup_{k=0}^{\infty}[10k, 10k+3)\right) \cup [4,6) \\ \frac{4}{(t-4)(t+3)(t+10)} & \text{if } t \in \bigcup_{k=0}^{\infty}\{10k+3\}, \quad \frac{1}{6} \quad \text{if } t = 0.. \end{cases}$$

Exercise 2.19. Answer:

$$f^{\Delta^2}(t) = \begin{cases} 6t+10 & \text{if } t \in (-2\mathbb{N}), \\ 9 & \text{if } t = 0, \\ 52t-4 & \text{if } t \in 3^{\mathbb{N}_0}. \end{cases}$$

Exercise 2.22. Answer: $\xi_2 \in \{-2,-1,0,1\}$, $\xi_1 \in \{-3,2,3\}$.
Exercise 2.23. *Hint.* Use the function $f(t) = t^3 + t^2$.
Exercise 2.27. Answer: Increasing on $\mathbb{T}$.
Exercise 2.28. Answer: Increasing on $(-\infty, 2-\sqrt{\frac{61}{6}}]_{\mathbb{T}} \cup [2+\sqrt{\frac{61}{6}}, \infty)_{\mathbb{T}}$; decreasing on $[2-\sqrt{\frac{61}{6}}, 2+\sqrt{\frac{61}{6}}]_{\mathbb{T}}$.
Exercise 2.29. Answer: Decreasing on $\mathbb{T}$.
Exercise 2.33. Answer:
1. f is convex for $t \geq 1$ and concave for $t \leq 1$.
2. f is convex for $t \in [-5,-4]_{\mathbb{T}} \cup [-3,\infty)_{\mathbb{T}}$ and concave for $t \in (-\infty,-5]_{\mathbb{T}} \cup [-4,-3]_{\mathbb{T}}$.

Exercise 2.34. Answer: Convex on $\mathbb{T}$.
Exercise 2.35. Answer: Convex on $\mathbb{T}$.
Exercise 2.36. Answer: Convex on $\mathbb{T}$.
Exercise 2.37. Answer: Concave on $\mathbb{T}$.
Exercise 2.38. Answer: Convex on $\mathbb{T}$.
Exercise 2.39. Answer: Concave for $t \in [0,4)_{\mathbb{T}}$, convex for $t \in (4,6]_{\mathbb{T}}$.
Exercise 2.40. Answer: Concave for $t \in (-\infty,-2]_{\mathbb{T}}$, convex for $t \in [-2,0]_{\mathbb{T}} \cup 3^{\mathbb{N}_0}$.
Exercise 2.42. Answer: Yes.
Exercise 2.43. Answer: Yes.
Exercise 2.44. Answer: Yes.
Exercise 2.45. Answer: Yes.
Exercise 2.46. Answer: Yes.
Exercise 2.47. Answer: No.
Exercise 2.48. Answer: Yes.
Exercise 2.49. Answer: No.
Exercise 2.50. Answer: No such α.
Exercise 2.51. Answer: $\alpha = 5$.
Exercise 2.52. Answer: Any $c \in [1,2]$.
Exercise 2.53. Answer: $-2\sin(t+1)\sin(t+1)^2$.

Exercise 2.55. Answer:

$$(f \circ g)^{\Delta}(t) = \begin{cases} -\frac{2(2t+3)}{(t^2+6t+10))(t^2+1)} & \text{if } t \text{ is right-scattered,} \\ -\frac{4t}{(1+t^2)^2} & \text{if } t \text{ is right-dense.} \end{cases}$$

Exercise 2.56. Answer:

$$(f \circ g)^{\Delta}(t) = \begin{cases} -\frac{t-1}{t^2}\left(\frac{2t^2-2t+1}{(8t-1)(t-1)} - \frac{1+t^2}{1+7t}\right) & \text{if } t \in \{-\frac{1}{n} : n \in \mathbb{N}\}, \\ \frac{7t^2+9t-6}{(1+7t)(8+7t)} & \text{if } t \in \mathbb{N}_0. \end{cases}$$

Exercise 2.57. Answer:

1. 3.
2. 5.
3. 1.
4. 0.
5. $\frac{1}{5}$.
6. −1.

Chapter 3

Exercise 3.1. Answer: No.
Exercise 3.2. Answer: Yes.
Exercise 3.3. Answer: No.
Exercise 3.4. Answer: Yes.
Exercise 3.5. Answer: Yes.
Exercise 3.6. Answer: No.
Exercise 3.7. Answer: The function f is delta predifferentiable with $D = \mathbb{T}\setminus\bigcup_{k=0}^{\infty}\{10k+3\}$.
Exercise 3.8. Answer: The function f is delta predifferentiable with $D = \mathbb{T}\setminus\{3,7\}$.
Exercise 3.9. Answer: The function f is delta predifferentiable with $D = \mathbb{T}\setminus\{-3\}$.
Exercise 3.19. *Solution.* Let $\phi(x) = |x|^{\alpha}$, $x \in \mathbb{R}$. Then applying Example 3.55, we get the desired result.
Exercise 3.20. *Solution.* Take arbitrary $\varepsilon > 0$. Since f is integrable on $[a,b]_{\mathbb{T}}$, there is a partition $P \in \mathscr{P}(a,b)_{\mathbb{T}}$ such that

$$U_{\mathbb{T}}(f,P) - L_{\mathbb{T}}(f,P) < \varepsilon. \tag{1}$$

Let $P_1 = P \cup \{c,d\}$. Then $P_1 \in \mathscr{P}(a,b)_{\mathbb{T}}$, and $P \subset P_1$. By Example 3.36 it follows that

$$L_{\mathbb{T}}(f,P) \le L_{\mathbb{T}}(f,P_1) \le U_{\mathbb{T}}(f,P_1) \le U_{\mathbb{T}}(f,P).$$

Hence from (1) we find

$$U_{\mathbb{T}}(f,P_1) - L_{\mathbb{T}}(f,P_1) \le U_{\mathbb{T}}(f,P) - L_{\mathbb{T}}(f,P) < \varepsilon.$$

Consider $P_2 \in \mathscr{P}(c,d)_{\mathbb{T}}$ consisting of all points of P_1 belonging to $[c,d]_{\mathbb{T}}$. Then $P_2 \subset P_1$, and

$$U_{\mathbb{T}}(f,P_2) \le U_{\mathbb{T}}(f,P_1),$$
$$L_{\mathbb{T}}(f,P_1) \le L_{\mathbb{T}}(f,P_2),$$

whereupon

$$U_{\mathbb{T}}(f,P_2) - L_{\mathbb{T}}(f,P_2) \le U_{\mathbb{T}}(f,P_1) - L_{\mathbb{T}}(f,P_1) < \varepsilon.$$

Hence by Example 3.40 we conclude that f is integrable on $[c,d]_{\mathbb{T}}$.

Exercise 3.21. *Solution.* If $c = 0$, then the statement is obvious. Suppose that $c \neq 0$. Take arbitrary $\varepsilon > 0$. Since f is Riemann integrable on $[a,b]_{\mathbb{T}}$, there is $\delta > 0$ such that for any $P \in \mathscr{P}_\delta(a,b)_{\mathbb{T}}$ given by

$$a = t_0 < t_1 < \cdots < t_n = b, \quad \xi_j \in [t_{j-1}, t_j)_{\mathbb{T}}, \quad j \in \{1,\ldots,n\},$$

we have

$$\left| S_{\mathbb{T}}(f,P) - \int_a^b f(t)\Delta t \right| < \frac{\varepsilon}{|c|}. \tag{2}$$

We have

$$S_{\mathbb{T}}(f,P) = \sum_{j=1}^n f(\xi_j)(t_j - t_{j-1}).$$

Then

$$S_{\mathbb{T}}(cf,P) = \sum_{j=1}^n (cf)(\xi_j)(t_j - t_{j-1}) = c\sum_{j=1}^n f(\xi_j)(t_j - t_{j-1}) = cS_{\mathbb{T}}(f,P).$$

Hence

$$\left| S_{\mathbb{T}}(cf,P) - c\int_a^b f(t)\Delta t \right| = \left| cS_{\mathbb{T}}(f,P) - c\int_a^b f(t)\Delta t \right|$$

$$
\begin{aligned}
&= |c| \left| S_{\mathbb{T}}(f,P) - \int_a^b f(t)\Delta t \right| \\
&< |c| \frac{\varepsilon}{|c|} \\
&= \varepsilon.
\end{aligned}
$$

Exercise 3.22. *Solution.* Take arbitrary $\varepsilon > 0$. Since f and g are Riemann integrable on $[a,b]_{\mathbb{T}}$, there is $\delta > 0$ such that for any $P \in \mathscr{P}_\delta(a,b)_{\mathbb{T}}$ given by

$$
a = t_0 < t_1 < \cdots < t_n = b, \quad \xi_j \in [t_{j-1}, t_j)_{\mathbb{T}}, \quad j \in \{1, \ldots, n\},
$$

we have

$$
\begin{aligned}
\left| S_{\mathbb{T}}(f,P) - \int_a^b f(t)\Delta t \right| &< \frac{\varepsilon}{2}, \\
\left| S_{\mathbb{T}}(g,P) - \int_a^b g(t)\Delta t \right| &< \frac{\varepsilon}{2}.
\end{aligned}
\tag{3}
$$

We have

$$
\begin{aligned}
S_{\mathbb{T}}(f,P) &= \sum_{j=1}^n f(\xi_j)(t_j - t_{j-1}), \\
S_{\mathbb{T}}(g,P) &= \sum_{j=1}^n g(\xi_j)(t_j - t_{j-1}).
\end{aligned}
$$

Then

$$
\begin{aligned}
S_{\mathbb{T}}(f+g,P) &= \sum_{j=1}^n (f+g)(\xi_j)(t_j - t_{j-1}) \\
&= \sum_{j=1}^n f(\xi_j)(t_j - t_{j-1}) + \sum_{j=1}^n g(\xi_j)(t_j - t_{j-1}) \\
&= S_{\mathbb{T}}(f,P) + S_{\mathbb{T}}(g,P).
\end{aligned}
$$

Hence

$$
\begin{aligned}
\left| S_{\mathbb{T}}(f+g,P) - \left(\int_a^b f(t)\Delta t + \int_a^b g(t)\Delta t \right) \right| &= \left| S_{\mathbb{T}}(f,P) + S_T(g,P) - \int_a^b f(t)\Delta t - \int_a^b g(t)\Delta t \right| \\
&= \left| S_{\mathbb{T}}(f,P) - \int_a^b f(t)\Delta t \right| + \left| S_{\mathbb{T}}(f,P) - \int_a^b g(t)\Delta t \right|
\end{aligned}
$$

$$< \frac{\varepsilon}{2} + \frac{\varepsilon}{2}$$
$$= \varepsilon.$$

Exercise 3.23. *Solution.* Firstly, observe that

$$fg = \frac{(f+g)^2 - (f-g)^2}{4}.$$

By Exercise 3.21 it follows that the function $-g : [a,b]_{\mathbb{T}} \to \mathbb{R}$ is integrable on $[a,b]_{\mathbb{T}}$. Now we conclude that the functions

$$f+g, \quad f-g : [a,b]_{\mathbb{T}} \to \mathbb{R}$$

are integrable on $[a,b]_{\mathbb{T}}$. By Exercise 3.19 we obtain that the function

$$(f+g)^2 + (f-g)^2 : [a,b]_{\mathbb{T}} \to \mathbb{R}$$

is integrable on $[a,b]_{\mathbb{T}}$. Applying Exercise 3.21, we get that the function

$$-(f-g)^2 : [a,b]_{\mathbb{T}} \to \mathbb{R}$$

is integrable on $[a,b]_{\mathbb{T}}$. By Exercise 3.22 we conclude that the function

$$(f+g)^2 - (f-g)^2 : [a,b]_{\mathbb{T}} \to \mathbb{R}$$

is integrable on $[a,b]_{\mathbb{T}}$. By Exercise 3.21 we find that the function

$$\frac{(f+g)^2 - (f-g)^2}{4} : [a,b]_{\mathbb{T}} \to \mathbb{R}$$

is integrable on $[a,b]_{\mathbb{T}}$, i. e., the function $fg : [a,b]_{\mathbb{T}} \to \mathbb{R}$ is integrable on $[a,b]_{\mathbb{T}}$.

Exercise 3.24. *Solution.* Since the function f is bounded on $[a,c]_{\mathbb{T}}$ and $[c,b]_{\mathbb{T}}$, we have that it is bounded on $[a,b]_{\mathbb{T}}$. Take arbitrary $\varepsilon > 0$. Because f is integrable on $[a,c]_{\mathbb{T}}$ and $[c,b]_{\mathbb{T}}$, there are partitions $P_1 \in \mathscr{P}(a,c)_{\mathbb{T}}$ and $P_2 \in \mathscr{P}(c,b)_{\mathbb{T}}$ such that

$$U_{\mathbb{T}}(f,P_1) - L_{\mathbb{T}}(f,P_1) < \frac{\varepsilon}{2},$$
$$U_{\mathbb{T}}(f,P_2) - L_{\mathbb{T}}(f,P_2) < \frac{\varepsilon}{2}.$$

Note that $P = \mathbb{P}_1 \cup P_2 \in \mathscr{P}(a,b)_{\mathbb{T}}$ and

$$U_{\mathbb{T}}(f,P) = U_{\mathbb{T}}(f,P_1) + U_{\mathbb{T}}(f,P_2),$$
$$L_{\mathbb{T}}(f,P) = L_{\mathbb{T}}(f,P_1) + L_{\mathbb{T}}(f,P_2).$$

Then

$$\begin{aligned}U_{\mathbb{T}}(f,P) - L_{\mathbb{T}}(f,P) &= U_{\mathbb{T}}(f,P_1) + U_{\mathbb{T}}(f,P_2)\\ &\quad - L_{\mathbb{T}}(f,P_1) + L_{\mathbb{T}}(f,P_2)\\ &= (U_{\mathbb{T}}(f,P_1) - L_{\mathbb{T}}(f,P_1)) + (U_{\mathbb{T}}(f,P_2) - L_{\mathbb{T}}(f,P_2))\\ &< \frac{\varepsilon}{2} + \frac{\varepsilon}{2}\\ &= \varepsilon.\end{aligned}$$

Applying Example 3.40, we conclude that f is integrable on $[a,b]_{\mathbb{T}}$. Moreover,

$$\begin{aligned}\int_a^b f(t)\Delta t &\le U_{\mathbb{T}}(f,P)\\ &= U_{\mathbb{T}}(f,P_1) + U_{\mathbb{T}}(f,P_2)\\ &< L_{\mathbb{T}}(f,P_1) + L_{\mathbb{T}}(f,P_2) + \varepsilon\\ &\le \int_a^c f(t)\Delta t + \int_c^b f(t)\Delta t + \varepsilon,\end{aligned}$$

and

$$\begin{aligned}\int_a^b f(t)\Delta t &\ge L_{\mathbb{T}}(f,P)\\ &= L_{\mathbb{T}}(f,P_1) + L_{\mathbb{T}}(f,P_2)\\ &> U_{\mathbb{T}}(f,P_1) + U_{\mathbb{T}}(f,P_2) - \varepsilon\\ &\ge \int_a^c f(t)\Delta t + \int_c^b f(t)\Delta t - \varepsilon,\end{aligned}$$

whereupon

$$\left|\int_a^b f(t)\Delta t - \left(\int_a^c f(t)\Delta t + \int_c^b f(t)\Delta t\right)\right| < \varepsilon.$$

Because $\varepsilon > 0$ was arbitrarily chosen, we get (3.19).

Exercise 3.25. *Solution.* Set

$$h(t) = g(t) - f(t), \quad t \in [a,b]_{\mathbb{T}}.$$

By Exercise 3.21 we have that $-g : [a,b]_{\mathbb{T}} \to \mathbb{R}$ is integrable on $[a,b]_{\mathbb{T}}$. By Exercise 3.22 we get that $g - f : [a,b]_{\mathbb{T}} \to \mathbb{R}$ is integrable on $[a,b]_{\mathbb{T}}$. Therefore $h : [a,b]_{\mathbb{T}} \to \mathbb{R}$ is integrable on $[a,b]_{\mathbb{T}}$. Next, we have

$$h(t) \ge 0, \quad t \in [a,b]_T,$$

and

$$U_{\mathbb{T}}(f,P) \geq 0,$$
$$L_{\mathbb{T}}(f,P) \geq 0$$

for all $P \in \mathscr{P}(a,b)_{\mathbb{T}}$. Therefore

$$\int_a^b h(t)\Delta t \geq 0.$$

By Exercise 3.22 we get

$$\int_a^b h(t)\Delta t = \int_a^b f(t)\Delta t + \int_a^b (-g(t))\Delta t.$$

Applying Exercise 3.21, we obtain

$$\int_a^b (-g(t))\Delta t = -\int_a^b g(t)\Delta t.$$

Therefore

$$\begin{aligned} 0 &\leq \int_a^b h(t)\Delta t \\ &= \int_a^b f(t)\Delta t - \int_a^b g(t)\Delta t, \end{aligned}$$

whereupon we get the desired result.

Exercise 3.26. *Solution.* Since $|\cdot| : [a,b]_{\mathbb{T}} \to \mathbb{R}$ is a continuous function, using Example 3.55, we conclude that $|f| : [a,b]_{\mathbb{T}} \to \mathbb{R}$ is integrable on $[a,b]_{\mathbb{T}}$. Next, we have the inequalities

$$\begin{aligned} -|f| &\leq f \\ &\leq |f|. \end{aligned}$$

Now applying Exercise 3.25, we get the inequalities

$$-\int_a^b |f(t)|\Delta t \leq \int_a^b f(t)\Delta t$$

$$\leq \int_a^b |f(t)|\Delta t,$$

whereupon we get the desired result.

Exercise 3.35. Answer:

1. $\frac{1}{2}$.
2. $\frac{1}{10}$.
3. $\frac{1}{4}$.

Exercise 3.37. Answer:

1. Divergent.
2. Divergent.
3. Divergent.
4. Convergent.
5. Convergent.
6. Divergent.

Exercise 3.38. Answer: Convergent.

Hint. Use

$$f(t) = \frac{1}{t+1}\operatorname{arctanh}\frac{t}{1-2t^2}, \quad g(t) = \frac{1}{t}\operatorname{arctanh}\frac{t}{1-2t^2},$$
$$(\operatorname{arctanh} t)^\Delta = \frac{1}{t}\operatorname{arctanh}\frac{t}{1-2t^2}.$$

Exercise 3.40. *Solution.* We have

$$\begin{aligned}\int_{t_0}^{\infty} g(t)\Delta t &= \sum_{k=0}^{\infty} g(t_k)\mu(t_k) \\ &\leq K\sum_{k=0}^{\infty} f(t_{k+1})\mu(t_k) \\ &\leq K\sum_{k=0}^{\infty} f(t_k)\mu(t_k) \\ &= K\int_{t_0}^{\infty} f(t)\Delta t \\ &< \infty.\end{aligned}$$

Exercise 3.41. *Solution.* Let $t_k = 2^k$, $k \in \mathbb{N}_0$. Then

$$t_{k+1} = 2^{k+1}, \quad \mu(t_k) = t_{k+1} - t_k = 2^{k+1} - 2^k, \quad k \in \mathbb{N}_0,$$

and

$$\int_1^{\infty} \frac{1}{t^p} \Delta t = \sum_{k=0}^{\infty} \frac{1}{t_k^p} \mu(t_k)$$
$$= \sum_{k=0}^{\infty} \frac{1}{2^{kp}} (2^{k+1} - 2^k)$$
$$= 2 \sum_{k=0}^{\infty} \frac{1}{2^{k(p-1)}} - \sum_{k=0}^{\infty} \frac{1}{2^{k(p-1)}}$$
$$= \sum_{k=0}^{\infty} \frac{1}{2^{k(p-1)}} \begin{cases} = \infty & \text{if } p \in [0,1], \\ < \infty & \text{if } p > 1. \end{cases}$$

Exercise 3.42. Answer:
1. Divergent.
2. Divergent.
3. Convergent.
4. Convergent.
5. Convergent.
6. Convergent.

Exercise 3.43. Answer: Convergent for $\alpha > -1$ and $\beta > -1$, divergent if $\alpha \le -1$ or $\beta \le -1$.

Exercise 3.48. *Solution.* Let

$$g(t) = \int_{t_0}^{t} h_k(t, \sigma(s)) h_m(s, t_0) \Delta s.$$

Then using the chain rule, we get

$$g^{\Delta}(t) = h_k(\sigma(t), \sigma(t)) h_m(\sigma(t), t_0)$$
$$+ \int_{t_0}^{t} h_{k-1}(t, \sigma(s)), h_m(s, t_0) \Delta s$$
$$= \int_{t_0}^{t} h_{k-1}(t, \sigma(s)) h_m(s, t_0) \Delta s,$$
$$g^{\Delta^2}(t) = h_{k-1}(\sigma(t), \sigma(t)) h_m(t, t_0)$$
$$+ \int_{t_0}^{t} h_{k-2}(t, \sigma(s)) h_m(s, t_0) \Delta s$$
$$= \int_{t_0}^{t} h_{k-2}(t, \sigma(s)) h_m(s) \Delta s,$$
$$\vdots$$

$$\begin{aligned} g^{\Delta^k}(t) &= \int_{t_0}^{t} h_m(s, t_0)\Delta s \\ &= h_{m+1}(t, t_0), \\ g^{\Delta^{k+1}}(t) &= h_{m+1}(t, t_0), \\ &\vdots \\ g^{\Delta^{k+m}}(t) &= h_{k+m+1}(t, t_0), \quad t, t_0 \in \mathbb{T}. \end{aligned}$$

Chapter 4

Exercise 4.1. Answer:

1. $$\begin{aligned} \mathrm{Re}_h(z) &= 0, \\ \mathrm{Im}_h(z) &= 0. \end{aligned}$$
2. $$\begin{aligned} \mathrm{Re}_h(z) &= 0, \\ \mathrm{Im}_h(z) &= -\frac{\pi}{12}. \end{aligned}$$
3. $$\begin{aligned} \mathrm{Re}_h(z) &= 0, \\ \mathrm{Im}_h(z) &= \frac{\pi}{48}. \end{aligned}$$
4. $$\begin{aligned} \mathrm{Re}_h(z) &= 0, \\ \mathrm{Im}_h(z) &= \frac{4\pi}{15}. \end{aligned}$$
5. $$\begin{aligned} \mathrm{Re}_h(z) &= 0, \\ \mathrm{Im}_h(z) &= -\frac{\pi}{8}. \end{aligned}$$

Exercise 4.2. Answer:

1. $30 + 2i$.
2. $16 + 6i$.
3. $13 + 14i$.
4. $3 + 7i$.
5. $34 + 66i$.

Exercise 4.3. Answer:

1. $-12 - 13i$.

2. $1560 - 3640i$.
3. $-1024 - 1848i$.
4. $-635 - 640i$.
5. 40.

Exercise 4.4. Answer:
1. $-\frac{214}{865} + \frac{7i}{865}$.
2. $-\frac{7}{25} - \frac{1}{25}i$.
3. $-\frac{17}{89} + \frac{1}{178}i$.
4. $-\frac{2}{5} - \frac{1}{5}i$.
5. $-\frac{33}{109} + \frac{1}{109}i$.

Exercise 4.5. Answer:
1. $-\frac{242}{865} + \frac{16}{865}i$.
2. $\frac{2}{25} - \frac{14}{25}i$.
3. $-\frac{39}{178} + \frac{9}{178}i$.
4. $-\frac{1}{5} - \frac{7}{5}i$.
5. $-\frac{44}{109} + \frac{74}{109}i$.

Exercise 4.6. Answer:
1. $\frac{379}{865} - \frac{1512}{865}i$.
2. $\frac{6}{25} + \frac{8}{25}i$.
3. $\frac{50}{89} - \frac{71}{178}i$.
4. $\frac{4}{5} + \frac{2}{5}i$.
5. $\frac{98}{109} - \frac{36}{109}i$.

Exercise 4.8. Answer: Delta regressive for $t \neq 2$.

Exercise 4.9. Answer: Delta regressive for $t \neq -10, 0, 9$.

Exercise 4.10. Answer: Delta regressive for $t \notin \bigcup_{k=0}^{\infty}\{10k + 3\}$.

Exercise 4.11. Answer: Delta regressive for $t \neq 11$.

Exercise 4.12. Answer: Delta regressive for $t \neq 2, 9$.

Exercise 4.13. Answer: $(f \oplus_\mu g)(t) = 2t^3 + 2t^2 - \frac{19}{2}t - \frac{11}{2}$, $t \in \mathbb{T}$.

Exercise 4.14. Answer:

$$(f \oplus_\mu g)(t) = \begin{cases} t^3 + \frac{7}{2}t^2 - \frac{137}{2}t - \frac{71}{2} & \text{if } t \in (-2\mathbb{N}_0) \\ \text{does not exist} & \text{if } t = 0 \\ \frac{t(-2t^2+161)}{162} & \text{if } t \in 3^{\mathbb{N}_0}. \end{cases}$$

Exercise 4.15. Answer:

$$(f \oplus_\mu g)(t) = \begin{cases} t^2 & \text{if } t \in \left(\bigcup_{k=0}^{\infty}[10k, 10k+3)\right) \cup [4, 6], \\ \text{does not exist} & \text{if } t \in \bigcup_{k=0}^{\infty}\{10k+3\}. \end{cases}$$

Exercise 4.16. Answer:

$$(f \oplus_\mu g)(t) = \begin{cases} -3 & \text{if } t = 0, \\ \frac{t(-10t+1209)}{1210} & \text{if } t \in 11^{\mathbb{N}_0}. \end{cases}$$

Exercise 4.17. Answer:

$$(f \oplus_\mu g)(t) = \begin{cases} 2 + t & \text{if } t \in [1,2) \cup [3,4) \cup [7,8), \\ \frac{-8t^2+72t-1}{72} & \text{if } t \in 9^{\mathbb{N}}, \\ \text{does not exist} & \text{if } t = 2, \\ 212 & \text{if } t = 4, \\ 35 & \text{if } t = 8. \end{cases}$$

Exercise 4.19. Answer: $\ominus_\mu g(t) = -\frac{t}{1+2t}$, $t \in \mathbb{T}$.
Exercise 4.20. Answer:

$$\ominus_\mu g(t) = \begin{cases} -\frac{t}{1+2t} & \text{if } t \in (-2\mathbb{N}), \\ 0 & \text{if } t = 0, \\ -\frac{t}{1+2t^2} & \text{if } t \in 3^{\mathbb{N}_0}. \end{cases}$$

Exercise 4.21. Answer:

$$\ominus_\mu g(t) = \begin{cases} -t & \text{if } t \in \left(\bigcup_{k=0}^{\infty}[10k, 10k+3)\right) \cup [4,6], \\ -\frac{t}{1+7t} & \text{if } t \in \bigcup_{k=0}^{\infty}\{10k+3\}; \quad -\frac{3}{4} \quad \text{if } t = 3; \quad -\frac{10}{41} \quad \text{if } t = 10. \end{cases}$$

Exercise 4.22. Answer:

$$\ominus_\mu g(t) = \begin{cases} 0 & \text{if } t = 0, \\ -\frac{t}{1+10t^2} & t \in 11^{\mathbb{N}_0}. \end{cases}$$

Exercise 4.23. Answer:

$$\ominus_\mu g(t) = \begin{cases} -t & \text{if } t \in [1,2) \cup [3,4) \cup [7,8), \\ -\frac{t}{1+8t} & \text{if } t \in 9^{\mathbb{N}}, \\ -\frac{2}{3} & \text{if } t = 2, \\ -\frac{4}{13} & \text{if } t = 4, \\ -\frac{8}{9} & \text{if } t = 8. \end{cases}$$

Exercise 4.25. Answer: $(f \ominus_\mu g)(t) = \frac{2t^2-t-11}{2(1+2t)}$, $t \in \mathbb{T}$.

Exercise 4.26. Answer:

$$(f \ominus_\mu g)(t) = \begin{cases} \frac{t^2+t-71}{2(1+t)} & \text{if } t \in (-2\mathbb{N}), \\ -1 & \text{if } t = 0, \\ -\frac{163t}{162(1+2t^2)} & \text{if } t \in 3^{\mathbb{N}_0}. \end{cases}$$

Exercise 4.27. Answer:

$$(f \ominus_\mu g)(t) = \begin{cases} t^2 - 2t & \text{if } t \in \left(\bigcup_{k=0}^{\infty}[10k, 10k+3)\right) \cup [4, 6], \\ -\frac{1}{7} & \text{if } t \in \bigcup_{k=0}^{\infty}\{10k+3\}. \end{cases}$$

Exercise 4.28. Answer:

$$(f \ominus_\mu g)(t) = \begin{cases} -3 & \text{if } t = 0, \\ -\frac{1211t}{1210(1+10t^2)} & \text{if } t \in 11^{\mathbb{N}_0}. \end{cases}$$

Exercise 4.29. Answer:

$$(f \ominus_\mu g)(t) = \begin{cases} 2 - t & \text{if } t \in [1, 2) \cup [3, 4) \cup [7, 8), \\ -\frac{1+72t}{72(1+8t^2)} & \text{if } t \in 9^{\mathbb{N}}, \\ -1 & \text{if } t = 2, \\ \frac{12}{13} & \text{if } t = 4, \\ -\frac{5}{9} & \text{if } t = 8. \end{cases}$$

Solutions, hints, and answers to problems

Chapter 1

Problem 1.1 Answer:

1. Yes.
2. No.
3. Yes.
4. Yes.
5. No.

Problem 1.2 Answer:

1.
$$\sigma(t) = \begin{cases} 4t & \text{if } t \in 4^{\mathbb{N}_0}, \\ t+5 & \text{if } t \in (-5\mathbb{N}), \\ 1 & \text{if } t = 0. \end{cases}$$

2.
$$\sigma(t) = \begin{cases} t & \text{if } t \in [-1,3) \cup [7,9) \cup [10,\infty), \\ 7 & \text{if } t = 3, \\ 10 & \text{if } t = 9. \end{cases}$$

3.
$$\sigma(t) = \begin{cases} \frac{t+3}{4} & \text{if } t \in \{1 - \frac{1}{4^n}\}_{n\in\mathbb{N}_0}, \\ 0 & \text{if } t = 0, \\ 2t & \text{if } t \in 2^{\mathbb{N}_0}. \end{cases}$$

4.
$$\sigma(t) = \begin{cases} t & \text{if } t \in \{2\} \cup [4,9) \cup \{10\}, \\ \frac{4-t}{3-t} & \text{if } t \in \{2 + \frac{1}{n}\}_{n\in\mathbb{N}, n\geq 2}, \\ 4 & \text{if } t = 3, \\ 10 & \text{if } t = 9, \\ 9 + \frac{t-9}{(\sqrt{t-9}-1)^2} & \text{if } t \in \{9 + \frac{1}{n^2}\}_{n\in\mathbb{N}, n\geq 2}. \end{cases}$$

5.
$$\sigma(t) = \begin{cases} t & \text{if } t \in [1,3), \\ 5 & \text{if } t = 3, \\ 5t & \text{if } t \in 5^{\mathbb{N}}. \end{cases}$$

https://doi.org/10.1515/9783112232088-006

Problem 1.3 Answer:

1.
$$\mu(t) = \begin{cases} 3t & \text{if } t \in 4^{\mathbb{N}_0}, \\ 5 & \text{if } t \in (-5\mathbb{N}), \\ 1 & \text{if } t = 0. \end{cases}$$

2.
$$\mu(t) = \begin{cases} 0 & \text{if } t \in [-1,3) \cup [7,9) \cup [10,\infty), \\ 4 & \text{if } t = 3, \\ 1 & \text{if } t = 9. \end{cases}$$

3.
$$\mu(t) = \begin{cases} \frac{3(1-t)}{4} & \text{if } t \in \{1 - \frac{1}{4^n}\}_{n\in\mathbb{N}_0}, \\ 0 & \text{if } t = 0, \\ t & \text{if } t \in 2^{\mathbb{N}_0}. \end{cases}$$

4.
$$\mu(t) = \begin{cases} 0 & \text{if } t \in \{2\} \cup [4,9) \cup \{10\}, \\ \frac{(t-2)^2}{3-t} & \text{if } t \in \{2 + \frac{1}{n}\}_{n\in\mathbb{N}}, \\ 1 & \text{if } t \in \{3,9\}, \\ (9-t)(1 - \frac{1}{(\sqrt{t-9}-1)^2}) & \text{if } t \in \{9 + \frac{1}{n^2}\}_{n\in\mathbb{N}}. \end{cases}$$

5.
$$\mu(t) = \begin{cases} 0 & \text{if } t \in [1,3), \\ 2 & \text{if } t = 3, \\ 4t & \text{if } t \in 5^{\mathbb{N}}. \end{cases}$$

Problem 1.4 *Solution.* We have

$$\mu(t_j) = t_{j+1} - t_j, \quad j \in \{1,2,\dots,p-1\},$$
$$\sigma(t_p) = t_p.$$

Then

$$\mu(t_j) = t_{j+1} - t_j, \quad j \in \{1,2,\dots,p\},$$

and

$$\mu(t_p) = 0.$$

Hence

$$\sum_{j=1}^{p}\mu(t_j) = \mu(t_1)+\mu(t_2)+\mu(t_3)+\cdots+\mu(t_{p-1})+\mu(t_p)$$
$$= t_2 - t_1 + t_3 - t_2 + t_4 - t_3 + \cdots + t_p - t_{p_1}$$
$$= t_p - t_1.$$

Problem 1.5 Answer:

1.
$$\rho(t) = \begin{cases} t-5 & \text{if } t \in (-5\mathbb{N}_0), \\ 0 & \text{if } t = 1, \\ \frac{t}{4} & \text{if } t \in 4^{\mathbb{N}}. \end{cases}$$

2.
$$\rho(t) = \begin{cases} t & \text{if } t \in (-1,3] \cup (7,9] \cup (10,\infty), \\ -1 & \text{if } t = -1, \\ 3 & \text{if } t = 7, \\ 9 & \text{if } t = 10. \end{cases}$$

3.
$$\rho(t) = \begin{cases} 0 & \text{if } t = 0, \\ 4t-3 & \text{if } t \in \{1-\frac{1}{4^n}\}_{n\in\mathbb{N},n\geq 2}, \\ 1 & \text{if } t = 1; \quad 0 \quad \text{if } t = \frac{3}{4}, \\ \frac{t}{2} & \text{if } t \in 2^{\mathbb{N}}. \end{cases}$$

4.
$$\rho(t) = \begin{cases} 2 & \text{if } t = 2, \\ \frac{3t-4}{t-1} & \text{if } t \in \{2+\frac{1}{n}\}_{n\in\mathbb{N}}, \\ 3 & \text{if } t = 4, \\ t & \text{if } t \in (4,9], \\ 9 + \frac{t-9}{(\sqrt{t-9}+1)^2} & \text{if } t \in \{9+\frac{1}{n^2}\}_{n\in\mathbb{N}}. \end{cases}$$

5.
$$\rho(t) = \begin{cases} t & \text{if } t \in [1,3], \\ 3 & \text{if } t = 5, \\ \frac{t}{5} & t \in 5^{\mathbb{N}}\backslash\{5\}. \end{cases}$$

Problem 1.6 Answer:

1.
$$\nu(t) = \begin{cases} 5 & \text{if } t \in (-5\mathbb{N}_0), \\ 1 & \text{if } t = 1, \\ \frac{3t}{4} & \text{if } t \in 4^{\mathbb{N}}. \end{cases}$$

2.

$$v(t)=\begin{cases}0 & \text{if } t\in[-1,3]\cup(7,9]\cup(10,\infty),\\ 4 & \text{if } t=7,\\ 1 & \text{if } t=10.\end{cases}$$

3.

$$v(t)=\begin{cases}0 & \text{if } t=0,\\ 3-3t & \text{if } t\in\{1-\frac{1}{4^n}\}_{n\in\mathbb{N},n\geq 2},\\ 0 & \text{if } t=1;\quad \frac{3}{4}\quad \text{if } t=\frac{3}{4},\\ \frac{t}{2} & \text{if } t\in 2^{\mathbb{N}}.\end{cases}$$

4.

$$v(t)=\begin{cases}0 & \text{if } t\in\{2\}\cup(4,9],\\ \frac{(t-2)^2}{t-1} & \text{if } t\in\{2+\frac{1}{n}\}_{n\in\mathbb{N}},\\ 1 & \text{if } t=4,\\ t-9-\frac{t-9}{(\sqrt{t-9}+1)^2} & \text{if } t\in\{9+\frac{1}{n^2}\}_{n\in\mathbb{N}}.\end{cases}$$

5.

$$v(t)=\begin{cases}0 & \text{if } t\in[1,3],\\ 2 & \text{if } t=5,\\ \frac{4t}{5} & t\in 5^{\mathbb{N}}\backslash\{5\}.\end{cases}$$

Problem 1.7 Answer: $t_p - t_1$.

Problem 1.8 Answer:

1. Any point is isolated.
2. $t\in(-1,3)\cup(7,9)\cup(10,\infty)$ is dense, $t=-1$ is right-dense, $t=3$ is left-dense and right-scattered, $t=7$ is left-scattered and right-dense, $t=9$ is left-dense and right-scattered, $t=10$ is left-scattered and right-dense.
3. $t=0$ is right-scattered, $t\in\{1-\frac{1}{4^n}\}_{n\in\mathbb{N}}$ is isolated, $t=1$ is left-dense and right-scattered, $t\in 2^{\mathbb{N}}$ is isolated.
4. $t=2$ is right-dense, $t=3$ is isolated, $t\in(4,9)$ is dense, $t=4$ is left-scattered and right-dense, $t=9$ is dense, $t=10$ is left-scattered, $t\in\{9+\frac{1}{n^2}\}_{n\in\mathbb{N},n\geq 2}$ is isolated.
5. $t\in(1,3)$ is dense, $t=1$ is right-dense, $t=3$ is left-dense and right-scattered, $t\in 5^{\mathbb{N}}$ is isolated.

Problem 1.9 Answer:

1.

$$\begin{aligned}[-10,16]_{\mathbb{T}} &= \{-10,-5,0,1,4,16\},\\ [-10,16)_{\mathbb{T}} &= \{-10,-5,0,1,4\},\\ (-10,16]_{\mathbb{T}} &= \{-5,0,1,4,16\},\\ (-10,16)_{\mathbb{T}} &= \{-5,0,1,4\}.\end{aligned}$$

2.

$$[0,11]_{\mathbb{T}} = [0,3] \cup [7,9] \cup [10,11],$$
$$[0,11)_{\mathbb{T}} = [0,3] \cup [7,9] \cup [10,11),$$
$$(0,11]_{\mathbb{T}} = (0,3] \cup [7,9] \cup [10,11],$$
$$(0,11)_{\mathbb{T}} = (0,3] \cup [7,9] \cup [10,11).$$

3.

$$\left[\frac{63}{64},8\right]_{\mathbb{T}} = \left\{1-\frac{1}{4^n}\right\}_{n\in\mathbb{N},n\geq 3} \cup \{1,2,4,8\},$$
$$\left[\frac{63}{64},8\right)_{\mathbb{T}} = \left\{1-\frac{1}{4^n}\right\}_{n\in\mathbb{N},n\geq 3} \cup \{1,2,4\},$$
$$\left(\frac{63}{64},8\right]_{\mathbb{T}} = \left\{1-\frac{1}{4^n}\right\}_{n\in\mathbb{N},n\geq 4} \cup \{1,2,4,8\},$$
$$\left(\frac{63}{64},8\right)_{\mathbb{T}} = \left\{1-\frac{1}{4^n}\right\}_{n\in\mathbb{N},n\geq 4} \cup \{1,2,4\}.$$

4.

$$[3,10]_{\mathbb{T}} = \{3\} \cup [4,9] \cup \left\{9+\frac{1}{n^2}\right\}_{n\in\mathbb{N}},$$
$$[3,10)_{\mathbb{T}} = \{3\} \cup [4,9] \cup \left\{9+\frac{1}{n^2}\right\}_{n\in\mathbb{N},n>2},$$
$$(3,10]_{\mathbb{T}} = [4,9] \cup \left\{9+\frac{1}{n^2}\right\}_{n\in\mathbb{N}},$$
$$(3,10)_{\mathbb{T}} = [4,9] \cup \left\{9+\frac{1}{n^2}\right\}_{n\in\mathbb{N},n\geq 2}.$$

5.

$$[3,125]_{\mathbb{T}} = \{3,5,25,125\},$$
$$[3,125)_{\mathbb{T}} = \{3,5,25\},$$
$$(3,125]_{\mathbb{T}} = \{5,25,125\},$$
$$(3,125)_{\mathbb{T}} = \{5,25\}.$$

Problem 1.10 Answer:

1.

$$\mathbb{T}^\kappa = \mathbb{T},$$
$$\mathbb{T}_\kappa = \mathbb{T}.$$

2.

$$\mathbb{T}^\kappa = \mathbb{T},$$
$$\mathbb{T}_\kappa = \mathbb{T}.$$

3.

$$\mathbb{T}^\kappa = \mathbb{T},$$

$$\mathbb{T}_\kappa = \left\{1 - \frac{1}{4^n}\right\}_{n\in\mathbb{N}_0} \cup 2^{\mathbb{N}_0}.$$

4.

$$\mathbb{T}^\kappa = \{2\} \cup \left\{2 + \frac{1}{n}\right\}_{n\in\mathbb{N}} \cup [4,9] \cup \left\{9 + \frac{1}{n^2}\right\}_{n\in\mathbb{N}, n\geq 2},$$
$$\mathbb{T}_\kappa = \mathbb{T}.$$

5.

$$\mathbb{T}^\kappa = \mathbb{T},$$
$$\mathbb{T}_\kappa = \mathbb{T}.$$

Problem 1.11 Answer:

1.

$$f^\sigma(t) = \begin{cases} \frac{6+t}{17+3t} & \text{if } t \in (-5\mathbb{N}), \\ \frac{2}{5} & \text{if } t = 0, \\ \frac{1+4t}{2+12t} & \text{if } t \in 4^{\mathbb{N}_0}, \end{cases}$$

$$f^{\sigma^2}(t) = \begin{cases} \frac{11+t}{32+3t} & \text{if } t \in (-5\mathbb{N}), \\ \frac{5}{14} & \text{if } t = 0, \\ \frac{1+16t}{2+48t} & \text{if } t \in 4^{\mathbb{N}_0}. \end{cases}$$

2.

$$f^\sigma(t) = \begin{cases} \frac{1+t}{2+3t} & \text{if } t \in [-1,3) \cup [7,9) \cup [10,\infty), \\ \frac{8}{23} & \text{if } t = 3, \\ \frac{11}{32} & \text{if } t = 9, \end{cases}$$

$$f^{\sigma^2}(t) = \begin{cases} \frac{1+t}{2+3t} & \text{if } t \in [-1,3) \cup [7,9) \cup [10,\infty), \\ \frac{8}{23} & \text{if } t = 3, \\ \frac{11}{32} & \text{if } t = 9. \end{cases}$$

3.

$$f^\sigma(t) = \begin{cases} \frac{7+t}{17+3t} & \text{if } t \in \{1 - \frac{1}{4^n}\}_{n\in\mathbb{N}_0}, \\ \frac{7}{17} & \text{if } t = 0, \\ \frac{1+2t}{2+6t} & \text{if } t \in 2^{\mathbb{N}_0}, \end{cases}$$

$$f^{\sigma^2}(t) = \begin{cases} \frac{31+t}{77+3t} & \text{if } t \in \{1 - \frac{1}{4^n}\}_{n\in\mathbb{N}_0}, \\ \frac{31}{77} & \text{if } t = 0, \\ \frac{1+4t}{2+12t} & \text{if } t \in 2^{\mathbb{N}_0}. \end{cases}$$

4.

$$f^{\sigma}(t)=\begin{cases}\frac{1+t}{2+3t} & \text{if } t\in\{2\}\cup[4,9]\cup\{10\},\\ \frac{7-2t}{18-5t} & \text{if } t\in\{2+\frac{1}{n}\}_{n\in\mathbb{N}_0},\\ \frac{5}{14} & \text{if } t=3,\\ \frac{11}{32} & \text{if } t=9,\\ \frac{10(\sqrt{t-9}-1)^2+(t-9)}{29(\sqrt{t-9}-1)^2+3(t-9)} & \text{if } t\in\{9+\frac{1}{n^2}\}_{n\in\mathbb{N}},\end{cases}$$

$$f^{\sigma^2}(t)=\begin{cases}\frac{1+t}{2+3t} & \text{if } t\in\{2\}\cup[4,9)\cup\{10\},\\ \frac{13-5t}{34-13t} & \text{if } t\in\{2+\frac{1}{n}\}_{n\in\mathbb{N}},\\ \frac{5}{14} & \text{if } t=3,\\ \frac{11}{32} & \text{if } t=9,\\ \frac{10+(t-9)(\sqrt{t-9}-1)^2}{29+3(t-9)(\sqrt{t-9}-1)^2} & \text{if } t\in\{9+\frac{1}{n^2}\}_{n\in\mathbb{N}}.\end{cases}$$

5.

$$f^{\sigma}(t)=\begin{cases}\frac{1+t}{2+3t} & \text{if } t\in[1,3),\\ \frac{6}{17} & \text{if } t=3,\\ \frac{1+5t}{2+15t} & \text{if } t\in 5^{\mathbb{N}},\end{cases}$$

$$f^{\sigma^2}(t)=\begin{cases}\frac{1+t}{2+3t} & \text{if } t\in[1,3),\\ \frac{26}{77} & \text{if } t=3,\\ \frac{1+25t}{2+75t} & \text{if } t\in 5^{\mathbb{N}}.\end{cases}$$

Problem 1.12 Answer:

1.

$$f^{\rho^2}(t)=\begin{cases}\sqrt{1+4(t-10)^2} & \text{if } t\in(-5\mathbb{N}),\\ \sqrt{401} & \text{if } t=0,\\ \sqrt{101} & \text{if } t=1,\\ \sqrt{1+\frac{t^2}{64}} & \text{if } t\in 4^{\mathbb{N}},\end{cases}$$

$$f^{\sigma\rho\sigma}(t)=\begin{cases}\sqrt{1+4(t+5)^2} & \text{if } t\in(-5\mathbb{N}),\\ \sqrt{5} & \text{if } t=0,\\ \sqrt{65} & \text{if } t=1,\\ \sqrt{1+64t^2} & \text{if } t\in 4^{\mathbb{N}}.\end{cases}$$

2.

$$f^{\rho^2}(t) = \begin{cases} \sqrt{1+4t^2} & \text{if } t \in (-1,3) \cup (7,9) \cup (10,\infty), \\ \sqrt{5} & \text{if } t = -1, \\ \sqrt{37} & \text{if } t = 3, \\ \sqrt{197} & \text{if } t = 7, \\ \sqrt{325} & \text{if } t = 9, \\ \sqrt{325} & \text{if } t = 10, \end{cases}$$

$$f^{\sigma\rho\sigma}(t) = \begin{cases} \sqrt{1+4t^2} & \text{if } t \in (-1,3) \cup (7,9) \cup (10,\infty), \\ \sqrt{5} & \text{if } t = -1, \\ \sqrt{197} & \text{if } t = 3, \\ \sqrt{197} & \text{if } t = 7, \\ \sqrt{401} & \text{if } t = 9, \\ \sqrt{401} & \text{if } t = 10. \end{cases}$$

3.

$$f^{\rho^2}(t) = \begin{cases} \sqrt{1+4(16t-15)^2} & \text{if } t \in \{1 - \frac{1}{4^n}\}_{n\in\mathbb{N}, n\geq 2}, \\ 1 & \text{if } t = 0; \quad 1 \quad \text{if } t = \frac{3}{4} \\ \sqrt{5} & \text{if } t = 1, \\ \sqrt{5} & \text{if } t = 2, \\ \sqrt{1+t^2} & \text{if } t \in 2^{\mathbb{N}},\ t > 2, \end{cases}$$

$$f^{\sigma\rho\sigma}(t) = \begin{cases} \sqrt{1+\frac{(3+t)^2}{4}} & \text{if } t \in \{1 - \frac{1}{4^n}\}_{n\in\mathbb{N}}, \\ \frac{\sqrt{13}}{2} & \text{if } t = 0, \\ \sqrt{17} & \text{if } t = 1, \\ \sqrt{65} & \text{if } t = 2, \\ \sqrt{1+16t^2} & \text{if } t \in 2^{\mathbb{N}},\ t > 2. \end{cases}$$

4.

$$f^{\rho^2}(t) = \begin{cases} \sqrt{17} & \text{if } t = 2, \\ \sqrt{1+4\left(\frac{5t-8}{2t-3}\right)^2} & \text{if } t \in \{2 + \frac{1}{n}\}_{n\in\mathbb{N}, n\geq 2}, \\ \frac{\sqrt{205}}{3} & \text{if } t = 3, \\ \sqrt{65} & \text{if } t = 4, \\ \sqrt{1+4t^2} & \text{if } t \in (4,9), \\ \sqrt{325} & \text{if } t = 9, \\ \sqrt{1+4\left(9+\frac{t-9}{(2\sqrt{t-9}+1)^2}\right)^2} & \text{if } t \in \{9 + \frac{1}{n^2}\}_{n\in\mathbb{N}, n\geq 2}, \\ \frac{\sqrt{26977}}{9} & \text{if } t = 10, \end{cases}$$

$$f^{\sigma\rho\sigma}(t)=\begin{cases}\sqrt{17} & \text{if } t=2,\\ \sqrt{1+4\left(\frac{4-t}{3-t}\right)^2} & \text{if } t\in\{2+\frac{1}{n}\}_{n\in\mathbb{N},n\geq 2},\\ \sqrt{65} & \text{if } t=3,\\ \sqrt{65} & \text{if } t=4,\\ \sqrt{1+4t^2} & \text{if } t\in(4,9),\\ \sqrt{325} & \text{if } t=9,\\ \sqrt{1+4\left(9+\frac{t-9}{(2\sqrt{t-9}+1)^2}\right)^2} & \text{if } t\in\{9+\frac{1}{n^2}\}_{n\in\mathbb{N},n\geq 2},\\ \sqrt{401} & \text{if } t=10.\end{cases}$$

5.

$$f^{\rho^2}(t)=\begin{cases}\sqrt{1+4t^2} & \text{if } t\in[1,3),\\ \sqrt{37} & \text{if } t=3,\\ \sqrt{37} & \text{if } t=5,\\ \sqrt{1+\frac{4t^2}{625}} & \text{if } t\in 5^{\mathbb{N}}\setminus\{5\},\end{cases}$$

$$f^{\sigma\rho\sigma}(t)=\begin{cases}\sqrt{1+4t^2} & \text{if } t\in[1,3),\\ \sqrt{101} & \text{if } t=3,\\ \sqrt{2501} & \text{if } t=5,\\ \sqrt{1+100t^2} & \text{if } t\in 5^{\mathbb{N}}\setminus\{5\}.\end{cases}$$

Problem 1.13 Answer:

1. $\frac{\sigma(t)}{9(\sigma(t)+3)}$.
2. -1.
3. $1+5\sigma(t)\rho(t)$.
4. -1.
5. $\frac{1}{t\sigma(t)\rho(t)}$.

Chapter 2

Problem 2.7. Answer: $2+t+(\sqrt[3]{t}+1)^3$.

Problem 2.9. Answer:

1. $3t-3$.
2. $\frac{6t^3+5t^2-3t}{(t+1)(2t+1)}$.
3. $\frac{2}{(t+1)(2t+1)}$.

Problem 2.10. Answer:

$$f^{\Delta}(t) = \begin{cases} -\frac{29}{(2+9t)(47+9t)} & \text{if } \in (-5\mathbb{N}), \\ -\frac{29}{22} & \text{if } t = 0, \\ -\frac{29}{2(2+9t)(1+18t)} & \text{if } t \in 4^{\mathbb{N}_0}. \end{cases}$$

Problem 2.11. Answer:

$$f^{\Delta}(t) = \begin{cases} 3t^2 + 1 & \text{if } t \in \{2\} \cup [4,9], \\ \frac{t^4-5t^3+4t^2-2t+25}{(t-3)^2} & \text{if } ty \in \{2 + \frac{1}{n}\}_{n\in\mathbb{N}, n\geq 2}, \\ 38 & \text{if } t = 3, \\ (a(t))^2 + ta(t) + t^2 + 1 & \text{if } t \in \{9 + \frac{1}{n^2}\}_{n\in\mathbb{N}}, \end{cases}$$

where

$$a(t) = 9 + \frac{t-9}{(\sqrt{t-9}-1)^2}, \quad t \in \left\{9 + \frac{1}{n^2}\right\}_{n\in\mathbb{N}}.$$

Problem 2.12. Answer:

$$f^{\Delta^2}(t) = \begin{cases} -\frac{2}{(t+2)(t+7)(t+12)} & \text{if } t \in (-5\mathbb{N}_0), \\ -\frac{1}{9} & \text{if } t = 0, \\ -\frac{5}{4(t+2)(2t+1)(8t+1)} & \text{if } t \in 4^{\mathbb{N}_0}. \end{cases}$$

Problem 2.13. Answer:

$$f^{\Delta^2}(t) = \begin{cases} \frac{6}{(1+t)^3} & \text{if } t \in [-1,3) \cup [7,9) \cup [10,\infty), \\ \frac{3}{256} & \text{if } t = 3, \\ \frac{3}{1210} & \text{if } t = 9. \end{cases}$$

Problem 2.14. Answer:

$$f^{\Delta^2}(t) = \begin{cases} \frac{5}{2} & \text{if } t \in \{1 - \frac{1}{4^n}\}_{n\in\mathbb{N}_0}, \\ 6 & \text{if } t \in 2^{\mathbb{N}_0}, \\ 10 & \text{if } t = 0. \end{cases}$$

Problem 2.15. Answer:

$$f^{\Delta^2}(t) = \begin{cases} -\frac{16}{(3+t)^3} & \text{if } t \in [1,3), \\ -\frac{11}{168} & \text{if } t = 3, \\ -\frac{45}{(3+t)(3+5t)(3+25t)} & \text{if } t \in 5^{\mathbb{N}}. \end{cases}$$

Problem 2.16. Answer: $12t^2 + 24ht + 14h^2$.

Problem 2.22. Answer:

1. Increasing on $\mathbb{T}$.
2. Increasing on $\mathbb{T}$.
3. Increasing on $\mathbb{T}$.

Problem 2.23. Answer: Decreasing on $\mathbb{T}$.

Problem 2.24. Answer: Increasing on $\mathbb{T}$.

Problem 2.25. Answer: Convex for $t \in (-\infty, -15]_{\mathbb{T}} \cup \{-5\}$, concave for $t \in [-10, -5)_{\mathbb{T}} \cup \{0\} \cup 4^{\mathbb{N}_0}$.

Problem 2.26. Answer: Convex on $\mathbb{T}$.

Problem 2.27. Answer: Convex on $\mathbb{T}$.

Problem 2.28. Answer: Concave on $\mathbb{T}$.

Problem 2.29. Answer: Convex on $\mathbb{T}$.

Problem 2.30. Answer: Yes.

Problem 2.31. Answer: Yes.

Problem 2.32. Answer: Yes.

Problem 2.33. Answer: Yes.

Problem 2.34. Answer: Yes.

Problem 2.35. Answer: No.

Problem 2.36. Answer: No.

Problem 2.37. Answer: No.

Problem 2.38. Answer: $a = -\frac{1}{2}$.

Problem 2.39. Answer: No such values.

Problem 2.40. Answer: $c = \sqrt{\frac{13}{2}}$.

Problem 2.41. Answer: $e^{t^3 + 3t^2 + 3t + 1} - e^{t^3}$.

Problem 2.43. Answer: $\frac{1}{2t+8}$.

Problem 2.44. Answer:

$$(f \circ g)^{\Delta}(t) = \begin{cases} -\frac{2}{(1+t)(4+t)} & \text{if } t \text{ is right-scattered,} \\ -\frac{2}{(1+t)^2} & \text{if } t \text{ is right-dense.} \end{cases}$$

Problem 2.46. Answer:

1. $\frac{1}{12}$.
2. -1.
3. 1.

Chapter 3

Problem 3.1. Answer: Yes.

Problem 3.2. Answer: Yes.

Problem 3.3. Answer: Yes.

Problem 3.4. Answer: No.

Problem 3.5. Answer: Yes.

Problem 3.6. Answer: The function f is delta predifferentiable with $D = \mathbb{T}\backslash\{7, 10\}$.

Problem 3.7. Answer: The function f is delta predifferentiable with $D = \mathbb{T}\backslash\{2, 4\}$.

Problem 3.18. Answer:

1. Divergent.
2. Divergent.
3. Divergent.
4. 2.

Problem 3.19. Answer:

1. Convergent.
2. Divergent.
3. Divergent.
4. Convergent.
5. Divergent.
6. Convergent.

Problem 3.20. Answer: Convergent.

Problem 3.22. Answer:

1. Convergent.
2. Convergent.
3. Divergent.
4. Divergent.
5. Convergent.
6. Divergent.

Problem 3.23. Answer:

1. Convergent for $\alpha > -2$, divergent for $\alpha \leq -2$.
2. Convergent for $\beta > -3$, divergent for $\beta \leq -3$.
3. Convergent for $\alpha > -2$ and $\beta > -4$, divergent if $\alpha \leq -2$ or $\beta \leq -4$.

Chapter 4

Problem 4.1. Answer:

1.

$$\mathrm{Re}_h(z) = \frac{1}{3},$$
$$\mathrm{Im}_h(z) = 0.$$

2.

$$\mathrm{Re}_h(z) = 0,$$
$$\mathrm{Im}_h(z) = \frac{31\pi}{32}.$$

3.

$$\mathrm{Re}_h(z) = 0,$$
$$\mathrm{Im}_h(z) = \frac{\pi}{4}.$$

4.

$$\mathrm{Re}_h(z) = \frac{\sqrt{34} - 1}{5},$$
$$\mathrm{Im}_h(z) = \frac{1}{5}\arctan\left(-\frac{5}{3}\right).$$

5.

$$\mathrm{Re}_h(z) = \frac{2\sqrt{5} - 1}{3},$$
$$\mathrm{Im}_h(z) = \frac{1}{3}\arctan(-2).$$

Problem 4.2. Answer:
1. $3 - 31i$.
2. $-12 + 53i$.
3. $24 - 3i$.
4. $27 + 33i$.
5. $407 + 12i$.

Problem 4.3. Answer:
1. $-24 - 26i$.
2. $365 + 720i$.
3. $3 + 262i$.
4. 7320.
5. $-29 - 59i$.

Problem 4.4. Answer:
1. $-\frac{107}{229} + \frac{1}{229}i$.
2. $-\frac{151}{457} - \frac{7}{457}i$.
3. $-\frac{9}{41} + \frac{1}{41}i$.
4. $-\frac{28}{61} - \frac{3}{61}i$.
5. $-\frac{506}{2536} + \frac{5}{1268}i$.

Problem 4.5. Answer:
1. $-\frac{103}{229} - \frac{29}{229}i$.
2. $-\frac{122}{457} - \frac{45}{457}i$.
3. $-\frac{2}{41} + \frac{23}{41}i$.
4. $-\frac{3}{61} - \frac{33}{61}i$.
5. $-\frac{225}{634} - \frac{27}{634}i$.

Problem 4.6. Answer:
1. $\frac{748}{229} - \frac{114}{229}i$.

2. $\frac{102}{457} + \frac{1064}{457}i$.
3. $\frac{8}{41} - \frac{10}{41}i$.
4. $\frac{47}{61} + \frac{90}{61}i$.
5. $\frac{495}{634} - \frac{2223}{634}i$.

Problem 4.7. Answer: Delta regressive for $t \neq 16$ and $t \notin (-5\mathbb{N})$.

Problem 4.8. Answer: Delta regressive for $t \neq 3, 9$.

Problem 4.9. Answer: Delta regressive for $t \neq 0$.

Problem 4.10. Answer: Delta regressive for any $t \in \mathbb{T}$.

Problem 4.11. Answer: Delta regressive for $t \neq 3, 9$.

Problem 4.12. Answer:

$$(f \oplus_\mu g)(t) = \begin{cases} \text{does not exist} & \text{if } t \in (-5\mathbb{N}), \\ 2 & \text{if } t = 0, \\ \frac{-3t^2+48t-1}{48} & \text{if } t \in 4^{\mathbb{N}_0}. \end{cases}$$

Problem 4.13. Answer:

$$(f \oplus_\mu g)(t) = \begin{cases} t^2 + 4t & \text{if } t \in [-1, 3) \cup [7, 9) \cup [10, \infty), \\ \text{does not exist} & \text{if } t = 3, \\ \text{does not exist} & \text{if } t = 9. \end{cases}$$

Problem 4.14. Answer:

$$(f \oplus_\mu g)(t) = \begin{cases} -12t^2 + 13t + 16 & \text{if } t \in \{1 - \frac{1}{4^n}\}_{n \in \mathbb{N}}, \\ \text{does not exist} & \text{if } t = 0, \\ t^4 + t^3 + t^2 + 2t & \text{if } t \in 2^{\mathbb{N}_0}. \end{cases}$$

Problem 4.15. Answer:

$$(f \oplus_\mu g)(t) = \begin{cases} 2t & \text{if } t \in [1, 3), \\ 24 & \text{if } t = 3, \\ 2t(1 + 2t^2) & \text{if } t \in 5^{\mathbb{N}}. \end{cases}$$

Problem 4.16. Answer:

$$(f \oplus_\mu g)(t) = \begin{cases} \frac{2t^2+2t+1}{2t+1} & \text{if } t \in \{2\} \cup [4, 9) \cup \{10\}, \\ t^2 + t + \frac{t^3(t-2)^2}{3-t} & \text{if } t \in \{2 + \frac{1}{n}\}_{n \in \mathbb{N}, n \geq 2}, \\ \text{does not exist} & \text{if } t = 3, 9, \\ 7 + t + \frac{7t(t-9)^{\frac{3}{2}}(2-\sqrt{t-9})}{(\sqrt{t-9}-1)^2} & \text{if } t \in \{9 + \frac{1}{n^2}\}_{n \in \mathbb{N}}. \end{cases}$$

Problem 4.17. Answer:

$$\ominus_\mu g(t) = \begin{cases} -\frac{t}{1+5t} & \text{if } t \in (-5\mathbb{N}), \\ 0 & \text{if } t = 0, \\ -\frac{t}{1+3t^2} & \text{if } t \in 4^{\mathbb{N}_0}. \end{cases}$$

Problem 4.18. Answer:

$$\ominus_\mu g(t) = \begin{cases} -t & \text{if } t \in [-1,3) \cup [7,9) \cup [10,\infty), \\ -\frac{3}{13} & \text{if } t = 3, \\ -\frac{9}{10} & \text{if } t = 9. \end{cases}$$

Problem 4.19. Answer:

$$\ominus_\mu g(t) = \begin{cases} \frac{4t}{3t^2-3t-4} & \text{if } t \in \{1 - \frac{1}{4^n}\}_{n\in\mathbb{N}_0}, \\ 0 & \text{if } t = 0, \\ -\frac{t}{1+t^2} & \text{if } t \in 2^{\mathbb{N}}. \end{cases}$$

Problem 4.20. Answer:

$$\ominus_\mu g(t) = \begin{cases} -t & \text{if } t \in [1,3), \\ -\frac{3}{7} & \text{if } t = 3, \\ -\frac{t}{1+4t^2} & \text{if } t \in 5^{\mathbb{N}}. \end{cases}$$

Problem 4.21. Answer:

$$\ominus_\mu g(t) = \begin{cases} -t & \text{if } t \in \{2\} \cup [4,9] \cup \{10\}, \\ -\frac{t(3-t)}{3-t+t(t-2)^2} & \text{if } t \in \{2 + \frac{1}{n}\}_{n\in\mathbb{N}, n\geq 2}, \\ -\frac{3}{4} & \text{if } t = 3, \\ -\frac{t(\sqrt{t-9}-1)^2}{(\sqrt{t-9}-1)^2+t(t-9)^{\frac{3}{2}}(2-\sqrt{t-9})} & \text{if } t \in \{9 + \frac{1}{n^2}\}_{n\in\mathbb{N}}. \end{cases}$$

Problem 4.22. Answer:

$$(f \ominus_\mu g)(t) = \begin{cases} -\frac{1+5t}{5(1+5t^2)} & \text{if } t \in (-5\mathbb{N}), \\ 2 & \text{if } t = 0, \\ -\frac{1+48t}{48(1+3t^2)} & \text{if } t \in 4^{\mathbb{N}_0}. \end{cases}$$

Problem 4.23. Answer:

$$(f \ominus_\mu g)(t) = \begin{cases} t^2 + 2t & \text{if } t \in [-1, 3) \cup [7, 9) \cup [10, \infty), \\ -\frac{1}{4} & \text{if } t = 3, \\ -1 & \text{if } t = 9. \end{cases}$$

Problem 4.24. Answer:

$$(f \ominus_\mu g)(t) = \begin{cases} \frac{4(16-t)}{-3t^2+3t+4} & \text{if } t \in \{1 - \frac{1}{4^n}\}_{n \in \mathbb{N}_0}, \\ -\frac{4}{3} & \text{if } t = 0, \\ \frac{t^2}{1+t^2} & \text{if } t \in 2^{\mathbb{N}_0}. \end{cases}$$

Problem 4.25. Answer: $(f \ominus_\mu g)(t) = 0$.

Problem 4.26. Answer:

$$(f \ominus_\mu g)(t) = \begin{cases} \frac{1-2t^2}{1+2t} & \text{if } t \in \{2\} \cup [4, 9) \cup \{10\}, \\ \frac{t(t-1)(3-t)}{3-t+t(t-2)^2} & \text{if } t \in \{2 + \frac{1}{n}\}_{n \in \mathbb{N}}, \\ \text{does not exist} & \text{if } t = 3, 9. \end{cases}$$

Index

https://doi.org/10.1515/9783112232088-007

www.ingramcontent.com/pod-product-compliance
Lightning Source LLC
Chambersburg PA
CBHW060948210326
41598CB00031B/4759
* 9 7 8 3 1 1 2 2 3 2 0 7 1 *